NEW APPLICATIONS OF
ELECTRON SPIN RESONANCE
Dating, Dosimetry and Microscopy

The frontispiece is the author's view of himself in this interdisciplinary research, as a traveler out of the solid state physics (crystalline lattice) to the interdisciplinary fields with geology and archaeology to find his fossilized cranium somewhere using ESR as a tool. It is a pleasant journey with many findings, sometimes gold mines in science, but also a "Winterreise" as a foreigner to both physicist and geologist community, which is inevitable in any interdisciplinary research. For geologists and archaeologists, conversely, a journey into solid state physics with this book will, I hope, bring about fruitful results in their own fields.

NEW APPLICATIONS OF
ELECTRON SPIN RESONANCE
Dating, Dosimetry and Microscopy

BY

Motoji Ikeya
Osaka University, Japan

Copy edited by

M. R. Zimmerman
Hahnemann University, USA

N. Whitehead
DSIR, New Zealand

World Scientific
Singapore • New Jersey • London • Hong Kong

Published by

World Scientific Publishing Co. Pte. Ltd.
5 Toh Tuck Link, Singapore 596224
USA office: 27 Warren Street, Suite 401-402, Hackensack, NJ 07601
UK office: 57 Shelton Street, Covent Garden, London WC2H 9HE

British Library Cataloguing-in-Publication Data
A catalogue record for this book is available from the British Library.

NEW APPLICATIONS OF ELECTRON SPIN RESONANCE
Dating, Dosimetry and Microscopy

ISBN-13   978-981-02-1199-8
ISBN-10   981-02-1199-6
ISBN-13   978-981-02-1200-1 (pbk)
ISBN-10   981-02-1200-3 (pbk)

Preface

This book describes applications of electron paramagnetic resonance (EPR) to dating, radiation dosimetry and microscopic imaging of the spatial distribution of paramagnetic defects. I will use, in the following, the term "electron spin resonance (ESR)" instead of "EPR" for simplicity and because it has been used widely in the field to be discussed here. ESR has become a standard analytical method as, for example, optical spectroscopy and very good commercial ESR spectrometers are available for its applications. Now it is also applied to develop a scientific method of radiation dosimetry, dating, and assessment of the past environment based on natural radiation effects, with applications to problems in geology, archaeology and forensic medicine. The recent development of magnetic resonance imaging (MRI) allows us to obtain the spatial distribution of spins.

ESR dating can cover the whole time range in the Quaternary, from hundreds to millions of years ago, far beyond the limit of radiocarbon (^{14}C) dating. The possibility of dating materials billions of years old is also noted. The lower limit of ESR dating may be reduced further as innovations in technology proceed. ESR dating is developing enormously and has the potential to develop even more if scientists become aware of the method and its limitations and apply it to solve problems in their respective fields.

Introductory books on ESR are intended for physics and chemistry students and generally involve mathematics and quantum physics beyond the understanding for those who are interested in geochronology. As a result, it is hard for them to understand what an electron spin is and why ESR can be used as a method of dating and radiation dosimetry. This book is directed to those who are interested in dating, dosimetry and assessments using ESR.

In this book, the basic principles of ESR and its application to dating and dosimetry are described without using complex quantum physics or mathematics. For those who need doses and dates, examples are given of typical ESR spectra from geological and biological materials used for ESR dating and dosimetry. A description of the physics and chemistry of relevant defects follows. Applications of ESR to several materials are presented with the aim of showing potential applications to people concerned with radiation dosimetry and to those in geology, anthropology, paleontology, geography, forensic science, history, archaeology and preservation science.

ESR has become a routine tool for physicists and chemists. There is the prospect that ESR dating could be done by geologists or archaeologists themselves, for their own purposes, with or without the help of physicists or chemists. This book on ESR dating will, it is hoped, be a source of information and *stimulate collaborative interdisciplinary work*. I have tried to reference all publications concerning ESR dating which will be established as the best dating method for the Quaternary when applied to a wide span of problems. The scope of ESR dating, dosimetry and imaging is expanding.

Readers who are familiar with dating methods and the basis of radiation effect physics can skip through Chapters 1 and 2. However, it may be instructive to examine the ESR spectra of geological samples dealt with in Chapter 2. A short description of pulsed ESR is given. Chapter 3 deals with the procedures of ESR dating, especially with a new additive dose method to obtain the age directly and Chapter 4 deals with radiation assessment in age derivation. Materials such as carbonate cave deposits, shells and corals are described in Chapters 5 and 6, and evaporites and apatite (including bones) are dealt with in Chapters 7 and Chapter 8. Applications to the age determination of fault formation and geothermal and volcanic studies based on defects in SiO_2 and silicate minerals appear in Chapters 9 and 10. Future ESR dating of ices (solid H_2O, CO_2, CH_4, etc.) in planetary science and applications to environmental sciences are discussed in Chapter 11. Chapter 12 deals with chemical ESR dating of organic materials in forensic science, history, preservation science, archaeology and coal and petroleum science. Chapter 13 is on ESR dosimetry for humans such as A–bomb survivors and residents close to the Chernobyl reactor accident and for monitoring irradiated foods. Chapter 14 covers ESR imaging with applications. Finally, appendices deal with a theory of defect formation, molecular orbitals, a portable ESR spectrometer, measurements of large objects and an in–vivo tooth dosimeter.

Forty years ago, I inherited a dream of my mother who hoped her naughty boy would be a scientist. Thirty years ago, I studied radiation effects in solids as a graduate student to seek the limit of my capability and realize her dream. Twenty years ago, fascinated by nature during my postdoctoral stay in University of North Carolina and depressed as a physicist in a local college back in Japan, I began amateur geology with ESR at the Akiyoshi cave and had a dream to write a science book dealing with an interdisciplinary field. Ten years ago, I started to write this book and soon found that my dream would not be realized without the help and dedication of many others including those of the publisher.

I sincerely thank Dr. M. R. Zimmerman and Dr. N. Whitehead for editing the manuscript during these ten years and several others acknowledged later for critical reading of chapters. Throughout this work, I owe much to my colleagues and coworkers as well as students and secretaries for their help and dedication. It must be added that my wife, Yoshiko (Okada) Ikeya (a biochemist) has tabulated ESR parameters and done most of the editing works for the last one year.

September 1993

Motoji Ikeya

Acknowledgement

In addition to my colleague scientists in this interdisciplinary field, I am grateful to the following persons who have contributed to this book and this field.

Late professors, T. Suita (Osaka Univ.), J. H. Crawford (North Carolina, USA) and H. Pick (Stuttgart, Germany) and Prof. N. Itoh (Nagoya Univ.) for introducing me to the field of physics on defects. Emeritus Prof. M. Date (President of Jpn. Phys. Soc.) and Prof. J. Kanamori (President of Osaka Univ.) for their support in the Dept. of Earth and Space Science founded in 1991, and Dr. H. Mitsudo (Okayama Ceramics Center) for his encouragement and supports.

My coworkers and friends in physics, geology and anthropology: Drs. W. C. Mallard (Georgia State Univ.), F. Goff (Los Alamos Nat. Lab.), A. K. Shinghvi (Phys. Res. Lab., Ahmedabad), W. L. McLaughlin (NIST, Washington) and J. Golson (ANU, Cannbera) for reading related chapters and correcting English.

My former staffs, Drs. T. Miki, A. Kai, secretaries Ms. J. Masaki and C. Kattoh (Yamaguchi Univ.) for their help and dedication, and Ms. R. Suehiro, Mr. A. Hochi and Dr. T. Miki for cartoons as an interdisciplinary language.

Coworkers Drs. A. Poulianos (Greece), J. L. Bischoff (USGS, Menlo Park), K. Ohmura (GEOL), M. Koba (Kansai Univ.), T. Nakata, S. Hoshi (Hiroshima Univ.), K. Tanaka (Inst. Elect. Power Ind.), T. Kobayashi (Hokuriku Univ.), J. Tatsumi (Kyoto Univ.), S. Okajima (Nagasaki Univ.), M. Iwasaki, T. Shimano (Ohu Univ.), Morinaga (Himeji Tech. Col.), Y. Tsuji (Jpn. Nat. Oil), H. Hara (Microdevice Co.) and A. Nakanishi (Sumitomo Sepc. Metal Co.) for discussions.

Staffs and students in Osaka University; Drs. M. Furusawa, S. Toyoda, C. Yamanaka, S. Ikeda, H. Ishii, and Mr. K. Ogoh, M. Hirai, Y. Tsukamoto, H. Miyamaru, K. Meguro, A. Hochi, H. Kohno, H. Sasaoka, T. Omura, M. Yamamoto and part time secretaries Ms. Y. Aoki, C. Mizushima and others.

Following organizations supported this work financially; Asahi Newspaper Co., and Yamada, Nissan and Kashio Science Foundations in addition to the Scientific Grants from the Ministry of Education of Japan.

For permission to reproduce figures and tables, I am indebted to the following publishers; Academic Press Ltd. (Table 2.3 and Fig. 9.1), American Institute of Physics (Fig. 2.22, 2.23, 7.5, 11.2, 11.5, 11.9), American Physical Society (Fig. 2.27), Blackwell Scientific Publications (Fig. 5.2), Conseil de l'Europe (Fig. 3.7, 4.4), Elsevier Science Publishers (Fig. 9.4, 10.15), Health Physics Society (Fig. 13.4), Longman Group UK (Fig. 6.2, 7.1), Macmillan Magazines Ltd. (Fig. 3.7, 8.15, 10.8), Material Science Society (Fig. 10.16), Pergamon Press (Tables 4.7, 10.2 and Fig. 5.8, 6.7, 7.6, 8.5, 9.14, 13.15), Royal Society of Chemistry (Fig. 11.1, 11.3), Springer-Verlag (Fig. 1.8, 2.16, 2.25, 7.1, 10.2, 10.13, A2.1–2.3), Springer-Verlag Tokyo (some figures from a book "*ESR Microscopy*" by M. Ikeya and T. Miki in Japanese) and Ionics Co. (from my book "*ESR Dating*").

On 2nd printing:

This book with tabulation of ESR parameters and spectra and almost complete references has been used as a guiding Atlas and a dictionary for radiation dosimetry and geological applications. After the publication of the former edition, seven years have passed. The rapidly developing fields seem to be matured, though breakthroughs are anticipated at the beginning of 21st century. Corrections of minor mistakes and some minor revisions and addition of new references up to middle of 2001 have been made in the 2nd printing. Table 9.4 for dangling bonds at the interface of Si and SiO_2 on Si-wafer would be useful in identifying new signals not only in semiconductor technology but also in geological materials. Main revisions are in Chapter 2 for use of pulse ESR, Chapter 9 to include recent works and references, Chapter 11 to include solid SO_2 at Io, and Chapter 13 on food irradiation monitoring and ESR dosimetry of tooth enamel. New accidents such as Techa River and the JCO critical accident in 2000 are briefly cited. Future revision will be made based on _International Symposium on ESR Dosimetry and Dating (2001-ESRDD-Osaka)_. Developments can be seen in the home page (http://quartz.ess.sci.osaka-u.ac.jp) for details. September 14, 2001.

Contents

Contents

Units, Constants and Notations

Units

a	:	Annum (=year) : ka = kiloannee, Ma = megaannee, Ga = gigaannee
Gy	:	Gray (SI unit of the absorbed radiation dose): 1 Gy = 1 J/kg = 100 rad
T	:	Tesla (SI unit of the magnetic flux density) : 1 T = 10^4 gauss

Constants

Avogadoro number	: N_A	$= 6.0221367 \times 10^{23}$ mol^{-1}
Bohr magneton	: β	$= 9.2740154 \times 10^{-24}$ J·T^{-1}
Nuclear Bohr magneton	: β_N	$= 5.0507866 \times 10^{-27}$ J·T^{-1}
Boltzman constant	: k_B	$= 1.380658 \times 10^{-23}$ J·K^{-1}
Planck constant	: h	$= 6.6260755 \times 10^{-34}$ J·s

Notations

Many abbreviations are used in this book but only those relevant to dating and dosimetry are listed below. Others are explained in each chapter. Some letters are doubly used, for example, T for the age and temperature. The readers are asked to allow the fuzziness and guess the meaning from the text.

A	:	hyperfine (hf) coupling constant
		A_{xx}, A_{yy} and A_{zz} for orthorhombic symmetry
		$A_{\parallel}$ and $A_{\perp}$ for axial symmetry
A_{shf}	:	superhyperfine (shf) coupling constant
AD	:	accident (or A-bomb) dose or archaeological dose
β	:	Bohr magneton
D	:	axial field splitting parameter for fine structure
D	:	annual dose rate of natural radiation
		D_α for α-rays, D_β for β-rays and D_γ for γ-rays
		D_{ex} for external and D_{in} for internal radiation
D'	:	artificial irradiation dose rate
E	:	orthorhombic field splitting parameter for fine structure
E	:	activation energy
E_z	:	Zeeman energy
ED	:	equivalent dose obtained by an additive dose method (cf. TD)
G-value	:	number of radicals formed per 100 eV
g factor	:	spectroscopic splitting factor
		g_N for nucleus, g_e for a free electron
		g_{xx}, g_{yy} and g_{zz} for orthorhombic symmetry
		$g_{\parallel}$ and $g_{\perp}$ for axial symmetry ($g_{zz} = g_{\parallel}$ and $g_{xx} = g_{yy} = g_{\perp}$)
		g_{iso} for an isotropic signal and g_{av} for average
Δg	:	g-shift caused by spin-orbit interaction

H	:	magnetic field; the unit is tesla (T) or mT
		H_o for resonance magnetic field $(h\nu = g\beta H_o)$
h	:	Planck constant
hf	:	hyperfine
I	:	nuclear spin quantum number
I	:	ESR signal intensity
K	:	rate constant of chemical reaction
		K_1 for formation and K_2 for decay
k–value	:	defect production efficiency by α–rays
λ	:	decay constant of radionuclides or second–order decay constant
M	:	magnetic quantum number of electron spin, S
m	:	magnetic quantum number of nuclear spin, I
N	:	number of atoms or traps
n	:	defect concentration
ν_o	:	preexponential factor
ν	:	frequency of electromagnetic wave
P	:	microwave power in mW
Q	:	radiation dose; $Q = D t$ (natural) or $D't'$ (artificial)
ρ	:	Density
S	:	electronic spin quantum number
$S(H)$	:	signal shape function
shf	:	superhyperfine
τ	:	lifetime of a defect
t	:	time
T	:	temperature
		T_a for annealing temperature
		T_g for geothermal temperature
T or T_{ESR}	:	ESR age
		T_d for the age considering ^{238}U–series disequilibrium
		T_e for the age for radioactive equilibrium
T_1	:	spin–lattice relaxation time
T_2	:	spin–spin relaxation time
$T_{1/2}$	:	half–life of radioactive elements
TD	:	total dose of natural radiation (cf. ED)

Chapter 1

Clocks of Elapsed Time

– The Place of ESR Dating –

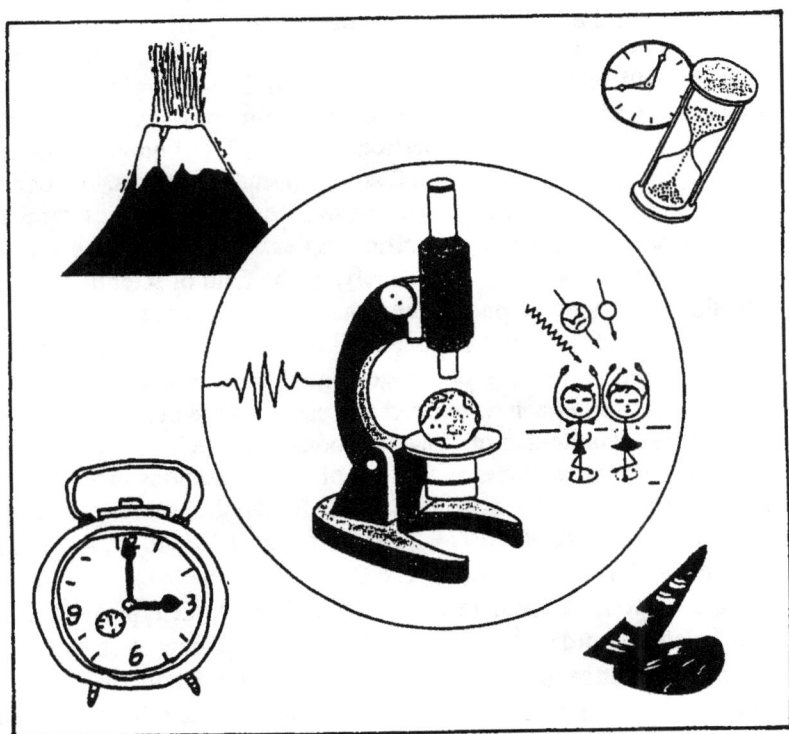

The principles of dating techniques in archaeology and geology are described briefly, to place electron spin resonance (ESR) dating among other dating methods. There are three general categories: (1) radioactive disintegration (based on half life, $T_{1/2}$), (2) accumulated radiation effect and (3) chemical reaction. ESR dating in earth sciences and archaeology belongs to category (2) and the one in history and forensic sciences to category (3).

1.1   Introduction

People age as time passes, forming histories of their own and interacting with their societies. History records important events with the time of their occurrences, whether it is a history of a person, a family, a community or a nation. Ancient cultures and societies which leave no written history can be traced back by archaeological studies. Ancient human remains studied by physical anthropologists improve our knowledge of the origin and development of early human beings. Animals and plants left as fossils in soil sediments are good stratigraphic markers and indicators of geologic ages. Tectonic movements, sea level changes and volcanic activities can also write the history of the earth and its inhabitants, since they can be the markers for important events in geological, prehistoric and protohistoric times.

Physics and chemistry have been involved in solving archaeological and geological problems. Several techniques from physics and chemistry have been applied to answering questions of "*why?*", "*how?*", "*where?*" and "*when?*" in studies of the natural and man–made objects. Geophysics and geochemistry in the earth sciences as well as archaeophysics (archaeometry) are examples of such scientific studies. Answering the question "**when?**", whether absolutely or relatively, is the field of scientific dating.

Physical or chemical phenomena that act as clocks for elapsed time and record it in some form have been utilized to find the age or the past passage of time for archaeological or geological objects and events. Several dating techniques have been developed and used and are being continuously refined for more accurate dating. Several books and articles are available on geochronology and archaeometry. Details of dating methods including limitations and applications to their respective fields should be consulted in appropriate textbooks (Aitken 1974, 1985, Fleming 1975, 1979, Rutter 1985, Zimmerman and Angel 1986, Zurer 1983).

In this chapter, the principles and methods for some typical dating are described briefly in order to understand the role of the new method, "**Electron Spin Resonance (ESR) Dating**". ESR is a method of microwave absorption spectroscopy used in physics and chemistry and is being used in earth sciences (Marfunin 1979). To understand the other dating techniques will be useful when we compare ESR ages with ages determined by other methods. First, the methods of dating, in which time is measured by the radioactivity decay curve (half–life), are described with examples of radiocarbon and uranium–thorium dating. Second, the methods utilizing the accumulation of natural radiation effects are presented. These involve fission track dating and thermoluminescence (TL) dating. ESR dating belongs to this category, though it is also a technique of chemical dating for organic materials. Third, the methods utilizing chemical reactions, and last, other methods are described briefly.

1.2 Radioisotope Dating

Radioactive elements emit $\alpha-$, $\beta-$ or $\gamma-$rays (He nuclei, electrons and electromagnetic waves, respectively) as shown in Figure 1.1 (a). Radioactive elements decay with a characteristic *half–life*, $T_{1/2}$, as

$$dN/dt = -\lambda N = -(0.693/T_{1/2})N , \tag{1.1}$$

or

$$N = N_0 \exp[-(0.693/T_{1/2})t] = N_0 (1/2)^{t/T_{1/2}} , \tag{1.2}$$

where N_0 and N are the numbers of the radioactive atoms present at time $t =$ 0 and at t, respectively. λ is the *decay constant* $(\lambda = 0.693/T_{1/2})$ and the radioactivity in disintegrations per second or per year is expressed as $-\lambda N$. The exponential decay of a radioactive element per unit of $T_{1/2}$ is shown in (b). The amount of radioactivity becomes extremely small as the time passes beyond several times the half–life of the radioisotope.

(a)   (b)

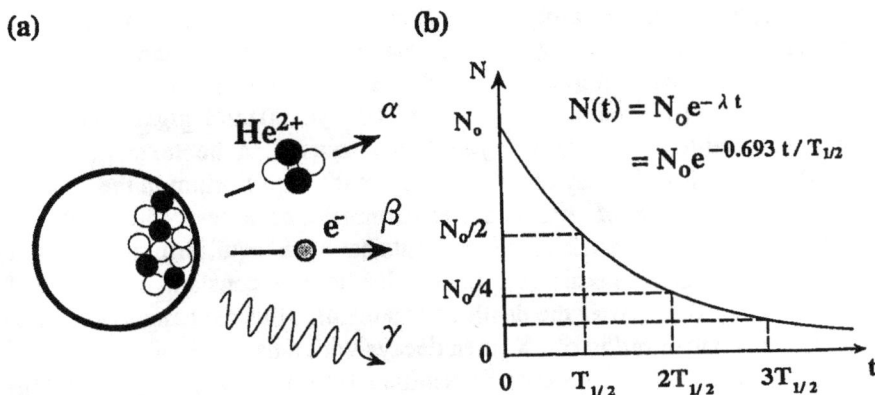

Figure 1.1 (a) Disintegration of a nucleus by the emission of $\alpha-$, $\beta-$ and $\gamma-$rays with a characteristic half–life $(T_{1/2})$. (b) The exponential decay of the number of nuclei is shown as a function of time, t, normalized by $T_{1/2}$.

As the decay of radioisotopes is a nuclear reaction deep in the nucleus, it is not affected by chemical reactions or the surrounding atomic structures. The development of nuclear technology has made the measurement of this radioactivity reliable. Consequently, radioisotope dating, especially radio-carbon dating has been accepted as a standard dating method among absolute dating techniques. Unless radioactive elements move in from the environment or leach out from the specimen, the radioactive content is a good indicator of elapsed time. Such an ideal sample is called a *"closed system"*.

1.2.1 Radiocarbon (^{14}C) Dating

This method, well accepted as a reliable dating technique, was proposed by W. E. Libby in 1946 and brought him a Nobel Prize in Chemistry. Naturally occurring carbon–14 ($^{14}_{6}C$) is produced by the nuclear reaction between nitrogen ($^{14}_{7}N$) and thermal neutrons ($^{1}_{0}n$) of cosmic ray origin in the upper atmosphere:

$$^{14}_{7}N \ + \ ^{1}_{0}n \ \Rightarrow \ ^{14}_{6}C \ + \ ^{1}_{1}H \ . \tag{1.3}$$

^{14}C decays with a half–life of 5,570 years by emitting a β–ray (electron) as

$$^{14}_{6}C \ \Rightarrow \ ^{14}_{7}N \ + \ e^{-} \ . \tag{1.4}$$

The sum of atomic mass numbers and atomic numbers is not changed before and after the nuclear reaction. The internationally preferred term for year, "annum [a]", is used for a half–life and an age along with **ka**, **Ma** and **Ga** for *kiloannee* (10^3 years), *megaannee* (10^6 years) and *gigaannee* (10^9 years), respectively. The term "year" is also used in the text.

Production and decay of ^{14}C in CO_2 reach equilibrium in the atmosphere. The presence of ^{14}C in our environment as a reservoir is shown schematically in Figure 1.2. Living plants have the equilibrium concentration of ^{14}C in their organic constituents due to the constant exchange of carbon during life. After the death of organisms, this exchange of carbon ceases. The concentration of ^{14}C then decays with a half–life of 5,570 years. The age after death can thus be determined by counting β–rays from the extracted carbon. The concentration of ^{14}C is usually much less than 1 part per billion (1 ppb) of the stable ^{12}C, but the radiation from ^{14}C can be measured by modern equipment.

It has been found, however, that the production rate of ^{14}C in the upper atmosphere is not constant due to a fluctuation in cosmic radiation.

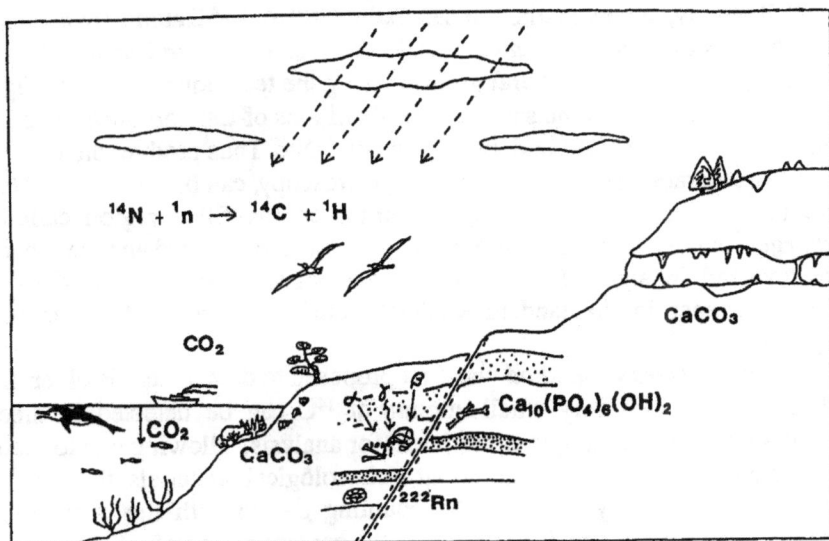

Figure 1.2 Radiocarbon (^{14}C) produced by cosmic ray–induced neutrons (^1n) in the upper atmosphere is kept in the reservoir of ocean, atmosphere, land and biosphere. Natural radiation, $\alpha-$, $\beta-$ and $\gamma-$rays are from ^{238}U, $^{232}Th-$ series disintegration and ^{40}K in soils and cosmic rays. Samples used for dating are indicated in the figure.

Comparison with *dendrochronology* (tree ring dating) has been the main source of calibration of radiocarbon ages with calendar dates. Correction of ^{14}C–ages must be made to derive an absolute age. It should be noted that errors expressed as "±" for ^{14}C–ages are statistical counting errors and do not necessarily mean that the true age lies within the error range given.

The time range of ^{14}C–dating is limited by the half–life of 5,570 years to from a few hundred years to about 40,000 ~ 50,000 years BP (before present). Radioactivity decreases exponentially, as is shown in Figure 1.1, until it becomes too low to be detected in old samples. A small amount of exchange of carbon from CO_2 in atmosphere or from water containing CO_2 results in a large error in the ^{14}C–age. Unexpectedly young ages have often been attributed to such modern carbon contamination. The operational assumption that the sample is a system isolated from the modern carbon environment (a closed system) may break down for old samples due to the very small amount of ^{14}C contaminated from the environment.

Recently, a new method of separating ^{14}N and ^{14}C atoms by means of a mass spectrometer and an *ion accelerator*, called a *tandem accelerator*, has been developed by several groups, using the technique shown in Figure 1.3. Bombardment of the sample with alkali ions of Cs^+ produces negative ions of $^{12}CH_2^-$, $^{13}CH^-$ and $^{14}C^-$, but not $^{14}N^-$. Thus production of $^{14}N^-$ ions, inseparable from $^{14}C^-$ by mass spectroscopy, can be avoided. Negative ions are accelerated through a thin foil or gas–filled region called a *"charge stripper"*, in which molecular ions are decomposed and the charges are reversed in sign. Thus, triple charged positive ions produced by the charge stripper in the tandem accelerator are led to the positive ion mass separator.

The *accelerator method* was first proposed to date materials older than 50,000 years since very small amounts of ^{14}C can be detected in theory. Small amount of the sample is sufficient for analysis, allowing one to obtain an adequate sample from important archaeological materials that are too precious to be totally sacrificed for counting β–rays with the conventional method. However, a major problem is the contamination of ^{14}C by modern carbon. The application to isotope separation of rare elements using accelerator mass spectroscopy will open up a vast field in archaeology and geology.

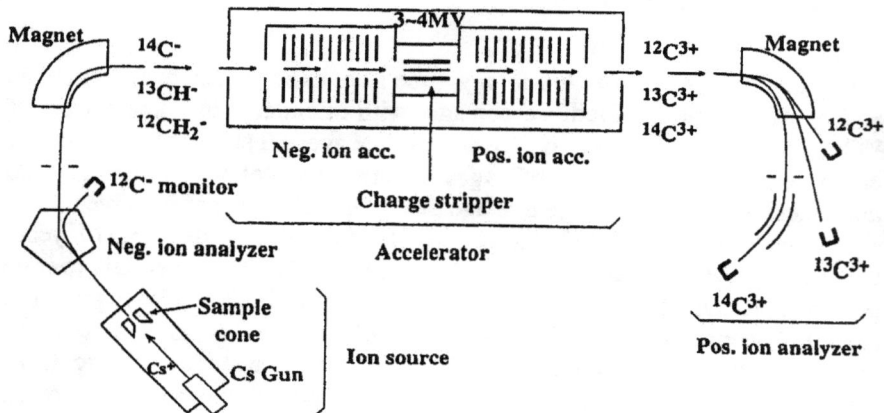

Figure 1.3 A tandem accelerator and mass analyzer system for ^{14}C dating. ^{14}N and negative ions $^{14}C^-$, $^{13}CH^-$ and $^{12}CH_2^-$ are separated at the negative ion analyzer on the left. The charge stripper in a tandem accelerator decomposes molecules and reverses the charges of carbon isotopes, which are separated by the positive ion mass analyzer to get $^{14}C^{3+}$ ions.

1.2.2 *Potassium–Argon* (^{40}K–^{40}Ar) *Dating*

Radioactive potassium–40 (^{40}K) present in nature is only 0.012% of stable potassium (^{41}K and ^{39}K). ^{40}K disintegrates into ^{40}Ca by emitting a β particle or into ^{40}Ar by capturing an electron as shown in Figure 1.4 (a). Gaseous argon is not incorporated when rocks involving potassium are formed. ^{40}Ar accumulates as a function of $[\exp(\lambda t) - 1]$ as time elapses due to the nuclear reaction of *electron capture* by ^{40}K. As the half life of ^{40}K is extremely long, 1.25×10^{10} years, the decay rate is almost constant over short periods. Thus, ^{40}Ar accumulates linearly as time elapses and one can detect ^{40}Ar by mass spectrometry. The lower age limit of this dating method is about half a million years in ordinary cases. Recent developments lower this limit considerably for some particular specimens. The age may be erroneous if argon is not well retained in the rock. The principle of other dating methods such as those involving $^{87}Rb/^{87}Sr$ is the same.

Inert helium gas is also formed by the α–*disintegration* of radioactive elements in the ^{238}U and ^{232}Th disintegration series. However, in most materials, helium–4 (^4He) is dissipated more rapidly than argon. Geothermal heating may cause the dissipation of helium or argon from minerals, making this dating method useless. However, the time clock is normally set to zero at the time of formation or eruption in samples of volcanic origin.

(a)

^{40}K (0.0118 %)

K-Capture
11%
1.3×10^9 a
88.8%
^{40}Ar

0.16%

^{40}Ca

^{40}Ar

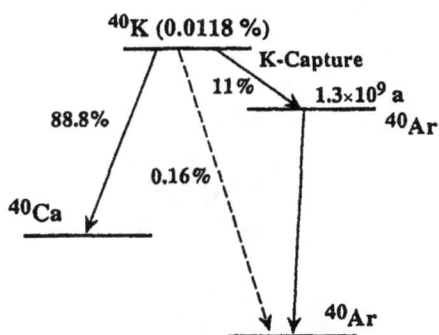

(b) **Radioactivity Dating Methods**
(Parent $\rightarrow$ Daughter)

Methods		$T_{1/2}$ [a]
P	$\rightarrow$ D	
^{40}K	$\rightarrow$ ^{40}Ar	1.25×10^9
^{87}Rb	$\rightarrow$ ^{87}Sr	4.88×10^{10}
^{147}Sm	$\rightarrow$ ^{147}Nd	1.06×10^{11}
^{176}Lu	$\rightarrow$ ^{176}Hf	3.60×10^{10}
^{187}Re	$\rightarrow$ ^{187}Os	4.56×10^{10}

$$D = P_o[\exp(\lambda t) - 1]$$
$$T = \log_e(1 + D/P_o)$$

Figure 1.4 (a) Radioactive decay of ^{40}K (parent: P) into ^{40}Ar (daughter: D) or ^{40}Ca. (b) The accumulation of the daughter D determined by a mass analyzer and the analysis of present parent number, P, are the basis of the dating method. In case of K–Ar dating, $^{40}Ar = {^{40}K}(\lambda_{Ar}/\lambda)[\exp(\lambda t) - 1]$, where λ_{Ar} is the decay constant of ^{40}K to ^{40}Ar and $\lambda = \lambda_{Ca} + \lambda_{Ar}$.

A convenient method of "**Isochron Dating**" has been developed to avoid the errors caused by the initial content of daughter elements, D_i in samples from the same rock or of separated minerals. The analyzed daughter contents are plotted against those of the parent (^{40}Ar versus ^{40}K for K–Ar dating). The slope gives the $[\exp(\lambda t) - 1]$ and the abscissa gives the initial ratio. An isochron dating in ESR dating is described in Chapter 4 (4.5.4).

1.2.3   *Uranium–Thorium* (^{230}Th/^{234}U) *Dating*

Several other dating techniques have been developed by utilizing differences in half–life or decay constant. Among these, **uranium–thorium** (^{230}Th/^{234}U) **dating** is a method utilizing the activity ratio due to *radioactive disequilibrium* of ^{238}U*–series disintegration* shown in Figure 1.5 (a). It can determine ages from about 10 ka to about 400 ka (Ivanovich and Harmon 1982). In a closed system like corals, ^{238}U and ^{234}U are taken in at the time of formation. Once ^{238}U disintegrates, ^{234}U is soon formed. The half–life of ^{234}U to disintegrate into ^{230}Th in radioactive equilibrium is 2.48×10^5 years, while that of ^{230}Th into ^{222}Rn is 7.52×10^4 years.

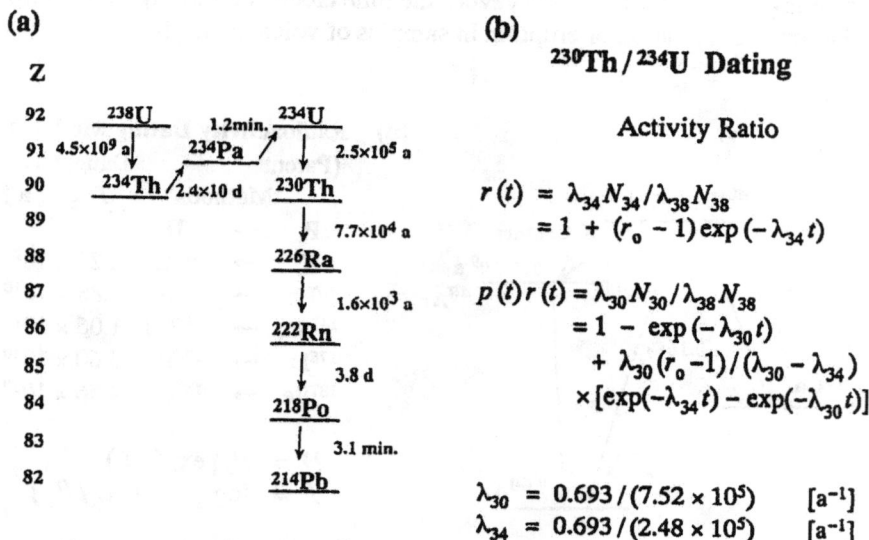

(a)

Z

92	^{238}U	1.2min.	^{234}U
91	4.5×10⁹ a↓	^{234}Pa ↗	↓ 2.5×10⁵ a
90	^{234}Th ↙ 2.4×10 d		^{230}Th
89			↓ 7.7×10⁴ a
88			^{226}Ra
87			↓ 1.6×10³ a
86			^{222}Rn
85			↓ 3.8 d
84			^{218}Po
83			↓ 3.1 min.
82			^{214}Pb

(b)

^{230}Th/^{234}U **Dating**

Activity Ratio

$$r(t) = \lambda_{34}N_{34}/\lambda_{38}N_{38}$$
$$= 1 + (r_0 - 1)\exp(-\lambda_{34}t)$$

$$p(t)\,r(t) = \lambda_{30}N_{30}/\lambda_{38}N_{38}$$
$$= 1 - \exp(-\lambda_{30}t)$$
$$+ \lambda_{30}(r_0-1)/(\lambda_{30}-\lambda_{34})$$
$$\times[\exp(-\lambda_{34}t) - \exp(-\lambda_{30}t)]$$

$$\lambda_{30} = 0.693/(7.52 \times 10^5) \quad [a^{-1}]$$
$$\lambda_{34} = 0.693/(2.48 \times 10^5) \quad [a^{-1}]$$

Figure 1.5 (a) ^{238}U–series radioisotopes with half–lives. The build–up of ^{230}Th (Ionium, Io) is the basis of uranium–thorium (^{230}Th/^{234}U) or ionium dating. (b) Activity ratios of ^{234}U/^{238}U and ^{230}Th/^{238}U with the decay constants, λ_{34} and λ_{30}.

^{230}Th which is initially absent is built up gradually from zero and approaches radiation equilibrium. As a result, the activity ratios of ^{234}U to ^{238}U and ^{230}Th to ^{234}U can be described as a function of age and the age can be determined. The derivation of the activity ratio is possible by solving the differential equations for decay of uranium–series radioactive elements shown in Figure 1.5 (b). Experimentally the activities of ^{230}Th, ^{234}U, etc., have been measured using *α–ray spectroscopy* after extensive radiochemical separation. Alternatively, *γ–ray photon spectroscopy* is used with a Ge(Li) detector or a pure Ge detector and a *pulse height analyzer*. The delayed build–up of ^{230}Th is discussed together with details of γ–ray spectroscopy for the assessment of the annual dose in Chapter 4.

^{230}Th/^{234}U dating has been successful for a closed system such as corals but not so successful for a partially open system such as shells and cave deposits. Leaching or uptake of parent (and/or daughter) elements results in erroneous ages. In γ–ray spectroscopy, it is necessary to grind a sample and compare the radioactivity with a standard so that the geometrical factors for the measurement become the same. In this respect, the measurement is destructive for the sample as in ^{14}C–dating. However, an anthropologically important hominid cranium from Arago, Tautavel in France, was dated non–destructively by calibrating the radioactivity response using a standard man–made model of the cranium with known radioactivity as shown in Figure 1.6 (Yokoyama and Nguyen 1981). The real and surrogate

(a)   (b)

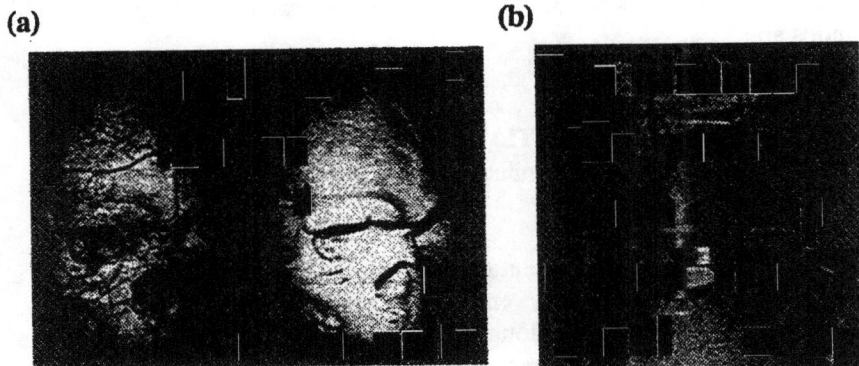

Figure 1.6   (a) A picture of the cranium excavated at Arago Cave, Tautavel in southern France, and its replica with a known amount of pitchblende as a radiation standard for ^{230}Th/^{234}U dating. (b) A low background γ–ray spectroscopy using a semiconductor Ge detector (courtesy by Y. Yokoyama at Low Radioactivity Lab., Gif–Sur–Yvette).

skulls were positioned in the same geometrical configuration in the liquid–nitrogen–cooled Ge(Li) detector. The ^{230}Th/^{234}U age of *Tautavel man* was 450 ± 150 ka, actually beyond the upper limit of this dating method.

Recently, a mass spectrometer has been used for the determination of ^{230}Th to expand the time range (Edwards *et al.* 1987). Although chemical separation is necessary and the use of an accelerator costs, this method will allow comparison with the results of ESR dating. Cross checking of ESR ages with ^{14}C– and ^{230}Th/^{234}U ages is frequently mentioned in this book.

1.3 Radiation Effect Dating

Radioactive elements emit α–particles, β–rays and γ–rays which produce atomic or electronic defects in solids as shown in Figure 1.7. This is called *"radiation effects"*. Particle radiation causes defects, vacancies and interstitials along a path called a *"track"*. Ionizing radiation such as β– and γ–rays or even light produces *"exciton"*, a pair of an electron and a hole (electron–deficient center) which leads either to recombination luminescence or to formation of an interstitial atom and a vacancy in insulating solid materials. Some of electrons and holes are trapped at the normal lattice site (self–trapping) or at the site of impurities and vacancies and are stabilized for millions of years.

One can date materials by observing the accumulated defects caused by radiation in solids. Any physical technique that can detect radiation effects can also be used as a method of dating in this category. Typical dating methods are:

(1) Fission track dating,
(2) Thermoluminescence (TL) dating and ESR dating, and
(3) Optically stimulated luminescence (OSL) dating.

1.3.1 *Fission Track Dating*

Uranium–238 (^{238}U) spontaneously fissions into two nuclei with a half–life of 1.01×10^{16} years, emitting a significant amount of energy. Atomic defects are produced along a narrow path – *"fission track"* – of the heavy fission fragments. The resultant tracks are $25 \sim 40$ Å in diameter and $10 \sim 20 \, \mu$m in length. An *"ion–explosion spike"* has been proposed as a model for formation of fission tracks (Fleischer *et al.* 1975). Positive ions densely produced in a cylindrical channel along the path are displaced into interstitial positions.

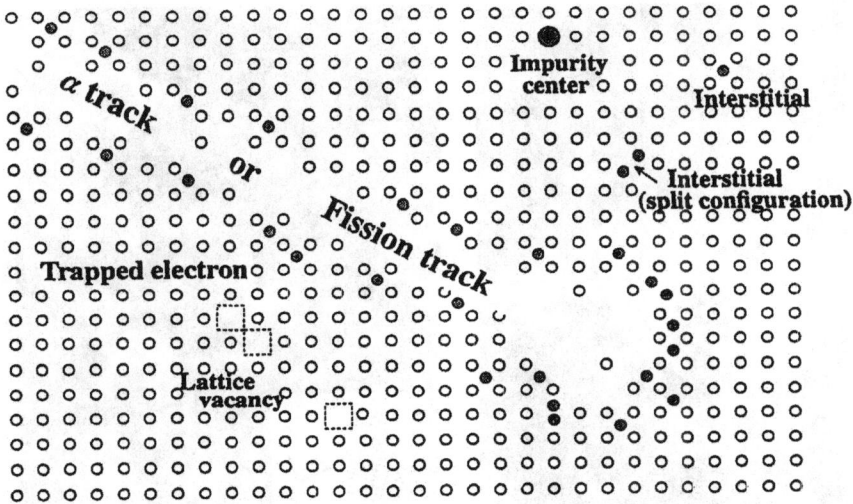

Figure 1.7 Radiation effects produced by radiation of α–particles and fission tracks. Point defects (vacancy and interstitial) and tracks are formed along the path of the charged particles. β– and γ–rays can produce point defects and trapped electrons (holes) in insulating (ceramic) materials.

A micrograph of fission tracks in *zircon* is shown in Figure 1.8 (a). The density of fission tracks in minerals is counted under an optical micro-scope after etching the surface of the specimen with acid or alkali. A fission track and point defects have been observed in zircon with electron micros-copy as shown in (b). The number of point defects may be counted directly with a more powerful high resolution microscope such as a scanning atomic force microscope, which will be called "**microscope dating**".

The content of ^{238}U can be assessed by irradiation of the specimen with *thermal neutrons* in a nuclear reactor, since the newly formed tracks due to the induced nuclear fission of the ^{235}U isotope are proportional to the ^{238}U content. The time range covered depends on the ^{238}U content and the degree of heating because the tracks (essentially atomic disorders or a locally high concentration of defects) are annealed by heating the specimen. This is a limitation of all methods based on radiation effects, though the phenomenon might be utilized for geothermal studies.

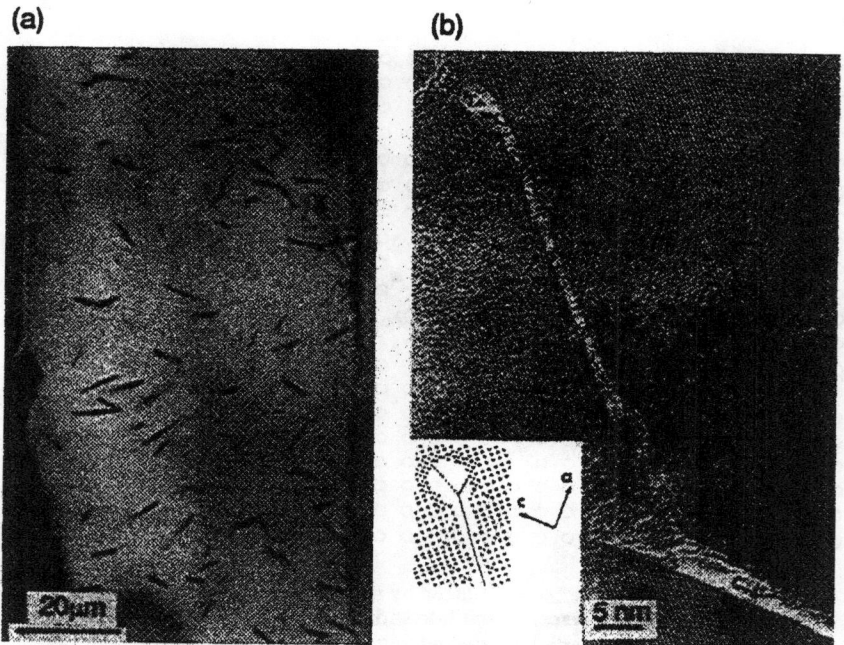

Figure 1.8 (a) A micrograph of fission tracks in a zircon ($ZrSiO_4$) crystal chemically etched with hydrofluoric acid (courtesy by Dr. M. Kasuya). (b) Electron microscopic (1 MeV) observation of fission tracks and point defects (arrowhead in the photograph) in zircon (Photograph by Yada *et al*. 1987).

Counting of tracks under a microscope is tedious. Computer process-ing of track images is developed but ambiguous due to the miscounting of surface scars, dislocation and microinclusion etch pits. Attempts to count bulk tracks may be made by ESR since electron spins can be measured in an amorphous state of the tracks. Decoration of tracks with some paramagnetic spin labeling compound is another way to count tracks with ESR though uniform decoration of only etch pits and not surface scars is difficult.

1.3.2 *Thermoluminescence (TL) and Electron Spin Resonance (ESR)*

Unpaired electrons and holes produced by natural radiation in minerals are trapped by impurities or lattice defects originally present or induced by radiation. These electrons or holes released from traps by laboratory heating recombine, emitting light; the process is called "**thermoluminescence (TL)**"

(Aitken 1985, Daniels *et al.* 1953, McKeever 1989). The emission intensity is proportional to the concentration of trapped electrons and therefore to natural radiation dose that the material has received.

Figure 1.9 illustrates the principle of TL and ESR dating. The paired electrons, indicated by a spinning boy and girl in (a), are split and ejected by natural ionizing radiation such as α–, β– and γ–rays in (b). The female unpaired electron is trapped by an impurity or inherent defect represented by a gangster in (c). TL is observed by heating the material and releasing the trapped electron (girl), from her captor to meet the hole (boy) in (d).

The apparatus used in TL dating is shown in Figure 1.10 (a). Samples are heated in vacuum or in an inert gas atmosphere and the emission of weak light is converted into an electric current with a photomultiplier. The TL light intensity is recorded as a function of time or temperature. Photon count- ing is now widely used to detect the weak light. Recently, a photodiode, which does not require a high voltage source, has been used in a small portable TL reader for student experiments (Ikeya *et al.* 1990).

In TL dating, the *TL light intensity* must be measured in one portion of the sample before and in other portions after artificial irradiation to assess the sensitivity to the radiation dose as shown in (b). This is called the "**additive dose method**" (see Chapter 3). The *total dose of natural radiation* (*TD*), usually called the *archaeological dose* (*AD*) or *equivalent dose* (*ED*), is deduced from the relationship between the TL intensity and the artificial irradiation dose. The age can be obtained by dividing the *TD* with the annual

Figure 1.9 Illustrating the principle of TL and ESR dating. (a) Paired elec- tron are spinning in opposite directions. (b) α–, β– or γ–rays knock off one of the paired electrons (girl) leaving the other (boy) at a hole. (c) The un- paired electron (girl) is trapped by a impurity (gangster). (d) Heating releases the electron (girl) and allows recombination with the hole (boy), emitting light called TL. ESR dating is a method of observing the trapped electron at (b) directly, by means of microwave absorption spectroscopy called ESR.

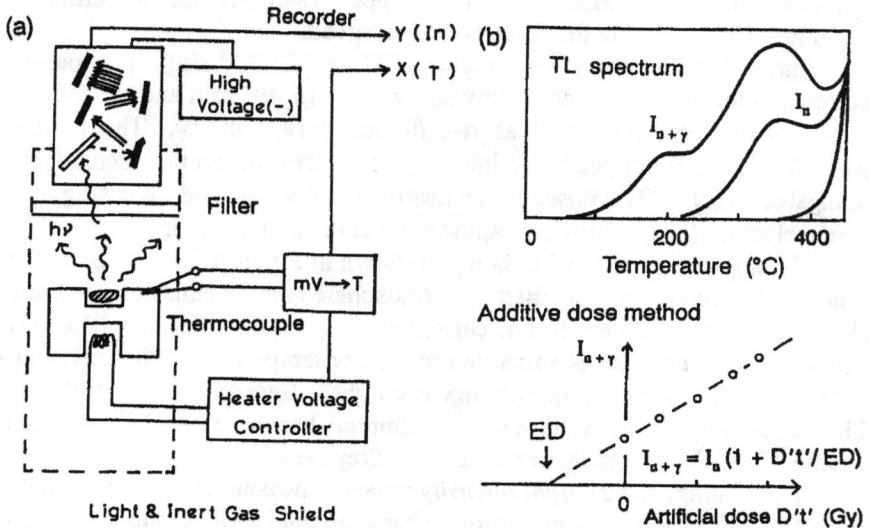

Figure 1.10 (a) Schematic drawing of TL dating equipment. (b) TL curves showing the measurement of the total dose of natural radiation (*TD*) or archaeological dose (*AD*) by the additive dose method.

dose rate of natural radiation. Chemical reactions, such as oxidation or decomposition of biological materials, produce spurious luminescence called *chemiluminescence* (CL). Absorbed oxygen causes CL in fossil bones, while mechanical grinding produces defects responsible for the emission of light called *tribo–thermoluminescence* (TTL). These are the particular difficulties in TL dating of bones and other biological materials.

Methods for TL dating of a number of inorganic materials have been developed considerably in the last decades. Reliable and systematic studies of TL dating have been directed to baked ceramics of archaeological remains and then to geologic loess, quartz, feldspars, zircon and other minerals (Aitken 1985, Fleming 1979). The age of mineralization or heating can be obtained from TL, and the application to geological materials has been studied (McDougall 1968, Sankaran *et al.* 1983), though some defects are not sufficiently stable for more than a few million years, the Quaternary Period. Recent studies are mostly *sediment dating* based on *sun–light bleaching* of defects, i.e., zeroing the TL clock by sunlight exposure (Wintle and Huntley 1979, Wintle and Proszynska 1982, Singhvi *et al.* 1982).

The principle, procedures and radiation dosimetry required for age derivation from the *TD* or *ED* in ESR dating are essentially identical to those in TL dating. ESR is a method of observing the trapped electrons in Figure 1.9 (c) directly, by measuring the microwave absorption, without allowing release and recombination with a hole. In contrast to TL dating, ESR is non–destructive, neither destroying the accumulated defects nor erasing the information contained. The time range covered by TL or ESR dating depends on the stability of the trapped electrons, i.e., the depth of the trap in the energy band scheme, or on the temperature at which TL occurs. In some materials, it may extend to a few million years far beyond the range of ^{14}C–dating (50,000 years BP) covering the whole Quaternary Period (Ikeya and Miki 1985, Ikeya 1987, Grün 1989).

1.3.3   *Optically Stimulated Luminescence (OSL)*

Among the physical methods that can detect lattice defects or unpaired electrons/holes, *optically stimulated luminescence* (OSL), *exoelectron emission* (EEE), *thermostimulated current* (TSC) and optical absorption have been studied to develop convenient radiation dosimeters. Each method has its own advantages and disadvantages if used as a method of dating. OSL was once widely used as a method of radiation dosimetry, using silver–activated phosphate glass rods called "Yokota glass". The method was developed as **"laser dating"** or **"OSL dating"** by using UV nitrogen laser and then applied to dating of carbonate deposits in caves (Ugumori and Ikeya 1980). Pulsed laser light allows the separation of luminescence with the emission wavelength and its lifetime as shown in Figure 1.11 (a). This allows detection of the radiation dose to the order of μGy (gray: 1 Gy = 1 J/kg).

Figure 1.11 (b) shows the processes electrons and holes undergo in the energy band structure scheme. Light is emitted when the trapped electron is excited by light from the ground state to the excited state. The atoms around the excited electron move and the lattice relaxes by emitting *phonons* (*lattice vibration*) leading to the *relaxed excited state*, from which the electron goes back to the ground state by emitting a *photon* or the light. This unrelaxed ground state shows the atomic configuration in a higher energy state than the real ground state. Hence, the photon energy of the emission is usually smaller than that of the excitation; the wavelength shifts to the red side. The difference is called the *"Stokes shift"*. The total population is not affected if only this process of photoluminescence is involved. However, the excited

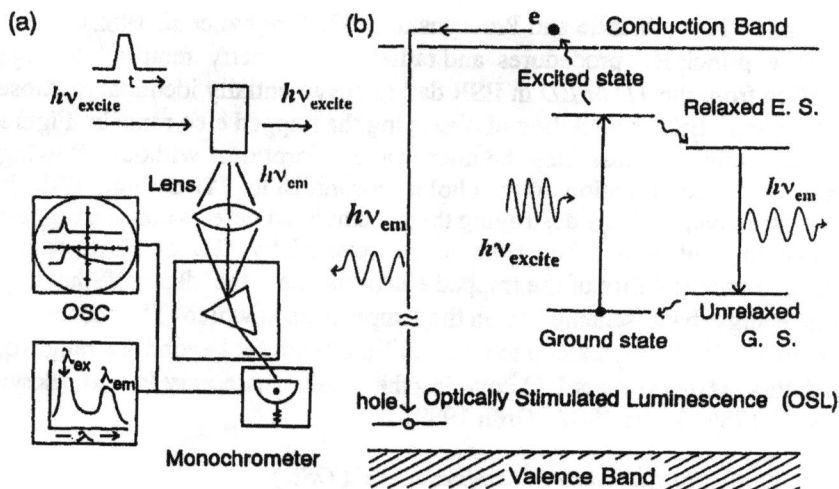

Figure 1.11 (a) Schematic drawing of *photoluminescence dating*. Mono-chromatic excitation light excites the electron (hole) centers created by natural radiation. The emission spectrum and the lifetime are measured. The emission intensity is proportional to the concentration of trapped electrons at the *ground state* (G.S.). (b) The electron in the *excited state* (E.S.) relaxes to the *relaxed excited state* (R.E.) and goes back to the unrelaxed ground state (U.G.) by emitting a photon. Lattice relaxation follows. Some electrons go to the conduction band from the R.E. thermally and recombine with a hole.

state is usually close to the conduction band. Some electrons are thermally ionized to the conduction band from the excited state, go to the hole center, and emit a photon by recombination, which leads to a decrease of the population. *UV- or laser-stimulated luminescence* of defects in some solids allows the dating of some minerals without heating the sample and emptying the trap, just as in ESR dating. A severe disadvantage of this method is that the sample needs to show OSL.

OSL dating has been developed and applied to quartz grains and feld-spars using *Ar laser, green light emitting diode* (LED) and *infrared light* (Aitken 1985, Huntley *et al.* 1985, Hütt *et al.* 1988). Many researchers involved in TL dating participated in this research, since the basic physics is the same as that of TL. Systematic studies with ESR will help to identify the defects and to understand the processes of OSL used for dating.

1.4   Chemical Dating
1.4.1   *Chemical Reactions and Analytical Dating*

A chemical reaction proceeds at a constant rate if the temperature, T, during the reaction is not varied, and this process can be the basis for the method of dating. The rate constant for a thermally activated process is generally of the form, $\exp(-E/k_B T)$ with *activation energy, E* and *Boltzman constant, k_B*. Hence, a small variation of T results in a large change of the rate constant. As chemical reactions depend very much on the surrounding environment such as temperature, ground water, the content of the relevant ions, chemical methods are often limited to auxiliary dating techniques.

The relative age determination of some fossils, especially fossil bones, can be made by assessing the content of fluorine, uranium, nitrogen and other elements. Nitrogen is leached out of bone, but conversely fluorine, uranium and others are gradually taken up from the environment. Thus, one can estimate age, at least relatively, by chemical analysis of these elements. ESR can be used as a tool for detecting paramagnetic Mn^{2+}, Fe^{3+} and Cu^{2+} ions in this analytical dating: old foraminifera contain a large amount of Mn^{2+} roughly proportional to the age (see Chapter 6).

The most famous application of analytical dating was in the refutation of the *Piltdown forgery*, a specimen long considered to be a fossil hominid representing a step in human evolution (Oakley 1949). Chemical analysis indicated that the bone was stained with an iron sulfate solution and oxidized with a dichromate solution. Apparently, the hoax was constructed from a fossil human cranium and modern ape mandible to fit the expected man–ape missing link.

Chemical reactions produce organic molecular fragments called "*radi-cals*" with an excess or a deficit of electrons similar to those produced by radiation. Such a molecule has an unpaired electron with *magnetic spin* properties. The number of such decomposed fragments with unpaired electron spin can be measured by ESR. A rough age might be estimated by solving the kinetic equations of radical formation. ESR detection of food oil oxidation has been used to determine the manufacturing date of foods such as fried potato chips using a digital ESR technique (Ikeya and Miki 1980). The extension to dating of dead bodies, papers, textiles and bloodstains is described in Chapter 12. Although ordinary ESR dating is based on radiation effects, ESR can be used as an auxiliary dating technique utilizing signals produced by chemical reactions.

1.4.2 *Amino Acid Racemization*

Amino acids are present in fossil shells, bones and teeth. Amino acids exist in two isomeric forms that have the same kind of atoms in the same proportion but rotate the plane of plane–polarized light in opposite directions. These isomers are called L– and D–*enantiomers*. Proteins in animals and plants consist solely of optically active L–type amino acids. After the death of the living organism, a chemical reaction called "racemization" converts L–amino acids to D–amino acids toward the equilibrium between the two amino acids. Hence, the amount of D–amino acids is tied to the age and the reaction is used to determine the age of biological materials after death, by studying the reaction rate at several temperatures (Abelson 1955). A known age obtained by another dating method for a sample buried under similar conditions is used to estimate the rate constant of the chemical reaction at the site. As various amino acids in proteins racemize at different rates, selection of appropriate amino acids allows the age determination beyond the limit of ^{14}C–dating. Fractionation of decomposed proteins increases the accuracy of the age (Matsu'ura and Ueta 1980).

As temperature variation results in an order of magnitude age difference, reliable ages are only obtained, for example, for fossil bones in a limestone cave where the temperature is nearly constant throughout the year (Bada and Deems 1975). However, temperature variation over long time periods, including the Ice Ages, should be considered even in such a cave. Application of this method to shell dating is considered more reliable as shells are nearly a closed system in contrast to the open system of fossil bones (Zurer 1983). The situation is the same for the ESR dating of shells and bones. Recently, the water regime has been found to affect the racemization reaction considerably. As peptide bonds in collagen are hydrolyzed by water, collagen changes into gelatin and easily leaches out of bones.

1.4.3 *Hydration of Obsidian*

A thin slice of *obsidian*, a natural volcanic glass which was widely used as prehistoric tools, shows a light rim under a polarizing microscope due to strain birefringence caused by water inclusion. The surface of obsidian incorporates water to the depth of a few μm due to diffusion from soil or atmosphere. This hydration rim can be used to date obsidian (Friedman and Smith 1959), since the hydration process began from a new surface when ancient man manufactured the tool. As water molecules diffuse inwards over time, the *hydration rim thickness* increases. The thickness can be estimated from the theoretical diffusion equation. The diffusion coefficient depends on the ambient temperature and environment.

Weathering is also a chemical process that can be used for relative dating purposes. Recently, the atomic movement of lithium, sodium and fluorine within obsidian and bone has been studied with the aid of nuclear resonance reactions and ion microanalysis. Diffusion profiles from the surface inward have been obtained for several elements. Diffusion of these atoms with time is a basis for dating and may be an auxiliary relative–dating technique similar to the hydration rim dating of obsidian.

1.5   Other Methods in Geochronology

1.5.1   Geomagnetism and Archaeomagnetism

The core of the earth is a hot liquid producing a current, and hence a magnetic field, due to self–rotation and precessional motion. The location and polarity of the north (N) and south (S) poles of the earth are not stable. Thus, 730,000 years ago they were reversed, an important event known as the Matsuyama–Brunhes boundary. The direction of the earth's magnetic field in ancient times was recorded in rocks when the hot rocks cooled. Baked clays and rocks also record such paleomagnetism. If one has a calibration curve for the variation of magnetic field direction with age, a sample which records ancient magnetic directions can be dated (Aitken 1974).

1.5.2   Isotopic Variation : $\delta^{18}O$ and $\delta^{13}C$

Changes in the stable oxygen isotopic ratio are related with paleotemperature and are used as a geothermometer. Water molecules with the light isotope of oxygen, $H_2^{16}O$, evaporate more easily than the heavier molecule, $H_2^{18}O$. The deviation from the Standard Mean Ocean Water (SMOW) is defined as $\delta^{18}O = [(^{18}O/^{16}O)_{sample}/(^{18}O/^{16}O)_{SMOW}) - 1]$ and expressed in per mill ($^o/_{oo}$). The relative ages of deep sea sediments have been determined with the relative isotopic variations of $\delta^{18}O$ (Shackleton and Opdyke 1973). ESR ages for marine carbonates and carbonate speleothems are often correlated with oxygen isotope ratios. The isotopic ratio of $^{13}C/^{12}C$ was compared with values of CO_2 released from the standard PDB (Belemnite from the Cretaceous Pee Dee Formation, South Carolina).

1.5.3   Dendrochronology

Counting the growth rings of trees and correlating the pattern of ring growth with climate have been the subject of extensive studies. The growth pattern is somewhat related to the cyclical nature of solar activity. Radiocarbon laboratories have found variation in ^{14}C production rate in the past and calibrated ^{14}C ages with dendrochronological data. Fitting the growth ring width of woods to known patterns can successfully tell us the age of prehistoric or historic construction materials. There seems to be some geophysical fluctuations in ^{14}C variation, as Japanese Yakusugi tree ring samples do not necessarily show the same ^{14}C variation as the Arizona specimens. Growth rings are formed by seasonal or daily change of the growth rates and can also be used with other biological materials like corals or shells. Calibration of the ESR age using the annual growth rings of corals just like tree rings (~ 10 mm/a) will be discussed in Chapters 3 and 6.

1.6   Summary

Physical and chemical principles have been described for understanding of conventional "*absolute*" dating methods. There are three categories of such dating methods: (1) decay of radioactivity, (2) natural radiation effects and (3) chemical reactions. ESR dating in geosciences belongs to category (2), since it utilizes natural radiation effects which have been accumulating

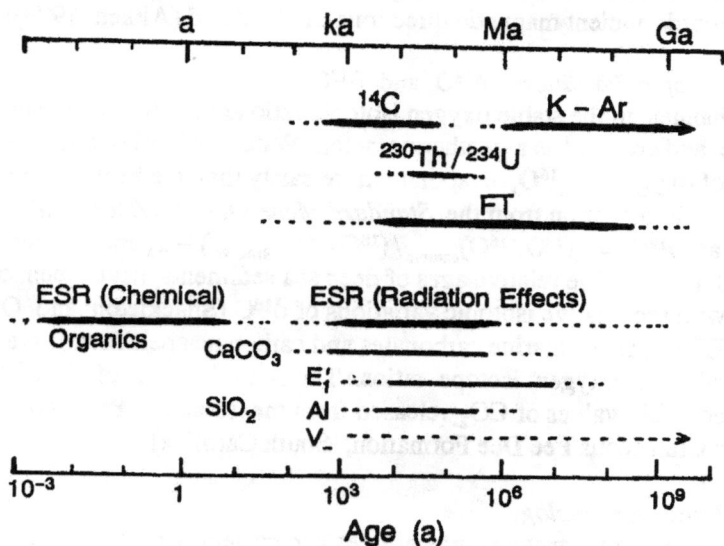

Figure 1.12   The time range of some dating methods.

since the formation of the material. Organic ESR dating in history and forensic sciences belongs to category (3). Since this book covers ESR dating and is not intended as a general text on dating in archaeology or geology, an extensive bibliography published in 20th Century refers the reader to other works.

The time ranges of coverage for major dating techniques are summarized in Figure A1.5. The upper limit of our ESR dating and TL dating depends on the lifetime of the particular signal or peaks (defects), i.e., on the material used and impurities as shown for carbonates and quartz. The lower limit depends on the sensitivity of the spectrometer and therefore will be improved considerably.

References

Abelson P.H. (1955): *Paleobiochem.* (Carnegie Inst., Washington) Yearbook **54**, 107-109.

Aitken M. J. (1974): *Physics and Archaeology*, 2nd Ed. (Clarendon Press, Oxford).

Aitken M. J. (1985): *Thermoluminescence Dating* (Academic Press, London).

Aitken M. J. (1998): *An Introduction to Optical Dating* (Oxford Univ. Press, Oxford).

Bada J.L. and Deems L. (1975): Accuracy of dates beyond the C14-dating limit using the aspartic acid racemization reaction. *Nature* **255**, 218.

Daniels F., Boyd C.A. and Saunders D.F. (1953): Thermoluminescence as a research tool. *Science* **117**, 343-349.

Edwards R.L., Chen J.H. and Ku T.L. (1987): Precise timing of the interglacial period from mass spectrometric determination. *Science* **236**, 1547-1553.

Fleischer R.L., Price P.B. and Walter R.M. (1975): *Nuclear Tracks in Solids: Principle & Application* (Univ. California Press, Berkley).

Fleming S.J. (1975): *Authenticity in Art* (The Institute of Physics, London and Bristol).

Fleming S.J. (1979): *Thermoluminescence Techniques in Archaeology* (Clarendon Press, Oxford).

Friedman I. and Smith R.L. (1959): The deuterium content of water in some volcanic glass. *Geochimica et Cosmochimica Acta* **15**, 218-228.

Grün R (1989): *Die ESR Alterbestimmung-methode* (Springer Verlag, Berlin) in German.

Grün R., Abeyratne M, Head J, et al (1999): AMS C-14 analysis of teeth from archaeological sites showing anomalous ESR dating results. *Quat. Sci. Rev.* **16**, 437-444.

Hennig G. and Grun R. (1983): ESR dating in Quaternary geology. *Quat. Sci. Rev.* **2**, 157-238.

Huntley D.J., Godfrey-Smith D.I. and Thewalt M.L.W. (1985): Optical dating of sediments. *Nature* **313**, 105-107.

Hutt G., Jaek I. and Tchonka J. (1988): Optical dating: K feldspars optical response stimulation spectra. *Quat. Sci. Rev.* **7**, 381-386.

Ikeya M. (1978): Electron spin resonance as a method of dating. *Archaeometry* **20**, 147-158.

Ikeya M. and Miki T. (1980): A new dating method with a digital ESR. *Naturwissen.* **67**, 191.

Ikeya (1986): *Dating and Age Determination of Biological Materials* Eds. Zimmerman M.R. and Angel L.(Croom Helm, London) Chapter 3 Electron Spin Resonance.

Ikeya M. and Miki T. (1985): *ESR Dating and Dosimetry* (Ionics, Tokyo).

Ikeya M. (1987): *Electron Spin Resonance Dating* (Ionics, Tokyo) in Japanese.

Ikeya M., Katakuse I. and Ichihara T. (1990): Portable thermoluminescence reader for dosimetry and dating in fields. *J. Nucl. Sci. Techn.* **27**, 188-190.

Ikeya M. (1994): ESR dating, dosimetry and microscopy- A trio to interdisciplinary fields as a wanderer from physics of defects. *Nucl. Instr. Meth.* **B 91**, 43-51.

Ikeya M. (1994): ESR (EPR) dating based on natural radiaation erffects. *Nucl. Geophys.* **8**, 201-224.

Ivanovich M. and Harmon R.S. (1982): *Uranium series disequilibrium application to environmental problems in the earth science* (Oxford Univ. Press, Oxford).

Jonas M. (1997): Concepts and methods of ESR dating. *Radiat. Meas.* **27**, 943-973.

Komura K. and sakanoue M. (1985): Gammma-ray spectroscopy for ESR dating. *ESR Dating and Dosimetry*, 9-18.

Libby W.F. (1946): Atmospheric helium three and radiocarbon from cosmic radiation. *Phys. Rev.* **69**, 671-672.

Marfunin A.S. (1979): *Spectroscopy, Luminescence and Radiation Centers in Minerals* (Springer Verlag, Berlin).

McDougall D.J. ed. (1968): *Thermoluminescence of Geological Materials* (Academic Press, London).

McKeever S.W.S. (1989): *Thermoluminescence of Solids* (Cambridge U.P., Cambridge).

Ohno K. (1984): ESR imaging. *Magn. Reson. Rev.* **11**, 275-130.

Poupeau G. and Rossi AM (1989): *Nuclear Methods of dating* eds. Roth E. and Poty B., Chapter 10. Electron Spin Resonance, 275-293.

Rink W.J. (1997): Electron spin resonance (ESR) dating and ESR applications in quaternary science and archaeometry. *Radiat. Meas.* **27**, 975-1025.

Rutter N.W. ed. (1985): *Dating Methods of Pleistocene Deposits and Their Problems* (Geoscience Canada), Reprints Series No.2, 73-87.

Sankaran A.V., Nambi S.V. and Sunta C.M. (1983): Progress of thermoluminescence research on geological materials. *Proc. Indian Nat. Sci. Acad.* 49, 18-112.

Shacklleton N.J. aand Oppdyke ND (1973): Oxygen isotope and paleomagnetic stratergraphy of equatorial Pacific core V28-238. *Quat. Res.* **3**, 39-55.

Singhvi A.K., Sharma Y.P. and Agrawal D.P. (1982): Thermoluminescence dating of sand dunes in Rajasthan, India. *Nature* **295**, 313-315.

Skinner A. (1985): The place of ESR dating among modern dating methods. *ESR Dating and Dosimetry*, 1-8.

Skinner A.R. (2000): ESR dating: is it still an 'experimental' technique? *Appl. Radiat. Isot.* **52**,1311-1316.

Ugumori T. and Ikeya M. (1980): Luminescence of $CaCO_3$ under N_2 laser excitation. Jpn. *J. Appl. Phys.* **19**, 459-465.

Wintle A. and Huntley D. J. (1979): Thermoluminescence dating of a deep-see ocean core. *Nature* **279**, 710-712.

Wintle A. and Proszynska H. (1982): TL dating of loess in Germany and Poland. *PACT* **9**, 547-554.

Yada K., Tanji T. and Sunagawa I. (1987): Radiation induced lattice defects in natural zircon $(ZrSiO_4)$ observed at atomic resolution. *Phys. Chem. Minerals* **14**, 197- 204.

Zimmerman M.R. and Angel L Eds. (1996): *Dating and Age Determination of Biological Materials* (Croom Helm, London) .

Zurer P. S. (1983): Archaeological Chemistry. *Chem. Eng. News* **21**, 26-44.

Chapter 2

Introduction to ESR

– What is Electron Spin Resonance (ESR) ? –

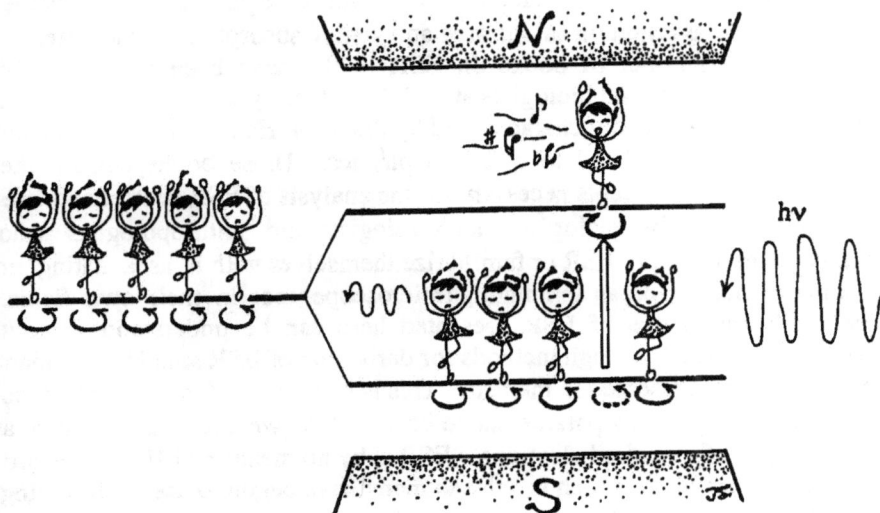

The principle of ESR. The unpaired electrons are excited to a high energy state under a magnetic field by the absorption of microwave "music". The excited electron changes its direction of spin and relaxes into the ground state by emitting phonons (song). Microwave absorption is measured as a function of the magnetic field by ESR spectroscopy.

2.1 Introduction

Electron spin resonance (ESR) or electron paramagnetic resonance (EPR) is a physical method of observing resonance absorption of microwave power by unpaired electron spins in a magnetic field. ESR was discovered by Zavoisky (1945) for transition metal ions in salts and developed after World War II using advances in microwave and solid state electronics. Purcell *et al.* (1946) and Bloch *et al.* (1946) discovered radiowave absorption by nuclear spins and developed a similar method, *nuclear magnetic resonance* (NMR). NMR is now widely used in chemistry and medicine. *Computerized tomography* NMR (CT–NMR) or *magnetic resonance imaging* (MRI), is widely used clinically because it is considered less hazardous than CT using X–rays. ESR or EPR, though less popular than NMR, is now used in radiation chemistry, biochemistry, solid state physics and medicine as a technique as common as infrared and visible absorption measurements.

A large number of books on ESR (EPR) have been published for physicists, chemists and biologists at both introductory and advanced levels. ESR spectra are usually characterized by the spin Hamiltonian parameters which are closely related to quantum physics. Those books contain the quantum physics concepts necessary for the analysis of ESR spectra and are not appropriate for geologists, archaeologists and anthropologists who simply want to apply ESR or familiarize themselves with it as a dating or dosimetry method or as an imaging (microscope) tool in their own fields. The basic knowledge of ESR presented here can be understood without quantum mechanics, though methods for derivation of ESR spin Hamiltonian parameters are presented in the text. This is because ESR signals must be identified using a few parameters to be sure that we are dealing with a specific type of unpaired electrons. ESR is by no means a difficult method for a non–specialist, which is why so many have begun to use ESR dating in addition to their own radiometric methods. A commercially available ESR spectrometer can be operated by anyone with one hour's training.

It should be noted that a single type of paramagnetic center gives a few derivative lines or peaks as a result of the averaging of anisotropic signals in polycrystalline materials. If the typical spectra shown in this book are experimentally observed for geological materials, it will not be necessary to derive the parameters again. It is necessary to remember that the signal magnitude is due to unpaired electrons produced by natural and artificial radiation. The signal amplitude is proportional to the *total accumulated dose of natural radiation (TD)* or the *equivalent dose (ED)* and is therefore related to the age. ESR spectra (resonance field intensity and signal shape) are

critically important in most studies, but the intensity of a particular radiation–induced signal for ESR dating is of interest here.

In this book, only the minimum knowledge necessary for understanding of ESR metrology is given with examples in geology, archaeology,
anthropology and earth and marine sciences. Those who know what ESR is
and what an ESR spectrum is may proceed to the next chapter dealing with
the principles of ESR dating and dosimetry. Those who need to know more
about ESR spectra and methods of analysis should consult standard ESR
textbooks (Carrington and McLachlan 1967, Marfunin 1979, Pake 1962,
Perry 1990, Slichter 1963, Stucki and Banwart 1980, Pilbrow 1990). The
classical picture of an electron spin is described first. Procedures to derive
parameters follow, with simple ESR spectra of minerals and fossils used for
ESR dating in later chapters. Applying the ESR dating by geologists or
archaeologists themselves, from their own viewpoint and considering their
own interests, would considerably advance the development of this method.

2.2 Electron Spins and Defects

2.2.1 *Electron Spin*

Classically an electron may be considered to be negatively charged,
rotating sphere. This rotation produces a circulating current in the direction
opposite to the rotation of the sphere, since the charge of an electron is
negative. The circular current produces a magnetic field in the vicinity of the
electron. Therefore, the electron spin (self rotation) can be considered
equivalent to a tiny magnet. The tiny electron spin magnets form magnetically neutral pairs in atoms and molecules as shown in Figure 2.1 (a). No net
magnetic moment is present due to the neutralizing effects of the pairing of
electron spins in most materials. A few specific atoms and ions that have
parallel spins are exceptions (atoms and molecules in a triplet state, and
transition metal and rare earth ions with d and f electrons, respectively).

Natural or artificial radiation ionizes atoms or molecules, i. e., breaks
the paired electron as shown in (b). When the ionized electron is trapped by
some other atoms, an *electron–excess atom* and an *electron–deficient atom*
are formed: the former and the latter, both with an unpaired electron, are
called "*trapped electron*" and "*trapped hole*" centers, respectively. They
now have net magnetic moment, μ_e due to the unpaired electron spins. An
unpaired electron may be considered as a tiny magnet as shown in (c).
Unpaired spins tend to align parallel to the direction of a magnetic field and
magnetize. Hence, the word "*electron paramagnetism*" is used for magnetic

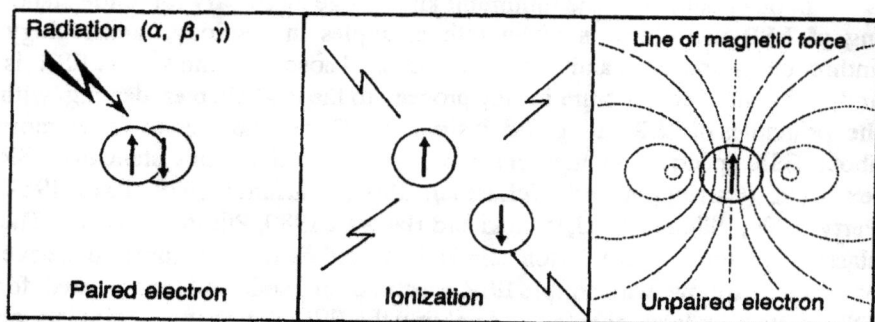

Radiation (α, β, γ)		Line of magnetic force
Paired electron	Ionization	Unpaired electron

Figure 2.1   (a) Nonmagnetic paired electrons and radiation, (b) Ionization and (c) Formation of an unpaired electron. The self rotation (spin) of a nega- tively charged sphere of an unpaired electron acts as a small magnet. Magnet- ic field is produced in the vicinity of an unpaired electron. Defects with un- paired electrons are called *paramagnetic defects*.

properties arising from unpaired electrons in a material. They are detectable with microwave absorption spectroscopy under an external magnetic field called "*Electron Paramagnetic Resonance* (EPR)" or   "*Electron   Spin Resonance* (ESR)" described in later sections.

2.2.2   *Color Center Defects*

Radiation ionizes atoms and molecules, producing free electrons. When a crystalline solid is irradiated, electrons and sometimes atoms are ejected from the lattice position, forming *vacancies* and *interstitials*. Alkali halides are ideal solids for experimental and theoretical studies of lattice defects since the large electronic band gaps give an optically transparent region. Indeed, these crystals are colored by radiation. The halogen anion sublattice is damaged by radiation in alkali halides (for example, NaCl) which have the simplest cubic crystal structure.

Models of color centers in NaCl are shown in Figure 2.2. An electron trapped at a negative ion vacancy is the simplest point defect called an "*F center*", which comes from the German word *Farbe* (color), since the crystal is colored by the presence of defects. The electron in the F center is excited from the ground state by absorption of light. The F center aggre- gates, forming the *F aggregate center*. A defect which has two F centers in the [110] direction is diamagnetic and called an "*M center*". This has optical absorption bands on the long wavelength side of the F center but has no net spin magnetic moment. *Metal colloids* are formed at high temperature due to the formation of Na^o from the F center and the F aggregate center.

An electron deficient Cl^o "*hole*" center combines at low temperature with a nearby Cl^- and becomes a molecular ion of Cl_2^- called a "V_k *center*". An interstitial Cl^o atom formed by leaving the F center out of the Cl^- site becomes a similar Cl_2^- molecule, which is called an "*H center*". These Cl_2^- centers become thermally mobile below room temperature and recombine with an electron center, emitting TL light. Some of these color centers absorb light at various wavelengths and can be detected magnetically since they have unpaired electron spins. Textbooks are available for the physics of color centers (Fowler 1968) and the structural analysis of point defects in solids using multiple magnetic resonance spectroscopy (Speath *et al.* 1992).

Some minerals that have been exposed to natural radiation have similar electron and hole centers. Geologists have estimated the age of zircon by simply looking at the color because the accumulation of electron and hole centers and the formation of the aggregates during a geological time produce various colors. Models of defects in geological minerals are described by Wertheim *et al.* (1971), Poole and Farach (1972) and Marfunin (1979).

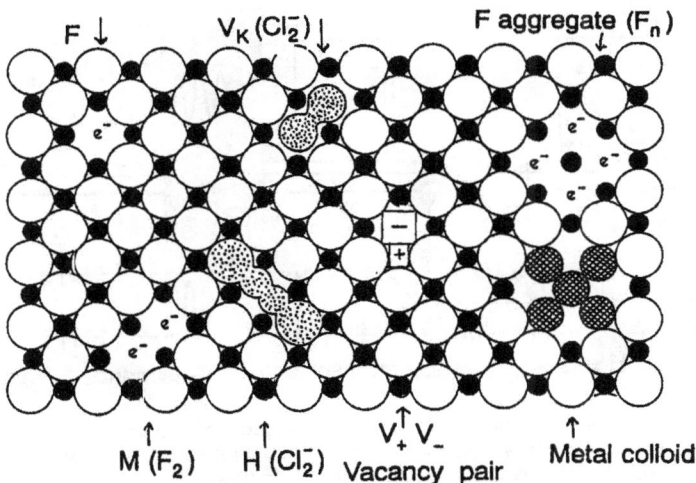

Figure 2.2 Lattice defects called "*color centers*" created by ionizing radiation in NaCl. Some have an unpaired electron and so are paramagnetic.

2.3   Electron Spin Resonance (ESR)

2.3.1 *Fundamental Principle of ESR*

When spinning electrons are placed in an external static magnetic field, the direction of the small magnet, i. e., the direction of the spin rotation which is initially random, becomes either the same as or opposite to that of the external magnetic field, as shown in Figure 2.3 (a) and (b).   These two situations are energetically different and are called *"up spin"* and *"down spin"* states. The electron spins excited from the lower level by the absorption of the microwave quanta flop to the lower level by the lattice vibration in a time called the *"spin-lattice relaxation time"* (T_1) as shown in (c). Flopping of the spins also occurs due to the interaction among spins and this flopping time is called the *"spin-spin relaxation time"* (T_2). If the microwave power is so high as to pump out the spins in the lower level or the relaxation time is too long for the population to be restored, a decrease in the signal intensity is observed. This is called *"saturation"*.

The magnetic properties of unpaired electrons are expressed by the following parameters:

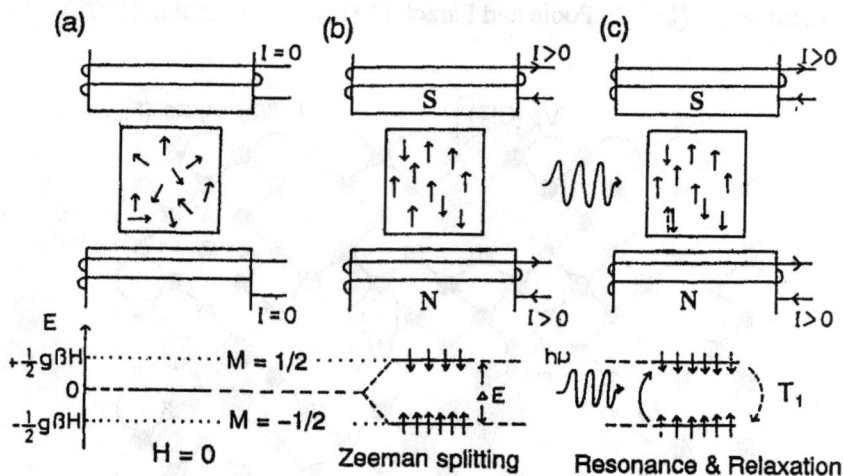

Figure 2.3 Energy separation of an unpaired electron spin under a magnetic field (Zeeman effect). (a) Random orientation of spins without current in the electromagnet, (b) a partial orientation under the magnetic field and (c) flipping of spins by microwave absorption and flopping of spins in the spin-lattice relaxation time, T_1.

Spin angular momentum, $S(h/2\pi)$ (*h : Planck constant*)
Spin quantum number, S : $S = 1/2$ for an electron.
Magnetic quantum number, M : $M = +1/2$ and $M = -1/2$ are allowed.
Bohr magneton, β : the basic unit of a small magnet for an electron spin.
Magnetic moment, μ_e : $\mu_e = -g\beta S$
Spectroscopic splitting factor, g : $g = 2.0023$ for a free electron.

The energy yielding different spin states under the external magnetic field H is known as the "*Zeeman effect*" and depends on H and the *magnetic moment* ($g\beta M$) of the electron. The *Zeeman energy*, $E_Z = -\mu_e H$, is

$$E_Z = g\beta HM , \qquad (2.1)$$

where H is expressed in **Tesla** (T ; 1 T = 10^4 gauss) or in **mT** (1 mT = 10 gauss). The energy level of an electron with $S = 1/2$ splits for $M = 1/2$ and $M = -1/2$ and is shown as a function of H in Figure 2.4 (a).

The direction of the spin is changed by the absorption of microwaves when the energy difference ($\Delta E_Z = g\beta H$) is equal to the quantum energy of an electromagnetic wave, $h\nu$. ΔE_Z is equal to the difference between two diagonal energy level lines in (a). Here, ν is the *frequency of the electro-magnetic wave*. This absorption of the electromagnetic wave (microwave) by the unpaired electrons is called "**electron spin resonance (ESR)**".

In solids, the g factors of most species do deviate slightly from that of the free electron (g_e) due to the contribution of orbital angular momentum via the magnetic *spin–orbit interaction* (see Section 2.4.2). Hence, the term, **electron paramagnetic resonance (EPR)** is preferred by physicists and chemists dealing with solid materials. However, the simple term, **ESR** is better understood by people in general in interdisciplinary fields.

A multi–spin system such as some paramagnetic ions or an exchange coupled pair of spins has a total spin quantum number larger than 1 ($S \geq 1$). In this case, M takes $2S + 1$ values from S, $S - 1$, $\cdots -S + 1$, $-S$. The g factor depends on the magnetic spin–orbit interaction ($\lambda L S$). Its accurate determination is the basis of modern analytic ESR (EPR) spectroscopy.

The resonance condition is represented by

$$g\beta H_o = h\nu , \qquad (2.2)$$

where H_o is the *resonance magnetic field*. One can obtain the resonance absorption by sweeping the magnetic field H, while maintaining the frequency ν constant. For electron spins with $g = 2.0$, ν is in the microwave frequency range when $H_o = 300 \sim 1000$ mT as indicated in Table 2.1.

Optical absorption is often measured in various scientific fields. The wavelength or frequency is swept to observe the optical transition between the electronic energy levels (see Figure 2.4). By analogy, it could be said that a material with unpaired electrons placed in an external magnetic field is "*colored*" at the microwave wavelength.

Table 2.1 The resonance magnetic field H_o for the signal at $g = 2.0$ at typical microwave frequency bands (wavelengths) using $h\nu = g\beta H_o$.

Band	Wavelength (cm)	ν [a] (GHz)	H_o [b] (mT)
L–band	20.0	1.5	53.5
S–band	9.4	3.2	114
X–band	3.2	9.5	339
K–band	1.2	25	892
Q–band	0.86	35	1,250

a: $\nu = g\beta H_o/h = 0.0140 \times g H_o$ [GHz]
b: $H_o = (h/\beta)(\nu/g) = 71.455 \times (\nu/g)$ [mT]

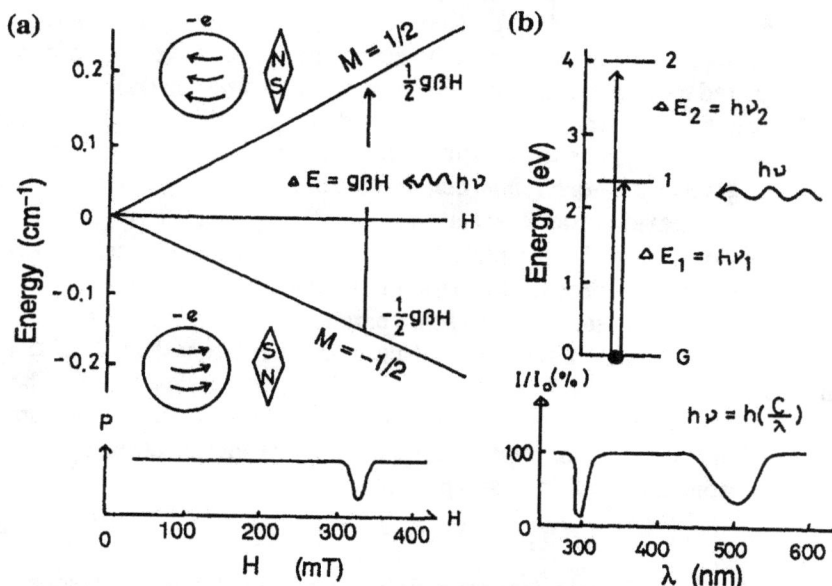

Figure 2.4 (a) Energy levels of an electron spin as a function of a magnetic field. The resonance by microwave absorption occurs at $H_o = h\nu/g\beta$. (b) Electronic ground and excited levels and a transition induced by an optical absorption: E [eV] $= 1239/\lambda$ (λ is a wavelength in [nm]).

2.3.2 ESR Spectrometer

ESR spectrometers are commercially available just as visible and infra-red optical spectrometers are. Figure 2.5 shows block diagrams of optical absorption and ESR spectrometers. In *optical spectrometers*, as shown in (a), the light is chopped so that a weak signal can be detected by electronically amplifying the detector output with the same frequency using a lock-in amplifier. Incident light with the intensity, I_o, is partially absorbed to give transmitted light with the intensity, I. In optical absorption the quantum energy of the electromagnetic wave, i.e., of the light, is changed by varying the wavelength to match the electronic energy separation. The optical absorbance, $\log_{10}(I_o/I)$ is recorded as a function of the wavelength.

In *ESR spectrometers* shown in (b), microwaves from an oscillator, called a Klystron or Gunn-diode, are led by a wave-guide to the resonant cavity in the static magnetic field produced by the electromagnet. The sample is inserted into the cavity, and the strength of the magnetic field is swept by slowly changing the current in the electromagnet coil. The absorption of the microwaves is measured using a detector diode.

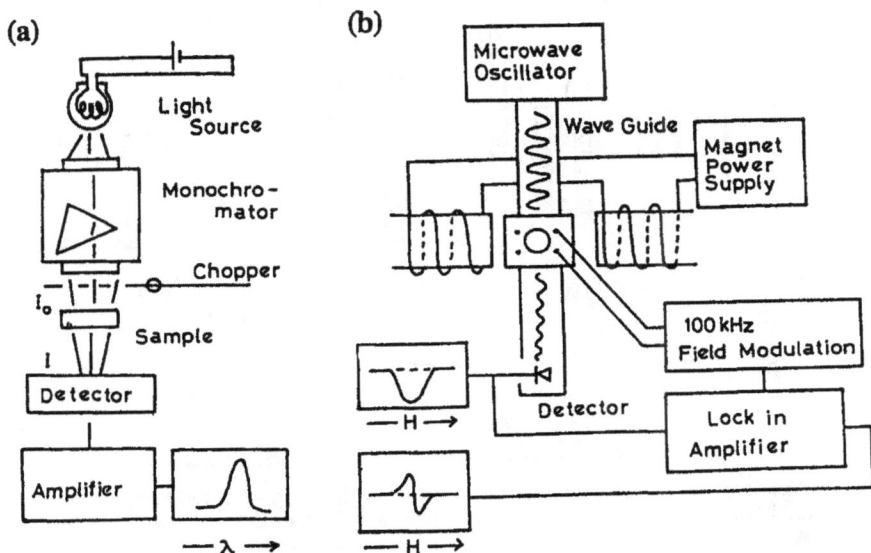

(a)

Light Source
Monochro-mator
Chopper
Sample
Detector
Amplifier

(b)

Microwave Oscillator
Wave Guide
Magnet Power Supply
100 kHz Field Modulation
Detector
Lock in Amplifier

Figure 2.5 Analogical illustration of spectrometers for (a) optical absorption and (b) ESR measurement.

Generally, the magnetic field is modulated by an additional 100 kHz sinusoidal magnetic field of small amplitude. Therefore, the microwave absorption is modulated by a high frequency of 100 kHz and synchronously amplified electronically (using a *lock–in amplifier system*). Modulation spectroscopy allows the detection of very small levels of the microwave absorption. The rectified output, which is displayed on the recorder, is a mathematically differentiated line of microwave absorption (dP/dH), i.e., the first derivative in Figure 2.6 (a) and (b). One may take the maximum–minimum magnitude as an index of the concentration of the unpaired electrons. *Modulation width* of the magnetic field should be less than the *line–width of the signal* as indicated in (c). In some cases, the second derivative is recorded to separate an overlapping broad line by taking the second harmonics. In some case, *double modulation* at different frequencies is adopted to study an additional effect (pressure, light, etc.) or spatial distribution (see Chapter 14).

If the width of the ESR signal is 0.1 mT, one can detect a minimum number of $10^9 \sim 10^{10}$ spins in the sample with an ordinary spectrometer.

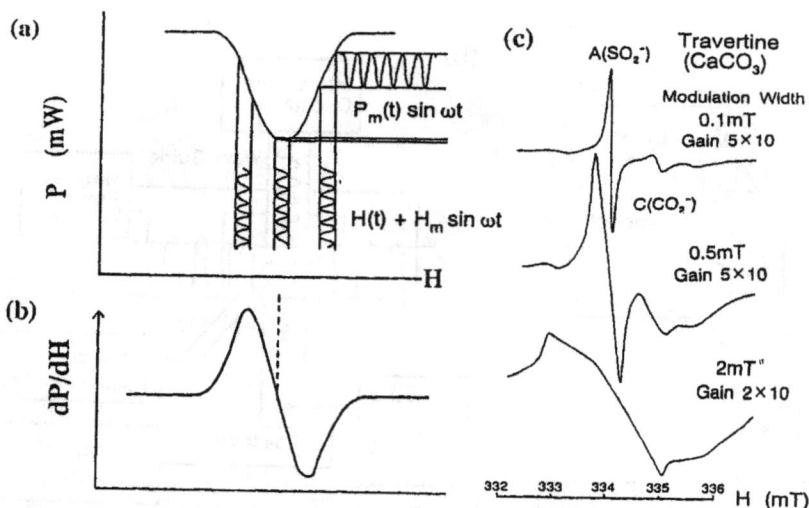

Figure 2.6 (a) The absorption curve and the magnetic field modulation method. (b) The ESR spectrum recorded in a form of the first derivative line as a function of the magnetic field, H. (c) ESR spectra of spring travertine (natural CaCO$_3$) with different modulation widths. Overmodulation larger than the linewidth of signals broadens the signals and distorts the line shape. Defects for the signals in CaCO$_3$ (SO$_2^-$ and CO$_2^-$) are described in Chapter 6.

This sensitivity is very high if you consider that the number of atoms per cm^3 in solids is of the order of $10^{22} \sim 10^{23}$. Room temperature measurements are generally sufficient for geological and archaeological samples.

Low temperature measurements give a sharp linewidth, increase the spin population in the lower state due to Boltzman factor, $e^{-g\beta H/k_B T}$ and so the signal to noise (S/N) ratio (about a factor of 4 increase by measuring at 77K). Freezing of water in a sample also reduces microwave power loss, which enhances the signal intensity. However, some signal intensity is saturated at low microwave power at a low temperature.

Microwave power level must be set to avoid the saturation of the signal intensity especially at low microwave power. It is a standard technique to separate overlapping signals from their different saturation behaviors by measuring the microwave power dependence. The ESR techniques and applications are described in the book by Alger (1973). An ESR spectrometer with a computer system is shown in Figure 2.7.

2.3.3 *Data Acquisition and Processing*

Data processing with analog to digital (A–D) conversion of the ESR signal has been done to obtain parameters and simulated spectra. Data processing for ESR dating consists of the improvement of the S/N ratio by repeated measurements and signal averaging. A simulation program for amorphous spectra using reported anisotropic parameters for a crystal helps to understand the complicated spectrum of the powder sample which is common in ESR dating and dosimetry. Data processing (subtraction of the

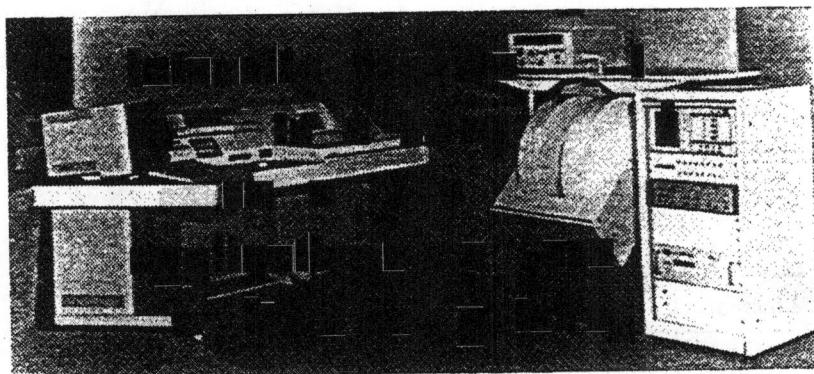

Figure 2.7 An ESR spectrometer with its personal computer system.

overlapping background signals, integration of a derivative spectrum for an absorption spectrum and double integration for the intensity determination, etc.) is useful for our purpose. It is convenient to compare a spectrum and its absolute intensity with those of standard samples of known localities and ages on CRT display. Monotonous measurements of similar samples are needed in dating and dosimetry. Hence, an automatic sample changer with a turn–table and sample cell gripper was manufactured and is described in Appendix 3. Both hardware and software should be developed for practical applications of ESR dating and dosimetry.

2.4   ESR Spectra and Parameters
2.4.1 Line shapes : Gaussian and Lorentzian Shapes
The linewidth and the lineshape are also useful information in ESR spectrum. The Gaussian and Lorentzian shapes can be seen in the shape of the derivative line. The relation between the relaxation time (T_2) and the line shapes, i.e., half–width ($\Delta H_{1/2}$), the maximum slope width (ΔH_{msl}) and the average second moment ($<\Delta H^2>^{1/2}$) are tabulated in Table 2.2.

Table 2.2 Absorption and the first derivatives of Gaussian and Lorentzian lineshapes and the half–width ($\Delta H_{1/2}$), maximum slope width (ΔH_{msl}) and the average second moment as indicated ($<\Delta H^2>^{1/2}$) with theoretical spin–spin relaxation time (T_2).

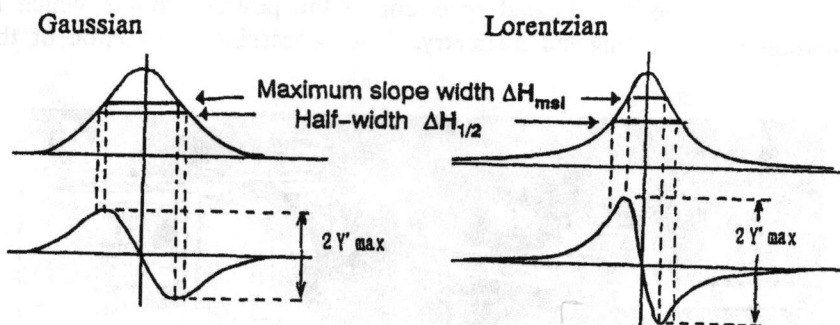

Gaussian   Lorentzian

Maximum slope width ΔH_{msl}
Half-width $\Delta H_{1/2}$

$2Y'$ max $2Y'$ max

Lineshape	$\Delta H_{1/2}$	ΔH_{msl}	$<\Delta H^2>^{1/2}_{av}$	$\dfrac{\Delta H_{1/2}}{\Delta H_{msl}}$	$\dfrac{\Delta H_{1/2}}{<\Delta H^2>^{1/2}_{av}}$
Gaussian	$\left(\dfrac{\ln 2}{\pi}\right)^{1/2}\dfrac{1}{T_2}$	$\dfrac{1}{\sqrt{2\pi}\,T_2}$	$\dfrac{1}{\sqrt{8\pi}\,T_2}$	1.177	2.355
Lorentzian	$\dfrac{1}{\pi T_2}$	$\dfrac{1}{\sqrt{3}\,\pi T_2}$	∞	1.732	0

2.4.2 g factor

The spectroscopic splitting factor, the g factor of an ESR signal, is an important parameter, since unpaired electrons in different environments have slightly different g factors, resulting in the appearance of signals for different centers at different magnetic field strengths. Classically, the slight g factor change due to the environment might be ascribed to the change of the effective negative charge by the spread of the electron wavefunction. The shift of g factor occurs indirectly through a magnetic interaction between the *spin magnetic moment* and *orbital magnetic moment* (*spin–orbit interaction*).

Figure 2.8 shows the energy diagram for two electron spins in different environments and thus they have different g factors, g_1 and g_2. As the magnetic moment and therefore the Zeeman energy is different for each electron spin, the slope of the energy level change is different. Since the incident microwave frequency is constant for an ESR spectrometer, the

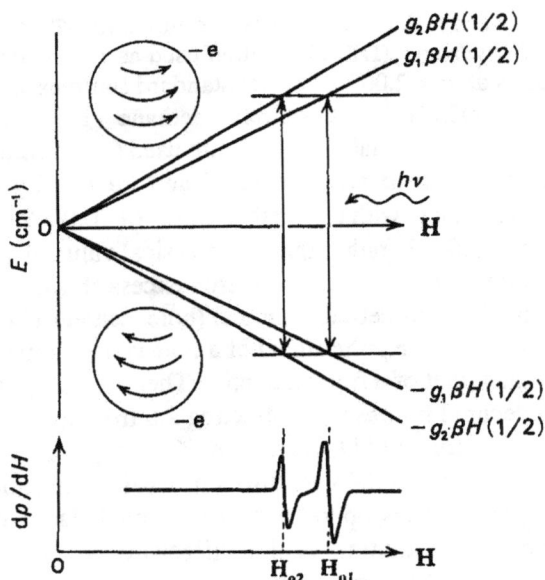

Figure 2.8 Energy levels of spins with two different g factors of g_1 and g_2. The magnitude of the magnetic moment or the interaction of the electrons with orbital magnetic moment through the spin–orbit interaction results in different g factors, i.e., different slopes in the energy level diagram. The absorption positions in the derivative form are shown below.

resonance occurs at magnetic fields H_{o1} and H_{o2}. The g factor is determined using Eq. (2.2) as

$$g = (h/\beta)(v/H) = 71.455 \times (v/H) \ . \tag{2.3}$$

The microwave frequency v (in GHz) and the magnetic field H (in mT) are
measured with a frequency counter and a fieldmeter, respectively. The g
factor of the unknown signal is determined using a standard signal with a
known g factor. If the resonance of a standard signal with g_1 and an
unknown signal with g_2 occurs at H_{o1} and H_{o2}, respectively (the magnetic
field separation $\Delta H = H_{o1} - H_{o2}$), the resonance condition gives

$$g_1 H_{o1} = g_2 H_{o2} = (hv/\beta) \ , \tag{2.4}$$

$$g_2 = g_1 H_{o1}/H_{o2} = g_1 H_{o1}/(H_{o1} - \Delta H)$$

$$= g_1/(1 - \Delta H/H_{o1}) \ . \tag{2.5}$$

Usually, the resonance field (H_o) falls around 340 mT for $v = 9.4$ GHz.
Diphenyl–picryl–hydrazyl (DPPH) is often used as a standard sample since
it has free radicals at $g = 2.0036$. Other standard samples are peroxyamine
disulfonate ($g = 2.00537$), tetracyanoxy methane ($g = 2.0026$) and tanol
($g = 2.0055$). The Mn^{2+} signals in MgO are used (see Section 2.4.4 (d)).

The g factors are generally around 2.0 and are usually determined to
the fourth decimal place with a probable error of $0.0002 \sim 0.0005$. Free
electrons have $g_e = 2.0023$ (rather than the classical value of 2.0000) due to
relativistic correction. A radical with electron excess shows a g factor smaller than g_e. An electron–deficient radical (hole center) shows a g factor
slightly larger than g_e i.e., the signal of a hole center appears at a lower
magnetic field than that of a free electron. There is a proposal to use the
third and fourth decimal figures by subtracting 2.0 from the g factor to identify a signal. The g factor would therefore be 23 for $g_e = 2.0023$. However,
signals of $g = 1.9975$ or $g = 2.43$ are difficult to call in this notation. Hence,
we describe the g factor using up to the fourth decimal place. It is possible to
obtain g factors of high resolution at high microwave frequencies of Q– and
K–bands ($v = 35$ and 25 GHz, respectively).

An ESR signal is identified by the g factor. The shift of the g factor
from the free electron value ($g_e = 2.0023$) is caused by the spin–orbit interaction, λLS, of the electron in the atomic orbital, where λ is the *spin–orbit
interaction parameter* and L is the *orbital magnetic moment* of the relevant
atom. The orbital magnetic moment is quenched in solids and the indirect

spin–orbit interaction by high energy state orbitals shift the g factor. For example, the g factor of Cu^{2+} can be written using perturbation theory as

$$g = g_e - \frac{8\lambda}{E_{ex} - E_G} , \tag{2.6}$$

where E_G and E_{ex} are the energies at the ground and excited state orbitals, respectively. The λ in a solid is usually about $10 \sim 20\%$ less than that of the free atom value obtained from atomic spectra.

2.4.3 *Anisotropic g factors and Random Orientation*

The g factor differs for electron or hole centers in crystalline materials when the direction of the magnetic field is changed. For the *orthorhombic symmetry*, the g tensor is expressed as g_{zz}, g_{xx}, g_{yy}. The *anisotropic g* factor of Cl_2^- in KCl considering molecular orbitals from halogen p electrons is described in the literature (Castner and Känzig 1957, Slichter 1963). The g factor for the magnetic field direction at intermediate angles θ and ϕ may be expressed as

$$g(\theta, \phi) = (g_{zz}^2 \cos^2\theta + g_{xx}^2 \sin^2\theta \cos^2\phi + g_{yy}^2 \sin^2\theta \sin^2\phi)^{1/2} . \tag{2.7}$$

For cases of *axial symmetry*, the g factor may be expressed as $g_\parallel$ and $g_\perp$, where $g_\parallel = g_{zz}$ and $g_\perp = g_{xx} = g_{yy}$. The g factor at θ is expressed as

$$g(\theta) = (g_\parallel^2 \cos^2\theta + g_\perp^2 \sin^2\theta)^{1/2} . \tag{2.8}$$

Figure 2.9 (a) shows an angular dependence of the g factors for a radical in irradiated $CaCO_3$. The g factor is angular independent for the direction of the magnetic field perpendicular to the [111] axis, if the crystal is rotated around the [111] axis. On the other hand, if the direction of the magnetic field is changed from perpendicular ($H_\perp$) to parallel ($H_\parallel$) to [111], g factor changes as shown in the figure. Thus, $g_\parallel$ and $g_\perp$ can be obtained from the spectrum and $g(\theta)$ is calculated using Eq. (2.8). Two–circle goniometer for the measurement of angular dependence of a single crystal is shown in (b).

When a defect with an anisotropic g factor makes a hindered rotation along the z–axis so as to average out the orthorhombic symmetry, the g factor can be expressed as

$$g_\parallel = g_{zz} , \qquad g_\perp = [(g_{xx}^2 + g_{yy}^2)/2]^{1/2} . \tag{2.9}$$

(a)   **(b)**

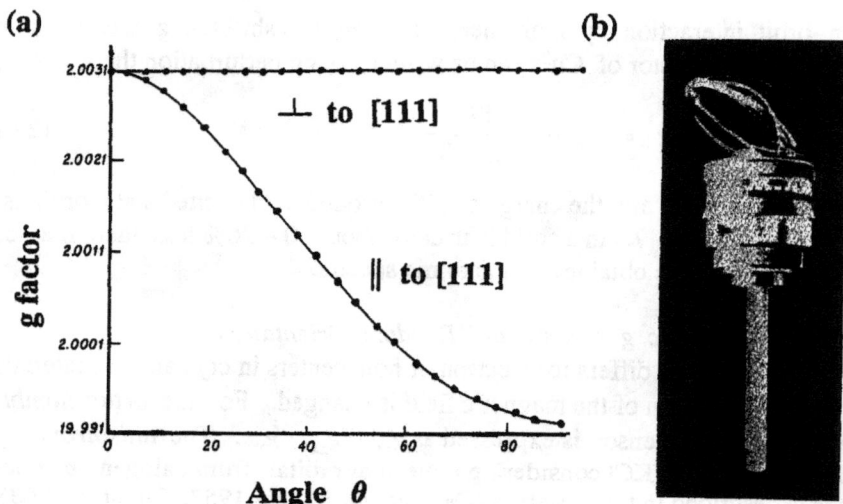

Figure 2.9 (a) Angular dependence of g factor as a function of the angle between the magnetic field and the crystalline c-axis for calcite (Rossi *et al.* 1985). (b) Two–circle ESR goniometer for the measurement of angular dependence of a single crystal (Kobayashi and Suhara 1985).

If the radical rotates rapidly and freely, the average g factor (g_{av}) for the signal with orthorhombic symmetry is written as

$$g_{av} = [(g_{zz}^2 + g_{xx}^2 + g_{yy}^2)/3]^{1/2} , \qquad (2.10)$$

while g_{av} for the signal with axial symmetry is written as

$$g_{av} = [(g_{\parallel}^2 + 2g_{\perp}^2)/3]^{1/2} . \qquad (2.11)$$

The molecular CO_2^- ion used in ESR dating of carbonates gives different signals due to the hindered rotation as described in Chapters 5 and 8.

A number of ESR studies on g factors have been made to calculate g values based on atomic or molecular orbitals if the atomic model of the defect or radical has been established. However, a model of a defect can not be drawn from the g factor alone.

2.4.4   *Hyperfine (hf) Structures*

(a)   *Hyperfine splitting* : The self–rotation of a positively charged nucleus (*nuclear spin*) produces a circular current and therefore a magnetic field in the vicinity, as for the case of an electron spin. A nuclear spin can also be regarded as a small magnet which interacts with electron spins magnetically. ESR is affected by the presence of a nuclear spin in the vicinity since the magnetic field induced by the nuclear spin also contributes to the Zeeman energy of the electron spin as shown in Figure 2.10. This is called the "*hyperfine (hf) interaction*" or the "*hf structure*" of the ESR signal.

The magnetic field induced by the nucleus differs slightly when the direction of the external magnetic field is changed relative to the axis of the p orbital mixture of the unpaired electron. We call the angular dependent part an "*anisotropic hf coupling*" and describe the *hf coupling constant*, A, as

$$A = A_s + A_p(3\cos^2\theta - 1) , \qquad (2.12)$$

where A_s and A_p are *isotropic* and *anisotropic hf coupling constants*, respectively, and θ is the angle between the direction of the magnetic field and the axis of p orbital. They are expressed as $A_\| = A_s + 2A_p$ and $A_\perp = A_s - A_p$ for $\theta = 0°$ and $90°$, respectively.

Physically, A_s is related to the probability of an electron cloud at the nucleus which has a magnetic moment of $\mu_N = g_N\beta_N I$, where g_N and β_N are the g factor and the Bohr magneton for the nucleus, respectively and I the *nuclear spin quantum number*. Isotropic A_s is called the "*Fermi contact*

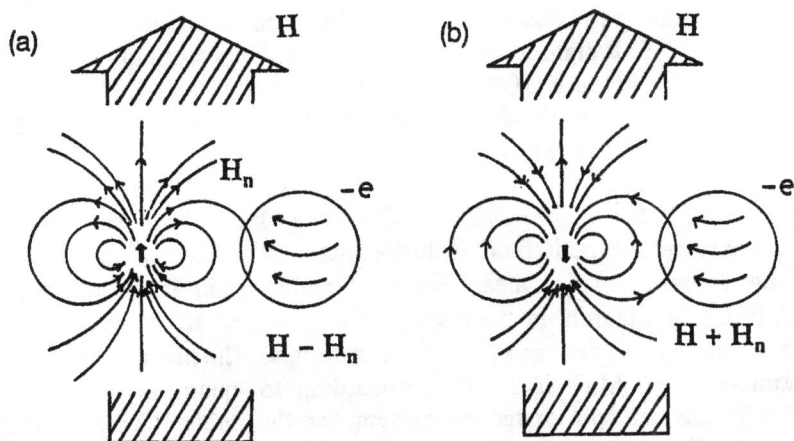

Figure 2.10   Hyperfine interaction due to nuclear spin ($m = 1/2$ and $m = -1/2$). A vector sum of the external magnetic field H and the magnetic field induced by the nuclear spin is the local field acting on the electron spin.   (a) The nuclear spin oriented to the field direction produces an opposite field to electron.   Resonance occurs at a more intense field.   (b) For the opposite direction of the nuclear spin, the resonance field shifts to a lower field.

interaction". A_p is due to the magnetic dipole–dipole interaction between the electron spin and the nuclear spin. These hf coupling constants are expressed as

$$A_s = (2\pi/3) g_N \beta_N \mid \Phi(0) \mid^2 , \tag{2.13}$$

and

$$A_p = (2/5) g_N \beta_N <r^{-3}> <3 \cos^2\theta - 1> , \tag{2.14}$$

where $\Phi(0)$ is the *electron wavefunction* at the center of the nucleus and r is the position of the electron relative to the nucleus. $<r^{-3}>$ and $<3\cos^2\theta - 1>$ are the averages using the spatial electron spin density.

Anisotropic hf constant is expressed using A_{xx}, A_{yy} and A_{zz} in a system of principal axes for the direction cosine of l, m and n as

$$A = A_{xx}^2 l^2 + A_{yy}^2 m^2 + A_{zz}^2 n^2 . \tag{2.15}$$

Anisotropic hf splitting is angle averaged for a powder spectrum (see 2.5.2).

(b) *Nuclear spins* :   Elements with nuclear spins are abundant. Theoretically computed estimates of isotropic (A_s) and anisotropic hf coupling constants (A_p) for $\mid \Phi(0) \mid^2 = 1$ of the elements relevant to ESR spectra are shown in Table 2.3. **Isotope ratios** in isotope geology can be studied in principle using hf lines of ESR signals if the intensity could be determined accurately to *per mill* (10^{-3}) or ppm (10^{-6}). Determination better than $1 \sim 0.1\%$ of the relative intensity is usually difficult due to the *S/N* ratio and overlapping signals with present day spectrometers.

(c) Mn^{2+} ($3d^5$: $S = 5/2$, $I = 5/2$) :   The energy level of the electron spin under an external magnetic field is further split by the magnetic field produced by the nuclear spin states. The most common and frequent example of hf splitting in minerals is the nuclear spin of Mn^{2+}. Mn^{2+} ions are involved in many biological and geological materials. The nuclear spin quantum number I of Mn^{2+} is $I = 5/2$. According to quantum physics, six $(2I + 1 = 6)$ nuclear spin states are present for the *nuclear spin magnetic quantum number*, m, in this case. Thus, possible m values are 5/2, 3/2, 1/2, –1/2, –3/2 and –5/2 or in quantum terms $m = I$ to $m = - I$. These give six different internal magnetic fields to the electron spin and so resonance occurs at six different external fields. The spectrum of Mn^{2+} constitutes six hf lines with a nearly equal intensity and separation of about $8 - 9$ mT.

Table 2.3 Hyperfine coupling constants, A of some elements.

Nucleus	Abundance (%)	Spin I ($\hbar$)	A_s (mT)	A_p (mT)
^{1}H	99.985	1/2	50.8	($A_{Ha} = 50.8\ \rho_{Ha}$)
^{10}B	19.9	3	24.0	1.27
^{11}B	80.1	3/2	72.0	3.79
^{13}C	1.10	1/2	111.0	6.48
^{14}N	99.634	1	55.0	3.41
^{15}N	0.366	1/2	77.1	4.79
^{17}O	0.038	5/2	160.0	10.26
^{19}F	100	1/2	1720.0	108.1
^{29}Si	4.67	1/2	121.8	3.08
^{31}P	100	1/2	363.0	20.48
^{33}S	0.75	3/2	97.0	5.63
^{35}Cl	75.77	3/2	166.5	10.02
^{37}Cl	24.23	3/2	138.5	8.34

Revised from Symons M. C. R. (1963): *Adv. Phys. Org.* **1**, 332.

(d) *Zeeman splitting of* Mn^{2+} ($S = 5/2$) : The electronic spin state of Mn^{2+} ($3d^5$, ^6S) is $S = 5/2$ since five d electrons are in the same direction according to Hund's rule. Six energy states with the *electron spin magnetic quantum number*, $M = 5/2$, $3/2$, $1/2$, $-1/2$, $-3/2$ and $-5/2$ arise due to the Zeeman effect ($E_M = g\beta HM$), which further split into six hf lines of the *nuclear spin magnetic quantum number*, $m = 5/2$, $3/2$, $1/2$, $-1/2$, $-3/2$ and $-5/2$ as shown in Figure 2.11. For the Zeeman energy larger than the hf energy (i.e., $g\beta H > AMm$), the $(2S + 1)(2I + 1) = 36$ energy levels are written as

$$E_{M, m} = g\beta HM + AMm .\qquad (2.16)$$

The allowed transitions are for $M \leftrightarrow M + 1$ and $m \leftrightarrow m$,

$$E_{M+1, m} - E_{M, m} = g\beta H + Am = h\nu .\qquad (2.17)$$

As the resonance field without hf splitting is expressed as $H_o = h\nu / g\beta$,

$$H = H_o - (A/g\beta)m .\qquad (2.18)$$

H depends only on m. For Mn^{2+}, six hf lines are observed with a separation of $A/g\beta = 8.69$ mT. The hf coupling constants expressed in the energy unit of 10^{-4} cm^{-1} or MHz are divided by the magnetic moment of an electron, $g\beta$, and thus $A/g\beta$ expressed in the magnetic field unit of mT is used as A value in this book. For literature values in the unit of **MHz** or 10^{-4} cm^{-1}, multiply

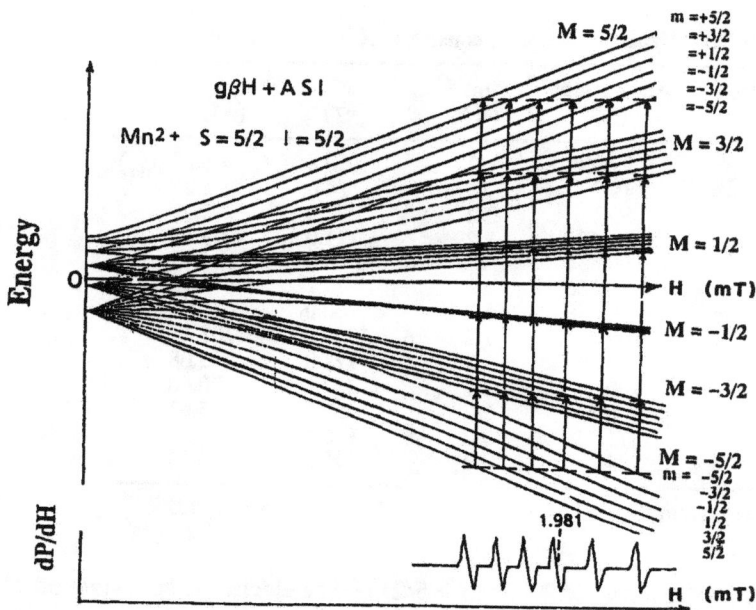

Figure 2.11 The energy levels of Mn^{2+} as a function of the external magnetic field H. Splitting into $2S + 1 = 6$ by the Zeeman effect for M and further into $2I + 1 = 6$ states for m by the hf magnetic field. The resonance fields for allowed transitions ($\Delta M = \pm 1$, $\Delta m = 0$) are the same for different M values when $A/g\beta < H_o$. Fine structure splitting occurs for a large A value, which is a part of the reason of the different linewidth of each hf line.

$0.07145/g$ or $0.21435/g$, respectively, to express them in the unit of mT.

The actual separation is slightly different in quantum physics due to so called "*second-order hf coupling*" or the "*off-diagonal terms*" in the spin Hamiltonian matrix (see other ESR textbooks). Instead of Eq. (2.18), the resonance field is given by second-order perturbation theory as

$$H = H_o - (A/g\beta)m - [(A/g\beta)^2/2H_o][I(I+1) - m^2 + m(2M-1)] \quad . \quad (2.19)$$

This is because energy level splitting by the Zeeman effect is not necessarily linear to the external magnetic field, especially at low magnetic fields as shown in Figure 2.11. The six hf lines split further into fine structures owing to different electron spin transitions of $M \leftrightarrow M+1$ at a frequency lower than 1.0 GHz. *Forbidden transition* signals associated with $M \leftrightarrow M+1$ and $m \leftrightarrow m+1$ are also observed on both sides of the hf lines although they are not indicated in the figure. Intensities of forbidden transition lines are very small except for the case of a large fine structure splitting.

(e) *g factors from standard signals of Mn²⁺ in MgO and of CuSO₄*: The g factor and hf splitting parameters can be obtained with the ESR spectrum of Mn^{2+} in MgO as demonstrated in Figure 2.12. The third and fourth lines of the six hf lines associated with Mn^{2+} appear at $g_3 = 2.0340$ and $g_4 = 1.9810$, respectively. The separation between the third and fourth line is **8.69 mT** at the X-band frequency and so signal separation from the 4th line, ΔH can be determined from the relative position of the signals. Thus, the g factor of a signal is written using Eq. (2.5) as

$$g = 1.981\, H_o/(H_o - \Delta H) \_ 1.981(1 + \Delta H/H_o), \qquad (2.20)$$

where $H_o = h\nu/g = 339$ mT for $\nu = 9.5$ GHz. One can use Eq. (2.20) for ESR signals that appear between the third and fourth lines of Mn^{2+} within the accuracy of 0.0002. The g factor at the center of the third and fourth lines is $g_{center} = \mathbf{2.0065}$ as one can obtain $\Delta H = 8.69/2 = 4.345$ mT. The $g = \mathbf{2.26}$ of Cu^{2+} in copper sulfate pentahydrate($\mathbf{CuSO_4 \cdot 5H_2O}$) is used as a standard for g factor and intensity calobration (Inoue et al., 2000). A slight change in ν_o does not affect the g factor

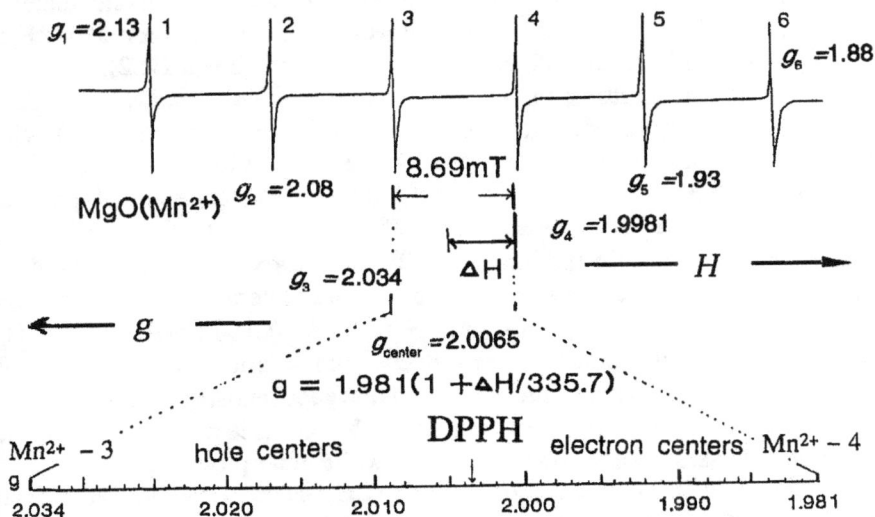

Figure 2.12 ESR spectrum of a standard sample of MgO (Mn^{2+}) and the derivation of a g factor using the g factor of the fourth line ($g_4 = 1.9810$) and 8.69 mT separation between the third and fourth lines.

much.   A convenient way is to make a "scale" with indication of the values of g factors for the full field range of 10 mT around $g = 2.0$. The standard sample of $MgO(Mn^{2+})$ is set at the position where the direction of the 100 kHz magnetic field modulation is opposite in phase relative to the sample position. So, the third and fourth lines are obtained as derivative lines with the phase opposite to the main signal of interest. It is desirable to use additional standards such as DPPH for accurate g factor.

2.4.5   Proton Hyperfine Structures : Equivalent Protons

In an organic radical, unpaired electrons interact with nearby protons or nitrogen (^{14}N, 99.63%, $I = 1$). An ESR signal is split into two lines of equal intensity (doublet) by hf coupling with the nuclear spin ($I = 1/2$) of a proton, usually with the splitting of $A = 50.8$ mT. Each line is split into two further lines if an additional proton is present. Figure 2.13 shows a typical example for the proton hf lines found in geological materials. The spectrum of fossil silicified wood taken from a seafloor shows a signal of two proton and three proton hf interactions. These triplet and quartet signals would presumably be associated with $\cdot CH_2COOH$ or $CH_3(CH_2)_n-\cdot CH_2$ radicals and methyl radicals ($\cdot CH_3$) formed by the decomposition of organic materials, respectively (Ikeya 1982). Methyl radicals are also observed in chert (Griffiths et al. 1982) and silica in geothermal areas (Ikeya 1982).

Theoretical splitting by the proton hf interaction is shown in Figure 2.14. As the coupling strength is nearly the same, three lines (triplet) with an intensity ratio of $1:2:1$ are observed and the ratio is $1:3:3:1$ for coupling with 3 equal protons (quartet). The hf interaction with 4 protons gives quintet with the intensity ratio of $1:4:6:4:1$ (Carrington and McLachlan 1967, Poole and Farach 1972). The septet spectrum observed in fossil horse molars and some shells is due to the hf interaction with 6 protons. This radical was identified as isopropyl radical, $(CH_3)_2C\cdot-R$, which is formed in synthetic valine–doped $CaCO_3$. The origin of organic radicals abundant in nature may be studied in organic geochemistry (Johns 1986).

The hf interaction of electron spins with protons is not necessarily the same in all cases. Some couple strongly, while other protons give weak hf magnetic fields. In the methanol radical ($\cdot CH_2OH$) observed in geothermal silica, the unpaired electron interacts with two protons equally and with one proton differently as shown in Figure 2.14 (b). Several organic molecules could be embedded in synthetic hydroxyapatite, $Ca_{10}(PO_4)_6(OH)_2$ and aragonite, $CaCO_3$. Ionizing radiation followed by thermal annealing produces organic radicals different from those created by radiation alone.

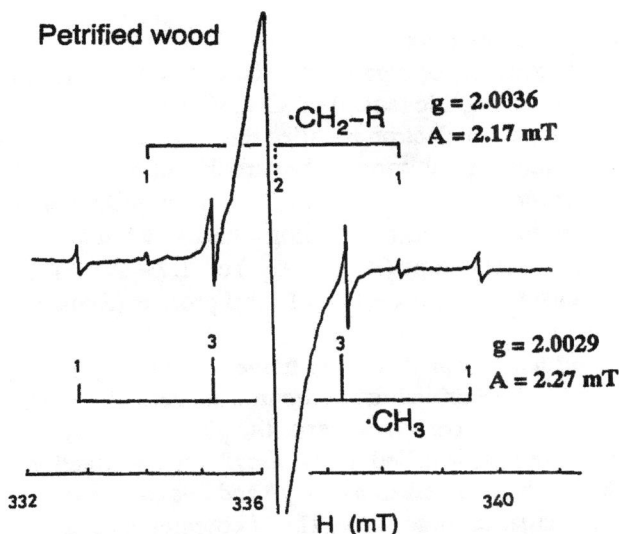

Figure 2.13 ESR spectrum of a fossil wood with proton hf lines of organic radicals, $\cdot CH_2-R$ and $CH_3\cdot$, together with an unresolved line of organic radicals at the central part.

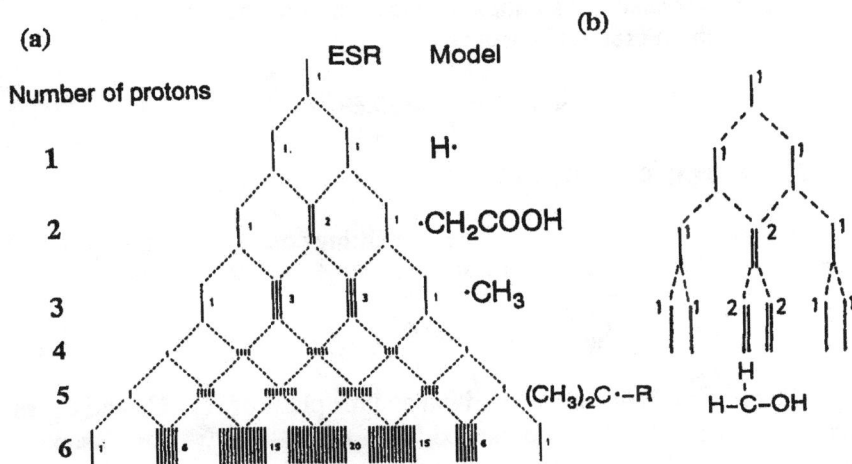

Figure 2.14 Theoretical splitting by the proton hf interaction. Equal contributions of 2, 3 and 4 protons with $I = 1/2$ give lines of different intensity ratios. A case of different hf coupling constant A within a molecule of methanol radical is indicated in (b).

2.4.6 *McConnell's Relation*

Proton hf structure for organic molecules has been studied extensively in chemistry. In an organic radical of $\cdot CH_\alpha-CH_\beta=$, a proton bonding to a carbon with an unpaired electron is called the "α *proton* (H_α)" and a proton bonding to a carbon neighboring to the carbon with an unpaired electron is called the "β *proton* (H_β)". It is difficult to distinguish between α and β protons by using the principal hf coupling constant A but a distinction can be made by the large anisotropy ($A_{max} - A_{min}$) of $1.7 \sim 2.2$ mT for α protons and the small anisotropy of $0.1 \sim 0.3$ mT for β protons (Symons 1963).

(a) α *proton* : π *electron density* : In the case of $-C_\alpha H_\alpha-$ type radicals, three bonds of the sp^2 orbital of C_α are in a plane. An unpaired electron enters the $p(\pi)$ orbital perpendicular to the plane as shown in Figure 2.15 (a). This type of radical is called a "π *radical*". If the direction of the $C_\alpha-H_\alpha$ bond and the p orbital are taken as the $y-$ and $z-$axes, respectively, the $x-$axis is perpendicular to these axes. The hf coupling constant is anisotropic and is given in mT as

$$A_x = -2.5\,\rho_{C\alpha} \quad , \quad A_y = -3.7\,\rho_{C\alpha} \quad , \quad A_z = -0.8\,\rho_{C\alpha} \quad , \tag{2.21}$$

where $\rho_{C\alpha}$ is the density of an unpaired electron in the p_x orbital of C_α. The average A for the α proton is given as

$$A_{av} = 2.25\,\rho_{C\alpha} = 50.68\,\rho_{H\alpha} \quad , \tag{2.22}$$

where $\rho_{H\alpha}$ is the spin density on H_α.

(b) β *proton* : The anisotropy of the hf splitting constant is small for the β proton far away from the unpaired electron. The A_{av} of the β proton is

$$A_{H\beta} = B\,\rho_{C\alpha}\cos^2\theta \quad , \tag{2.23}$$

where θ is the angle between the p orbital and the plane of H_β, C_β and C_α as shown in Figure 2.15 (b) and the constant $B = 4 \sim 5$ mT. For instance, *methyl group* that is freely rotating shows average $A = B\,\rho_{C\alpha} <\cos^2\theta> = B/2$ $= 2.25$ mT. ESR spectra of organic radicals are detailed in the literature (Symons 1963, Carrington and McLachlan 1967).

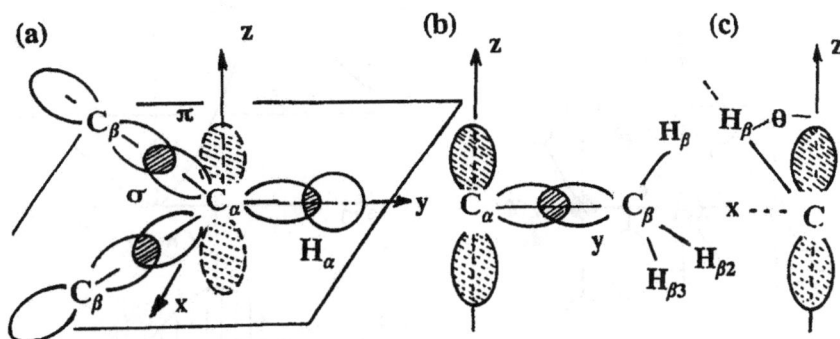

Figure 2.15 (a) The axes for anisotropic hf coupling of α proton with an unpaired electron in the p (π) orbital (hatched by dashed and dotted lines). (b) The hf coupling of β proton with an unpaired electron in the α carbon. (c) The projection of the $x-y$ plane from the z direction to indicate the angle θ for one of the β protons.

2.4.7 Superhyperfine (shf) Structures

When an unpaired electron interacts with nuclear spins of the *nearest neighboring* (n.n) atoms or ions, or in some cases the *next nearest neighbor* (n.n.n) or third shell ions, resonance lines are split into hf structures due to the magnetic field produced by the neighboring atom's nucleus. These are called "*superhyperfine (shf) structures*". This splitting is mostly unresolved and contributes to the broadening of the ESR signal. Characteristic features of organic radicals in minerals are the stable signal at room temperature and the sharp resonance lines, whereas those observed in radiation chemistry have a broad linewidth and are mostly stable at lower temperatures. Organic radicals embedded in silica or carbonate minerals are not surrounded by molecules with H which broaden the signal due to shf interaction.

Figure 2.16 (a) shows a model of the F center with the first n.n shell of six alkali ions and the second n.n.n shell of twelve halogen ions in alkali halides. A derivative line of the F center is shown schematically in (b) to-gether with the theoretical position and intensity of the shf lines from the first and second shells. Two isotopes ^{35}Cl (75.77%) and ^{37}Cl (24.23%) have the same nuclear spin ($I = 3/2$) with different nuclear magnetic moments. The broad linewidth of the F center in NaCl or KCl is due to the unresolved shf structures from Na$^+$ or K$^+$ nuclei and also from $^{35}Cl^-$ and $^{37}Cl^-$ nuclei. Resolved shf structures due to $^{19}F^-$ ($I = 1/2$, 100%) and Li$^+$ (^6Li; $I = 1$, 7.5% and ^7Li; $I = 3/2$, 92.5%) are observed for the F center in a LiF crystal.

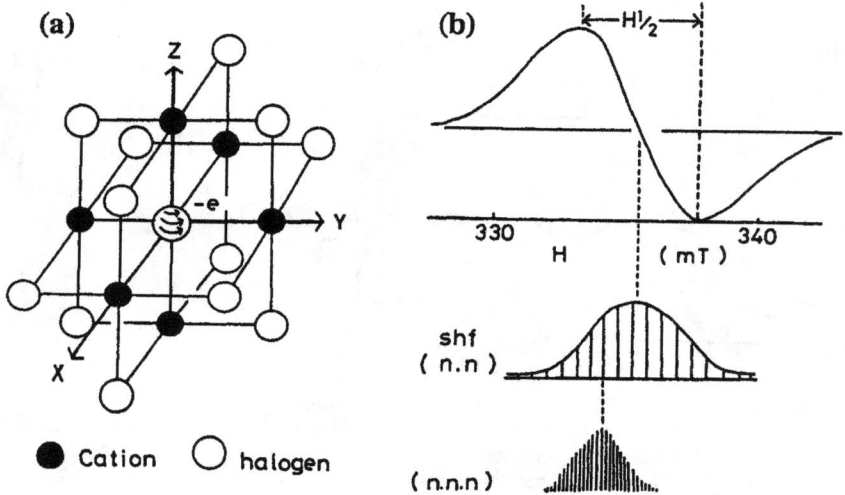

Figure 2.16   (a) A model of the F center in alkali halides and (b) ESR spectrum of the F center with the width due to shf by the nearest neighbor (n.n) alkali and next nearest neighbor (n.n.n) halogen ions (Seidel 1961).

Point defects such as F centers are common in solids as elementary products of radiation effects. The linewidth of the spectrum is dominated by the hf interaction.   F centers are sometimes so close to each other as to inter-act magnetically in α–ray track regions in a neutron–irradiated LiF crystal. In case of geological materials, we have a chance to observe such defects (F aggregate type center) in the region of α–recoil or fission tracks. In such a case, the linewidth may be broadened.

The *shf coupling constant*, A_{shf}, in a crystalline material has an aniso-tropy similar to the hf coupling constant and is written as

$$A_{shf} = a + b\,(3\cos^2\theta - 1) \ , \tag{2.24}$$

where a and b are *isotropic* and *anisotropic shf coupling constants*, respec-tively. Both a and b for ions in the n.n and n.n.n shells around the unpaired electron in the F center are obtained with an *electron nuclear double reso-nance* (ENDOR) technique. The isotropic a represents the electron probabili-ty at the surrounding nucleus (*Fermi contact term*) and the anisotropic b indicates the *dipolar interaction* of the electron spin with the nuclear spin.

A typical ESR spectrum of *interstitial hydrogen atoms* (H_i^o) in a KF crystal shown in Figure 2.17 indicates a hf doublet by a proton (H, $I = 1/2$) and shf structures due to four $^{19}F^-$ ions nearby having $I = 1/2$. As the four F^- ions interact equally with the electron spins for H along [100], the total nuclear spin of $I = 1/2 \times 4 = 2$ leads to $2I + 1 = 5$ lines with an intensity ratio of $1 : 4 : 6 : 4 : 1$ and the separation of A_{shf}. The shf due to the four K^+ (^{39}K, 93.26% and ^{41}K, 6.73%) both with $I = 3/2$ is not resolved due to the small nuclear magnetic moment.

Atomic hydrogen centers and their association with impurities and defects in alkali halides have been studied in the field of color center physics. They are called *"U centers"* since optical absorption bands are observed in the ultraviolet and violet regions. The U center is diamagnetic H⁻ substituting for a halogen anion, the U_1 *center* is H⁻ in the interstitial position (H_i^-) and the U_2 *center* is an interstitial H° (H_i^o) (Speath 1966, Seidel and Wolf 1968). Stabilization of H° by an impurity like Li⁺ at an off–center site in KCl is observed still at low temperature. In the case of geological materials, ESR spectra of hydrogen atoms are not observed, as they become mobile well below room temperature. However, in some minerals like hydroxyapatite, such centers are observed soon after artificial γ–irradiation.

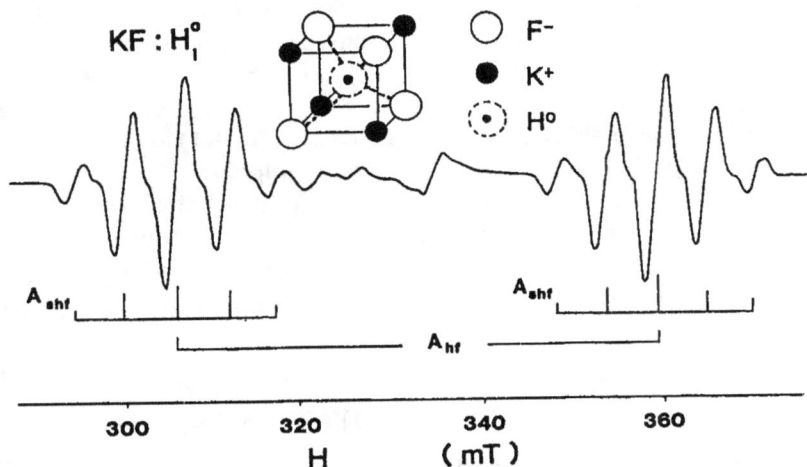

Figure 2.17   ESR spectrum of an interstitial atomic hydrogen centers (H_i^o) in a KF crystal for H parallel to the [100] direction. Resonance lines by hf splitting $(A_{hf}$ by a proton with $I = 1/2)$ are further split into five lines by shf due to nearby four F⁻ ions with $I = 1/2$.

2.4.8   Fine Structures

Triplet states ($S = 1$) are formed by the *spin–spin exchange coupling* of two unpaired electrons in close proximity. In transition metal ions, the spin quantum number S is larger than 1 due to the alignment of 3d electron spins. Rare earth ions have 4f electrons whose spins orient to the same direction to have a maximum S according to *Hund's rule*. The coupling of the electron spins with the orbital magnetic moment leads to an indirect spin–spin coupling. The spin energy state under external magnetic field depends on the angle between the direction of the crystalline field and the magnetic field.

Transition metal ions have from one to ten d electrons. Mn^{2+} and Fe^{3+} have five d electrons and the total spin is five times the half spin ($S = 5/2$). The fine structure splitting due to different $M \rightarrow M+1$ transitions is observed when the ion is in a quadratic term of the *crystalline static electric field* such as an axial electric field ($V = D'z^2$), an orthorhombic electric field [$V = D'z^2 + E'(x^2 - y^2)$] or a strong cubic (quadratic term) crystalline field [$V(r) = (a_c/6)(x^4 + y^4 + z^4)$].

$$V(x, y, z) = D'z^2 + E'(x^2 - y^2) , \qquad (2.25)$$

where D' and E' are parameters for an axial and orthorhombic crystalline field, respectively and a'_c is a cubic field crystalline field parameter.

The energy levels split into ($2S + 1$) lines for an axial field parameter D as shown in Figure 2.18 for (a) $S = 1$ and (b) $S = 5/2$. Resonance occurs between different magnetic spin quantum numbers, i.e., $M \leftrightarrow M \pm 1$ ($\Delta M = \pm 1$). Since M takes S, $S-1$, $\cdots\cdots$, $-S$, $2S$ angle dependent resonance lines have different intensity ratios of [$S(S + 1) - M(M + 1)$].

The energy state is described by "*spin Hamiltonian*" $DS_z^2 + E(S_x^2 - S_y^2)$ using operators S_x, S_y and S_z, where ESR parameters D and E are called "*zero field splitting parameters*" related with the above D' and E', respectively. When $D \neq 0$ and $E = 0$, the fine structure splitting is called *axial field splitting*. When $E \neq 0$, it is called *orthorhombic field splitting*.

(a) *Axial field* (DS_z^2) : The general features of the energy are described by

$$E = g\beta HM + D(3\cos^2\theta - 1)(M^2 - S(S+1)/3) , \qquad (2.26)$$

and therefore for the transition of $M \leftrightarrow M+1$,

$$h\nu = g\beta H_o = g\beta H + D(3\cos^2\theta - 1)(2M + 1) ,$$
$$H = H_o - (D/g\beta)(3\cos^2\theta - 1)(2M + 1) . \qquad (2.27)$$

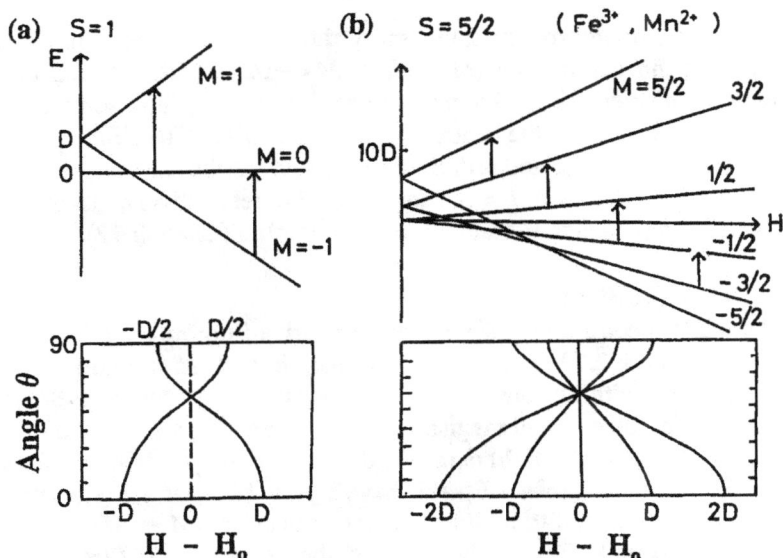

Figure 2.18 Fine structure levels as a function of the magnetic field for (a) $S = 1$ and (b) $S = 5/2$ with the angular dependence of fine structure splitting.

The *unit* of D (cm^{-1}) is practically expressed as mT by dividing D by $g\beta$ (multiply $0.21435/g$ to convert the unit of 10^{-4} cm^{-1} to mT). We call this angle dependent splitting *"fine structures"* or *"axial field splitting"*.

No fine structure occurs for the transition $M = -1/2 \leftrightarrow M = 1/2$ or for an unpaired electron with $S = 1/2$ as is seen from Eq. (2.27). For a triplet ground state of two unpaired electrons ($S = 1$), the fine structure is a doublet and for $S = 5/2$ like Mn^{2+} or Fe^{3+}, quintet fine structures.

(b) *Orthorhombic field* $[DS_z^2 + E(S_x^2 - S_y^2)]$: So long as the energy splitting of spins by the orthorhombic crystalline field via spin–orbit interaction is smaller than the Zeeman energy, the splitting of the resonance field can be described by perturbation theory. The resonance field for new D and E in the unit of mT (i.e., $D/g\beta$ and $E/g\beta$, respectively) is written using $E_{M+1} - E_M = g\beta H_0$ for a general case of θ and ϕ as

$$H = H_0 - [(2M + 1)/2][D(3\cos^2\theta - 1) + 3E\cos2\phi\sin^2\theta]$$

$$- (\sin^2\theta/2H_0)[(D - E\cos2\phi)^2\cos^2\theta + E^2\sin^22\phi][24M(M + 1) - 4S(S + 1) + 9]$$

$$- (1/8H_0)\{[D\sin^2\theta + E\cos2\phi(1 + \cos^2\theta)]^2 + 4E^2\cos^2\theta\sin^22\phi\}$$

$$\times [2S(S + 1) - 6M(M + 1) - 3] . \tag{2.28}$$

In powder or amorphous state, most of the fine structures are averaged out except for the central line $(M = 1/2 \leftrightarrow M = -1/2)$ which is almost angular independent of orientation. Even these transitions are split in some cases as shown in Figure 2.12 for the central six lines of Mn^{2+}. The ESR spectra of Fe^{3+} in the strong axial and orthorhombic fields at the X-band frequency (9.5 GHz) shows signals at $g = 6.0$ and $g = 4.27$, respectively, as observed in clay minerals (Chapter 10) and blood hemoglobin (Chapter 12).

2.4.9   NaCl $(Mn^{2+} : F^-)$

An ESR spectrum of Mn^{2+} associated with a fluorine ion (F^-) in NaCl is shown in Figure 2.19 as an example of fine, hf and shf structures. Mn^{2+} ions form a pair with F^- ions in an NaCl crystal doubly doped with Mn^{2+} and F^-. When the external magnetic field is along a crystal axis of [100], two sites for the magnetic field parallel $(\theta = 0)$ and perpendicular $(\theta = 90°)$ to the direction of the $Mn^{2+} - F^-$ axis have a probability of 1:2 (site symmetry). Fine structure splitting lines for the transition $(M \leftrightarrow M+1)$ are described using Eq. (2.27) as the shift of the field by $2D(2M + 1)$ and $-D(2M + 1)$ for $\theta = 0°$ and $90°$, respectively. For Mn^{2+} with $S = 5/2$, $M = -5/2, -3/2, -1/2, 1/2$ and $3/2$. The intensity for $\theta = 90°$ is small when the

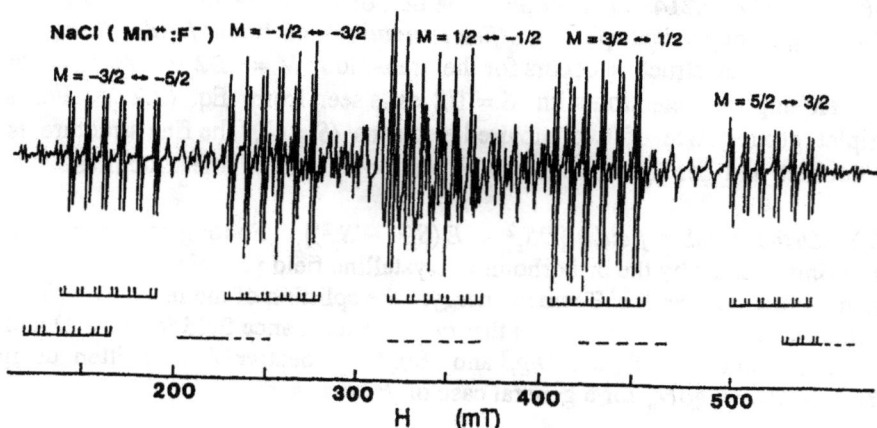

Figure 2.19 ESR spectrum of NaCl doubly doped with Mn^{2+} and F^- for the magnetic field along the [100] direction. The association of Mn^{2+} with F^- resolves six hf lines of Mn^{2+} $(I = 5/2)$ into a doublet due to the shf interaction of F^- $(I = 1/2)$. Five groups of lines are due to fine structure splittings of Mn^{2+} $(S = 5/2)$ for $\theta = 0°$.

D value is large. These fine structures split into six lines due to hf splitting of Mn^{2+} nucleus ($I = 5/2$) and further into two by shf splitting due to nearby $^{19}F^-$ ($I = 1/2$). The shf splitting due to $^{35}Cl^-$ and $^{37}Cl^-$ ($I = 3/2$) is not resolved. The angular dependence of the signal positions can be analyzed using Eq. (2.28). If the sample is rotated from $H \parallel [100]$ to $[010]$ via $[110]$ in the (001) plane, one site has always $\theta = 90°$, while others change from $0°$ to $90°$ and $90°$ to $0°$.

2.5 Powder Spectrum
2.5.1 Averaging of g factors

Most materials for ESR dating are polycrystalline or powder. Fossil bones or shells consist of microcrystalline materials oriented either randomly or preferentially along a particular direction. The angular dependent ESR signals appear at different magnetic field positions and superimpose to produce a broad spectrum (Kneubühl 1960). Figure 2.20 indicates the probability that the principal axis of randomly oriented microcrystals makes an angle of θ with the magnetic field direction. The *probability*, $P(\theta)$, is proportional to a fraction of $\sin\theta\, d\theta$,

$$P(\theta)\, d\theta = \sin\theta\, d\theta .$$ (2.29)

The *signal shape function*, $S(H)$, is then obtained as

$$S(H) = \sin\theta\, d\theta / dH ,$$ (2.30)

where
$$\int_{-\infty}^{\infty} S(H)\, dH = 1 .$$ (2.31)

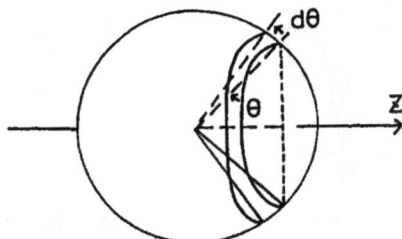

Figure 2.20 Probability for anisotropic centers to make an angle θ with the direction of the external magnetic field H for the analysis of powder pattern lineshapes. An angular ring is seen between the angles θ and $\theta + d\theta$ with the area $2\pi r^2 \sin\theta\, d\theta$ (Poole and Farach 1972).

For the case of an *anisotropic g factor*, the resonance position changes as a function of θ (Figure 2.21(a), Eq. (2.7) and Eq. (2.8)). The magnetic fields of $H_{\parallel}$ and $H_{\perp}$ correspond to g factors of $g_{\parallel}$ and $g_{\perp}$ for the field parallel ($\parallel$) and perpendicular ($\perp$) to the principal axis of the crystal, respectively. The signal shape function $S(H)$ which is partly proportional to $\sin\theta$ is shown as a function of the magnetic field H in Figure 2.21 (c), using the angular dependence of the resonance line in (a) and the probability $P(\theta)$ in (b). In (c), the broken curve is the theoretical one for a sharp signal and the solid curve is obtained by taking the linewidth into account. The derivative form is shown in (d). It should be noted that the maximum and minimum

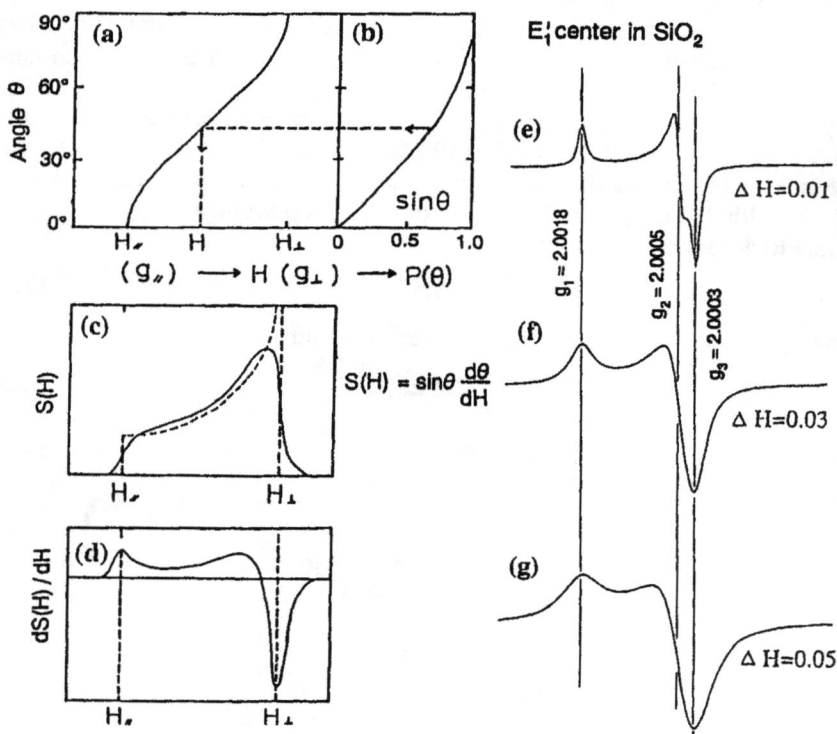

Figure 2.21   Powder pattern line shape due to g factor anisotropy.   (a) Angular dependence of the resonance line, (b) the probability $P(\theta)$ that the anisotropic center makes an angle θ to the field, (c) the signal shape function $S(H)$ (broken curve) and the one taking the finite linewidth into account (solid curve), and (d) the derivative line of the ESR signal (Lefebve and Maruani 1965). (e), (f) and (g) Calculated powder spectra of the E_1' center in SiO_2 for different linewidths (ΔH in mT) assuming a Lorentzian lineshape.

points of the derivative spectrum correspond roughly to $g_{\parallel}$ and $g_{\perp}$, respectively. Calculated spectra with anisotropic g factors, g_1, g_2 and g_3 of the E'_1 center (an electron at an oxygen vacancy) in SiO_2 with the indicated linewidth (ΔH) are given in (e), (f) and (g).

2.5.2 Averaging of Fine and Hyperfine Structures

Fine structure splitting shows an angular dependence of form $-D(3\cos^2\theta - 1)$ as in Eq. (2.27). The shift of the field for a triplet state ($S = 1$, $M = 0 \leftrightarrow M = 1$) is $-D$ for $\theta = 0°$ (the z direction), while the shift is $+D/2$ for $\theta = 90°$ (x and y directions). For $M = -1 \leftrightarrow M = 0$, they are $+D$ and $-D/2$ for $\theta = 0°$ and $90°$, respectively. Figure 2.22 shows the powder spectra for fine structure splitting for the axial and orthorhombic crystalline

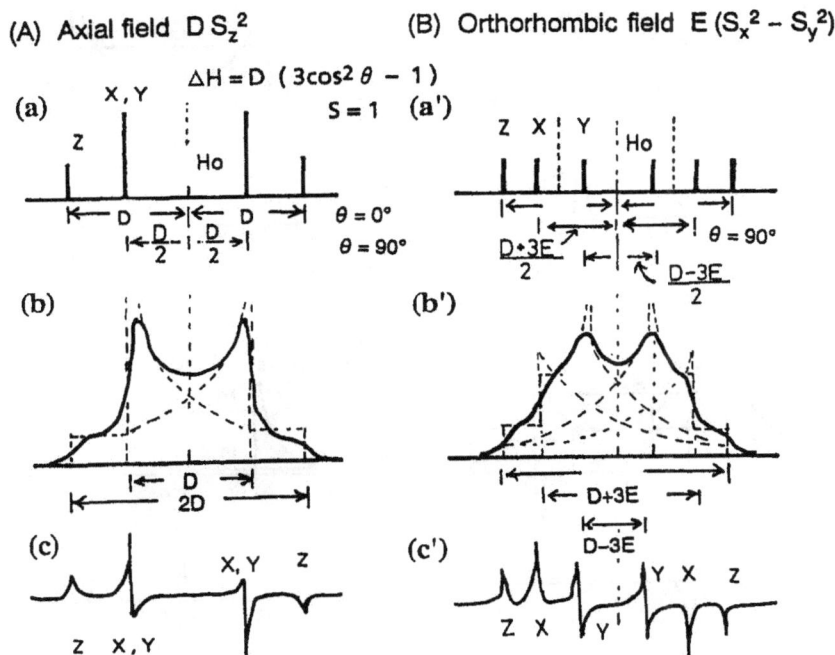

Figure 2.22 Powder spectra for fine structure splitting. (A) An axial field and (B) an orthorhombic field. (a) The signal position and intensity, (b) lineshape integrated for all angles for $S = 1$ and (c) a derivative line (Lefebve and Maruani 1965).

fields. The overall average of the theoretical signal shape function is shown in (b) where the solid curve is the calculated one by taking the linewidth into account. It is noted that the parameters for fine structures can be derived roughly from such a powder spectrum. The ESR spectrum for the orthorhombic field shows splitting for X and Y as shown in (c).

Powder spectra for anisotropic hf structures are shown in Figure 2.23. The theoretical signal intensity and the derivative line considering the linewidth are shown for different magnitudes of $A_\parallel$ and $A_\perp$ for the axial symmetry in (a) and (b), and for those of A_1, A_2 and A_3 for the orthorhombic symmetry in (c). The triplet hf for $I = 1$ is shown in (d). Anisotropic hf structures of the powder signal smear out the angular dependence.

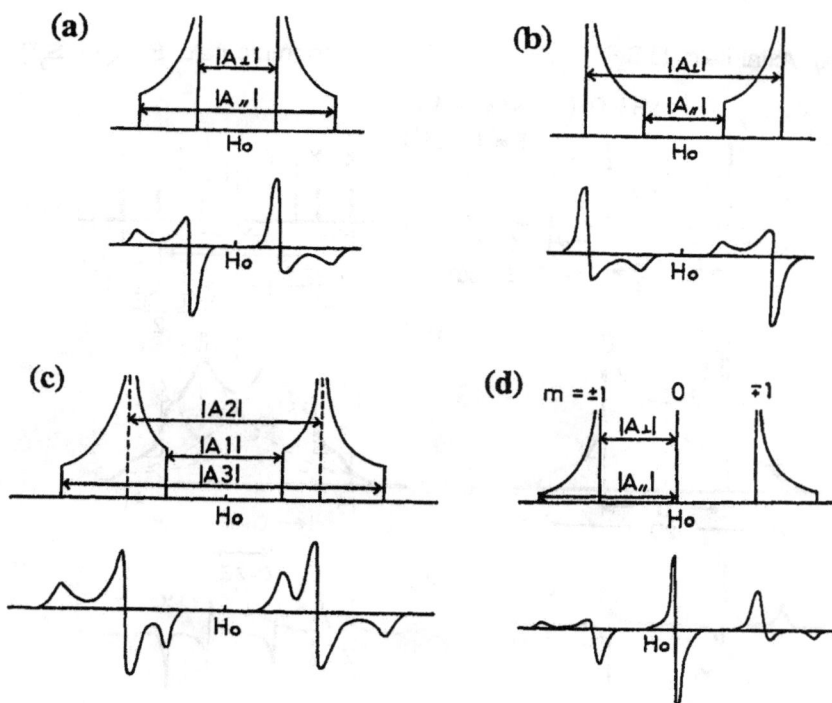

Figure 2.23   Theoretical powder spectra of anisotropic hf structures for (a) $A_\perp < A_\parallel$, (b) $A_\perp > A_\parallel$, (c) $A_1 < A_2 < A_3$ and (d) triplet state ($I = 1$) (Lefebve and Maruani 1965).

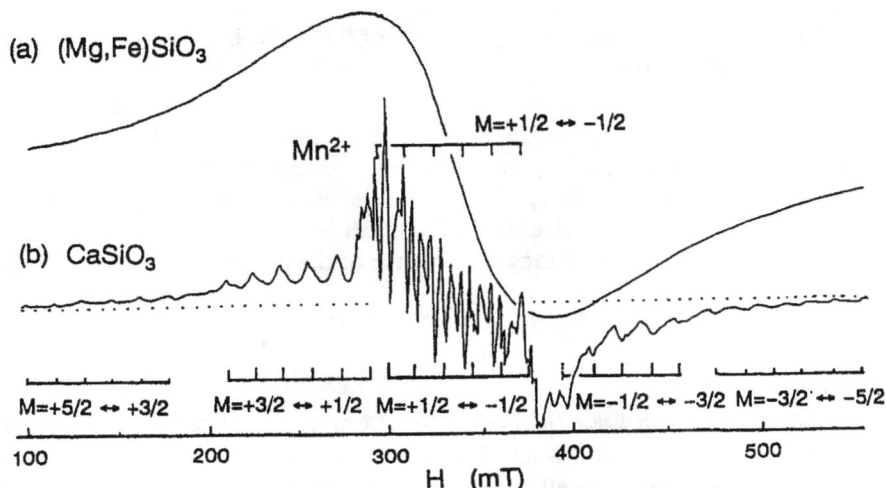

Figure 2.24   (a) One broad line is observed for (Mg, Fe)SiO$_3$ due to dipolar and exchange interactions of Mn^{2+} and Fe^{3+}. (b) ESR spectrum of Mn^{2+} in pyroxene (CaSiO$_3$). The allowed and forbidden lines of hf splittings are associated with $\Delta M = \pm 1$, $\Delta m = 0$ and $\Delta M = \pm 1$, $\Delta m = \pm 1$, respectively.

If the concentration of paramagnetic centers is high, the *dipole–dipole* interaction broadens the signal linewidth and the *exchange interaction* changes the lineshape into the Lorentzian shape as shown in Figure 2.24 (a). It must also be noted that the $M = 1/2 \leftrightarrow M = -1/2$ transition for $\theta = 0°$ and $\theta = 90°$ gives fields at different magnetic intensities due to second–order terms, complicating the spectrum further. However, the lines at both ends indicate the averaged signals of fine structure associated with $M = \pm 3/2 \leftrightarrow \pm 5/2$ and $M = \pm 1/2 \leftrightarrow \pm 3/2$.

An ESR spectrum of Mn^{2+} in pyroxene is shown in (b) as an example of fine structure, though further splitting into hf lines makes the analysis difficult. Weak signals of fine structures are observed on both wings of the spectrum. Complicated signal lines due to the allowed and forbidden hf lines are intense. Anisotropy of the hf structure for Mn^{2+} is usually very small.

2.6   Signal Saturation, Relaxation and ENDOR

2.6.1   *Relaxation Time and Saturation*

We have described in Section 2.2 how ESR is caused by the transition of electron spins from a lower to a higher energy level. The Zeeman energy for the transition of a microwave quantum of 9 GHz is about 10^{-5} eV. This is about $1/10^3$ of thermal energy ($k_B T = 0.025$ eV) at 300 K. Therefore, the difference in the number of spins in the lower and upper levels is actually small. Electron spins in the upper level fall to the lower level by emitting *phonons* or transferring energy to the *lattice vibration* with a time called the "*spin–lattice relaxation time, T_1*". If the *microwave power P* is intense enough to pump the lower electron to the upper level, the upper level population, N_2, is reduced considering the balance of the power as

$$dN_2/dt \;=\; cP - N_2/T_1 \;,$$ (2.32)

where c is a constant.

One way to separate the ESR signal from a complicated overlapping spectrum is to measure at low and high microwave power levels. The intensity of some signals saturates at low microwave power but that of others increases in proportion to $P^{1/2}$ up to high microwave power. Appropriate setting of power levels is therefore necessary for the ESR measurement. Examples of the microwave power dependence are shown in each chapter.

2.6.2 *Electron Nuclear Double Resonance (ENDOR)*

Shf interactions with the first nearest neighbor (n.n) ions, next n.n (n.n.n) ions and so on can be resolved by the method of *electron nuclear double resonance* (ENDOR). One can observe the *nuclear resonance* caused by the neighboring ions at *radio frequencies* through the change of the ESR spectrum of unpaired electrons at the microwave frequency. Figure 2.25 (a) shows a schematic illustration of the ENDOR principle for an unpaired electron interacting with a nuclear spin $I = 1/2$. The electron spin states are $M = \pm 1/2$. The ESR absorption occurs at two magnetic field intensities.

In ENDOR, ESR absorption from $M = -1/2$ to $M = 1/2$ is saturated at a high microwave power. The radio frequency wave associated with the nuclear hf coupling is then added to cause transitions between two nuclear magnetic spin states, m. Microwave absorption then occurs, since the saturated balance is destroyed at the resonant radio frequency. One can thus resolve hf splitting by a particular nucleus which creates a broad linewidth. ENDOR of the F center has been studied extensively (Seidel and Wolf 1968). ENDOR should also be used to identify the model of ESR signals in geological materials to give a sound basis for ESR dating.

Figure 2.25 (b) shows the model of an interstitial hydrogen atom (H_i°), sometimes called the U_2 *center*, in KCl. A part of the ENDOR spectrum of

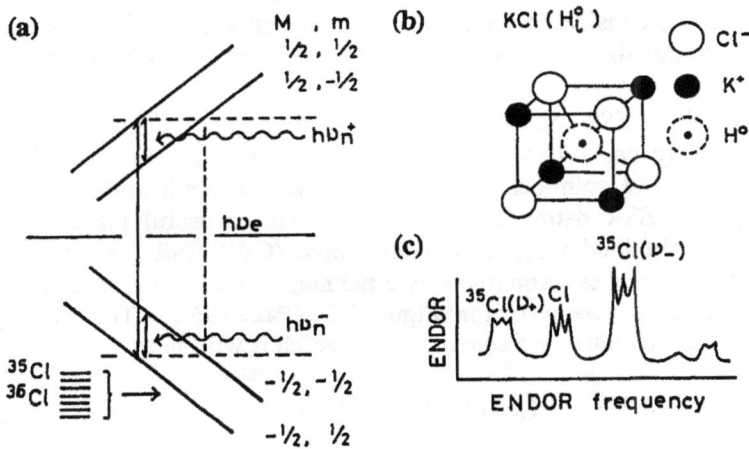

Figure 2.25 (a) Energy levels of an interstitial hydrogen atom, H_i^o in KCl with the indication of ESR resonant transitions by the microwave quanta of $h\nu_e$ and the ENDOR transition by the radiowave of $h\nu_{n+}$ and $h\nu_{n-}$. (b) The model of H^o in KCl. (c) ENDOR spectrum shows shf structures by Cl^- nuclear spins (revised from Speath 1966).

the centers is shown in (c). Shf structures by the n.n and n.n.n ions are clearly resolved by ENDOR, from which one can determine the spread of the unpaired electron spin density on neighboring ions (Speath 1966). The atomic arrangement and impurity association in the n.n and n.n.n can be described as well, as if one had seen the neighboring atoms around the un-paired electron from inside the material. However, the method requires a sufficient number of unpaired electrons having some specific neighboring atoms with a nuclear spin and so high–power source of radiowave at a specif-ic frequency region. Hence, ENDOR, though used in physics, has not been applied to geological materials except for some mineral crystals. A good standard general ENDOR spectrometer is not available. The contribution of ENDOR to mineral physics would be enormous if it were widely used.

We still have plenty of problems in ESR with geological and archaeo-logical materials containing several different impurities. Some might be solved with ENDOR in the future, but most of the problems could be solved by using geological and archaeological criteria to choose better samples and by analyzing impurities chemically. A solid state physics approach to syn-thesize a material doped with impurities gives us an insight on the models of defects produced in nature. Our understanding of ESR dating of carbonates has expanded considerably since the first work in 1975 by Ikeya and models of the ESR signals such as CO_2^-, SO_3^- and SO_2^- are being clarified.

2.6.3   Pulsed ESR : Spin–Echo

(a)   *Rotating coordinate and pulsed microwave* :   A brief description of pulsed ESR and spin–echo is given here since the application of this new technique to ESR dating and dosimetry produces useful information that cannot be obtained with continuous wave (CW) ESR.   Spin magnetic moments that precess around the external magnetic field are described using a *rotating coordinate* system in Figure 2.26 (Pake 1962). The spin system is rotating at the *Larmor frequency*, ω, which is written as

$$\omega = (g\beta/h)H = \gamma H , \qquad (2.33)$$

where γ is the *gyromagnetic ratio, $g\beta/h$*.   The microwave magnetic field is given as

$$H_1 \sin\omega t = H_1(e^{i\omega t} - e^{-i\omega t})/2 , \qquad (2.34)$$

where H_1 is the intensity of the microwave magnetic field and therefore is proportional to $P^{1/2}$. The microwave magnetic field becomes "*static*" in a rotating frame along the x' direction.

The spin behavior at off–resonance is described as the precession of the *magnetic moment*, μ, around the *effective magnetic field*, H_{eff}, in the rotating coordinate as shown in Figure 2.26 (a). The precession is around the x'–axis at the resonance condition since $H_{eff} = H_1$ as shown in (b). The spin oriented to the z direction precesses toward the $-z$ direction around H_1 in (c). The energy loss due to lattice phonons acts to take the magnetic moment back to the z direction. The *spin–lattice relaxation time* T_1 is defined as the relaxation time to orient the spin back to the z direction and is called "*longitudinal relaxation*".

One method to orient spins to the $-z$ direction is to apply an intense microwave H_1 for a period of $t = \pi/2\omega_1 = \pi/2\gamma H_1$. This pulse is called the "π *pulse*" or "$180°$ *pulse*" as the magnetization is directed to the $-z$ direction. One needs an intense H_1 or P to produce a pulse with a width shorter than T_1. If we could monitor the magnetization along the z direction, it would be recovered as $-M_z + 2M_z(1 - e^{-t/T_1})$, where M_z is the net magnetization along the z direction. Although ESR signals overlap, they have different T_1.

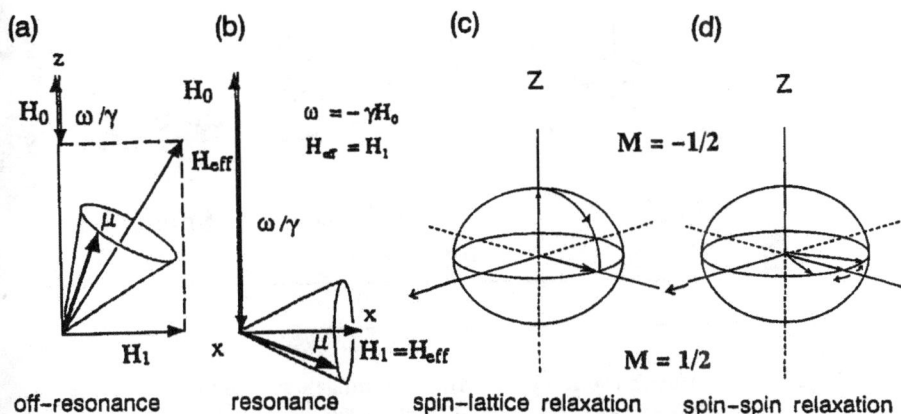

Figure 2.26 (a) A coordinate system rotating at a frequency ω. The precessing spin feels an effective magnetic field, H_{eff}, at off-resonance. (b) H_{eff} is along the x' direction at resonance and spins precess around the x'-axis in the rotating frame (Pake 1962). (c) Measurements of spin-lattice relaxation time T_1 by pulse magnetization to the $-z$ direction using the π (180°) pulse with pulse width of $t = \pi/2\gamma H_1$. Relaxation to the $+z$ direction is called spin-lattice relaxation. (d) Spin-spin relaxation after the $\pi/2$ pulse which brings the magnetization to the $x'-y'$ plane of the rotating coordinate.

Separation of overlapping signals is possible by selecting T_1 (Kevan and Bowman 1989) and this will contribute considerably in determining the concentration of defects in geological and archaeological materials.

(b) *Spin-spin relaxation ($\pi/2$ and π pulses)*: The *spin-spin relaxation time, T_2,* can be obtained by Hahn's two echo method. The spin system is oriented to the $-y'$ direction around the x' direction in the rotating coordinate with a $\pi/2$ (90°) pulse with a width of $t = \pi/4\omega_1 = \pi/4\gamma H_1$ as shown in Figure 2.26 (d) or Figure 2.27 (a). If the spin-spin interaction is large, the phase is disturbed. The spin directions diffuse in the $x'-y'$ plane as shown in Figure 2.27 (b). When the second pulse (π (180°) pulse) with a width of $t = \pi/2\omega_1 = \pi/2\gamma H_1$ is applied at time τ after the first pulse, spins are oriented to the y' direction as shown in (c). Spins further diffuse in the rotating coordinate and the magnetization along the $-y'$ direction is formed in time τ after the second pulse as shown in (d) and (e). The microwave power is

(a)   (b)   (c)   (d)   (e)

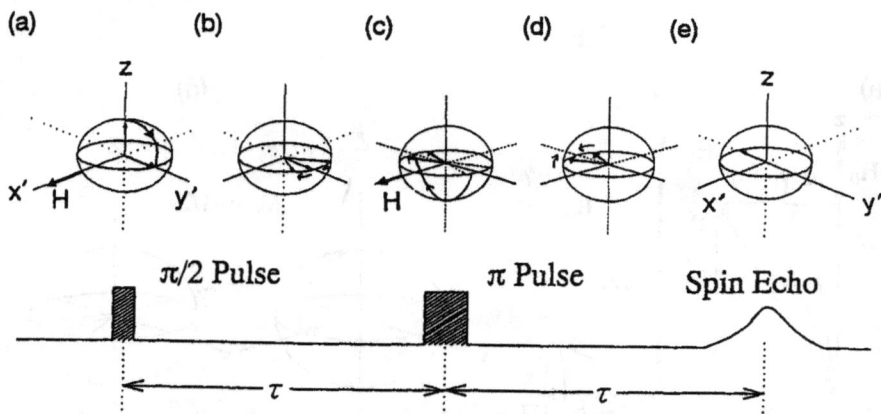

π/2 Pulse   π Pulse   Spin Echo

Figure 2.27   Hahn's two π/2 pulses for the coordinate system rotating at the angular frequency ω = 2πf. (a) Spin rotation around the x'-axis by π/2 (90°) pulse with the width $t = (\pi/2)/(\gamma H)$, (b) spin dephasing due to spin–spin relaxation, (c) π (180°) pulse after the time τ brings into the $-y'$ direction, (d) inphasing due to spin–spin relaxation and (e) spin echo for the spin magnetization along the y' direction (Hahn 1950).

detected at that time as a *spin–echo*. If the time t between the first and the second pulses is varied, the *echo intensity* (I/I_o) after the second pulse decays as a function of τ in the form of e^{-t/T_2}. A series of pulses are created with a computer to obtain the echo intensity automatically as a function of τ. The spin–echo intensity is obtained for granitic quartz grains irradiated by neutrons or by γ–rays after mechanical crushing as well as for quartz grains from a geological fault as shown in Figure 2.28 (Ikeya *et al.* 1992). The T_2 of quartz grains from a geological fault is about 30 μs just between 50 μs for those irradiated with neutrons and 15 μs for those crushed and γ–irradiated.

   The T_2 is a good indicator of the local concentration of spins. ESR signals of an unpaired electron at an oxygen center (E_1' center in SiO_2) give different T_2 depending on whether they are formed by neutrons randomly or by γ–rays after crushing. The results are due to the localization of defects created randomly or around dislocations. Plastic deformation or distortion creates the oxygen vacancy (or interstitial trapping site) around dislocations leading to localized the E_1' center by additional γ–irradiation. If one could separate the localized E_1' center around dislocations, the elapsed time after the fault movement may be determined.

2.6.4   *Fourier Transform (FT–) ESR*

Pulsed electromagnetic waves give a transient transition from the low to the high energy level. They have components of other frequencies close to their frequency, if the pulse shape is Fourier transformed mathematically. Hence, neither sweeping of the frequency nor the external magnetic field is necessary in FT–NMR equipped with a computer. Signal averaging of the transient response by repeated measurements in a short time gives a good *S/N* ratio. Separation of signals with different relaxation times is possible. A variety of pulsed NMR techniques have been developed for which L. R. Ernst was given the 1990 Nobel Prize in chemistry. Spin–echo observed as an echo of the pulsed electromagnetic wave is also often tried in ESR, as in NMR. A pulsed FT–ESR has different characteristics from CW–ESR.

Pulsed ESR or FT–ESR spectrometers are available by using digital microprocessors and sophisticated techniques of microwave pulse generation and detection systems. Various techniques developed in pulsed NMR can be used in pulsed ESR and allow different radiation–induced centers to be separated by means of their relaxation behavior; however, the sensitivity of pulsed ESR is still not comparable with that of CW–ESR at present though accumulation of data by repeated measurements is used. ESR imaging to

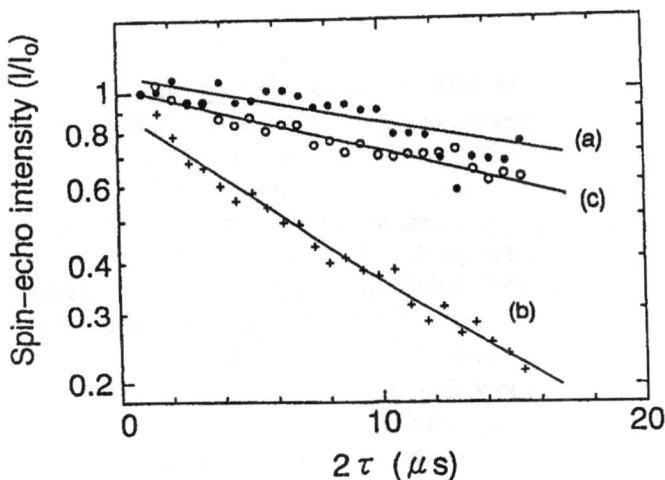

Figure 2.28   The exponential decay of the spin–echo intensity as a function of the time τ between the $\pi/2$ and π pulses for quartz grains ($e^{-\tau/T2}$). T_2 of the E_1' center were obtained from the intensity decay down to $1/e = 0.3678$ of the initial value. (a) Granitic quartz grains irradiated with neutrons, (b) those crushed prior to γ–irradiation and (c) quartz grains from a geological fault.

detect the spatial distribution of unpaired electron spin concentrations and ESR microscopy to detect microscopic distribution of spins (see Chapter 14) will be much improved by the use of FT–ESR. The technique will be a powerful tool in future for geological and archaeological materials.

2.6.5 *Vector Transfer Detection*

The spectral shape of some signals changes as the detection phase of the lock–in–amplifier to magnetic field modulation is varied. For the E_1' center, this is because the relaxation time is about $10\,\mu s$ for a modulation frequency of 100 kHz. Vector transfer ESR (EPR) by varying the phase of a lock–in–amplifier for a modulation spectroscopy is used for detecting radicals in coal samples and SiO_2 (Watari *et al.* 1985). Information on the spin localization can be obtained for centers with relaxation times close to the inverse of the modulation frequency. Although the method is somewhat classic and only applicable to specific cases, simple measurements can be done with CW–ESR or EPR apparatus. Selection of the E_1' center signal in quartz (SiO_2) fraction of granite and rhyolite has been made by suppressing the signal of the oxygen hole center (OHC) using this method (Shimoyama and Rink 1992).

2.7   Summary

The principles of ESR and some ESR spectra have been presented together with examples using geological materials. ESR parameters, the *g* factor, hyperfine (hf), superhyperfine (shf) and fine structures are used to identify the unpaired spin state in the material. This identification assures us that we are dealing with the same ESR signals as have been measured by other investigators and helps them to extend the work later.

All we need in ESR dating and dosimetry is to measure the intensity of an ESR signal and its enhancement by artificial irradiation or by the passage of time as described in the next chapter. Detailed knowledge of ESR is not necessary if samples show ESR signals which have already been identified.

Later chapters deal with typical ESR spectra and parameters (spin Hamiltonian parameters) in the format of a guiding atlas for ESR dating. The detailed description of spectra may be required when reassignment of the ESR signal must be made or when a new ESR signal for dating and dosimetry is discovered. This chapter will have to be rewritten completely when FT–ESR or pulsed ESR is used routinely in ESR dating, dosimetry and imaging (microscopy). Future development of ESR metrology is anticipated as our technology develops. Vast fields remains unexplored. A new field of "Quantum Geophysics" is expected in an interdisciplinary field.

(2nd edition): After writing this summary, almost a decade has passed. Several books on EPR (Mabbs and Collison, 1992; Weil *et al.*, 1994; Eaton et al.,1998; Poole, 1996) and pulse ESR (Schweiger and Jeschke, 2001) were published. However, the price of pulse ESR and its maintenance cost is so expensive that ordinary geoscientists cannot afford to buy the spectrometer. Furtherworse, the sensitivity (S/N ratio) is still two or three orders of the magnitude smaller that the conventional CW-spectrometer even with signal averaging.

New applications of pulse ESR demonstrated that the intense overlapping signal of Mn^{2+} or Fe^{3+}, both are common impurities in geological materials, can be erased by selecting the relaxation time and observing the *field sweep electron spin echo (ESE)* (Kohno et al.,1996a), but practical applications cannot be done for young samples of interests. Microwave power amplitude modulation also separate some signal (Kohno et al., 1996b). We need technological breakthroughs which reduces the costs and enhance the sensitivity of a pulse ESR spectrometer so that this method is used to solve problems in dating, dosimetry and imaging.

References

Alger R. S. (1973): *Electron Paramagnetic Resonance: Techniques and Applications* (Wiley & Sons, New York).

Bloch F., Hansen W. W. and Packard M. (1946): Nuclear Induction. *Phys. Rev.* **69**, 127.

Castner T. G. and Kaenzig W. (1957): The electronic structure of V-centers, *J. Phys. Chem. Solids* **3**, 178-195.

Carrington A. and McLachlan A. D. (1967): *Introduction to Magnetic Resonance* (Harper & Row, New York).

Eaton G.R., Eaton S. and Salikhov K.M. (1998): Foundations of Modern EPR (World Scientific, Singapore) pp.818.

Fowler W. B. (1968): *Physics of Color Centers* (Academic Press, New York)

Griffiths D. R., Robins G. V., Seeley N. J., Chandra H., McNeil A. C. and Symons M. C. R. (1982): Trapped methyl radicals in chert. *Nature* **300**, 435-436.

Hahn E. L. (1953): Spin echoes. *Phys. Rev.* **80**, 580-594.

Ikeya M (1975): Dating a stalactite by electron paramagnetic resonance. *Nature* **255**,48.

Ikeya M. (1982): Electron spin resonance of petrified woods for geological age assessment. *Jpn. J. Appl. Phys.* **21**, L28-30.

Ikeya M., Kohno H., Toyoda S. and Mizuta Y. (1992): Spin-spin relaxation time of E' centers in neutron irradiated quartz (SiO_2) and fault gouge. *Jpn. J. Appl. Phys.* **31**, L1539-L1541.

Inoue K., Hirai M., Ikeya M. and Yim W. (2000): Calibration method for ESR signal intensity of calcite shells. *Appl. Magn. Reson.* **19**, 255-269.

Johns R. B. ed. (1986): *Biological Markers in the Sedimentary Record* (Elsevier Sci. Pub., New York).

Kevan L. and Bowman M. K. (1989): *Modern Pulsed and Continuous Wave Electron Spin Resonance* (John Wiley & Sons, New York).

Kneub hl F. K. (1960): Lineshapes of electron paramagnetic resonance signals produced by powders, glasses and viscous liquids. *J. Chem. Phys.* **33**, 1074-1078.

Kobayashi T. and Suhara M. (1985): Single crystal ESR dating : method and apparatus. *ESR Dating and Dosimetry* 293-298.

Kohno H., Yamanaka C. and Ikeya M. (1996a): Effects od alpha-irradiation and pulsed ESR measurements of evaporites. *Appl. Radiat. Iso.* **47**, 1459-1463.

Kohno H. and Yamanaka C. (1996b): Amplitude modulation of eESR for signal separation using the difference of microwave power dependence. *Adv. ESR Appl.* **11**, 10-16.

Lefebve R. and Maruani J. (1965): Use of computer program in the interpretation of electron paramagnetic resonance spectra of dilute radicals in amorphous solid samples. *J. Chem. Phys.* 42, 1480-1502.

Mabbs FE and Collison D. (1992): *Electron Paramagnetic Resonance of Transition Compounds* (Elsevier, Amsterdam) pp1326.

Marfunin A. S. (1979): *Spectroscopy, Luminescence and Radiation Centers in Minerals* (Springer-Verlag, Berlin).

Pake G. E. (1962): *Paramagnetic Resonance* (Benjamin Inc., New York).

Perry D. L. ed. (1990): *Surface Chemistry of Geologic Materials* (VCH, Weinheim).

Pilbrow J. R. (1990): *Transition Ion Electron Paramagnetic Resonance* (Oxford Univ.).

Poole C. P. and Farach H. A. (1972): *The Theory of Magnetic Resonance* (Wiley & Sons, New York).

Poole C. P.(1996): *Electron Spin Resonance:* a comprehensive treatise on experimental technique (2nd edition) (Dover , New York), pp780.

Purcell E. M., Torrey H. C. and Pound R. V. (1946):   Resonance absorption by nuclear magnetic moments in solids. *Phys. Rev.* **69**, 37.

Rossi A., Poupeau G. and Danon J. (1985): On some paramagnetic species induced in natural calcites by α- and γ-ray irradiations. *ESR Dating and Dosimetry* 77-85.

Schweiger A. and Jeschke G. (2001): *Principles of Pulse Electron Paramagnetic Resonance* (Oxford Uni. Press, Oxford) pp.578.

Seidel H. (1961): Electron-nuclear-double-resonance spectra of F-centers in alkali halides single crystals. *Z. Physik* **165**, 218-238.

Shimoyama Y. and Rink W. J. (1992): Vector EPR detection of dating quartz. Adv. ESR Appl. 8, 2-8.

Slichter C. P. (1963): Principle of Magnetic Resonance (Harper & Row, New York).

Speath J. M. (1966): ENDOR of U2 centers. *Z. Physik* **192**, 107-111.

Speath J.M., Niklas J.R. and Bartram, R.H. (1992): *Structural Analysis of Point Defects in Solids* (Springer-Verlag, Berlin).

Stucki J. W. and Banwart W. L. (1980): *Advanced Chemical Methods for Soils and Clay Minerals Research* (Reidel Publishing Co., Boston).

Symons M. C. R. (1963): Identification of organic free radicals by electron spin resonance. *Adv. Phys. Org. Chem.* **1**, 283-363.

Watari H., Shimoyama Y. and Matsuki K. (1985): Development of vector saturation transfer ESR. *ESR Dating and Dosimetry* 299-306.

Weil JA, Bolton JR and Wertz JE (1994): Electron Paramagnetic Resonance - Elementary Theory and Practicak Applications (John Wiley & Sons Inc, New York).

Wertheim G.K., Hausmann A. and Sander W. (1971): *The Electronic Structure of Point Defects* (North Holland Co., Amsterdam).

Zavoisky E. (1945): Spin-magnetic resonance in paramagnetics. *J. Phys.* **9**, 245.

Chapter 3

ESR Dating and Dosimetry

– Principles and Procedures –

Principles and procedures of ESR dating and dosimetry are described together with the events which set the clock time of radiation effects to zero in ESR dating. A brief history of ESR dating and dosimetry is also described.

3.1 Principles of ESR Dating and Dosimetry

Natural radiation from radionuclides including uranium (^{238}U and ^{235}U), thorium (^{232}Th) and their daughter radionuclides, and potassium (^{40}K) in the environment or inside an archaeological or geological material produces electronic and atomic defects in the material. The radiation produces free (unpaired) electrons, some of which may be trapped by impurities as well as by inherent defects. They are often quite stable and accumulate with time. These unpaired electrons may be in either the organic or inorganic portion and can be detected by ESR. The ESR signal intensity is proportional to the spin concentration and so to the total dose, i.e., the radiation dose rate and the time elapsed after their formation or an event, such as heating, which affects the spin concentration.

Thermoluminescence (TL) is observed when the trapped electrons (or holes) are released thermally and recombine with holes (or electrons) at an elevated temperature (see Figure 1.9). On the other hand, ESR detects unpaired electrons created by α–, β– and γ–rays, as is again shown in Figure 3.1. The direction of the unpaired electron spin (self–rotation), represented by a dancing girl and boy is changed by microwave absorption under a magnetic field as shown in (c) and (d). The concentration of the unpaired electrons can be determined directly by microwave absorption as the ESR signal intensity. These unpaired electrons trapped by an impurity or left at a hole are simply called "defects" in this book. The difference between TL and ESR dating is only in the detection method of defects from which "*total dose of natural radiation (TD)*" is obtained.

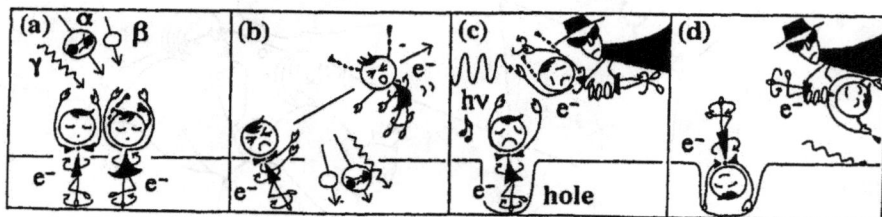

Figure 3.1 The principle of ESR dating. (a) Paired electrons are spinning in opposite directions, (b) ionization by α–, β– or γ–rays knocks off one of the paired electrons (girl) leaving the other (boy) in a hole, (c) the unpaired electron (girl) is trapped by an impurity (gangster), and (d) the spin direction of unpaired electrons is changed by microwave absorption. The concentration of unpaired electrons in proportion to the *TD* or *ED* is determined with ESR.

The *TD* is called the "*archaeological dose (AD)*" in archaeological applications and may be called the "*geological dose (GD)*" in geological dating. "*Accumulated dose of natural radiation (AD)*" and "*equivalent dose (ED)*" are also used. For ESR dating we propose "*TD*" as a general term among these synonyms. We use the term "*ED*" for indicating experimentally obtained equivalent dose in this book. The radiation qualities or the energy spectra of the artificial and natural radiation are not the same and the defect production efficiency depends on the radiation quality. The *ED* is expressed as the dose equivalent to that of $\alpha-$, $\beta-$ or $\gamma-$rays used for the artificial irradiation. Hence, the obtained *ED* by artificial irradiation is not necessarily equal to the *TD* derived from natural radiation.

The *TD* depends on the age and "*annual dose rate of natural radiation (D)*". For the determination of ESR or TL ages from the *TD* or *ED*, separate assessment of the average *D* is necessary. The ESR age, T_{ESR}, can be obtained by dividing the *TD* or *ED* with *D*, i.e.,

$$T_{ESR} = TD/D \quad \text{or} \quad T_{ESR} = ED/D \ . \tag{3.1}$$

Both ESR and TL can be used as methods of radiation dosimetry since their signal intensities are proportional to *TD* or *ED*. TL dosimetry has been widely used in radiation facilities, while ESR dosimetry has not yet because of the absence of an appropriate dosimeter element and a low–cost light–weight reliable spectrometer. Such a spectrometer will increase the usage of ESR dating. The advantage that ESR measurements have is that they can be done without heating the material, which allows us to use tissue equivalent organic materials as a dosimeter element and biological materials to determine accident dose. Radiation dose of irradiated foodstuffs for preservation can be monitored with ESR.

In this chapter, the theory and standard procedures of ESR dating and dosimetry are described together with their limitation. A brief historical development of ESR dating and some technical notes are given at the end of this chapter. It must be noted again that radiation assessment described in Chapter 4 is necessary to obtain the ESR age. Obtaining the *TD* or *ED* is ESR dosimetry and only a part of ESR dating.

3.2 Procedures for ESR Dating

3.2.1. Additive Dose Method

An ESR signal intensity does not indicate the radiation dose or the age unless the intensity (defect concentration) is calibrated to the dose. If one could know the defect concentration at a certain future time, it would be possible to estimate the past passage of time from the present concentration. In conventional TL dating (Aitken 1985) and ordinary ESR dating (Zeller *et al*. 1967), known additive doses of artificial irradiation are used to produce additional defects which can be used to calibrate the concentration within a sample and to determine the production yield of defects. In other words, artificial irradiation usually by γ–rays from a ^{60}Co or a ^{137}Cs source is used as a "*time machine*" to let the defect concentration proceed to a future state. This method is called the "*additive dose method*".

The signal intensity is enhanced as a function of the *absorbed dose of artificial irradiation*, Q, which equals the product of the *additive dose rate*, D' and *irradiation time*, t', in a laboratory, i.e., $Q = D't'$. The *ED* is determined from the enhancement of the signal intensity, assuming a linear relationship between the defect concentration and *ED*. This is written as

$$I(Q)/I_0 = (ED + Q)/ED \ . \tag{3.2}$$

On rearranging, this becomes

$$I(Q) = I_0(1 + Q/ED) \ , \tag{3.3}$$

where I_0 and $I(Q)$ are the observed *signal intensities* before and after irradiation, respectively. Figure 3.2 (a) shows that the *ED* is obtained by linear extrapolation of the data points to the zero ordinate using the least–square fitting method.

The growth curve may be fitted to a simple saturation curve,

$$I(Q) = I_s(1 - e^{-(Q + ED)/SD}) \ , \tag{3.4}$$

where I_s is the *saturation intensity*. *SD* is the *saturation dose* in a laboratory irradiation and equal to the product of the additive dose rate D' and the *saturation lifetime*, τ_s' in a laboratory. The saturation lifetime τ_s in nature is described later in the theoretical analysis of the growth curve (see Section 3.5). Using the relation $Q = D't'$ and $SD = D'\tau_s'$, Eq. (3.4) becomes

$$I(t') = I_s(1 - e^{-(t' + ED/D')/\tau_s'}) \ . \tag{3.5}$$

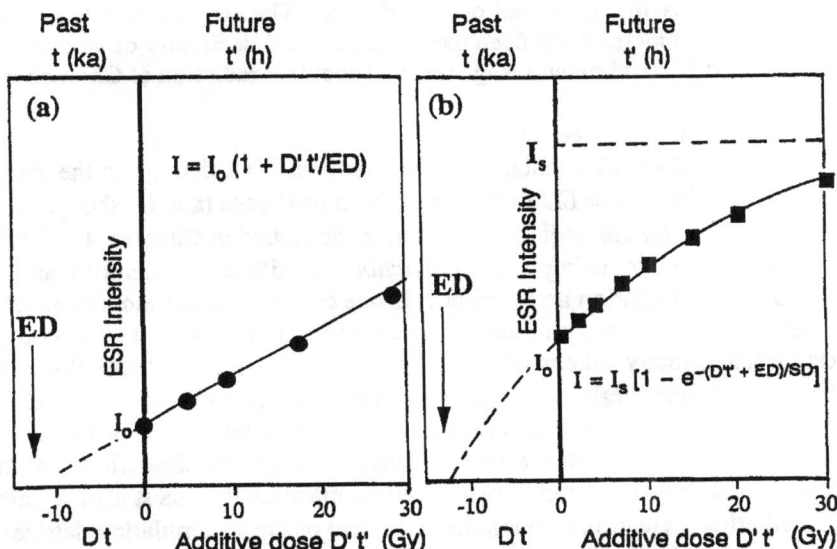

Figure 3.2 An additive dose method. The enhancement of ESR signal intensities by an additive artificial irradiation with the dose Q [Gy] ($Q = D't'$). The ED is obtained by extrapolation back to the zero ordinate. Additive irradiation is a kind of *"time machine"* to proceed an event to the future. (a) A linear growth of the signal intensity and least–square fitting to a line. (b) Saturation growth fitting to a saturation curve of Eq. (3.4).

The saturation intensity I_s is best fitted with a computer by iteration using the initial expected value. This is made using a semi–logarithmic plot,

$$\log_e (1 - I(Q)/I_s) = -(Q + ED)/SD . \qquad (3.6)$$

The radiation intervals should be exponentially increased to fit the data to the saturation curve (Grün and MacDonald 1989). However, the accuracy in the high dose region affects the ED greatly and the statistical error estimation of the ED is difficult in the best fitting (Wu 1986, Berger *et al*. 1987). The ED obtained is generally smaller than that determined by linear least–square fitting. In some cases, fitting two saturation curves is necessary for two different types of defects.

If the dose rate D (usually in mGy/a; see chapter 4) is obtained, the age T_{ESR} may be expressed in [ka] using Eq. (3.1) as T_{ESR} [ka] = ED [Gy]/D [mGy/a]. The ED and D estimation for geological and archaeological als constitutes the main part of ESR dating. The estimation of *ED* by the additive dose method has been used also for ESR dosimetry of atomic bomb radiation at Hiroshima and Nagasaki and accident radiation at Chernobyl.

3.2.2.   *"Wait and See" Method*

The *TD* or *ED* is obtained in conventional ESR dating. In the conversion from the *TD* or *ED* to the age, the annual dose rate *D* should be obtained based on the radiation assessment described in Chapter 4. Is there any alternative to using a *"time machine"* (artificial irradiation) and thus avoiding the radiation assessment ? If one could know the increment of the defect concentration with time and know how it has grown, the age could be estimated directly (Ikeya and Miki 1980a). If we let time flow, the natural passage of time might enable us to determine the past passage of time. We therefore measure the initial defect concentration with ESR, wait patiently and measure again after a certain time. We let the time flow naturally toward the future instead of using a "time machine". This is also a method of radiation detection through measurement of the accumulating lattice defects produced by radiation and is called the "wait and see method".

In the new method of *digital ESR dating*, the computer–averaged signal intensity obtained by repeated measurements is given as a digital number. The ESR measurement is repeated after the passage of time for a few minutes, hours, days or even a year. The signal intensity of the defects produced by natural radiation is enhanced from I_0 to $I(t)$ after a time, t. If the rate of defect production is nearly constant, the defect concentration is proportional to the passage of time. Thus, Eq. (3.3) is rewritten using time t and age T_{ESR} as

$$I(t) = I_0(1 + t/T_{ESR}) \; . \tag{3.7}$$

The "wait and see method" is shown schematically in Figure 3.3. For example, one can determine the age (time after manufacturing) of the potato chips from the accumulating radical concentration by fitting the results to a growth curve. The resulting age of about 50 days determined by linear extrapolation, is shortened to 40 days when a t^2 term is included. Fitting to a saturation curve gives 38 days. These dates are close to the 35 days obtained from the printed date on the package. The slight discrepancy is partly due to the non–linear kinetics of the chemical reaction as well as to changes in the environment (temperature) and the initial oil oxidation at the time of manufacturing (zero age) (see Chapter 12).

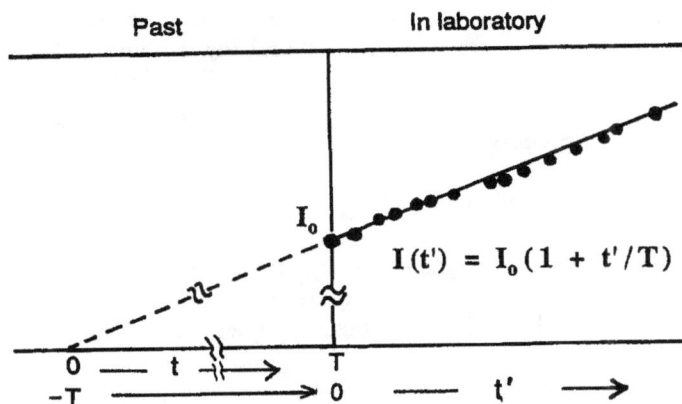

Figure 3.3   *"Wait and see"* method in ESR dating, where the growth of the signal intensity is followed as a function of the waiting time *t*. The age T_{ESR} is directly obtained by extrapolation to the zero intensity. Day by day increase of the signal intensity of lipid peroxide radicals in potato chips is shown. Extrapolation of the increase gives the age (the manufactured time), roughly in agreement with the printed date on the envelope.

The assumptions that the increase in the defect concentration is linear from the beginning and that the natural radiation dose rate is constant must be studied carefully. If the growth behavior is known, the growth can be fitted to a known function or approximated by a simple saturation curve or a polynomial including a t^2 term.   Archaeological and geological samples should be kept at the excavation site between measurements to maintain a constant radiation environment.

The stabilities of the magnetic field and microwave power of an ESR spectrometer are fractions of 10^{-7} and 10^{-6}, respectively.   Fluctuation of the signal intensity due to sensitivity and stability of the signal may mostly be avoided by averaging repeatedly measured intensity ratios of the sample signal to the standard Mn^{2+} signals. Thus, the increase of the signal intensity has been detected with a precision of $\sim 10^{-4}$. Even so, a few days are necessary to detect an increase of the intensity in a sample of a thousand years old.   Progress in digital technology will make this method practical for archaeological or geological materials. This method can be used for rapidly accumulating defects, as is the case of chemically induced lipid peroxide radicals in foods such as potato chips and coffee beans (see Chapter 12).

3.2.3   *Accelerated Passage of Time for Chemical Reactions*

Chemical reactions are accelerated by raising the temperature. In the ESR context, this is equivalent to an irreversible time machine that speeds up the time passage. The rate constant, K, of a chemical reaction for a thermally activated process is written as

$$K = \nu_0 \exp(-E/k_B T) \;, \tag{3.8}$$

where ν_0 is a *pre−exponential factor*, E the *activation energy*, k_B the *Boltzman constant* and T the *absolute temperature*. Chemical ESR dating of organic materials in historical and forensic samples is given in Chapter 12.

3.2.4   *ESR Dendrochronology*

The "wait and see" method requires appreciable waiting time. Natural phenomena recorded in specimens in the form of annual growth band like tree rings can be used to calibrate the ESR signal intensity. If one could accurately determine the defect concentration at several positions within the sample, the age can be determined in principle without radiation assessment. This method is called the "**ESR dendrochronology**". There are plenty of materials that have annual growth lines which may also be used to assess the paleo−climate and acid rainfalls (see Chapter 6).

Figure 3.4 shows an example of dendrochronological ESR dating using a large coral with annual growth bands. The ESR signal intensity at $g = 2.0007$ due to CO_2^- is increased as a function of the annual growth band number from the outside. The age ($T_{ESR} = 2{,}930$ a) is obtained directly by extrapolating the linear growth of the signal intensity to the zero ordinate. Although the distribution of uranium in the sample and ^{238}U−series disequilibrium could modify the age, the calculated age is close to the ^{14}C−age of the coral.

3.2.5   *Plateau Method*

The *ED* is usually obtained by least−square fitting of the enhancement of signal intensity after artificial irradiation, in accordance with Eq. (3.3). In TL dating, the reliability is increased by using TL intensities at several temperatures and obtaining the *ED* as a function of temperature. The con-stant *ED* obtained around the TL peak temperature region is called the "plateau". In ESR measurements, the signal intensity is recorded as a func-tion of the *magnetic field*, H. The *signal intensity*, $I_0(H)$, at different magnetic fields is enhanced to $I(H, Q)$ by an additive irradiation dose of Q

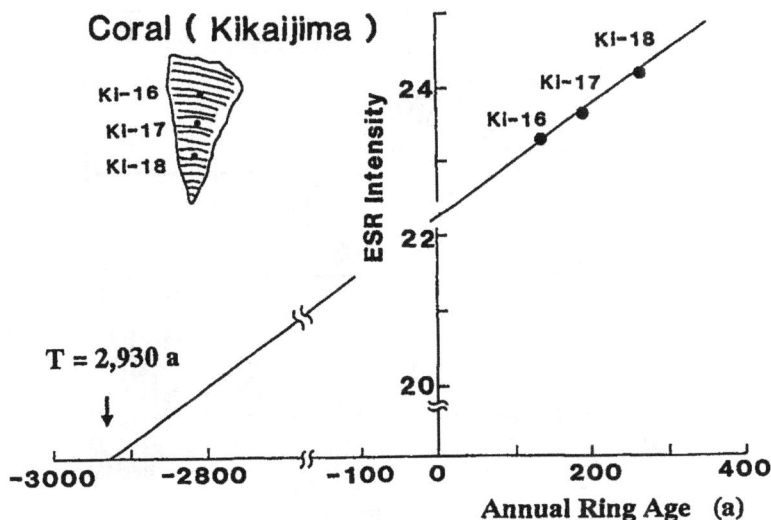

Figure 3.4   Dendrochronological ESR dating of a large coral with annual growth bands. The signal intensity plotted as a function of the year intervals is extrapolated back to zero to obtain the age, T_{ESR} = 2,930 a.

[Gy] according to Eq. (3.3), which is rewritten as

$$I(H, Q) = I_0(H)(1 + Q/ED(H)) .   \qquad (3.9)$$

The *ED* is thus obtained as a function of *H* by the additive dose method, just as the TL plateau is found as a function of temperature. A typical example of the use of the ESR plateau method is shown for hydroxyapatite in Figure 3.5 (Miki and Ikeya 1985). Computer processing gives a reliable *ED* over the wide range of magnetic field strengths. This method can also be applied to the direct age determination in the "wait and see" method. In that case, the plateau of the age $T_{ESR}(H)$ is obtained against the magnetic field.

If several signals overlap in the ESR spectrum and their intensities are enhanced linearly with radiation dose, a meaningful *ED* may be obtained. However, if one or some of the signals are unstable and partially annealed in nature, the *ED* plateau can not be obtained in the region of unstable signals. Caution is needed in interpreting the plateau since the TL plateau involves thermal decay of one type of defects and conversion to others during TL measurement, while the ESR plateau does not involves thermal annealing.

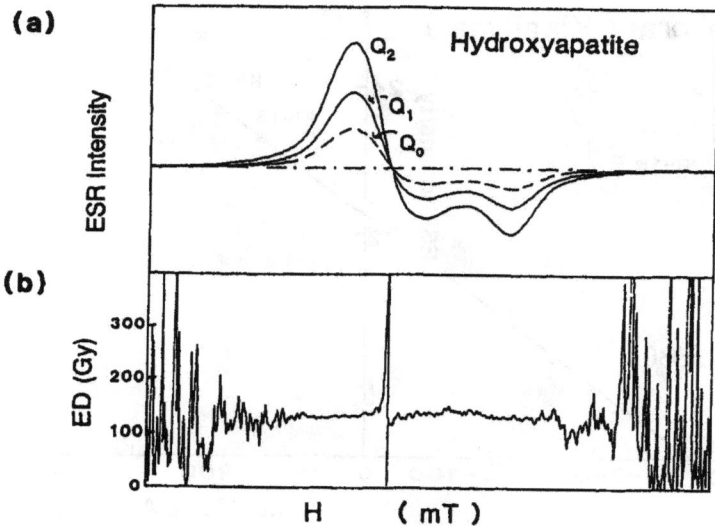

(a)

(b)

Figure 3.5 A plateau method in ESR dating. (a) ESR spectra of hydroxyapatite before (Q_o) and after artificial irradiation (Q_1 and Q_2). (b) A computer best fit of the growth of signal intensity at different magnetic fields H to a linear growth or saturation curve leads to ED's as a function of H.

Application to an integrated spectrum also gives a plateau (Hütt *et al.* 1985), as does each hyperfine line of the Al center in quartz (Yokoyama *et al.* 1985).

3.2.6 *Direct Age Determination from a Theoretical Growth Curve*

A new additive dose method is proposed to obtain the ED or the age directly without extrapolating the growth curve of the signal intensity. The theoretical growth curve at the *artificial irradiation* dose rate gives the defect production efficiency from the initial growth and the interaction volume from the saturation behavior as described later in Section 3.6. A growth curve in nature for a geologically long period at the *natural radiation* dose rate can be calculated theoretically by considering first– and second–order annealing (see Section 3.5). An exact age T_{ESR} or TD which is larger than the ED obtained by extrapolation is determined by solving the general equation of defect formation (see Appendix A1.2).

3.3 Scope of ESR Dating : Events Which Set the Clock to Zero

The signal intensity of radiation–induced defects is related to the ESR age and is considered to be zero for a sample of zero age. The following events destroy defects in materials and set the clock time to zero.

3.3.1. *Crystallization*

ESR dating of crystallization has been done for a stalactite in a lime-stone cave (Ikeya 1975). The signal intensity of the recrystallized carbonate is essentially zero at the surface of the speleothem where the age is very young, while the old interior carbonate shows intense signals of defects. The possibility of ESR dating has thus been proved by the stratigraphic evidence (see Chapter 5). Biological materials such as shells, corals and bones also belong to this category, as crystallization proceeds in the living organisms (see Chapter 6). Partial recrystallization of the materials can give mislead-ingly young ages. Other minerals are also objects of ESR dating of crystalli-zation (see Chapters 7, 8 and 9).

3.3.2. *Action of Heat*

The firing of natural clay minerals used in pottery anneals out defects in quartz and feldspars and provides the basis for dating pottery with TL. Defects produced by natural radiation are annealed by heating and the clock time is set to zero at the time of firing. The radiation absorbed since then has produced the defects. In this way, ESR dating can be applied to rocks and minerals heated by ancient man or by geothermal events such as volcanic eruptions (see Chapter 10).

3.3.3. *Optical Bleaching*

Shining light on materials leads to *optical bleaching* of defects. This phenomenon has been used in TL dating of deep–sea materials and loess which were once suspended in air and exposed to sunlight. Defects close to the surface contribute more to TL signals than those inside due to the self-absorption of the TL light. In ESR dating, however, defects in the entire material are measured. Exposure to an ordinary fluorescent light for a short time does not lead to an appreciable amount of optical bleaching unless the material is transparent or contains some light sensitive defects. The sunlight effect on the decay of the ESR signal intensity at $g = 1.997$ (Ge center) of quartz grains (SiO_2) is observed though not so much for the signal at $g = 2.0009$ (E'_1 center). An attempt to date sediments based on this mechanism has been made using the Al and Ge centers in SiO_2 (see Chapter 9).

3.3.4. *Mechanical Bleaching or Annealing*

Plastic deformation produces a movement of atomic dislocations and/or introduces new dislocations in solids. Trapped unpaired electrons which are released by plastic deformation recombine with holes, producing the luminescence called *"deformation luminescence* (DL)". Defects which have optical absorption, i.e., *"color centers"* in ionic crystals, are bleached by plastic deformation; the phenomenon is called *"mechanical bleaching"*. This is the basis of dating fault and glacier movements, which produce sand formation and sedimentation (see Chapter 9). If the temperature at the local site is actually raised by the frictional heat, the mechanism is similar to the thermal annealing in a heating event as described previously.

It should be noted that new types of defects are also introduced by plastic deformation. These defects give *"exoelectron emission* (EEE)" or thermoluminescence called *"tribothermoluminescence* (TTL)" on heating the mineral. This is the reason why such defects must be removed by etching with acids in TL and ESR dating of the material which has been ground to equal sized grains during sample preparation.

3.3.5 *Surface Etching in Nature*

Defects created by α–rays or fission fragments at the surface of a mineral are easily etched by acid or acidic water. The high concentration of ^{234}U relative to ^{238}U in ground water is due to etching of the region around α–recoils. Such etching is an additional cause of zeroing the age. The etching speed might vary depending on the acidity of ground water.

Biological etching, etching in a worm that swallowed soils, might occur for surface defects. This hypothesis has not yet been confirmed experimentally. Preliminary studies indicate a low content of the E' center in mud and worm soils and easy etching of defects in quartz grains irradiated by He^+ from an accelerator. These observations are the subject of future studies.

3.4   Natural Annealing : Lifetime of Defects

Annealing of defects occurs thermally in a geologically long period through the recombination of *electron centers* and *hole centers*, and hence these defects have a finite lifetime. The lifetime of the defect or radical restricts the upper limit of applicability of ESR dating. It depends on the properties of the defect and temperature. Two types of recombination are considered; one is a correlated recombination leading to *first–order kinetics* and the other is an uncorrelated type leading to *second–order kinetics*.

In theoretical consideration of natural annealing of defects, the follow-ing parameters are used.

n :   Defect concentration in cm^{-3} or signal intensity at time t

n_0 :   Initial defect concentration or signal intensity for a laboratory time $t' = 0$ or for the age, $t = T_{ESR}$

t_a :   Annealing time of isochronal annealing, experimentally $t_a = 900$ s here

T :   Absolute temperature

T_a :   Annealing temperature at which n is reduced to $1/e$ (i.e., 0.36788 times) of the previous value for a certain isochronal annealing time t_a

T_{a10}:   Annealing temperature at which n is reduced by 10 % for a time t_a

τ :   Lifetime of defects at the temperature T when $n = n_0/e = n_0 \times 0.36788$

τ_i :   Lifetime at the i-th step annealing at the temperature T_i

ν_0 :   Frequency (or pre-exponential factor), assumed to be 3×10^{12} s^{-1}

λ :   Rate constant for the second-order decay

k_B :   Boltzman constant ($k_B = 1.380662 \times 10^{-23}$ J/K)

E :   Activation energy in eV.

(a) Correlated recombination

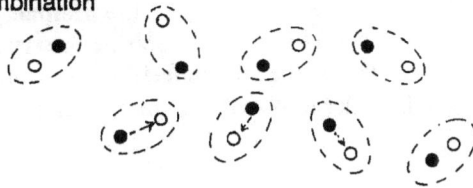

(b) First-order decay   (c)

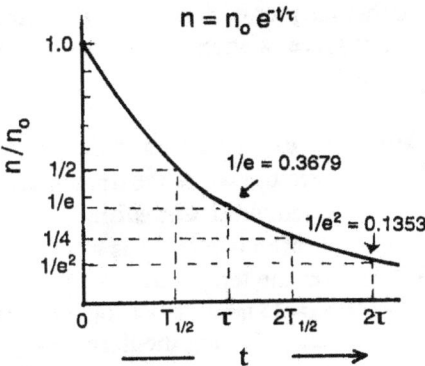

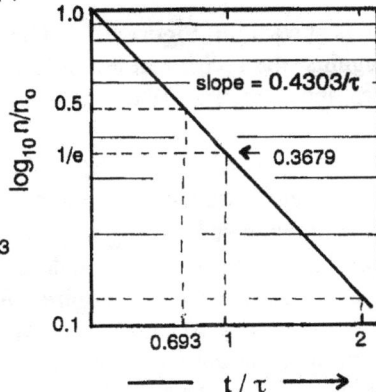

Figure 3.6   (a) Recombination of a correlated pair of an electron and a hole (vacancy and interstitial) leading to first-order annealing.   (b) The exponential decay of the defect concentration n as a function of time τ.   (c) A semilogarithmic plot of the first-order decay gives a straight line with a slope of $0.4303/\tau$

3.4.1. First–Order Annealing

(a) *Isothermal annealing* : An annealing experiment measuring the decay of the defect concentration with time while keeping the specimen at a certain constant annealing temperature is called "**isothermal annealing**". In the first–order kinetics, an electron and a hole which are separated, but which still form a separated pair may recombine with each other as shown in Figure 3.6 (a). Since the concentration of electron centers is equal to that of the correlated pair, it decreases at a rate proportional to the concentration, n. This is written as

$$dn/dt = -n/\tau . \tag{3.10}$$

On integrating, this becomes

$$n = n_0 \exp(-t/\tau) . \tag{3.11}$$

This is shown in Figure 3.6 (b) and (c). A semilogarithmic plot of the ratio n/n_0 against the annealing time t gives the lifetime τ from the slope.

Since the lifetime τ at room temperature is extremely long for defects used in ESR dating, the isothermal annealing experiment is carried out at high temperatures in order to obtain τ in a laboratory. Theoretically τ is written as

$$\tau = (1/\nu_0) \exp(E/k_B T) , \tag{3.12}$$

This is shown in Figure 3.7. The Arrhenius plot of experimentally obtained τ against the reciprocal temperature T^{-1} gives a straight line from which τ at ambient temperature can be deduced.

(b) *Isochronal annealing* : An annealing experiment in which a speci- men is kept for a fixed time at a fixed temperature while the temperature is raised in each succeeding step is called "**isochronal annealing**" or "*pulse annealing*". In this case, the defect concentration is plotted as a function of annealing temperature. Figure 3.8 illustrates the temperature versus time t relation in isochronal annealing for $t_a = 900$ s ($= 15$ min). Isochronal annealing curves for carbonate deposits have been calculated using the lifetimes shown in Figure 3.7. The defect concentration n at the i–th step annealing is reduced to $n_i = n_{i-1} e^{-t_a/\tau_i}$, where τ_i is the lifetime at the i–th temperature T_i. The signal intensity is then calculated as

$$n_i/n_0 = e^{-t_a/\tau_1} \cdot e^{-t_a/\tau_2} \cdots e^{-t_a/\tau_i} = e^{-t_a \Sigma 1/\tau} . \tag{3.13}$$

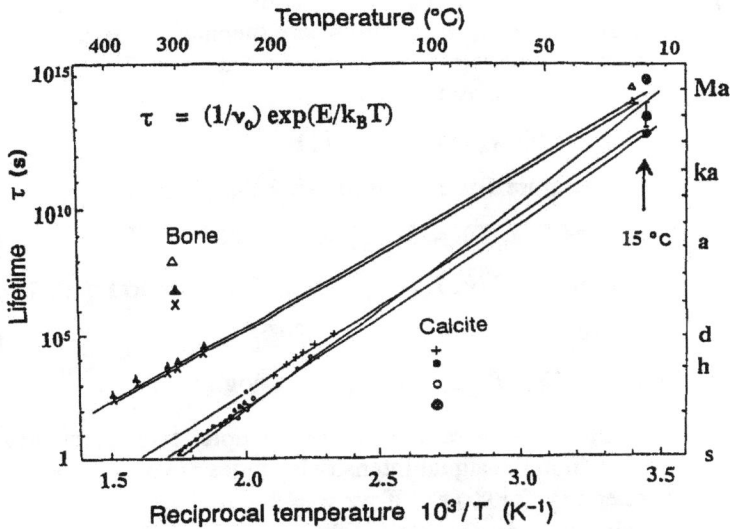

Figure 3.7 The lifetimes obtained from isothermal annealing experiments are plotted in a logarithmic scale versus the reciprocal temperature. Extrapolation of the straight line gives the lifetime at ambient temperature. Data for calcite and bones are shown (Yokoyama *et al.* 1983, Debenham 1983).

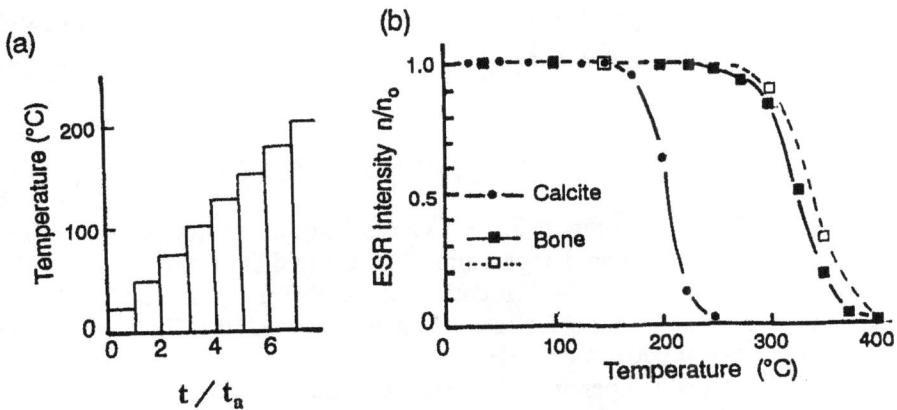

Figure 3.8 Isochronal annealing. (a) Time dependence of the temperature and (b) calculated curve for the pulse annealing.

The number of steps and different t_a give different isochronal annealing curves. If we designate the temperature where the intensity is reduced to $1/e$ of the initial value for a certain fixed annealing time t_a as T_a, then the lifetime τ is equal to t_a and is expressed as

$$t_a = (1/\nu_0) \exp(E/k_B T_a) \ . \tag{3.14}$$

Therefore, τ at temperature T is expressed using Eq. (3.12) and (3.14) as

$$\tau = (1/\nu_0) \exp(E/k_B T) = (1/\nu_0) [\exp(E/k_B T_a)]^{T_a/T} \ . \tag{3.15}$$

Applying the relation $\exp(E/k_B T_a) = t_a \nu_0$ from Eq. (3.14) to Eq. (3.15),

or

$$\tau = (1/\nu_0)(t_a \nu_0)^{T_a/T} = t_a (t_a \nu_0)^{T_a/T - 1} \ , \tag{3.16}$$

$$\log_{10}\tau = (T_a/T - 1) \log_{10}(t_a \nu_0) + \log_{10} t_a \ . \tag{3.17}$$

When the annealing time t_a is set to 900 s for example, this means that T_a is the temperature at which the signal intensity becomes $1/e$ of the previous one by isochronal annealing for 900 s. If we $\nu_0 = 3 \times 10^{12}$ s^{-1}, then, Eq. (3.17) is rewritten as a simple linear function of $\log_{10}\tau$ versus T_a, i.e.,

$$\log_{10}\tau = (15.431/T)T_a - 12.477 \qquad (t_a = 900 \text{ s}) \ . \tag{3.18}$$

Practically, one can use the temperature T_{a10} where the intensity I_i is reduced by 10% from the previous one I_{i-1} (i.e., $I_i/I_{i-1} = 0.9$) rather than to $1/e$. Then, using $e^{-t_a/\tau_{a10}} = 0.9$, i.e., $\tau_{a10} = 9.5 t_a$, Eq. (3.14) and (3.16) become

and

$$9.5 t_a = (1/\nu_0) \exp(E/k_B T_{a10}) \ , \tag{3.14'}$$

$$\tau = (1/\nu_0)(9.5 t_a \nu_0)^{T_{a10}/T} \ , \tag{3.16'}$$

respectively, and finally, Eq. (3.17) becomes

$$\log_{10}\tau = (16.409/T)T_{a10} - 12.477 \qquad (t_a = 900 \text{ s}) \ . \tag{3.19}$$

This relation is shown in Figure 3.9 which can be used to estimate the stability of defects at an ambient temperature of 15°C (bold line) and geothermal temperatures (solid lines) in ESR dating of geothermally altered minerals (see Chapter 10). The defects stable up to $T_{a10} = 250$°C in laboratory annealing for $t_a = 900$ s are stable up to 40 Ma at 15°C but stable only for 50 ka at geothermal temperature $T = 50$°C as indicated by dashed arrows in the figure.

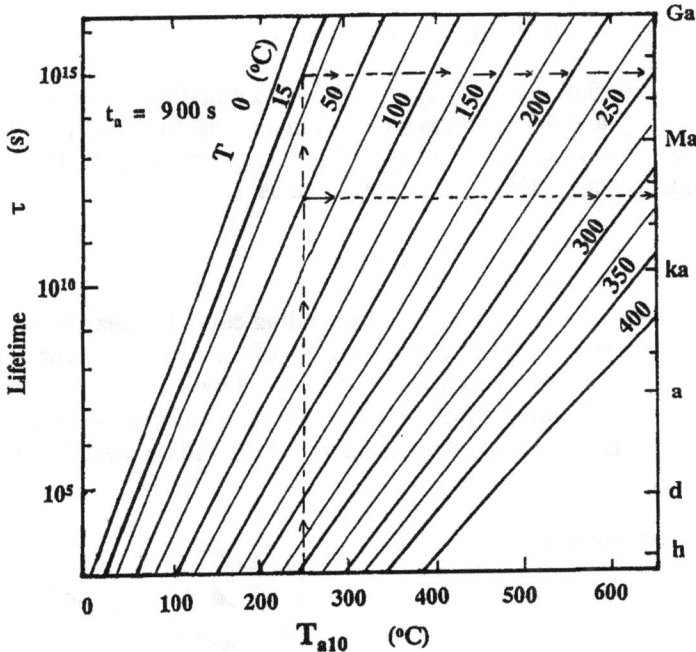

Figure 3.9 The lifetime of a defect at several temperatures (T) as a function of T_{a10} for $t_a = 900$ s. The stability of defects that anneal at $T_{a10} = 250°C$ is indicated by dashed arrows at $T = 15°C$ (40 Ma) and 50°C (50 ka).

3.4.2   Second-Order Annealing

Sometimes, an isothermal annealing curve does not fit a simple exponential decay. If the recombination of electron and hole type centers is uncorrelated, i.e., both centers move freely to recombine mostly with other partners as shown in Figure 3.10 (a), the decay rate is proportional to the product of the concentration of both electron and hole centers, i.e., n^2. This type of process obeys second-order decay kinetics and may be written as

$$dn/dt = -\lambda n^2 = -n/(1/\lambda n) , \qquad (3.20)$$

which indicates that the apparent lifetime τ in the sense of the first-order decay ($dn/dt = -n/\tau$) is expressed as

$$\tau = 1/\lambda n . \qquad (3.21)$$

The τ increases as the concentration decreases. On integrating Eq. (3.18),

$$n = n_0/(1 + \lambda n_0 t) \; . \tag{3.22}$$

The concentration n is 1/2, 1/3, 1/4 and 1/5 with increasing time of $1/\lambda n_0$, $2/\lambda n_0$, $3/\lambda n_0$ and $4/\lambda n_0$ as shown in (b) in contrast to the 1/2, 1/4, 1/8 and 1/16 with increasing time of $t_{1/2}$, $2t_{1/2}$, $3t_{1/2}$ and $4t_{1/2}$ in the first–order exponential decay. Eq. (3.19) is rewritten as

$$1/n = 1/n_0 + \lambda t \; . \tag{3.22'}$$

Thus if these conditions apply, the annealing curve is linear when the reciprocal concentration is plotted as a function of the annealing time and the rate constant λ may be obtained from the slope as shown in (c).

These first– and second–order annealing processes may coexist in nature and form a mixed–type annealing which will be described in Appendix A1.1.

(a) Uncorrelated recombination

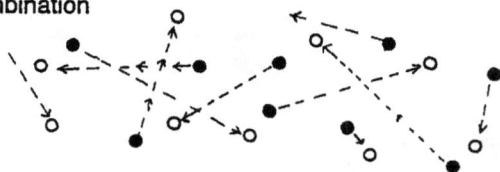

(b) Second–order decay   (c)

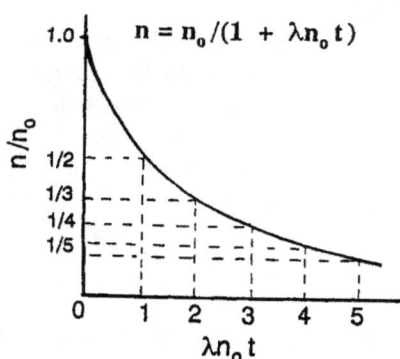

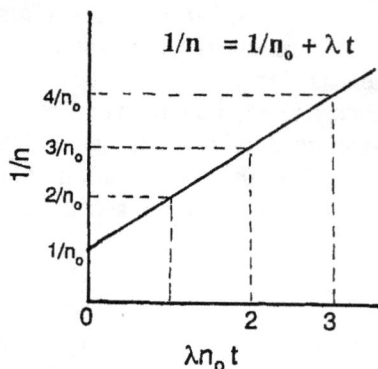

Figure 3.10 (a) Recombination of uncorrelated pairs of defects (electron and hole or vacancy and interstitial). (b) Second–order decay as a function of time normalized by $1/\lambda n_0$. (c) Plot of $1/n$ versus t gives λ from the slope and $1/n_0$ at the abscissa.

3.5   Growth Curve : Formation and Decay of Defects

3.5.1   *Insensitive Volume around a Defect*

The change in the formation yield is often neglected in the analysis of defect formation. Interaction of defects with impurities or other defects occurs in a high concentration region. The defect production yield is reduced by an enhanced rate of recombination if the material is exposed to radiation of the order of $1 \sim 10$ kGy. The signal growth curve following artificial irradiation thus shows a tendency to saturate at a high dose region. The reason why α–rays produce defects with less efficiency at the same radiation dose (energy) as β–rays is also due to the interaction among defects in the α–ray track and the track of an α–recoil atom, which will be described in Appendix A1.4. The defect formation is described with the concept of "*interaction volume*" or "*inactive volume around a defect*" (Itoh 1972, Ikeya 1986, Barabas *et al.* 1988). The saturation of signal growth due to the interaction among defects in an additive irradiation experiment should be distinguished from that due to the lifetime of defects in nature. The following parameters are used in addition to those described in the previous section:

D, D' : Dose rate of natural radiation and artificial irradiation, respectively.

t, t' : Time in nature and in additive artificial irradiation, respectively.

I, I_0, I_s : Signal intensities at time t and time $t' = 0$ ($t = T_{ESR}$) and saturation intensity, respectively.

a : Defect creation efficiency [$cm^{-3}\,Gy^{-1}$] expressed in terms of G (the number of radicals formed per 100 eV) and the density of the material, ρ, as

$$a = 6.25 \times 10^{13}\,G\rho \qquad (3.23)$$

$a = 1.7 \times 10^{13}\,cm^{-3}\,Gy^{-1}$ when $G = 0.1$ and $\rho = 2.6$.

b : Interaction volume defined as the number of lattice sites insensitive to radiation as schematically shown in Figure 3.11 (usually $b = 10^3 \sim 10^5$). Neither interstitials nor vacancies are stabilized since the lattice is distorted in this volume.

N_0 : The number of the available lattice site expressed as

$$N_0 = 6 \times 10^{23} \times \rho / MW \qquad (3.24)$$

$N_0 = 2.6 \times 10^{22}\,/cm^3$ for the *molecular weight*, $MW = 60$ and the density, $\rho = 2.6$.

τ_s, τ_s' : Saturation lifetime defined as the time required to fill all sites with the volume b at the natural dose rate D and the artificial one D', respectively.

τ_e : **Effective lifetime,** $1/\tau_e = 1/\tau + 1/\tau_s$.

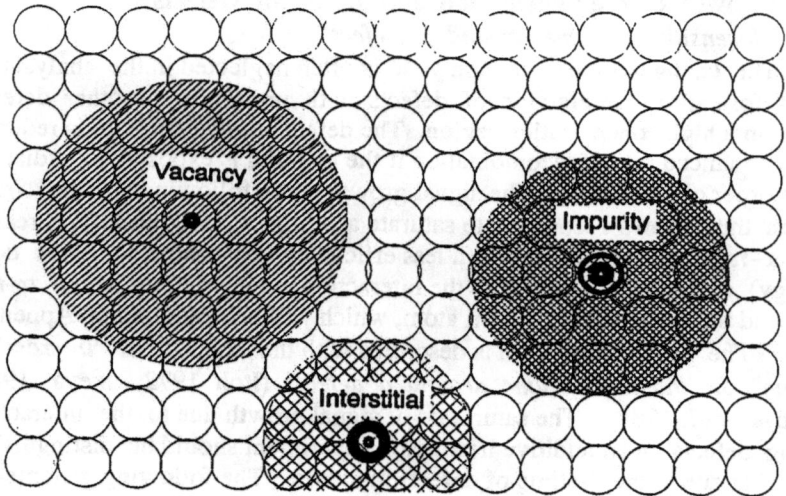

Figure 3.11 Saturation of the ESR signal intensity at a high dose region due to interaction among defects. The insensitive volumes around defects such as a vacancy, an interstitial and a substitutional impurity are indicated by shaded circles. Small open circles are atoms at the lattice site.

The defect concentration n is described by a rate equation of the defect formation during additive irradiation at the dose rate D' as

$$dn/dt = aD'(N_o - bn)/N_o$$
$$= aD' - n/[N_o/(abD')] = aD' - n/\tau'_s . \qquad (3.25)$$

This equation is somewhat analogous to the first–order annealing case with a *lifetime* τ_s' in artificial irradiation at D'. The *saturation lifetime* τ_s in nature at the dose rate D is a time required to fill all of the lattice sites with volumes b and it is obtained from the saturation curve by the additive dose method. Both lifetimes are written as

$$\tau'_s = N_o/abD' , \quad \tau_s = N_o/abD . \qquad (3.26)$$

The solution of Eq. (3.24) is

$$n = (N_o/b)[1 - e^{-(ab/N_o)D't'}] . \qquad (3.27)$$

This is rewritten using the relation $D't' = Dt$ and Eq. (3.26) as

$$n = aD'\tau_s'(1 - e^{-t/\tau_s'})$$

$$= aD\tau_s(1 - e^{-D't/D\tau_s}) \ . \tag{3.28}$$

The *saturation of growth*, $n = n_s(1 - e^{-1}) = 0.63212n_s$ is observed at the saturation dose where $abD't'/N_o = 1$ or $D't' = N_o/ab$ which is set to a saturation lifetime τ_s in nature using $D\tau_s = D'\tau_s'$.

If typical values of N_o (= 2.6×10^{22} cm^{-3}), a (= 1.7×10^{13} cm^{-3} Gy) and D (= 1 mGy/a) are assumed and the experimental saturation dose $D't' = 5 \times 10^5$ Gy is obtained, the following values are calculated:

The saturation concentration : $N_o/b = aD't' = 8.5 \times 10^{18}$ cm^{-3}
The inactive volume : $b = 3 \times 10^3$
The saturation lifetime : $\tau_s = (N_o/b)/aD = D't'/D = 5 \times 10^8$ a

The saturation of growth restricts the upper age limit of ESR dating to about several times of τ_s.

3.5.2   Formation and First–Order Decay in Nature

(a) *Analysis of growth curve* : Formation and decay of defects in nature are described using the lifetime τ as

$$dn/dt = aD(1 - bn/N_o) - n/\tau \ . \tag{3.29}$$

This indicates a saturation curve due to equilibrium between formation and decay if b is zero. If we use the relation $N_o = abD\tau_s$, then

$$dn/dt = aD - n(1/\tau_s + 1/\tau)$$

$$= aD - n/\tau_e \ , \tag{3.30}$$

where the *effective lifetime* τ_e is expressed as

$$1/\tau_e = 1/\tau + 1/\tau_s \ . \tag{3.31}$$

Then the solution of Eq. (3.29) is

$$n = aD\tau_e(1 - e^{-t/\tau_e}) \ . \tag{3.32}$$

Hence the growth shows a tendency to saturate for ages older than the life-time, reaching the steady state at ages more than a few times the lifetime. The saturation value equals $aD\tau_e$. If artificial additive irradiation is done at a dose rate D', higher than the natural dose rate D, the enhancement of the signal intensity is observed since the new saturation value becomes $aD'\tau_e'$.

If we substitute Eq. (3.31) into Eq. (3.32), then

$$n = [aD\tau\tau_s/(\tau + \tau_s)][1 - e^{-t/\tau} e^{-t/\tau_s}] \quad . \tag{3.33}$$

If we normalize the defect concentration with the saturation concentration obtained by artificial irradiation and express it as the signal intensity ratio, then

$$I/I_s = [(\tau/\tau_s)/(1 + \tau/\tau_s)][1 - e^{-t/\tau} e^{-t/\tau_s}] \quad . \tag{3.34}$$

This is a saturation curve including both τ and τ_s. Figure 3.12 (a) shows the calculated growth as a function of time t/τ_s using τ/τ_s as a parameter. The signal intensity I normalized by the saturation intensity $I_s = aD\tau_s$ is shown for convenience. The additive dose method gives the growth curve for $\tau/\tau_s = \infty$ from which the ED is obtained. This curve is drawn from the point of I_0 at $t' = 0$. Then the exact TD is obtained. Figure 3.12 (b) shows such curves of additive dose in a laboratory, all of which are shifted ones of the

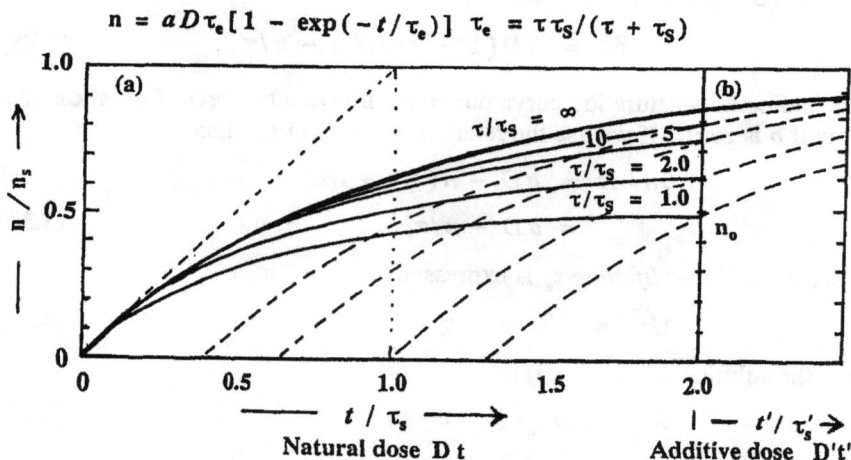

Figure 3.12 (a) Growth curves of the signal intensity I/I_s for defects with the first–order lifetime τ and the saturation lifetime τ_s due to the interaction volume b. I_s is the saturation intensity ($I_s = N_s/b = aD\tau_s$). (b) Growth by further irradiation in a laboratory (dashed curves).

saturation curve drawn by the thick solid line. Thus, one can get the saturation lifetime τ_s of the defects at the dose rate D' and the defect production
efficiency a from the growth curve by the additive dose in a laboratory.

The solution of Eq. (3.29) with a new boundary condition of $n'(t') = n_0 = n(T_{ESR})$ at $t' = 0$ is

$$n'(t') = n_0 e^{-t'/\tau_e} + aD'\tau_e(1 - e^{-t'/\tau_e}) \ . \qquad (3.35)$$

It must be noted that the first term is usually constant $(= n_0)$ for the short
time irradiation of $t'/\tau_e \ll 1$. This is shown in Figure 3.13 for $D = 1$ mGy/a
and different D'. Additive irradiation at much higher dose rate theoretically
enhances n_0 unless $\tau/\tau_s \gg 1$. The decrease of n is observed for $D' < D$ due
to the decay.

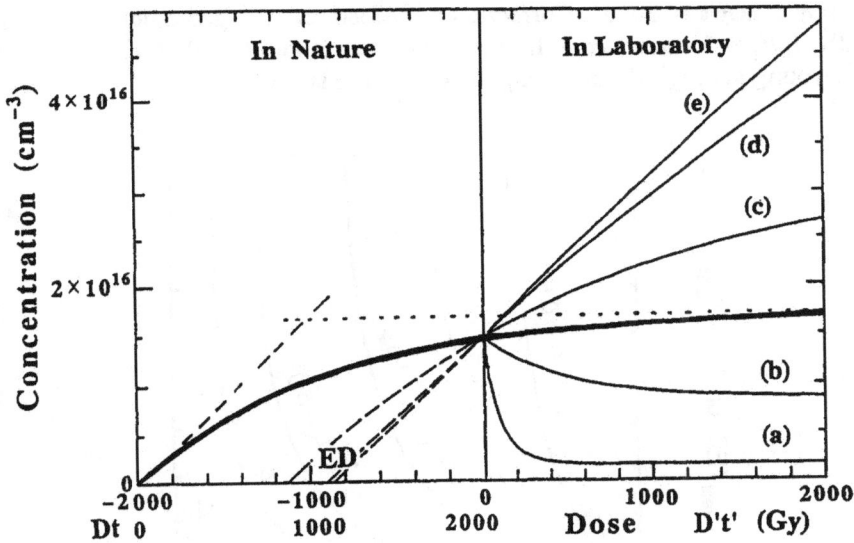

Figure 3.13   Formation of defects by natural radiation and the first-order
decay over the lifetime lead to the saturation level $aD\tau$ at $D = 1$ mGy/a (bold
curve). The growth is enhanced or reduced depending on the artificial dose
rate $D' = 1.0 \times 10^{-4}$ (a), 5.0×10^{-4} (b), 2.0×10^{-3} (c), 1.0×10^{-2} (d) and
1.0 Gy/a (solid curves). The dashed lines show extrapolation. $\tau_e = 10^6$ a and
$a = 1.7 \times 10^{13}$ cm^{-3} Gy^{-1} were used for calculation.

(b) *Correction factor* (TD_{real} / ED) : One can determine the real age, T_{real} and the age, T_{add} at an additive dose rate D' from the saturation curve of Eq. (3.32) and Eq. (3.28), respectively.

$$T_{\text{real}} = \tau_e \log(1 - n_0/aD\tau_e) , \qquad (3.36)$$

$$T_{\text{add}} = \tau_s \log(1 - n_0/aD\tau_s) . \qquad (3.37)$$

The ratio $n_0/aD\tau_s = n_0/aD'\tau_s'$ is equal to the signal intensity ratio I_0/I_s. Hence, the ratio TD_{real}/ED or $T_{\text{real}}/T_{\text{add}}$ may be written as

$$TD_{\text{real}}/ED = \frac{1}{1 + \tau_s/\tau} \cdot \frac{\log[1 - (I_0/I_s)(1 + \tau_s/\tau)]}{\log(1 - I_0/I_s)} . \qquad (3.38)$$

This is shown in Figure 3.14. If τ is comparable to τ_s, $TD \approx ED$ is obtained only for the weak initial intensity $I_0 \ll I_s$. In most materials, $\tau_s = TD_s/D \approx$ 10 kGy/1 mGy/a $\approx 10^7$ a. Correction is needed for old geological materials with $I_0/I_s \approx 1$ and $\tau \leq \tau_s$. In the case of $\tau \gg \tau_s$ and $I_0 \ll I_s$ (archaeological and young geological materials), no correction is necessary.

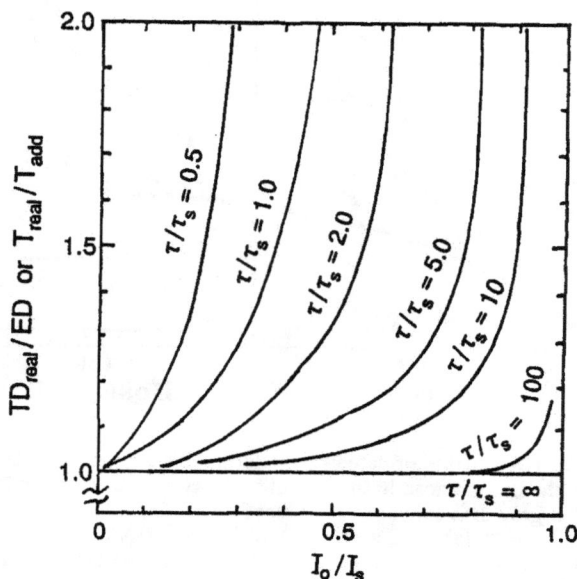

Figure 3.14 A calculated additive dose correction factor of TD_{real}/ED as a function of the degree of saturation (the ratio I_0/I_s) using a saturation curve for the first-order natural annealing.

3.5.3   *Formation and Second–Order Decay in Nature*

Uncorrelated pairs of electrons and holes recombine according to the second–order kinetics. The rate equation for defect formation and second–order natural decay is written as

$$dn/dt = aD - \lambda n^2 \qquad . \qquad (3.39)$$

The saturation curve is obtained in the form of

$$n = (aD/\lambda)^{1/2} \tanh[(aD\lambda)^{1/2}t] \; . \qquad (3.40)$$

A linear increase of the defect concentration with the dose is clear as $n = aDt$ for a small dose region, $(aD\lambda)^{1/2} t \ll 1$. The fitting of a growth curve to Eq. (3.40) must be made for the second–order decay which is common in nature. When Eq. (3.39) is solved with the boundary condition, $t = T_{ESR}$ and $n = n_0$ at the high additive dose rate of D', the solution is

$$n = (aD'/\lambda)^{1/2} \; \frac{n_0 + (aD'/\lambda)^{1/2} \tanh[(aD'\lambda)^{1/2}(t-T)]}{n_0 \tanh[(aD'\lambda)^{1/2}(t-T)] + (aD'/\lambda)^{1/2}} \; . \qquad (3.41)$$

The fitting of the growth curve to this equation should be made. In practice, however, the defects used in ESR dating and dosimetry are stable enough for a time t' in a laboratory. It is noted that we have neither included the interaction volume b nor the effective lifetime τ_e in the second–order kinetics. A general equation including first– and second–order mixed annealing as well as the analytical age equation is described in Appendix A1.2.

3.6   Additive Dose Method without Extrapolation : Exact Solution

Although we extrapolate the growth curve back to the zero ordinate in the conventional additive dose method in both ESR and TL dating, this is actually not correct. What one should do is to draw a theoretical growth curve in nature from that obtained in a laboratory. This has neither been done in dating with ESR nor with TL in spite of a long history of the latter method in dating and dosimetry. A simple hypothesis was that we could simulate the natural dose rate D with 10^8 times higher dose rate D'. The followings are the procedures to construct a natural growth curve and to obtain the age (Ikeya 1992).

(1)   The fitting of the experimental growth curve obtained by the additive dose $D't'$ to theoretical curves of Eq. (3.27) leads to the ED but not to

the TD_{real}. Note $t' \ll \tau$ and $t' \ll 1/\lambda n$.

(2) The parameters a and b are obtained from a saturation curve of Eq. (3.27) by the additive dose method ($I_s \propto N_0/b = aD'\tau_s' = aD\tau_s$ and the initial growth of $aD't'$), while τ and λ are obtained from annealing experiments and extrapolation in the Arrhenius plot according to Eq. (3.11) or Eq. (3.12) and Eq. (3.23), respectively.

(3) The growth curve is calculated using Eq. (A1.4) and parameters a, b, τ, λ and D.

(4) The age T_{ESR} is obtained by equating $n(T_{ESR}) = n_0$ using Eq. (A1.4) with a computer or by using Eq. (A1.6).

These procedures are carried out for OH· radicals in solid H_2O in Chapter 11. Unless second–order annealing is dominant, the relative concentration of defects or the signal intensity relative to its saturation intensity I/I_s is sufficient to determine the age for defects with a known τ. Figure 3.15 shows theoretical curves for samples of different ages or growth stages in the first– and second–order mixed annealing. Solid curves are obtained for $0 < t < T$ and $t' > 0$ and dashed curves are extrapolations of natural and laboratory growth curves according to the theoretical formula. The use of the conventional additive dose method to obtain ED by extrapolating the growth curve is allowed so long as the growth is still linear ($I/I_s \ll 1$) and the obtained age is much smaller than the effective lifetime τ_e or the second–order annealing lifetime ($1/\lambda n$). The errors in annealing parameters do not affect ED much so long as the curve does not reach a saturation level.

3.7 Time Range and Samples

3.7.1 *Time Range : Upper and Lower Limits*

An *upper limit* of billions of years for quartz grains has been proposed based on the correlation between geological ages and the signal intensity for the E' center (an electron at an oxygen vacancy) created by α–recoils in quartz grains (Odom and Rink 1988, Grün 1989a). Similar trials have been made by correlating TL intensity to geologic age of minerals. If the defects have been formed by α–rays or α–recoil atoms, the growth should be logarithmic and not saturated as described in Appendix A1.4. Quartz grains contain only a few ppb of uranium in the SiO_2 lattice and the γ–rays from the external environment have been simply assumed not to create oxygen vacancies. This will be discussed in Chapter 9, dealing with defects in SiO_2.

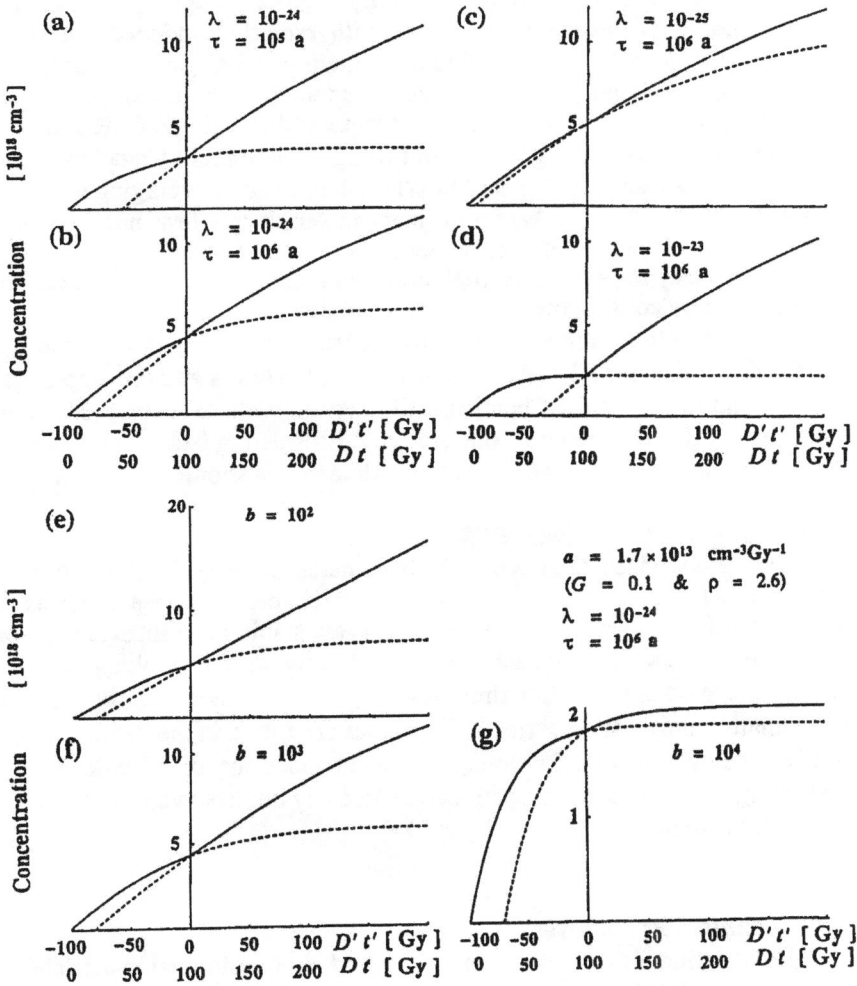

Figure 3.15 A new additive dose method of fitting to theoretical curves of exact solution without extrapolating the growth curve in a laboratory. Defect formation by natural radiation in the past at the dose rate D and by artificial irradiation at the dose rate D' are shown by solid curves. Dashed curves are theoretically extrapolated ones. Typical values for the defect production efficiency, $a = 1.7 \times 10^{13}$ cm^{-3} Gy^{-1} and the interaction volume, $b = 10^3$ atoms are used in (a) – (d), while other values including annealing parameters τ and λ are indicated in the figures.

The *lower limit* of the age range in ESR dating depends on the signal to noise (S/N) ratio and interfering impurity signals. Impurity signals may overlap and mask the signal associated with radiation–induced defects. If one improves the S/N ratio by a signal–averaging technique and subtraction of an impurity signal, the lower limit is reduced to about 10 years (TD = 10 mGy) for carbonate fossils. Thus, it is possible to extend ESR dating to quite recent events. The general limit using commercially available spectrometers (sensitivity : 10^{10} spins/0.1 mT) without signal averaging is about a thousand years (TD = 1 Gy) for ordinary materials or a few hundred years for corals. Cooling the detector or electronic and microwave circuits will be an orthodox way to get a better S/N ratio. In fact, the diode characteristics are improved by cooling the detector diode (Ikeya 1993).

The case for dating young materials based on chemical reactions is described in Chapter 12, where the lower limit is hours and the upper limit is a few hundred years. Chemical ESR dating, such as dating of potato-chips, is a different method from ordinary ESR dating based on dosimetry of natural radiation. The age limits for both methods should not be confused.

3.7.2 Samples Dated with ESR

Samples whose TD's have so far been determined by ESR are listed in Table 3.1. Details are given in the referred chapters. These samples are suitable for ESR dating because the defects are stable for a sufficiently long time or the clock has been set to zero according to the geological events described previously. If the time resetting is not made clearly, the age determination may not be correct. Examples are found in the dating of fossil woods or clay minerals involving chemical reactions of silicification or weathering which did not happen instantaneously but has even continued to the present time.

3.8 Technical Procedures

ESR dating of specific geological and archaeological materials are detailed in the following chapters. General technical information is briefly outlined here. Since ESR dating is still under development and experimental conditions depend greatly on materials and involved impurities, standard procedures have been not established although some recommendations, suggestions and minimum requirements, unified notations, *etc.*, are discussed by those involved in ESR dating. The methods described here are just routine time saving procedures practiced in our laboratory.

Table 3.1 Materials used for ESR dating and dosimetry.

Materials	Chemical Form	Applications	Chapters
Carbonates	$CaCO_3$	stalactite, travertine spring deposit, etc.	5
Biocarbonates	$CaCO_3$	coral, shell, foraminifera egg shell, calcrete	6
Sulfates	$CaSO_4$ $CaSO_4 \cdot 2H_2O$	anhydrite gypsum (desert deposit, cave deposit)	7
Bicarbonates	$NaHCO_3$	saline lake deposit	7
Phosphates	$Ca_5(PO_4)_3X$ $CaHPO_4$	hydroxyapatite, tooth, bone phosphate nodule	8, 13 8
Silica	SiO_2	geological fault volcanic rock, altered rock	9 10
Silicates	$ZrSiO_4$	zircon feldspars, clay minerals	10
Ice & Dry Ice	H_2O, CO_2	comet, solid H_2O and CO_2	11
Organics	radicals * radicals metal ions *	food, crop, leather, paper alanine, sugar mummy, blood	12 13 12

* ESR signals produced by chemical reactions rather than natural radiation.

3.8.1 Sampling and Preparation

(a) *Sampling* : Samples which have been _recrystallized_ should be avoided unless you are interested in the age of recrystallization. Recrystallization always produces a younger age than expected. Note that brown iron minerals included in crystalline materials give signals which interfere with ESR signals of radiation–induced defects.

If possible, avoid samples whose _radiation environment_ might have changed during the burial time, for example, by groundwater radioactivity. Samples close to wind–blown or water–deposited soils make the dose rate assessment difficult due to non–uniform radiation both in space and time.

Avoid samples which have been exposed to _light_ and wrap them with aluminum foil to avoid excessive sunlight. Avoid samples previously examined by _X–rays_ or _electron microscopy_ since these processes produce defects. Even the light intensity of a surface microscope is high enough to destroy some defects. Avoid samples that might have been _heated_ unless you are interested in the time after the heating event.

(b)   *Sample preparation* :   First, _wash off_ any attached dirt with water in an ultrasonic bath.   A slight acid treatment may be necessary to remove all the dirt and surface carbonate deposits.   A magnetic separator should be used to remove ferromagnetic minerals. Etch the sample with weak acid to remove the surface material of grains or small particles attached to large grains.   Do not heat the sample or grind it into a fine powder.   Use water cooled saws for cutting to avoid heating.   Note that grinding reduces the content of unpaired electron spins and creates defects which trap electrons produced by irradiation.   An early interlaboratory comparison project for carbonate speleothems milled into fine powder failed because of the selection of inappropriate samples and the grinding effect (Hennig *et al.* 1983).

A sample _grain size_ of $100 \sim 250 \, \mu m$ is preferable for an ordinary experiment. Grains of $40 \sim 80 \, \mu m$ are used for fault dating.   Small grains of less than $100 \, \mu m$ have a tendency to stick to the quartz sample holder. Samples of large size crystalline materials show angular–dependent signals and may vary in density due to the degree of packing.

Use a _fixed amount_ of the sample, for example, 200 or 300 mg for carbonate if available.

3.8.2 *ESR Measurements*

(a) *Sample setting* :   Wait half an hour until the spectrometer is well stabilized.   A constant temperature reduces the drift of the sensitivity.   To obtain high sensitivity, use a cylindrical cavity with the TE_{011} mode rather than a rectangular one with the TE_{102} mode.

Note that all sample holders are not exactly the same in size or shape. Use the same holder in the same orientation relative to the magnetic field by marking the holder in order to obtain an accurate intensity measurement.   A standard quartz sample holder with an inner diameter of 5 mm is convenient for ordinary use.   Repeat the measurement while reinserting the holder into the microwave cavity to check the reproducibility of the signal intensity.   A broken piece of solid sample may be measured but the same orientation and position within the cavity must be used each time.

(b)   *Conditions for ESR measurements* :   Optimum conditions of ESR measurement of dating signal are described in each chapter.   Conditions for ESR measurements are listed in Table 3.2. Detailed conditions used should

Table 3.2 Conditions for ESR measurements

Microwave		
Frequency	υ	X–band (9.5 GHz)
Power	P	Appropriate power from the power dependence. (Simply 1 mW if the signal is not saturated)
Modulation field		
Frequency	f_m	100 kHz (commercial apparatus), not 80 Hz.
Width	ΔH_m	0.05 ~ 0.5 mT (or close to the peak to peak if no signal overlaps)
Amplifier		
Gain	G	$1 \sim 10,000 \times 10$, Confirm the intensity at two gain setting.
Time constant	τ	0.1 or 0.3 s (If noisy, 1 or 3 s with slow field sweep)
Magnetic field H		
Center field	H_o	340 or 345 (335) mT at $g = 2.0$. Calibrate with DPPH ($g = 2.0036$) or Mn^{2+}.
Sweep width	ΔH	± 5 mT or appropriate
Recorder & Computer		
Time constant	τ	Shorter than τ of the amplifier.
Data acquisition		If the signal intensity is accurately determined, any software is all right.
Temperature		Room temperature for ordinary signals. Low temperature (77 K) for particular signals.

The appropriate condition depends on materials as will be mentioned in each chapter. Conditions used for measurement should be described in ESR papers for reproduction by other researchers.

be recorded.

Microwave frequency : The X–band frequency (9.5 GHz) is commonly used. Higher microwave frequencies of K–band (25 GHz) and Q–band (35 GHz) can be used to obtain better resolution of signals with different g factors but the reproducibility of the signal intensity is poor due to difficulty in precise sample setting. Overall preference for ESR dating will be at an X–band frequency. Since hyperfine splitting is not affected greatly by microwave frequency, one can distinguish signals due to different g factors from hf signals.

Microwave power : The signal intensity depends on the microwave power. For new materials, microwave power dependence should be meas–

ured. Use the microwave power where the signal intensity does not reach to saturation. The saturation behavior is different for different signals. If several signals are overlapping, separation of the dating signal is necessary. This is done by saturating other signals at an appropriate microwave power.

Modulation field : Commercial spectrometers have 100 kHz field modulation. Select the modulation width less than the linewidth of the signal of interest. If it is too small, the signal intensity is small. If it is larger than the linewidth, the signal becomes broad. Overlapping signals are well resolved with a small modulation width.

Amplifier gain : Select the *"time constant"* of the amplifier considering the S/N ratio. Check the linearity of the amplifier and recorder for two *"amplifier gain"* settings. Measure the same standard specimen of MgO (Mn) to check the sensitivity and reproducibility of the intensity. Note that the intensity of the Mn^{2+} signal is slightly reduced by γ-irradiation and thus internal calibration may give an underestimated ED .

Magnetic field sweep : Record the *"center field"* and the *"range"* of the magnetic field sweep in mT. Use a slow scan of 8 to 15 min and response time of 0.1 to 0.1 s with an appropriate 100 kHz modulation width for a series of experiments. If the region of interest is small, one can shorten the scanning time to 4 or 8 min. If the sweep range is large, a slower scan with a short time constant must be used depending on the S/N ratio.

Low temperature measurements : A sharp linewidth produced due to the relaxation time as well as a high sensitivity due to reduced microwave loss is a bonus for low temperature measurements. Microwave loss is reduced because the involved moisture is frozen. The difference in the spin population between the lower and upper Zeeman levels is increased at low temperatures leading to the enhancement of the signal intensity.   Some signals such as the Al and Ti centers in SiO_2 appear at low temperature while others disappear due to the prevention of hindered rotation as in the case of CO_2^- in carbonates.

(c)   *Data processing with a computer* : If a computer is attached, one can obtain the ED at different magnetic fields by using the amplitude from the base line (*ED-plateau methods*). Softwares to subtract the overlapping signals and the base line are available so that the concentration is obtained by doubly integrating the derivative spectrum. The absolute concentration should be calculated using standard samples. Simulation of powder spectra can be made using anisotropic parameters of ESR spin Hamiltonian (g, A, D, E, etc.).

3.8.3 *Artificial Irradiation*

(a) *Albedo effect* : Irradiate samples sandwiched between PMMA plates or placed in a thick glass holder (about 1 mm). The surface of the sample must be subjected to secondary electrons to the same extent as the inside, and to do this the sample should be covered with another material of the same density.   Otherwise, the dose is generally lower at the surface than inside. This can be the cause of an inaccurate absorption dose and results in a fluctu-ation of the intensity – dose response curve. X–*rays should not be used* because the energy dependence is appreciable in the tens of kV region.

(b) *High dose rate effect* : Avoid a higher dose rate than 100 Gy/h (Ikeya 1985).   A higher dose rate creates fewer electron centers as is known from TL dating studies (Groom *et al.* 1978). This *"high dose rate effect"* leads to an erroneously high *ED*. The effect is not due to the temperature rise by γ– ray heating but presumably due to the interaction among electrons and holes during irradiation. The sample holder should never be irradiated.

3.9 Brief History of ESR Dating and Dosimetry
3.9.1 *Development of ESR Dating and Dosimetry*

Natural radiation effects have been studied by various physical detec-tion techniques. The possible use of TL in studying radiation effects on geological samples was suggested in 1953 by Daniels *et al.* and this tech-nique was then applied to archaeological ceramics in 1974 by M. J. Aitken. Electron micrographs of defect trails created by fission fragments of ^{235}U led to the development of fission track detection by chemical etching and its application to dating as described in Chapter 1.

ESR technology capable of detecting many paramagnetic ions in solids was developed in the 1950's and 1960's with interpretation based on quantum physics. Early studies on free radical detection in coal (Ubersfeld *et al.* 1954) and smoky quartz (O'Brien 1955) indicated that ESR could be applied to geological materials especially petroleum and clay minerals. Duchesne *et al.* (1961) tried to use ESR as a dating method for coal, lignite, peat and petroleum and discussed the effect of natural radioactivity and geothermal heat. Additive artificial irradiation was tried for the first time in this pioneer-ing work to calibrate the production yield of electron spins. The annual radiation dose rate was estimated from the content of uranium. However, the radiation effect was unfortunately masked by geothermal organic decom-position of the coal.

The use of ESR techniques incorporating the additive dose method used in TL dating was then suggested for dating of ancient geological materials (Zeller *et al*. 1967, Zeller 1968). However, ESR dating of old geological materials could not give meaningful ages in the 1960's. No controlled study with standard samples of distinctly different ages was carried out even for relative ESR dating at that time.

While the main stream of TL dating studies was directed towards archaeological ceramic materials of known origins using quartz and feldspar grains, there were some attempts to use ESR for dating of natural quartz grains in 1969 by McMorris but the ESR sensitivity was too low. The study of a lunar rock was made to detect ferromagnetic resonance of iron particles created by solar winds (Tsay *et al*. 1972).

ESR dating was then applied to Quaternary secondary deposits of carbonate, stalactites from the Japanese Akiyoshi cave (Ikeya 1975). Dose rates were determined later using TL dosimeters and the ^{238}U content measured by γ-ray spectroscopy. Radiation equilibrium was assumed in these early studies but the importance of ^{238}U-series radiation disequilibrium, which had been pointed out in TL studies, was later taken into account in ESR dating (see Chapter 4).

Dating of carbonates and bones in anthropological studies was carried out using the dose rate tuned to fit known ages in different environments (Ikeya and Miki 1980b). Absolute ESR ages of bones from Petralona in Greek, Tautavel in France and Heidelberg in Germany were still younger than those some anthropologists expected. A precise study of radiation assessment has established ESR dating of tooth enamel (Grün and Invernati 1985). Impure carbonate cave deposits related to anthropological remains as well as impurity centers intentionally doped were studied with ESR. Some radioisotopes were doped into minerals to study the internal radiation dose. Most ESR dating studies up to the middle 1980's were on secondary deposits of calcite with impurities. Work on carbonates was extended to the dating of fossil shells in sediments and to planktonic foraminifera in deep-sea sediment cores. Secondary deposits of $NaHCO_3$ and $NaCl$ in lake sediments and of marine gypsum have also been studied (see Chapters 5, 6 mad 7).

In archaeological applications, ESR dating of burnt flints was made (Robins *et al*. 1978), while unburnt geological flint was dated to be 300 Ma with the additive dose method using fast neutrons from a nuclear reactor (Garrison *et al*. 1981). However, it is not clear whether a meaningful age was successfully obtained from these studies.

Application of ESR to fault dating in structural geology was carried out using quartz grains in the fault gouge with mechanical bleaching of defects used as a mechanism of zeroing the age (Ikeya et al. 1982). These investigations used defects associated with Al, Ti and Ge impurities in SiO_2 (see Chapter 9). Application to geothermal silicates has also been done for the assessment of geothermal heat sources. Sediments baked by lava flows, volcanic ash and volcanic rocks were dated (see Chapter 10). Sediment dating based on the sunlight bleaching mechanism were reported.

ESR dosimetry using alanine as a dosimeter was reported (Bradshaw et al. 1962) and then alanine pellets, prepared using paraffin and polystyrene, were developed as dosimeter elements. In situ observation of radiation dose in a biological material was tried using a rat tail (Swartz 1965). ESR dosimetry of atomic bomb radiation for survivors at Hiroshima and Nagasaki was made using tooth enamel extracted upon dental treatment (Ikeya et al. 1984) and extended to Chernobyl, Techa River, and Toukai JCO-critical accidents. Radiation dose was studied using sugar-coating tablets and shell buttons. Monitoring of radiation dose in irradiated foods such as the bones of chicken, frog and fish have been studied (Chapter 13). A low-cost portable ESR spectrometer was developed using a permanent magnet, Nd-B-Fe (see Appendix 3).

3.9.2 Symposia and Proceedings up to 2001

The applicability of ESR techniques was first recognized as a potential method of dating similar to Tl for paleo-anthropology in French CNRS Congress at Tautavel in 1981. The _Specialist Seminar (SS) on TL Dating_ included ESR from its 3rd symposium. The proceeding volumes are listed below for references.

1982	3rd SS on TL & ESR Dating	Helsingor, Denmark	PACT 9 (1983)
1984	4th SS on TL & ESR Dating	Worm, Germany	Nuclear Tracks 14 (1985)
1987	5th SS on TL & ESR Dating	Cambridge, UK	Quatern.Sci. Rev.7 (1988).
1990	6th SS on TL & ESR Dating	Clermont, France	Radiat. Meas. 18 (1991),
1993	7th SS on TL & ESR Dating	Krems, Austria	Quatn. Geochr. 13 (1994).
			Radiat. Meas. 23 (1994)
1996	8th Intrn. Conf. Luminescence and ESR Dating		Quatn. Geochr.16 (1997)
		Canberra, Australia	Radiat. Meas. 27 (1997)
1999	9th I.C. ESR and Optical Dating	Rome, Italy	Quatn.Geochr. 19 (2000).
			Radiat. Meas. 32 (2000)

I organizes a small symposium specified as _The First International Symposium on ESR Dating_, including dosimetry, at Ube-Akiyoshi, Yamaguchi University, Japan, in 1985 simply because I could not afford to organize the SS and invite specialists of optical dating in a small collge. A Proceeding Book, _ESR Dating and Dosimetry_ (Ikeya and Miki ed. 1985) was published with a complete bibliography up to 1985. Dosimetry people joined and continued _International. Symp. on ESR Dosimetry and Applications._ Researchers worked hard and the fields were stimulated by two Intern. Symposia,which developed the method rapidly at the early stage of this new field. Followings are Symposia Proceedings with the volume and number and publication years for references.

1985	1st *ESR Dating (including dosimetry)*	
	Akiyoshi-Yamaguchi, Japan	*ESR Dating and Dosimetry* (1985)
1988	2nd *ESR Dosimetry and Applications*	
	GFS, Munchen Germany	*Appl. Radiat. Isot.* **40-10/12** (1989)
1991	3rd *ESRDA* NIST, Maryland, USA	*Appl. Radiat. Isot.* **44-1/2** (1993)
1995	4th *ESRDA* GFS Munchen Germany	*Appl. Radiat. Isot.* **47-11/12** (1997)
1998	5th ESRDA Obninsk, Russia	*Appl. Radiat. Isot.* **52-5** (2000)
2001	(ESR Dosimetry & Dating, Osaka, Japan)	*ESR Appl. Metrol.* **17** (2001)
2002	6th ESRDA Denver, USA	*Appl. Radiat. Isot.*

A series of Workshops on ESR Applied Metrology have been held every year in Japan since 1986 and extended to an *Intern. Symp. on ESR Dosimetry and Dating (2001-ESRDD-Osaka)* as a satellite meeting before the *3rd Asia-Pacific EPR Symposium* with main emphasis on new prospects at the start of 21st century. **Breakthroughs** were (1) *Diamagnetic to paramagnetic (D-P) Convertion* by low temperature γ -or UV-irradiation such as H_2O_2 in water and irradiated foods. (2) Triplet state detection by optical excitation for diamagnetic defects. (3) The last land on earth for ESR dating would be Antarctic samples. (4) Future outer planet worlds await us.

Proceedings are available as CD at a low cost as in the home page of Quantum Geophysics Laboratory: *http://quartz.ess.sci.osaka-u.ac.jp/*

Several review articles (Ikeya 1988, Gr n 1989b, Poupeau and Rossi 1989) and books on ESR dating in Japanese (Ikeya 1987) and in German (Grun 1989c) as well as books on the ESR microscope (imaging) in Japanese (Ohno 1991, Ikeya and Miki 1992) were published and finally an Atlas book with ESR spectra and tabulated parameter was published as the former edition (Ikeya, 1993).

3.10 Summary

The general method of ESR dating utilizing natural radiation effects are described with its limitations. The *TD* or *ED* related to the age is obtained from the ESR signal intensity and its enhancement by artificial irradiation. The stability of defects and saturation of defect concentration are considered according to the defect models in solid state physics A new additive dose method to obtain the age or the real *TD* from the *ED* directly was proposed based on the first- and second-order kinetics. Various materials have been used for ESR dating and applications to practical use in various fields have increased considerably. A vast area for the application of ESR dating and dosimetry is still there.

References

Aitken M. J. (1985): *Thermoluminescent Dating* (Academic Press, London).
Aitken M. J. (1998): *An Introduction to Optical Dating* (Oxford Univ. Press, Oxford).
Barabas M., Bach A. and Mangini A. (1988): An analytical model for the growth of ESR

signals. *Nucl. Tracks* **14**, 231-235.

Berger G. W., Lockhart R. A. and Kuo J. (1987): Regression and error analysis applied to the dose - response curves in thermoluminescence dating. *Nucl.Tracks* **13**, 177-184.

Bradshaw W. W., Cadena D. G., Crawford E. W. and Spetzler H. A. (1962): The use of alanine as a solid state dosimeter. *Radia. Res.* **171**, 11.

Daniels F., Boyd C.A. and Saunders D. F. (1953): Thermoluminescence as a research tool. *Science* **117**, 343-349.

Debenham N. C. (1983): Reliability of thermoluminescence dating of stalagmite calcite. *Nature* **304**, 154-156.

Desrosiers M. F. (1990): Assessing radiation dose to food. *Nature* **345**, 485.

Duchesne J., Depireux J. and van der Kaa J. M. (1961): Origin of free radicals in carbonaceous rocks. *Geochimica et Cosmochimica Acta*, **23**, 209-218.

Garrison E. G., Rowlett R. M., Cowan D. L. and Holroyd L. V. (1981): ESR dating of ancient flints. *Nature* **290**, 44-45.

Groom P. J., Durrani S. A., Kazal K. A. R. and McKeever S. W. S. (1978): The dose rate dependence of TL response and sensitivity in quartz. *PACT J.* **2**, 200-210.

Gr n R. and Invernati C. (1985): Uranium accumulation in teeth and its effect on ESR dating - A detailed study of a mammoth tooth. *Nucl. Tracks* **10**, 867-877.

Gr n R. (1989a): ESR dating for the early Earth. *Nature* **338**, 543-544.

Gr n R. (1989b): Electron Spin Resonance (ESR) Dating. *Quat. Intern.* **1**, 65-109.

Gr n R. (1989c): *ESR Alterbestimmungs-methode* (Springer-Verlag, Berlin) in German.

Gr n R. and MacDonald P. D. M. (1989): Non-linear fitting of TL/ESR dose-response curves. *Appl. Radiat. Isot.* **40**, 1077-1080.

Gr n R. and Rhodes E. J. (1991): On the selection of dose points for saturating exponential ESR/TL dose response curves. *Ancient TL* **9**, 40-46.

Gr n R. (1992): Suggestions for minimum requirements for reporting ESR age estimates. *Ancient TL* **10**, 37-41.

Hennig G. J., Gr n R. and Br nnacker K. (1983): Interlaboratory comparison project of ESR dating, phase I 1982. *PACT J.* **9**, 447-452.

H tt G., Molodkov A., Kessel H. and Raujas A. (1985): ESR dating of subfossil Holocene shells in Estonia. *Nucl. Tracks* **10**, 891-898.

Ikeya M. (1975): Dating a stalactite by electron paramagnetic resonance. *Nature* **255**, 48-50.

Ikeya M. and Miki T. (1980a): A new dating method with a digital ESR. *Naturwissenschaften* **67**, 191-19.

Ikeya M. and Miki T. (1980b): Electron spin resonance dating of animal and human bones. *Science* **207**, 977-979.

Ikeya M., Miki T. and Tanaka K. (1982): Dating of a fault by ESR on intrafault materials. *Science* **215**, 1392-1393.

Ikeya M., Miyajima J. and Okajima S. (1984): ESR dosimetry for atomic bomb survivors using shell buttons and tooth enamel. *Jpn. J. Appl. Phys.* **23**, L699-L710.

Ikeya M. (1985): *Dating Methods of Pleistocene Deposits* ed. Rutter N. W. (Geoscience Canada, No.2), Chapt 9 Electron Spin Resonance, 73-87.

Ikeya M. and Miki T. (1985): ESR Dating and Dosimetry (Ionics, Tokyo).

Ikeya M. (1986): *Dating and Age Determination of Biological Materials.* ed. Zimmerman M. R. and Angel S. L. (Croom Helm, London). Chapt. 3 Electron Spin Resonance.

Ikeya M. (1987): *Electron Spin Resonance Dating* (Ionics, Tokyo) in Japanese.

Ikeya M. (1988): Dating and radiation dosimetry with electron spin resonance (ESR). *Magn. Reson. Rev.* **13**, 91-134.

Ikeya M. (1992): A theoretical growth curve of defect formation for ESR and thermoluminescence dating. *Jpn. J. Appl. Phys.* **31**, L1618-L1620.

Ikeya M. and Miki T. (1992): ESR Microscopy - New Development of Applied Electron
Spin Resonance - (Springer-Tokyo, Tokyo) in Japanese.

Ikeya M. (1993): From earth to space: ESR dosimetry moves toward the 21st century.
Appl. Radiat. Isot. **44**, 1-5.

Itoh N. and Stoneham A.M. (2001): *Materials Modification by Electron Excitation* (Cambridge Univ. Press, Cambridge).

McMorris D. W. (1969): Trapped-electron dating: ESR studies. *Nature* **222**, 870-871.

Miki T. and Ikeya M. (1985): A plateau method for total dose evaluation in ESR dating
with a digital data processing. *Nucl. Tracks* **10**, 913-919.

O'Brien M. C. (1955): The structure of color centers in smoky quartz. *Proc. Roy. Soc.
(London)* **231A**, 404-414.

Odom A. L. and Rink W. J. (1988): Natural accumulation of Schottky-Frenkel defects :
implications for a quartz geochronometer. *Geology* **17**, 55-58.

Ohno K. ed. (1991): *ESR Imaging* (IPC Press, Tokyo) in Japanese.

Poupeau G. and Rossi A. M. (1989): *Nuclear Methods of Dating* ed. Roth E. and Poty E.
(CEA, Paris), Chapt. 10 Electron spin resonance dating, 275-293.

Regulla D.F. and Deffner U. (1982): Progress in alanine/ESR transfer dosimetry in high
dose dosimetry. IAEA-SM-272/ 39, 221-235.

Robins G. V., Seeley N. J., McNeil D. A. C. and Symons M. R. C. (1978): Identification
of ancient heat treatment in flint artifacts by ESR spectroscopy. *Nature* **276**, 703-704.

Romanyukha AA, Ignatiev EA, Degteva MO, Kozheurov VP, Wieser A. and Jacob P. (1996):
Radiation dose from Ural region. *Nature* **381**, 199-200.

Shimokawa K. and Imai N. (1985): ESR dating of quartz in tuff and tephra. ESR *Dating
and Dosimetry*, 181-185.

Skinner A.R. (2000): ESR dating: is it still an 'experimental' technique? *Appl.Radiat.
Isot.* **52**, 1311-1316.

Swartz H. M. (1965): Long lived electron spin resonance in rats irradiated at room temperature. *Radiat. Res.* **24**, 579-583.

Tanaka T., Sawada S. and Itoh T. (1985): ESR dating of late Pleistocene near shore and
terrace sands in southern Kanto, Japan. *ESR Dating and Dosimetry*, 275-280.

Tsay F. D., Chan S. I. and Manatt S. L. (1972): Electron paramagnetic resonance of radiation damage in a lunar rock. *Nature* **237**, 121-122.

Ubersfeld J., Etienne A. and Combrisson J. (1954): Paramagnetic resonance: a new property of coal like materials. *Nature* **174**, 614.

Yokoyama Y., Quaegebeur J.P., Bibron R. and Leger C. (1983): ESR dating of paleolithic calcite: thermal annealing experiment & trapped electron lifetime. *PACT* **9**,
372-379.

Yokoyama Y., Falgueres C. and Quaegebeur J.P. (1985): ESR dating of quartz from Quaternary sediments: first attempt. *Nucl. Tracks* **10**, 921-928.

Wu C. F. J. (1986): Jacknife, bootstrap and other resampling methods. *Annals of Statis-
tics* **14**, 1261-1294.

Zeller E. J., Levy P. W. and Mattern P. L. (1967): Geological dating by ESR. *Proc. Symp.
Radioactive Dating and Method of Low Level Counting*, 531-540.

Zeller E. J. (1968): Use of electron spin resonance for measurement of natural radiation
damage. *Thermoluminescence of Geological Materials*, ed. MacDougall D. J. (Academic Press, London), 271-279.

Chapter 4

Assessment of Radiation Dose

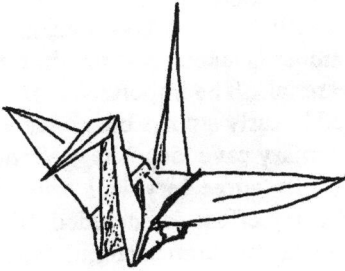

" Mr. Turtle and Ms. Crane ! How old are you ? "

According to an old Japanese saying, a crane lives for one thousand years and a turtle lives for ten thousands years. Assessment of radioactivities and environmental radiation makes the conversion of the equivalent dose (*ED*) or total dose of natural radiation (*TD*) into an ESR age possible.

4.1   Introduction

ESR dating described in this book belongs to a category of dating techniques utilizing radiation effects induced by natural radiation similar to fission track (FT) and thermoluminescence (TL) dating. ESR dating owes much to these two dating techniques, especially to TL dating in assessment of the annual dose rate and calibration procedures. Ordinary TL and ESR dating methods measure the *equivalent dose* (*ED*) or the *total dose of natural radiation* (*TD*) as described in Chapter 3. Extensive studies have evaluated the annual radiation dose in TL dating of pottery, assuming radioactive equilibrium of the uranium–238 (^{238}U) and thorium–232 (^{232}Th) series disintegration (Aitken 1985, Fleming 1979, Bell 1979).

Radioactive disequilibrium, especially in the ^{238}U–series, is common even in a closed system where only uranium is taken up at the time zero and daughter nuclides are created in the material. The importance of this disequilibrium in ESR dating was neglected in early studies but considered later following the TL dating study of Quaternary cave deposits (Wintle 1978). ESR ages for closed system corals and shells agree very well with ^{14}C–ages and ^{230}Th/^{234}U ages if the disequilibrium effect is included (Ikeya and Ohmura 1983). A model of constant (linear) uranium accumulation has been proposed to obtain the age for open system bones or teeth.

This chapter introduces the elementary physics of environmental radioactivity necessary for deduction of the age from the *ED*. The annual dose rate equations for both radioactive equilibrium and disequilibrium are given for the contents of ^{238}U, ^{232}Th and potassium–40 (^{40}K) in a finite or infinite medium. The *TD*'s (*ED* is only used for an experimental *TD* in this book) are given in an analytical form as a function of age and the ^{238}U content. Equations are presented for the initial incorporation of radioactive elements such as radium–226 (^{226}Ra) and ^{230}Th. Proposals to calculate the theoretical *TD* from the present radioactivity for a closed system and to study the dependence on the ^{238}U content (isochron method) are described with the experimental method to determine the internal and external doses considering the range of α–, β– and γ–rays.

The book on "*TL Dating*" by Aitken (1985) is recommended for the study of radiation assessment because of the similarity of ESR dating to TL dating. Sampling in a field should be made, hopefully in a simple radiation environment, considering the radiation assessment at a later stage.

4.2 Dosimetric Quantities and Units

The most important dosimetric quantity is the "**absorbed dose**", which is defined as the mean energy imparted by an ionizing radiation to a material of unit mass (ICRP 1983, Brodsky 1976). The special name for the unit of Joule/kg, [J/kg], for the absorbed dose is "**gray [Gy]**". The Gy should not be confused with Giga (10^9) years in earth science. Radiation energy is usually expressed in MeV. The conversion of MeV per kg of material to Joule/kg and Gy is: 1 MeV/kg $= 1.6 \times 10^{-19} \times 10^6$ J/kg $= 1.6 \times 10^{-13}$ Gy. The *TD* or *ED* referred to in the previous chapter should be in Gy. The special old unit, rad, in erg/g (1 rad $= 10^{-2}$ Gy) will not be used here (Brodsky 1978).

The radiation quantity for an additive artificial irradiation of γ–rays should be expressed strictly as absorbed dose in Gy. Another quantity conventionally used is the "**exposure**". The exposure is defined as the absolute value of the total charge of ions (of one sign) produced in air of the unit mass when all the secondary electrons are liberated by the γ–ray photons. The unit of exposure is "**Coulomb/kg [C/kg]**". The conventional unit, roentgen [R], corresponds to 2.58×10^{-4} C/kg. One may say that an additive artificial irradiation was made in a photon field of an exposure in C/kg.

The exposure can be converted into the absorbed dose in a medium using a **conversion factor** f when electronic equilibrium exists (Attix *et al.* 1986). The conversion factor is expressed as

$$f = 33.7 \, (\mu_{en}/\rho)_{med} / (\mu_{en}/\rho)_{air} \quad [Gy/(C/kg))] \; , \qquad (4.1)$$

where μ is the *mass absorption coefficient* and ρ is the *density*. $(\mu_{en}/\rho)_{med}$ and $(\mu_{en}/\rho)_{air}$ are the *mass energy absorption coefficients* for the electromagnetic radiation (photons) in the medium and air, respectively and depend on the energy of photons. Tables 4.1 and Table 4.2 show (μ_{en}/ρ) and f, respectively for various media.

Table 4.1 Mass energy absorption coefficient (μ_{en}/ρ) [cm^2/g] for several media.

hv[MeV]	Air	H_2O	Bone	Muscle	SiO_2	$CaCO_3$
1.0	0.0278	0.0309	0.0295	0.0306	0.0278	0.0278
1.5	0.0254	0.0282	0.0270	0.0280	0.0254	0.0254

Table 4.2   The conversion factor, f, to the absorbed dose of gray [Gy]*
from the exposure in Coulomb/kg, [C/kg].

f	Materials	Air	H_2O	Bone	Muscle	SiO_2	$CaCO_3$
C/kg	f(1.0Mev)	33.7	37.5	35.8	37.1	33.7	33.7
	f(1.5Mev)	33.7	37.4	35.8	37.1	33.7	33.7
Gy	$f'=f_m/f_{H_2O}$	0.90	1.00	0.96	0.99	0.90	0.90

* The absorption dose in SI unit is gray [Gy] (1 Gy = 1 J/kg = 100 rad).

The effect of absorbed dose on biological materials for different kinds of radiation (α-, β-, γ-rays and neutrons) with different energies and qualities is described with a unit of "Sievert [Sv]", which is equal to gray [Gy] for γ-rays. *Relative biological effect* (RBE) for the radiation energy must be multiplied by the absorbed dose in gray [Gy] to obtain the dose in [Sv]. The dose and the dose rate in radiation monitoring are expressed in [Sv] and [Sv/h]. The background natural radiation dose in a building is around $0.05 \sim 0.1 \mu Sv/h$. The annual natural γ-ray dose rate, D, is about $0.1 \mu Sv/h = 0.876$ mSv/a $= 0.876$ mGy/a using **1 a = 8,760 h = 3.1536 $\times$ 10^7 s**.

4.3   Annual Dose Rate : Radioactive Equilibrium

4.3.1   *Natural Radioactivity*

In nature, there are radioactive elements in three major nuclear disintegration series. Tables 4.3 and 4.4 show the energies of α-, β- and γ-rays (E_α, E_β and E_γ, respectively) and the half-lives for nuclides of the 238U-series and 232Th-series, respectively (Lederer and Shirley 1978, Nambi and Aitken 1986, Liritzis and Kokkoris 1992). The third 235U-series is not shown as the contribution from these nuclides is small. Beside these radioactivities, 40K which makes up 0.0117% of natural potassium decays into 40Ar (11%) and 40Ca (89%), and 87Rb whose atomic abundance is 27.8% decays by emitting β-rays into 87Sr (see Figure 1.4).

238U disintegrates by emitting α-particle in addition to the spontaneous fission which causes fission track in insulating solid minerals. As the half-life of 238U is extremely long ($T_{1/2} = 4.468$ Ga), the decay rate, $\lambda_{38} N_{38} = (0.693/T_{1/2}) N_{38}$, may be considered to be constant over a period of 10^8 a. The decay from 234Th to 234U is relatively rapid, but 234U decays to 230Th with the half-life of 2.45×10^5 a. The delayed build-up of 230Th is taken

Table 4.3   Disintegration, average energies of α-, β- and γ-rays and half-lives ($T_{1/2}$) of isotopes in ^{238}U–series.

Nucleus Z	Decay	$T_{1/2}$	Energy (MeV)		
			E_α [a] (b_α)	E_β [a] (b_β)	E_γ [a] (b_γ)
92 ^{238}U	α	4.468×10^9 a	4.198 (77) 4.149 (23)	0.00815	0.00136
90 ^{234}Th	β	24.10 d		0.0506 (73) 0.0249 (19)	0.00935
91 ^{234}Pa	β	1.17 min		0.8253 (98)	0.01880
92 ^{234}U	α	2.45×10^5 a	4.773 (72) 4.721 (28)	0.0110	0.00172
90 ^{230}Th	α	7.70×10^4 a	4.688 (76) 4.621 (23)	0.0127	0.00154
88 ^{226}Ra	α	1,602 a	4.785 (94.4) 4.602 (5.6)	0.0034	0.00674
86 ^{222}Rn	α	3.8235 d	5.490 (99.9)		0.5100 (0.00078)
84 ^{218}Po	α, β	3.05 min	6.003 (100)	0.0705 (0.0002)	0.0000
82 ^{214}Pb	β	26.8 min		0.2072 (48) 0.2274 (42)	0.2486
83 ^{214}Bi	α, β	19.9 min		0.6482	0.6093 (46)
84 ^{214}Po	α	1.64×10^{-4} s	7.685 (100)		0.00008
82 ^{210}Pb	β	22.3 a		0.0042 (80) 0.0161 (20)	0.0130 (11)
83 ^{210}Bi	α, β	5.01 d		0.3889 (100)	
84 ^{210}Po	α	138.4 d	5.297 (100)		
82 ^{206}Pb		stable			

$$E_{\text{total } \alpha, \beta, \gamma} \text{ [b]} = \Sigma\, b_z E_{z, \alpha, \beta, \gamma} = \quad 42.806\ (18.33) \quad 2.270\ (0.913) \quad 1.753\ (0.0395)$$

a) Only major fractions are given as b and their abundance (%) in parentheses. Minors are not shown although the total energy includes all. Details are tabulated by Liritzis and Kokkoris (1992).
     b_α, b_β and b_γ are branching ratios of α–, β– and γ–rays in %.
b) Total energy in parentheses is for pre–radon, i.e., 100% ^{222}Rn loss.

into account in the calculation of the *TD*. Except for ^{226}Ra, half–lives of the residual elements are again short indicating early radioactive equilibrium. Since gaseous ^{222}Rn often emanates from a sample, the rate of ^{222}Rn loss must be considered in the dose rate calculation.

Table 4.4 Disintegration, average energies of α-, β- and γ-rays and half-lives ($T_{1/2}$) for isotopes in ^{232}Th–series.

Nucleus Z ^{A}Element	Decay	$T_{1/2}$	Energy (MeV)			
			E_α [a]	(b_α)	E_β	E_γ
90 ^{232}Th	α	1.41×10^{10} a	4.010 (77)		0.0104	0.00130
			3.952 (23)			
88 ^{228}Ra	β	5.75 a	–		0.0144	
89 ^{228}Ac	β	6.31 h	–		0.4516	0.92870
90 ^{228}Th	α	1.913 a	5.396		0.0184	0.00322
88 ^{224}Ra	α	3.66 d	5.674		0.0021	0.00989
86 ^{220}Rn	α	55.6 s	6.282			0.54970 [b]
84 ^{216}Po	α	0.15 s			6.779	0.80600 [b]
82 ^{212}Pb	β	10.64 h	–		0.1702	0.14810
83 ^{212}Bi	α, β	60.6 min	2.172		0.4667	0.18460
84 ^{212}Po (64.07%)	α	0.307 μs	5.633		–	–
81 ^{208}Tl (35.93%)	β	3.07 min			0.2147	1.20589
82 ^{208}Pb		stable				

$$E_{\text{total } \alpha, \beta, \gamma} \,^{c)} = \Sigma\, b_z E_{z, \alpha, \beta, \gamma} = \begin{array}{ccc} 35.9323 & 1.3462 & 2.4820 \\ (15.0663) & (0.4969) & (0.9431) \end{array}$$

Liritzis and Kokkoris (1992): Note that ^{208}Tl is not included in their table though included in the dose rate calculation in Table 4.5.

a)　see Table 4.3.
b)　Average values $b_\gamma E_\gamma$ are 0.00070 and 0.00002, respectively.
c)　Total energy in parentheses is for pre–thoron, i.e., 100% ^{220}Rn loss.

The disintegration of ^{232}Th will reach equilibrium to the end of the ^{232}Th–series, ^{208}Pb, in a short time as the decay times of the daughter elements are all short in comparison with ages of geological or anthropological interest. Unless gaseous theron (^{220}Rn) emanates from the material, all the energy involved in the disintegration series would be dissipated to the material. The dose rate can be calculated assuming that ^{232}Th disintegrates nearly at a constant rate. In other words, the ^{232}Th–series may be in radioactive equilibrium. Similarly, the dose rate from ^{40}K would be nearly constant as the half–life is very long. Thus, the energies released from ^{232}Th and its daughters as well as from ^{40}K are calculated from the present concentrations of ^{232}Th and ^{40}K using the calculated dose rate already familiar from TL dating. Some energy values in Tables 4.3 and 4.4 are unnecessarily shown up to the 4th or 5th digit since the accurate energy is needed in α– and γ–ray spectroscopy.

4.3.2   *Annual Dose Rate*

The annual dose rate D is calculated using the radiation energy of the i-th disintegration element, E_i in MeV, the decay rate, λ_i in a^{-1} and the number of element, N_i, per kg of the material as

$$D = (1.60218 \times 10^{-19} \text{ J/eV})(10^6 \text{ eV})(\Sigma \lambda_i N_i E_i) \times 10^3 \text{ [mGy/a]}$$

$$= 1.60218 \times 10^{-10} (\Sigma \lambda_i N_i E_i) \text{ [mGy/a]} . \qquad (4.2)$$

The number of disintegration per year, $\lambda_i N_i$, is calculated using the half-life $T_{1/2}$ as $\lambda_i N_i = (\ln 2/T_{1/2\,i})N_i = (0.69315/T_{1/2\,i})N_i$ ($T_{1/2}$ of each nuclide is shown in Tables 4.3 and 4.4). For the **radioactive equilibrium** of ^{238}U, $\lambda_i N_i = \lambda_{i+1} N_{i+1} = \cdots = \lambda_{38} N_{38}$, where λ_{38} is the decay rate and N_{38} is the number of ^{238}U per kg of the material. Hence the dose rate is written as

$$D = 1.60218 \times 10^{-10} \lambda_{38} N_{38} \Sigma E_i \text{ [mGy/a]} . \qquad (4.3)$$

For example, for 1 ppm ^{238}U, i.e., 1 mg per kg of the material,
1) $N_{38} = 6.02214 \times 10^{23} \times (10^{-3}/238.05) = 2.5300 \times 10^{18}$ and
       $\lambda_{38} = 0.69315/(4.468 \times 10^9 \text{ a}) = 1.55136 \times 10^{-10}$ a^{-1} are obtained.
2)   Then, $\underline{D = 0.062879 \times \Sigma E_i}$ is used to calculate the dose rate.

For 1 ppm ^{232}Th, $\underline{D = 0.020514 \times \Sigma E_i}$ is used.

The energies for α-, β- and γ-rays of the ^{238}U- and ^{232}Th-series disintegrations are shown in Tables 4.3 and 4.4. The dose rates of α-, β- and γ-rays (D_α, D_β and D_γ, respectively) per 1 ppm or 1% of radioactive elements calculated based on these energies are shown in Table 4.5. The dose rates are calculated using this table according to following equations,

$$D_\alpha = C_U D_{U-\alpha} + C_{Th} D_{Th-\alpha} , \qquad (4.4)$$

$$D_\beta = C_U D_{U-\beta} + C_{Th} D_{Th-\beta} + C_{K,Rb} D_{K,Rb-\beta} , \qquad (4.5)$$

$$D_\gamma = C_U D_{U-\gamma} + C_{Th} D_{Th-\gamma} + C_K D_{K-\gamma} , \qquad (4.6)$$

where, for example, C_U is the concentration and $D_{U-\alpha}$ is the α-ray dose rate of ^{238}U. Since gaseous ^{222}Rn in the ^{238}U-series and ^{220}Rn (thoron) in the ^{232}Th-series may emanate from the sample, these tables include the case for 100% loss of ^{222}Rn and ^{220}Rn in parenthesis. The energy from ^{222}Rn or ^{220}Rn down to the end of decay is not counted in the calculation for the   case

Table 4.5   The annual dose rates for α–, β– and γ–rays for radioactive equilibrium of 238U- and 232Th-series disintegration and for natural potassium and rubidiuma).

Disintegration	D_α [b]	D_β [b]	D_γ [b] (mGy/a)
U (1ppm)	2.7920 (1.2040)	0.14660 (0.05907)	0.11130 (0.00395)
UO_3 (1ppm)	2.3195 (1.0006)	0.12181 (0.04909)	0.09259 (0.00324)
^{238}U (1ppm)	2.6916 (1.1528)	0.14273 (0.05739)	0.10207 (0.00248)
^{232}Th (1ppm)	0.7371 (0.3091)	0.02762 (0.01019)	0.05092 (0.01935)
K_2O (1 %)	-	0.67805	0.20287
Rb_2O (1ppm)	-	0.00047	-
K_2O (1%) & Rb_2O (50ppm)	-	0.70170	0.20287
^{238}U - ^{234}Pa	0.26321	0.05568	0.001856
^{234}U	0.29899	0.00069	0.000108
^{230}Th - ^{206}Pb	2.1294 (0.5906)	0.08636 (0.0010)	0.10824 (0.00052)
^{226}Ra - ^{206}Pb	1.8388 (0.3001)	0.08556 (0.0002)	0.10815 (0.00042)
^{222}Rn ^{206}Pb	1.5387 (-)	0.08534 (-)	0.10772 (-)
^{235}U - ^{207}Pb	0.11518(0.05786)	0.00468(0.00202)	0.001857 (0.001429)

a) The dose rates are based on the data presented by Liritzis and Kokkoris (1992) and revised by Ogoh et al. (1993). Further revision made the table closer to that by Nambi and Aitken (1986) against the reported discrepancy of 2~3%.

b) Numbers in parentheses are the dose rates, D' for 100% loss of ^{222}Rn or ^{220}Rn. The factors of 0.8322 and 0.8788 are multiplied for the dose rate of 1ppm of UO_3 and $ThO2$, and the dose rates are divided by 0.8301 and 0.9158 for 1% of K and Rb, respectively.

of 100% loss. In the case of Nloss% loss, the dose rate is written as

$$D_N = D(1 - N_{loss}/100) + D'(N_{loss}/100), \qquad (4.7)$$

where D, D_N and D' are dose rates for 0%, N_{loss}% and 100% radon loss, respectively.   The partial dose rates in ^{238}U-series disequilibrium described in Section 4.5.2 are also given in the table for calculation of losses and uptakes. Note that the dose rate is different for 1 ppm of U, UO_3 and ^{238}U which depends on the method among uranium analyses.

4.4 Radiation Dose Assessment

4.4.1 *Ranges and quality of α–, β – and γ– rays*

α– β– and γ–rays create lattice defects in the crystalline lattice in addition to the ionization of paired electrons. The radiation effect is not only the production of unpaired electrons but also involves the formation of atomic defects that stabilize the unpaired electrons. In other words, the trap described as a gangster who traps the girl (electron) in the cartoon in Figure 3.1 may be created. For the radiation dose assessment for both internal and external doses, the range and quality of α-, β- and γ-rays are important as well as the defect production efficiency, especially of α-rays. The ranges of α-, β- and γ-rays for material with density $\rho = 2.6$ are shown in Figure 4.1 (a) as a function of energy. The range indicated is the distance where the radiation intensity is decreased to $1/e = 0.3679$ since the intensity is expressed as $I = I_o e^{-x/R}$, where x and R are the distance and range, respectively. In other word, the intensity is decreased to $0.1353 I_o$ and to $0.05 I_o$ by the penetration of $2R$ and $3R$, respectively. Therefore, practically, the maximum penetration distance is a few times the range.

(a) *Alpha−rays* : The range of 4 ~5 MeV α-rays is 15 ~20 μm as shown in Figure 4.1 (a) and so the maximum penetration depth is 40 ~60 μm. Alpha-rays produce lattice defects partly by energy transfer through elastic collisions but mostly through inelastic collisions due to ionization along their path. A narrow region with a thin diameter of 2 ~ 10 nm along a range of 10 ~ 20 μm is damaged particularly at the end of the range, where the energy loss rate per unit length, dE/dx, (stopping power) is high.

The α-recoil atom, the daughter atom after the α-ray emission has an appreciable recoil energy which produces lattice defects. The track by the α-recoil atom may be quite short as the atomic weight is usually about 200. The crystalline lattice at the site of the α-recoil track may be changed locally into an amorphous state along the track.

In insulating materials, the defect creation energy needed is about 40 eV (a few times the band gap of the material) and therefore a 4 MeV α-particle would create about 10^5 defects along its track in the material. The local concentration of lattice defects may reach about 10^5 defects/10^{-14} cm^3 = 10^{19} defects/cm^3, in which region the mutual interaction reduces the defect creation efficiency (α-ray efficiency: k-value) as is known in the saturation behavior of the growth curve by γ-irradiation described in Chapter 3. A part of the site within the range may be close to an amorphous state due to heavy damage. Hence, most of the electrons and holes recombine and only a small fractions remain as stable centers. The defect production efficiency by α-rays relative to β- and γ-rays (k-value) should always be considered in TL and ESR dating.

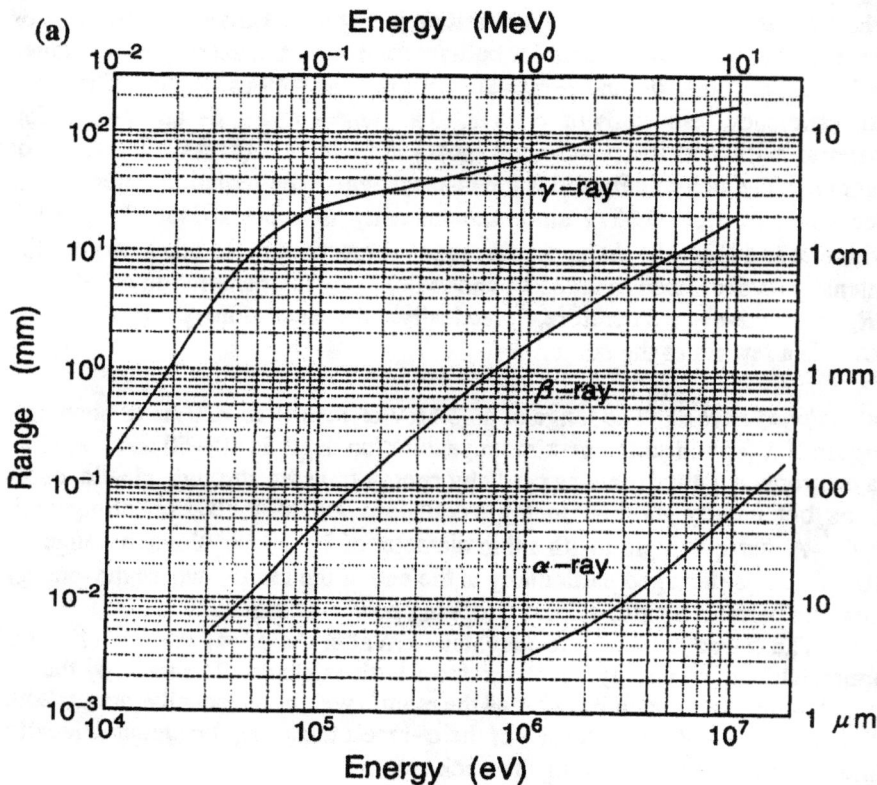

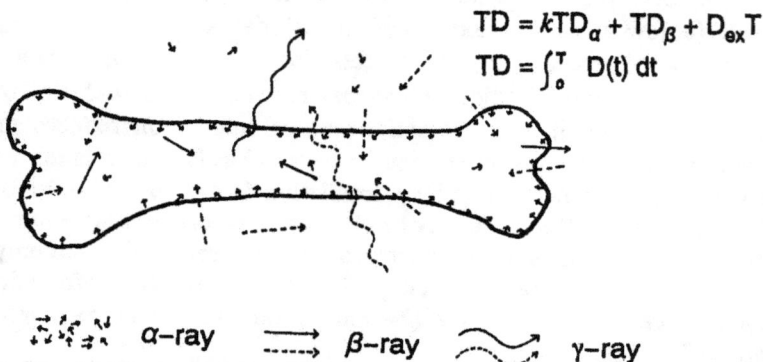

Figure 4.1 (a) The ranges of α-, β- and γ-rays as a function of energy in MeV. The energies of natural radiation are indicated. (b) A schematic illustration of the ranges of α-, β- and γ-rays for both internal (solid arrows) and external radiation (dashed arrows) in a bone buried in a sediment.

(b)   *Beta–rays* :   Energetic electrons (β–rays) ionize the material.   The range of β–rays around 0.5 ~1.0 MeV is 0.7 ~1.5 mm and so the maximum penetration depth is 2 –4 mm.   Experimental and theoretical estimates for the attenuation within the sample and from the soil indicate that the removal of the outer 2 mm is adequate to avoid the influence of β–radiation from the surrounding soil.   The defect production efficiencies by the absorbed energy of β– and γ–rays are regarded as nearly the same and become larger below 50 keV since some inner shell (K–shell) of specific atoms in the constituent solid are ionized.

(c)   *Gamma–rays* :   The range of γ–rays of 1 ~2 MeV is 6 ~9 cm and so the maximum penetration depth is 20 ~30 cm.   Natural radiation with different energy spectra gives different average conversion factors f described in Section 4.2.   This is especially true for bones or TLD elements of $CaSO_4$, where the defect production efficiency for γ–rays of low energies (< 50 keV) is a few times higher than for those of 1.0 MeV.   At the moment, we simply derive an *ED* energy corresponding to the radiation quality of the artificial irradiation.   Gamma–rays from a ^{60}Co source should be used rather than low energy X–rays because the *TD* obtained by X–rays is markedly dependent on the X–ray energy and different from that by γ–rays.

4.4.2   *Quality Effect : k–values for α–Rays*

Considering the efficiencies of stable defect production by α–, β– and γ–rays, the effective annual dose rate may be written as

$$D = k_\alpha D_\alpha + k_\beta D_\beta + k_\gamma D_\gamma , \qquad (4.8)$$

where k_α, k_β and k_γ are defect production efficiencies relative to those of the artificial irradiation.   We usually disregard the k–value for β– and γ–rays ($k_\beta = k_\gamma = 1$) and leave only k_α as the k–value.   Then, Eq. (4.8) becomes

$$D = kD_\alpha + D_\beta + D_\gamma . \qquad (4.9)$$

If α–rays are used as an additive artificial dose to get the ED_α, the ratio of ED_α to that by γ–rays gives the k–value, i.e.,

$$k = ED_\gamma / ED_\alpha , \qquad (4.10)$$

which is equal to the ratio of defect production efficiencies, a_α / a_γ.   It is noted that the low efficiency of defect formation by α-rays gives a large ED_α by additive α-irradiation. Since the k-value is the ratio of the defect creation efficiency, $a_\alpha/a_\gamma = G_\alpha/G_\gamma$ using G value (the number of radicals produced per 100 eV), it should not depend on the defect concentration in principle. However, experimentally $k = a_\alpha/a_\gamma (1 - bn/N_0)$ is measured and given as the k-value (see Section 3.5.1 in Chapter 3). Hence, the k-value appears to depend on the defect concentration.

In TL dating, the k_α for a heavily damaged region along an α-track is about 0.2 due to the high rate of recombination. The situation would be similar for ESR dating though it may not necessarily be of the same magnitude. Retrapping processes of electrons released by heating during TL measurement are involved in the k-value for TL, while only the production yield gives the difference in ESR dating (Nambi 1979).

Most k-values have been obtained from ED's by the conventional additive dose method using Eq. (4.10) (Lyons 1988). Annealing of α-tracks or α-recoil tracks may be dominant in nature, while measurements were usually made soon after additive α-irradiation. Hence, the ESR signal intensity enhancement by α-rays may be overestimated due to inclusion of unstable defects. Actual ED_α may be larger than that leading still to small k-values (see Eq. (4.10)). If dissolution of the α-ray damaged region occurs by chemical etching in nature, the apparent k-value would be close to zero. If the locally amorphous damaged part recrystallizes, the k-value may even become negative.

ESR studies of carbonates doped with α-emitter Po or ^{238}U have been made to evaluate the internal α-ray effects (Apers et al. 1981, Hennig and Grün 1983). Much smaller k-values have been reported for several materials as shown in Table 4.6. In most cases, especially when the internal dose rate is not predominant, the uncertainty in the k-value will contribnte less than 2% to the uncertainty in the age estimate, and in younger samples probably less than 1% (Lyons and Brennan 1991).

4.4.3 Internal and External Doses

The absorbed dose for materials buried in a sediment, must be treated for each of α-, β- and γ-rays considering their ranges and the sample size. The maximum penetration depth (effective range) of α- and β-rays is 40–60 μm and 2–4 mm, respectively, while that of γ-rays is 20–30 cm as described previously and illustrated in Figures 4.1 (b). Hence, the effective volume around the finite-sized sample contributing to radiation damage is different for α-, β- and γ-rays.

Table 4.6 Alpha–ray efficiency, k–value, so far reported for ESR dating.

Material	Signal	k–value	Reference
Speleothem	$g_C = 2.0007$	0.052 ± 0.006	Lyons & Brennan 1991
Coral	$g_C = 2.0007$	0.15 ± 0.05 [a]	Ikeya & Ohmura 1983
		0.06 ± 0.01	Radtke & Grün 1988
		0.05 ± 0.01	Grün *et al.* 1992
		0.007 [b]	Kohno & Ikeya (unpublished)
Foraminifera	$g_C = 2.0007$	0.08 ± 0.02	Mudelsee *et al.* 1992
	$g_B = 2.0036$	0.1 ± 0.02	Mudelsee *et al.* 1992
Tooth enamel	$g_C = 2.0018$	0.15 ± 0.02	DeCanniere *et al.* 1986

These k–values are obtained from the ratio of ED_γ / ED_α except those: a) empirically by tuning ED to ^{14}C–ages and b) experimentally obtained from the ratio of G_α / G_γ.

Alpha–rays :   The internal α–dose rate (D_α) may be used except for the surface few μm where external α–rays cause damage. The surface few μm is low in weight and may be removed by washing with acid. The internal α–ray dose can be calculated from the content of ^{238}U and ^{230}Th.

Beta–rays :   The external β–rays produce radiation damage at the surface 1 mm, while a part of internal β–rays goes out of the superficial 1 mm of the sample. Hence the average dose rate of both the internal and external β–dose rate must be used for the surface $0.5 \sim 1$ mm of the sample.   If this part is removed, only the internal dose rate (D_β) may be used.

Gamma–rays :   The range of natural γ–rays is about 10 cm.   If the outside environment is uniform and infinite (more than 30 cm from the sampling position) and the radioactivity is in equilibrium, one can use the γ–ray dose rate (D_γ) calculated from the contents of ^{238}U, ^{232}Th and ^{40}K in the outside sediment and cosmic ray component (D_{cos}) as external dose rate (D_{ex}). If this is not possible, one must embed TLD to estimate the D_{ex} which includes both D_γ and D_{cos}.

Whether the sample is considered to be *infinite* or *finite* depends on the volume of the sample and the surrounding environment:

1) If radioactive elements exist uniformly in the sample and matrix radiation energy emitted by the sample is equal to the absorbed dose by the sample, we might consider that the *sample size is infinite*. In this case, only the internal dose can be used.

2) If the surrounding materials within the effective γ-ray range of 30 cm from the sample have different concentrations of U, Th and K, we might consider that the *sample size is finite*. In this case, the external D_γ has to be considered for the D_{ex}. The external dose is $D_{ex}T_{ESR}$.

These procedures are given as a flow diagram in Figure 4.2.

Quartz grains in a large granite or in a clay or sand sediment or a coral sample from a massive coral may be considered to be in an infinite medium. The surface of the quartz has to be removed for α-ray dose estimation. Same quartz grains in a pottery, or a solitary coral or a shell in a sediment should be considered to have a finite size. In this case, the D_{ex} should be used rather than the D_γ of the material.

For example, the TD for shells or bones embedded in a sandy sediment is calculated in two ways:

1) The TD from inside the material should be considered for both α- and β-rays, i.e., $TD_{in} = kTD_\alpha + TD_\beta$.
2) The TD from the outside environment (TD_{ex}) should be considered for γ-rays, i.e., $TD_{ex} = D_{ex}T_{ESR} = (D_\gamma + D_{cos})T_{ESR}$.
3) Therefore, the TD is obtained by

$$TD = TD_{in} + TD_{ex} = kTD_\alpha + TD_\beta + (D_\gamma + D_{cos})T_{ESR} \quad . \quad (4.11)$$

The dose rates D_α, D_β and D_γ vary depending on the local environment and also on radioactive disequilibrium. Therefore, *correction factors* and *radioactive disequilibrium* should be introduced to obtain the accurate age as described in following sections.

4.4.4   *Dose Correction Factors*

The external dose or dose rate decreases with the penetration depth. Dose attenuation factors for α-, β- and γ-rays as well as cosmic rays have been used by researchers in TL dating (Aitken *et al.* 1985, Bell 1980, Mejdahl 1979). Attenuation curves considering the stopping power of radiation or electron energy degradation are averaged over the volume for both thin plate and grain samples.

(a)   *Alpha–dose attenuation* : Alpha–dose attenuation factors in a mineral with the density, $\rho = 2.7$ g/cm^3 were calculated for a thin plane and a spherical grain by averaging the dose over the volume with appropriate thickness or

START

C_U, C_{Th}, K, D_{ex}

yes — Infinite — no

yes — Equil — no

no — Equil — yes

D_α, D_β
Eq.(4.4)
Eq.(4.5)

U-series disequil TD(t)
TD_α, TD_β, TD_γ
Eq.(4.28)

D_α, D_β
Eq.(4.4)
Eq.(4.5)

$D = kD_\alpha + D_\beta + D_\gamma$
Eq.(4.9)

$TD = kTD_\alpha + TD_\beta + TD_\gamma$
Eq.(4.28)

$TD = kTD_\alpha + TD_\beta + D_{ex}T$
Eq.(4.38)

$D = kD_\alpha + D_\beta + D_{ex}$
Eq.(4.9)

$T = ED/D$

$TD(t) = ED$: Graphics
$T_i = T_{i-1} + (ED - TD_{i-1})/D$
Eq.(4.29)

$T = ED/D$

T_i — no — $\Delta T < 1$ — no — T_i

yes

ESR Age

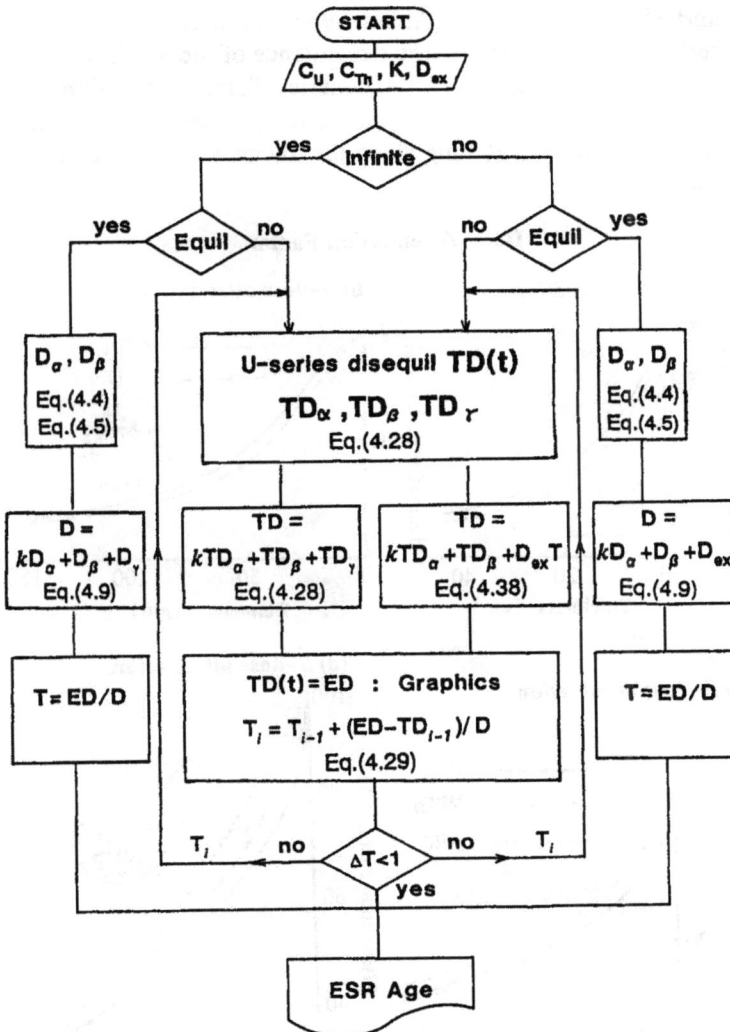

Figure 4.2 A flow diagram for the assessment of the *TD* in ESR dating for an infinite or finite medium considering radioactive equilibrium and uranium-series disequilibrium. The D_α, D_β and D_γ are obtained using Eq. (4.4), (4.5) and (4.6), respectively. The D_{ex} is measured with TLD or with a field surveymeter, or obtained as the sum of the D_γ and D_{cos}. The external D_γ can be calculated from the contents of ^{238}U, ^{232}Th and ^{40}K in the environment. D_{cos} depends on the altitude and height (usually, 0.25 mGy/a in Japan).

diameter (Bell 1980). The results are shown in Figure 4.3 (a) and (b) for ^{232}Th– and ^{238}U–series decays (Aitken 1985, Grün 1989). This effect was reevaluated considering the energy dependence of the track length by α–particles for both ^{238}U–series and ^{232}Th–series disintegration (Brennan *et al.* 1991). The effect of surface etching on the α–ray dose was given as a function of the grain size, but etching along the tracks was not considered in the calculation.

Dose Attenuation Factor

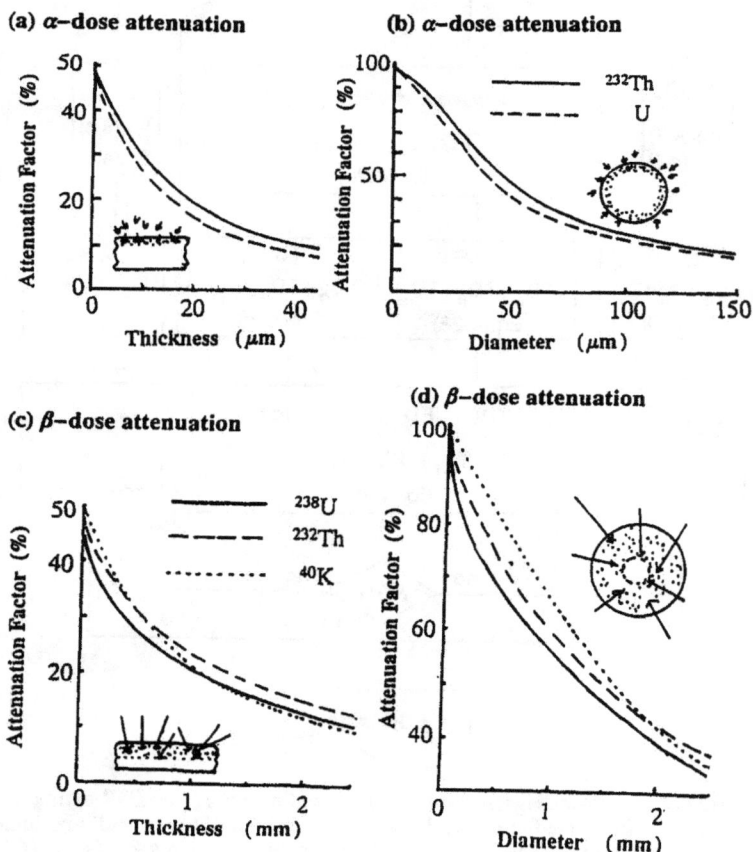

Figure 4.3 (a) and (b) Alpha–dose attenuation factor for the ^{238}U– and ^{232}Th–series for a thin plane and a spherical grain, respectively (Bell 1980). (c) and (d) Beta–dose attenuation factors for β–rays and IC electrons from ^{238}U– and ^{232}Th–series radioactive elements and ^{40}K (Mejdahl 1979).

(b)   *Beta—dose attenuation* : Quartz grains contain no radioactive elements. The average β–dose rate for quartz in a clay matrix differs from the
infinite matrix dose rate due to the attenuation of the β–rays within the grain.
Beta—rays and internal conversion (IC) electrons arising out of a matrix with
uniformly distributed radionuclides lose energy in the sample. The latter
contributes for the β–dose rate, being 13 and 7% of the whole in the case of
^{234}Th– and ^{238}U–series, respectively. The absorbed fraction for a spherical
non–radioactive region, has been calculated theoretically based on radiation
physics involving energy loss straggling and secondary bremsstrahlung.

The β–ray attenuation factor for external radiation, Att_β, is given as

$$Att_\beta = 0.15 \times (V_s / V_T) , \tag{4.12}$$

where V_T is the total volume of the sample and V_s is the volume adjacent to
the surface to a depth corresponding to the effective β–ray range. The effective β–ray range, R_β (g/cm^2), is given as

$$R_\beta = 0.75 \times 0.11 \times (1 + 22.4 E^2)^{1/2} - 1 , \tag{4.13}$$

where E is the maximum β–ray energy in MeV. For small samples,

$$Att_\beta = (1 - e^{-\mu d})/\mu d , \tag{4.14}$$

where d (g/cm^2) is the sample thickness and μ (cm^2/g) is the attenuation coefficient for β–rays (Yokoyama *et al.* 1982).

Figure 4.3 (c) and (d) show the relative attenuation factor for β–rays
from ^{40}K and both β–rays and IC electrons from ^{232}Th– and ^{238}U–series
radioactive elements for the refinement of the β–ray dose rate (Mejdahl
1979). It is about 90% for quartz grains of 0.2 mm in diameter.

The relative attenuation factor is virtually independent of the water
content as the stopping power of the clay is almost identical with that of
quartz. Since the attenuation factor depends naturally on the energy of electrons or nuclides, the radioactive disequilibrium affects the absorbed β–dose
fraction and so the attenuation factor. This effect is about 10% for 1 mm
grains and is almost negligible for the β–dose rate. Attenuation factors are
summarized in the book by Aitken (1985) and in the review article by Grün
(1989), and they are further discussed by Lyons and Brennan (1989).

(c)   *Cosmic ray corrections* : Cosmic rays have to be considered for the
dose rate estimation. The dominant portion of the hard (high energy)
component of cosmic rays is muons which compose 75% of all particles at

the sea level. Figure 4.4 (a) shows the depth profile of cosmic rays. The cosmic ray dose relative to that at the sea level (zero altitude) is shown in (b) for three different latitudes (Prescott and Stephan 1982). Attenuation by soils has to be calculated for geological and archaeological materials embedded in soil sediments. Cosmic ray dose rate is higher at a high altitude since the air attenuation of cosmic rays is small at a high altitude. Thus, modification is needed at a high mountain. Attenuation factors deep inside caves and tunnels measured with a NaI (Tl) detector are shown in (c) (Komura and Sakanoue 1985).

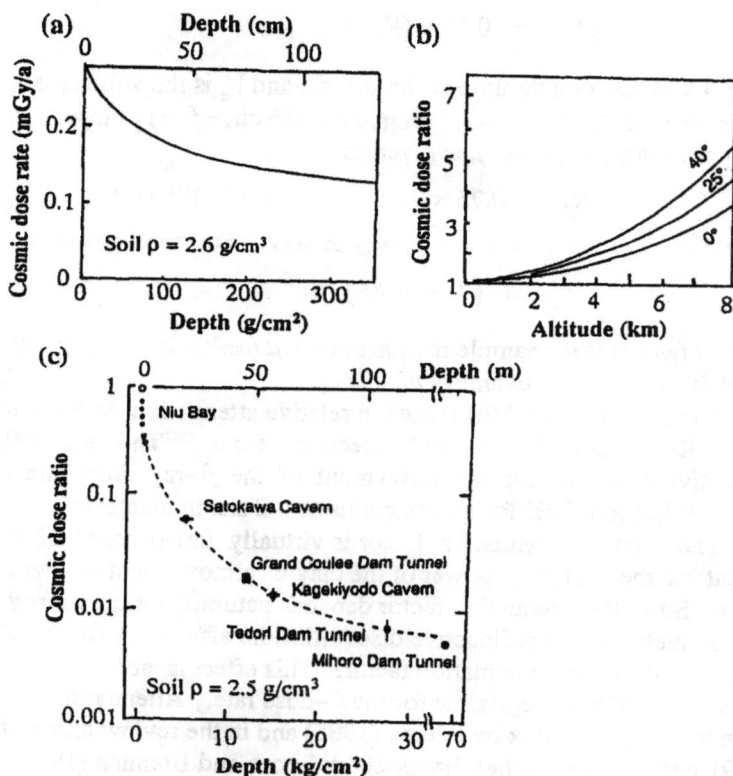

Figure 4.4 (a) Attenuation of cosmic rays with depth and (b) cosmic ray dose relative to that at the sea level for three different latitudes (Prescott and Stephan 1982). (c) Measured attenuation of cosmic rays for soils from caves and tunnels. Intensity is normalized at sea level (modified after Komura and Sakanoue 1985).

(d)   *Water content* : The external annual dose rate D_{ex} may be calculated from the contents of ^{238}U, ^{232}Th and ^{40}K, using the values of D_α, D_β and D_γ in Table 4.5. The effect of moisture or water at the excavation site must be taken into account if the materials were dried before analysis. The dose rate for $\alpha-$, $\beta-$ and γ–rays (D_α', D_β' and D_γ', respectively) for an infinite medium of pottery with the water content of W % in weight can be written as,

$$D_\alpha' = D_\alpha / [1 + 1.49W/(100 - W)] , \qquad (4.15)$$

$$D_\beta' = D_\beta / [1 + 1.25W/(100 - W)] , \qquad (4.16)$$

$$D_\gamma' = D_\gamma / [1 + 1.14W/(100 - W)] , \qquad (4.17)$$

where D_α, D_β and D_γ are calculated as the annual dose rate for no water and W denotes the water content in the surrounding soils. These are often quoted from Bowman's Ph.D. thesis (1976) in the literature (Aitken 1985).

4.5   Radioactive Disequilibrium

4.5.1   *Initial Incorporation*

The initial incorporation of radioactive elements such as ^{226}Ra and ^{230}Th causes ^{238}U–series disequilibrium. The time dependent radiation dose rate due to decay of the incorporated radioactivity is written as

$$D_A(t) = D_A e^{-\lambda t} , \qquad (4.18)$$

where D_A is the dose rate of the radioactive element with a mass number A. If the daughters decay in a short time, we might include their dose rate into D_A as is shown for D_{26} for ^{226}Ra down to ^{206}Pb in the decay series. The TD_A can be calculated by integrating the dose rate as

$$TD_A = (D_A/\lambda)(1 - e^{-\lambda T}) . \qquad (4.19)$$

This is shown in Figure 4.5 (a). The effect of radioactivity leaching on the TD can similarly be assessed if the process of leaching is known.

If one assesses the TD due to the activity of the daughter elements, the time dependent dose rate due to the time dependent concentration of the radionuclides, $C(t)$, should be calculated. It is written as

$$TD = \int C(t) D_A e^{-\lambda T} dt . \qquad (4.20)$$

This is shown in Figure 4.5 (b).

Alpha–ray disintegrations produce recoil daughter nuclei which leach out of the material (Kigoshi 1971). Objects for ESR dating do not necessarily represent a closed system where no radioactivity comes in or goes out. Leaching and incorporation of radioactive elements are sometimes the cause of disequilibrium aside from the delayed build–up of ^{230}Th. Yokoyama et al. (1982) assessed the radiation dose for each radioactive element confirming leaching and incorporation of radioactivity, using α–ray and γ–ray spectroscopy. Since we do not know when the leaching of radioactive elements occurred, the calculation is more or less the same as that using the present concentration of ^{238}U as described in the next section.

4.5.2 ^{238}U–Series Disequilibrium : Closed System

The TD of α–, β– and γ–rays for ^{238}U–series disequilibrium has been calculated (Wintle 1978, Yokoyama et al. 1982, Ikeya 1982, Gosler and Hercman 1988). The delayed build–up of ^{230}Th and time–dependent annual dose rate, $D(t)$, has been derived for ESR dating of shells and corals (Ikeya and Ohmura 1983, 1984). The TD was estimated by integrating the $D(t)$.

The $D(t)$ for a closed system of ^{238}U–series disintegration at the time t after the incorporation of uranium may be written using the activities of ^{238}U, ^{234}U and ^{230}Th as

$$D(t) = \sum D_i \lambda_i N_i , \tag{4.21}$$

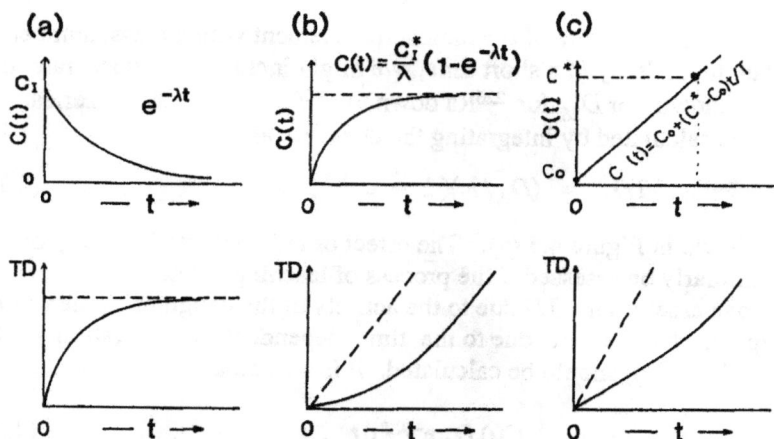

Figure 4.5   Schematic illustration for the TD versus age T relation for radiation from (a) decay of one type of radioactivity, (b) a daughter nuclide that reaches an equilibrium and (c) a constant uptake ratio of the element.

or

$$D(t) = D_{38}(\lambda_{38}N_{38}) + D_{34}(\lambda_{34}N_{34}) + D_{30}(\lambda_{30}N_{30}) \; , \quad (4.22)$$

where λ_{38}, λ_{34} and λ_{30} are decay constants and N_{38}, N_{34} and N_{30} are the number of atoms for ^{238}U, ^{234}U and ^{230}Th, respectively. D_{38}, D_{34} and D_{30} represent the effective energy released per decay from ^{238}U to ^{234}U, ^{234}U to ^{230}Th and ^{230}Th to stable ^{206}Pb, respectively, as shown in Table 4.5.

The equation is rewritten using the activity ratio of $r = {}^{234}U/{}^{238}U$ and $p = {}^{230}Th/{}^{234}U$ as shown in Figure 1.5 (b), i.e.,

$$r = \lambda_{34}N_{34}/\lambda_{38}N_{38} \quad \text{and} \quad p = \lambda_{30}N_{30}/\lambda_{34}N_{34} \; , \quad (4.23)$$

as

$$D(t) = \lambda_{38}N_{38}[D_{38} + rD_{34} + pD_{30}] \; , \quad (4.24)$$

or

$$D(t) = D + D_{34}(r-1) + D_{30}(pr-1) \; , \quad (4.25)$$

where $D = D_{38} + D_{34} + D_{30}$ is the annual dose rate for radioactive equilibrium. The activity ratios, $r(t)$ and $p(t)$, are given as

$$r(t) = 1 + (r_0 - 1)e^{-\lambda_{34}t} \; , \quad (4.26)$$

$$p(t)r(t) = 1 - e^{\lambda_{30}t} + (r_0 - 1)\lambda_{30}(e^{-\lambda_{34}t} - e^{-\lambda_{30}t})/(\lambda_{30} - \lambda_{34}) \; , \quad (4.27)$$

where r_0 is the initial activity ratio of $^{234}U/{}^{238}U$. The TD can be calculated by integrating $D(t)$ from $t = 0$ to $t = T$ as

$$
\begin{aligned}
TD(T) &= \int_0^T D(t)\,dt \\
&= DT + D_{34}(r_0 - 1)(1 - e^{-\lambda_{34}T})/\lambda_{34} - D_{30}\{(1 - e^{\lambda_{30}T})/\lambda_{30} \\
&\quad - (1/\lambda_{34})(r_0 - 1)[1 - (\lambda_{30}e^{-\lambda_{34}T} - \lambda_{34}e^{-\lambda_{34}T})/(\lambda_{30} - \lambda_{34})]\} \; . \quad (4.28)
\end{aligned}
$$

The TD is calculated taking ^{230}Th build-up into account as follows;
1) First, the age T_e, assuming radioactive equilibrium, is calculated from the ED and the dose rate D using the relation of $T_0 = T_e = ED/D$.
2) Next, the TD_i is calculated with Eq. (4.28) using the T_{i-1} (the first T_i).
3) Then, the difference $ED - TD_i$ is divided by D.
4) The age is adjusted to

$$T_i = T_{i-1} + (ED - TD_{i-1})/D \; . \quad (4.29)$$

5) The procedure is repeated until the increment $(ED - TD_{i-1})/D$ becomes sufficiently small or the T_i converges by iteration (Ikeya 1986).

The results of a generalized TD calculation per 1 ppm of ^{238}U are shown in Figures 4.6 and 4.7 in a normal and logarithmic scale, respectively, for $k = 0$ and 0.1. If the effect of the uranium–series radioactivity is dominant, one can estimate an approximate age from the ED obtained by the additive dose method. The ED/C_U may be regarded as TD/C_U and the age can be obtained using the figure.

4.5.3 Open System Disequilibrium : Uranium Accumulation

Incorporation of uranium and other radioactive elements is not a simple process. Uranium accumulates rapidly from the environment in a short time at the first stage to a certain level following organic decomposition and then slowly increases linearly as a function of time (Seitz and Taylor 1974).

(a) Early uptake model : The content of uranium is considered to be constant as shown in Figure 4.8 (a) if uptake of uranium at the initial stage occurs in a short time relative to the age. The TD evaluation for a closed system can be applied in this case.

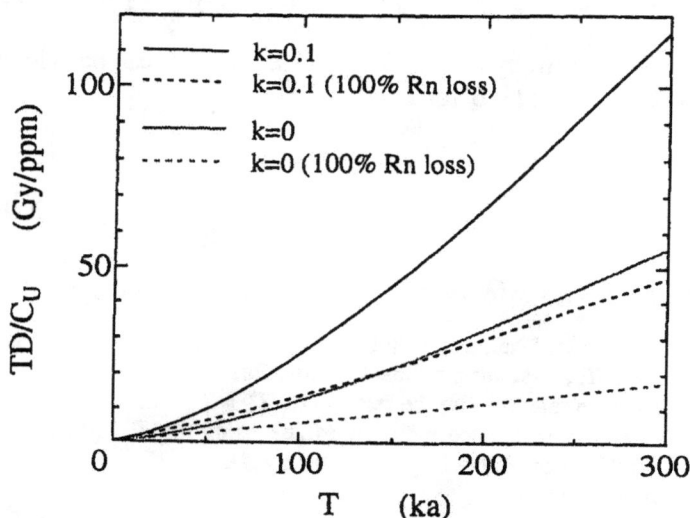

Figure 4.6 TD versus age T relation for 1 ppm of ^{238}U due to ^{238}U–series disequilibrium based on Eq. (4.28) for $k = 0$ and 0.1. The dashed straight lines are for 100% Rn loss.

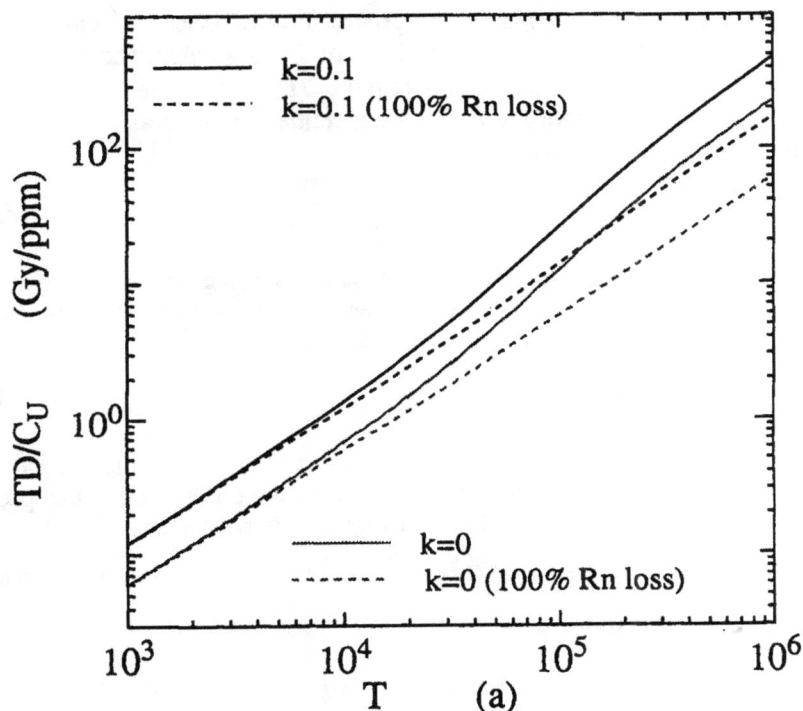

Figure 4.7 Logarithmic relationship of the TD versus age T for 1 ppm of ^{238}U due to the ^{238}U–series disequilibrium for $k = 0.1$ and 1.0. The dashed lines are for 100% Rn loss.

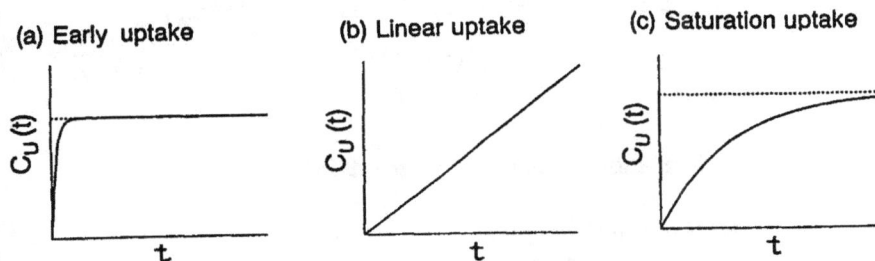

Figure 4.8 Models of uranium uptake. (a) An early uptake (constant C_U), (b) a linear accumulation (uptake at a constant rate), $C_U(t) = (C_U^\bullet/T)t$ and (c) a saturation uptake, $C_U(t) = C_{Us}(1 - e^{-ct})$.

(b) *Linear (constant) uptake model* : Uptake of uranium at a constant rate as shown in Figure 4.8 (b) may be a process for old bones containing a large amount of uranium. A simple hypothesis of linear uranium accumulation for a fossil bone was introduced to obtain the TD versus age relation (Ikeya 1982). The increment of the TD by the increment of the uranium concentration at time t is written as

$$\mathrm{d}\,TD \;=\; \mathrm{d}\,C_U(t)\,TD\,(T-t) \; , \qquad\qquad (4.30)$$

where $TD(t)$ is a function of the total dose considering uranium–series disequilibrium. Thus, the TD_a considering uranium accumulation is written using the initial condition of $C_U(0) = 0$ and $TD(0) = 0$ as

$$TD_a \;=\; \int TD\,(T-t)\,C_U(t)\,\mathrm{d}t \; . \qquad\qquad (4.31)$$

The initial concentration is negligibly small and the concentration may be assumed to have increased linearly from zero at the burial time to the present uranium concentration, $C_U{}^*$. The concentration is written as

$$C_U(t) \;=\; C_U{}^*\,t/T \; , \qquad\qquad (4.32)$$

and therefore, for linear uranium accumulation,

$$TD_a(T) \;=\; \int (C_U{}^*/T)\,TD(t)\,\mathrm{d}t \; . \qquad\qquad (4.33)$$

The calculated TD_a for 1 ppm of $C_U{}^*$ is given by integrating the $TD(T)$ in Eq. (4.28) as

$$TD_a(T) = D\,T/2 + (D_{34}/\lambda_{34})(r_0 - 1)[\,1 - (1 - e^{-\lambda_{34}T})/\lambda_{34}\,T\,]$$
$$- (D_{30}/\lambda_{30})[\,1 - (1 - e^{-\lambda_{30}T})/\lambda_{30}\,T\,] - [\,D_{30}(r_0 - 1)/\lambda_{34}\,]$$
$$\{1 - [\lambda_{30}{}^2(1 - e^{-\lambda_{34}T}) - \lambda_{34}{}^2(1 - e^{-\lambda_{30}T})]/[\lambda_{30}\lambda_{34}\,(\lambda_{30} - \lambda_{34})\,T\,]\} \; .$$

$$(4.34)$$

The TD_a versus T relation is given in Figure 4.9 for $C_U{}^* = 1$ ppm.

(c) *Other models of uranium uptake*
     Saturation uptake model :   The rate of uranium uptake may be reduced as the C_U increases. If the time–dependent content $C_U(t)$ changes according to a saturation function, it may be expressed as

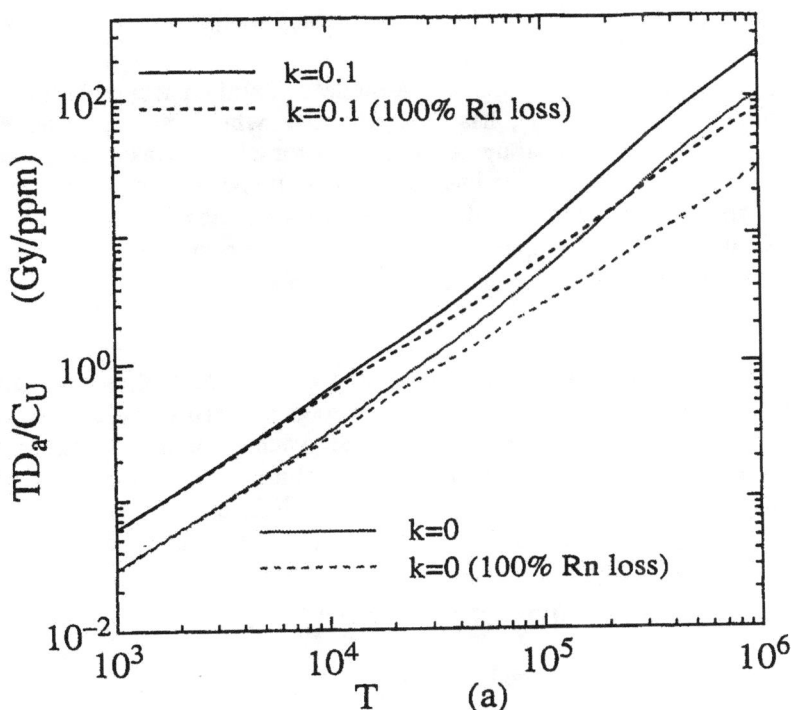

Figure 4.9 Calculated curve of the *TD* versus *T* relation for a linear uranium accumulation during the burial time of *T* years taking ^{238}U–series disequilibrium into account. The present uranium concentration $C_U^* = 1$ ppm.

$$C_U(t) = C_{Us}(1 - e^{-ct}) , \qquad (4.35)$$

where C_{Us} is a saturation value and c is an unknown parameter, both of which depend on the uranium content in the environment and the temperature. The C_U^* may be written as

$$C_U^* = C_U(T) = C_{Us}(1 - e^{-cT}) , \qquad (4.36)$$

as shown in Figure 4.8 (c). The constant uptake model considering the uranium site leads to the equation. The TD_a can be calculated using Eq. (4.31). If the second–order annealing effect of the α–track has to be considered as described in Chapter 3, Eq. (4.33) becomes for the α–ray dose as

$$TD_{\alpha \alpha}(T) = \int D_{\alpha} C_U(t) TD(T-t')/[1 + \ln(T - t')] dt' \quad , \tag{4.37}$$

Polynominal $(t/T)^m$ model : A model of uranium accumulation using a polynomial form, $(t/T)^m$, has been proposed, where $m = 0$ for an early uptake, $m = 1$ for a linear uptake and $m > 1$ for a late uptake (Grün _et al._ 1988). However, a polynominal equation has no physical meaning and is not an appropriate function to describe the accumulation process. The best way would be to use samples which do not involve ambiguous processes. Objects close to a closed system should be sampled.

4.5.4 ESR Isochron Dating

The internal dose can be estimated from the content of radioactive elements. The difficulty in assessing the average external radiation dose can be avoided in principle by the method of ESR isochron dating (Karakostano-glou and Schwarcz 1983), which follows isochron dating procedures often used in the Rb–Sr method. If the dose from ^{232}Th and K_2O is negligible (i.e., $D_{Th, K} T_{ESR} = 0$) as in the case of cave deposits, the _TD_ is expressed as a simple function of C_U,

$$TD = C_U(kTD_{\alpha} + TD_{\beta}) + D_{ex} T_{ESR} \quad . \tag{4.38}$$

When the _TD_ is plotted against the C_U in Figure 4.10, the external dose $D_{ex} T_{ESR}$ is obtained from the point of intersection of the _y_–axis, and the internal dose $(kTD_{\alpha} + TD_{\beta})$ is obtained from the slope. The internal dose may be calculated considering ^{238}U–series disequilibrium given by Eq. (4.28) or in some cases the linear uranium uptake model given by Eq. (4.34). The method was not successful for carbonate deposits since the C_U is more or less the same.

ESR isochron plot of _TD_ against the internal dose rate instead of the C_U was introduced to obtain the external dose for tooth enamel samples. The age T_{ESR} is determined from the slope and $D_{ex} T_{ESR}$ from the abscissa since

$$TD = D_{in} T_{ESR} + D_{ex} T_{ESR} \quad . \tag{4.39}$$

The dose from the dentine to enamel was also included as the D_{in}. Thus, archaeological tooth samples lacking sediments for the D_{ex} determination can be dated (Blackwell and Schwarcz 1993, Blackwell _et al._ 1993). The dose rate is usually time–dependent due to ^{238}U–series disequilibrium. It would be better to use Eq. (4.39) to obtain the T_{ESR} from the slope.

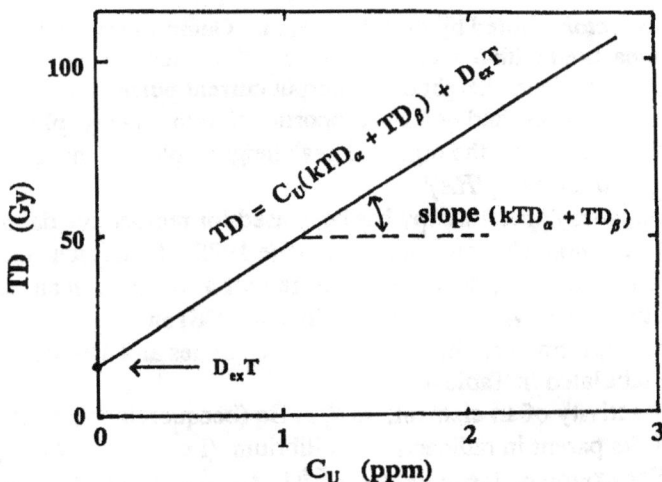

Figure 4.10 Principle of ESR isochron dating. ED's for samples of the same age are plotted as a function of C_U. The slope gives the internal dose and the abscissa gives $D_{ex} T_{ESR}$.

4.6 Methods used to Measure Dose

4.6.1 *Measurements of Radioactive Elements*

Several methods are employed to determine the contents of ^{238}U, ^{232}Th and ^{40}K in geological and archaeological materials and to assess the annual dose rate for the deduction of ESR ages from the experimental ED. Each method has its own merits and convenience. Experimental errors in the dose rate evaluation directly affect the ESR age. To confirm the overall accuracy of the analytical method, the result obtained should be checked by a separate radiation measurement.

Radioactive disequilibrium can be checked by measuring the activities of a parent and a daughter nuclide. The rate of ^{222}Rn loss, N_{loss}, can be determined from the activity ratio of ^{226}Ra and ^{218}Po or ^{214}Pb. The delayed build-up of ^{230}Th can be studied by measuring ^{234}U and ^{230}Th.

(a) *Gamma-ray spectroscopy and neutron activation analysis* : Detection of low background γ-rays has been developed considerably in the last three decades due to the development of semiconductor detectors and digital technology. Gamma-ray photons create electron and hole pairs in photon detectors such as a coaxial *highly pure germanium (HPGe) detector*, a planar *Ge low energy photon spectroscopy (Ge-LEPS)* and a *lithium-doped Ge*

[Ge(Li)] detector cooled by liquid nitrogen. Gamma–ray peaks above 200
keV are measured with a Ge(Li) detector, while those below 100 keV with
Ge–LEPS.   The peak height of the output current pulse due to the created
charges of electrons and holes is proportional to the γ–ray photon energy.
The apparatus that sorts the digitized peak heights, photon energies is called a
"pulse height analyzer (PHA)".

Gamma–ray spectroscopy has been used for radioactive dating such as
^{230}Th/^{234}U method (Komura and Sakanoue 1985, Ivanovich and Harmon
1982).   A typical γ–ray spectrum measured for a stalactite in an early stage
of ESR dating is shown in Figure 4.11 (Ikeya 1978) and the selected energies
in the γ–ray spectra used for evaluation of U–series and Th–series elements
and K are tabulated in Table 4.7.

The activity of an element, $\lambda_i N_i$ in Bq (becquerel; counts/s), is equal
to that of its parent in radioactive equilibrium (i.e., $\lambda_i N_i = \lambda_{i-4} N_{i-4}$ for α
decay).   For example, the activity of ^{238}U, $\lambda_{38} N_{38}$, can be determined from
the γ–ray spectrum of ^{234}Pa since $\lambda_{38} N_{38} = \lambda_{34} N_{34}$.   If the activity of ^{226}Ra
is not equal to that of ^{222}Rn, it is considered that ^{222}Rn has emanated from

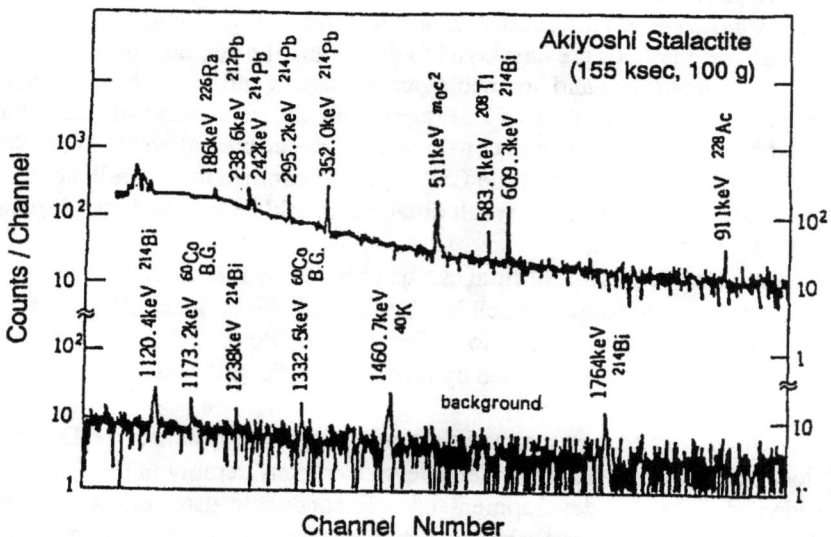

Figure 4.11 Gamma–ray spectrum of 100 gram stalactite with a Ge–Li detector (courtesy by Dr. K. Miyake at Kobe Marine College).
Table 4.7 Selected γ–ray lines and nuclides used for the determination of

Table 4.7 Selected γ–ray lines and nuclides used for the determination of ^{238}U, ^{232}Th and ^{40}K compiled from published data.

Element	Energy (keV)	Nuclide	Branching ratio (%)	Coincident photons
Uranium (U)	13.0, 13.3 *	(X–rays)		
	25.6 *	^{231}U		
	46.5	^{210}Pb	63.3	
	63.3	^{234}Th	3.83	0.00
	67.7	^{230}Th		
	92.4	^{234}Th	5.39	0.00
	92.8	^{234}Th		
	185.72	^{235}U	2.67	0.00
	186.05	^{230}Th	0.09	0.00
	186.10	^{226}Ra	3.50	0.00
	295.09	^{214}Pb	18.59	0.00
	351.87	^{214}Pb	36.5	0.01
	609.31	^{214}Bi	44.8	0.95
	1120.27	^{214}Bi	14.8	1.00
	1764.49	^{214}Bi	15.36	0.02
Thorium (Th)	**238.63**	^{212}Pb	43.65	0.00
	241.0	^{224}Ra	4.04	0.00
	383.32	^{228}Ac	11.25	0.24
	583.19	^{208}Tl	30.55	1.45
	911.21	^{228}Ac	26.6	0.18
	968.97	^{228}Ac	16.17	0.18
	2614.53	^{208}Tl	35.86	1.39
Potassium (K)	**1460.83**	^{40}K	10.67	0.00

after Guibert and Schvoerer (1991) and 91/92 Detectors & Instruments for Nuclear Spectroscopy, EG & G ORTEC.

the sample. The rate of ^{222}Rn loss, N_{loss} in %, can thus be determined as

$$N_{loss} = (1 - \lambda_{22}N_{22}/\lambda_{26}N_{26}) \times 100.$$

Gamma–ray peaks to obtain the contents of ^{238}U, ^{232}Th and ^{40}K as well as the rate of radon (^{222}Rn) or thoron (^{220}Rn) loss are as follows:

1) The ^{238}U content is estimated from the ^{234}Th peak at 63.3 keV.
2) The ^{230}Th content from the ^{230}Th peak at 67.7 keV. ^{238}U–series disequilibrium can be determined from the ratio of $^{230}Th/^{238}U$.
3) The ^{232}Th content from the ^{228}Ac peaks at 383.32 and 911.21 keV.
4) The rate of radon(^{222}Rn) loss from the ^{214}Pb (351.84 keV) or ^{214}Bi (609.31 keV) and ^{226}Ra (186.10 keV).
5) The rate of thoron (^{220}Rn) loss from the ^{212}Pb (238.63 keV) and

^{228}Ac (911.21 keV).

6) ^{40}K is directly estimated from the peak at 1461 keV.

Detection limits of these elements are 0.2 ~ 0.3 ppm for U and Th and 0.02% for K using 10 g sample, 32 mmϕ ×10 mm Ge–LEPS and Ge(Li) with 16% efficiency for the measuring time of 10^4 min (Komura and Sakanoue 1985).
    In addition to these, the following peaks are observed:

$e^+ + e^-$	: *Positron annihilation*	: 511 keV.
	(Positrons are created by photons with energies higher than 1.022 keV)	
^{137}Cs	: *A–bomb or accident nuclide*	: <u>31.8, 32.2, 36.4</u> and 661.6 keV
^{60}Co	: *Involved in modern iron*	: 1173 and 1332.5 keV

    Gamma–ray spectroscopy is also used for the detection of γ–ray peaks in *"neutron activation analysis (NAA)"* to determine the content of elements such as Mn and Ag that have a large *cross section* for neutron activation.

(b)   *Alpha–ray spectroscopy : rate of radon loss* : Measurement of radionuclides necessary for annual dose rate estimation has been carried out by α–ray spectroscopy after radiochemical separation of radionuclides.   Figure 4.12 shows α–ray spectra of (a) uranium isotopes and (b) ^{230}Th in the ^{238}U–series as well as the standard ^{228}Th and its daughters. The contents of ^{238}U and ^{234}U can be determined to give the ratios of ^{234}U/^{238}U and ^{230}Th/^{234}U.
    The ratios of ^{226}Ra/^{238}U and ^{210}Pb/^{238}U indicate the degree of ^{222}Rn loss and the α–recoil dissolution of daughter elements over 100 years (five times the half–life of ^{210}Pb) in the embedded site. The ratios of ^{226}Ra/^{238}U and ^{214}Pb/^{238}U indicate the rate of *radon loss during the measurements*.

(c)   *Track detection (fission tracks and α–tracks)* : The uranium content in a small amount of sample (~ 20 mg) can be determined with *fission track* detection.   The sample sandwiched by mica plates is irradiated by neutrons. The induced fission causes tracks on the mica. Chemical etching with hydrofluoric acid allows observation of the number of fission tracks and then the content of ^{235}U (hence ^{238}U) based on the number of incident neutrons (*fluence*).   Details are described in Chapter 1.
    The number of α–rays emitted from the sample can be counted directly with a *solid state nuclear track detector (SSNTD)* using cellulose nitrate (CN) films for an estimation of the dose rate.   Kodak–Pate type LR–115 film consists of a 12 ~ 13 μm red–colored CN layer on a 100 μm polyester support.

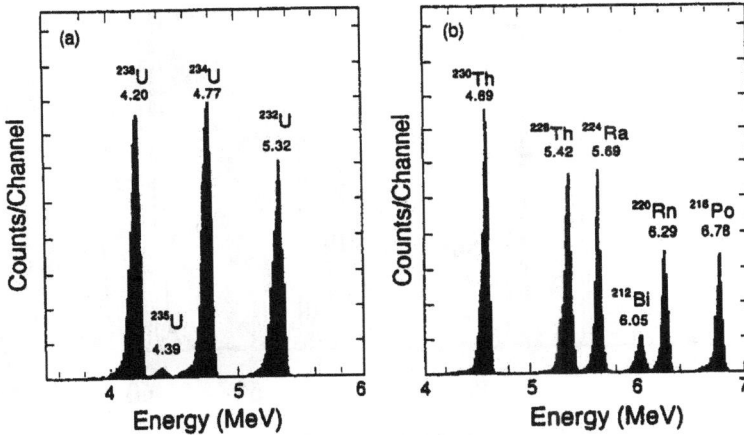

Figure 4.12   Alpha–ray spectra of radiochemically separated (a) uranium isotopes with added standard ^{232}U, and (b) ^{230}Th and the daughter elements (^{224}Ra, ^{220}Rn, ^{216}Po and ^{212}Bi) of added standard ^{228}Th (courtesy by Dr. K. Komura at Low Level Radioactive Laboratory, Kanazawa University).

The film is directly attached to the sample, kept in an evacuated container for a few months and then etched with 6N NaOH at 60°C to observe the tracks.   The total counts from ^{238}U– and ^{232}Th–series disintegration are thus obtained. A plastic film CR–39 (allylglycal–carbonate) is also used.

(d) *Mass spectrum* : *Secondary ion mass spectrometry* (SIMS) or an *ion microanalyzer* (IMA) can be used for the analysis of constituent and impurity atoms in materials. The principle is to bombard and sputter the surface of the material with accelerated ions. The sputtered ions are then led to a mass analyzer system. The sensitivity of SIMS is extremely high and allows the detection of impurities at ppb (10^{-9}) levels.   Figure 4.13 shows a SIMS spectrum of Choukoutien bones. As the sensitivity is high, almost all elements including peaks of uranium and uranium oxide ions are detected .

(e) *Chemical analysis* : Determination of ^{238}U, ^{232}Th and K_2O contents by chemical analysis is the method used in many laboratories. Chemical analysis is generally sensitive enough to determine at ppm levels.   Flame spectroscopy is famous for determining the potassium content. The content of uranium and thorium is determined by optical absorption at 660 and 766.5 nm, respectively, after the chemical separation of the elements.

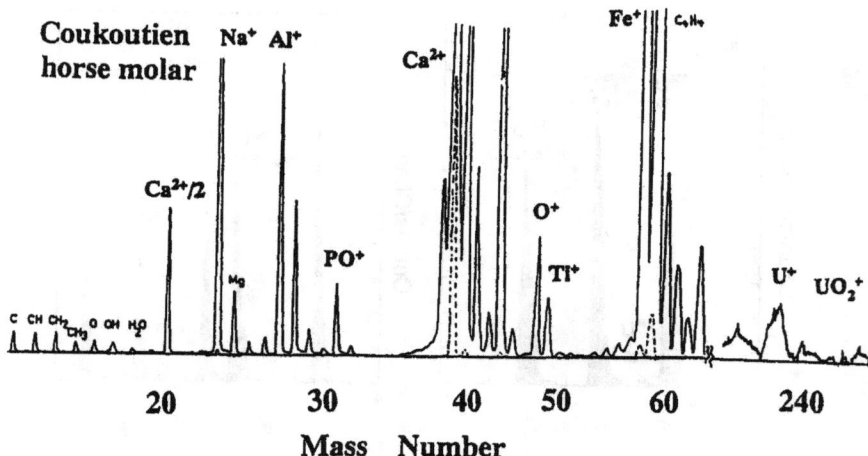

Figure 4.13   SIMS (secondary ion mass spectrometry) of Choukoutien bones for uranium detection. Oxygen ions bombard and sputter the sample. The mass numbers of UO and UO_2 are used for the determination of uranium content (courtesy by C. Ueda and J. Okano, Osaka university).

4.6.2   *External Radiation Measurement*

(a)   *Thermoluminescence dosimeter (TLD)* :   Commercial TLD elements are $CaSO_4$ doped with a rare earth element, Tm. Sometimes, LiF phospher is used as a TLD element. The TLD should be exposed in situ for at least a year to average out variation, mainly the water content in the sediment or soil. Water absorbs γ–rays and affects the γ–ray dose rate. The external dose rate D_{ex} at the sample position should be measured before a large–scale excavation, since the disturbed soils in the environment affect the dose rate. The TLD should be set in a 1 mm Cu–pipe (corresponding to 0.37 mm Pb shielding) to eliminate β–radiation. In ordinary measurements at a homogeneous radiation environment such as a large coral, the result obtained by TLD gives an overall good concordance with the γ–ray dose rate estimated from the content of ^{238}U, ^{232}Th and K and the cosmic ray dose rate.

(b)   *Field γ–ray spectrometer* :   Although a portable γ–ray surveymeter used in radiation facilities is manufactured to detect much higher dose rate than the background natural radiation, some good apparatus may be used at an ordinary environment where natural radiation dose rate is not so low as in a limestone cave. Recently, a scintillation–type surveymeter using a semiconductor photondetector became available. Gamma–ray spectra are also measured in–situ at the sampling site. The measured D_{ex} at that time does not necessarily agree with the dose rate by TLD which is the accumulated and averaged one. Problems are noted by comparing the dose rate obtained by TLD and a field γ–ray spectrometer (Lyons *et al.* 1993). An inhomogeneous radiation environment such as a cave had better be avoided in such a comparison.

(c) *Solid state dosimeter* :   A semiconductor diode–type detector operated by a battery is popular as a pocket dosimeter in monitoring the accumulated dose.   One can detect the background radiation dose in a few hours.   It should be noted that all these measurements are for the D_{ex} at that moment, and that variation in the water content is not averaged out.

4.6.3   *Constancy of Annual Dose Rate*

The environmental γ–ray dose rate depends on the content of ^{238}U, ^{232}Th and K_2O in soils and rocks in the vicinity (Eisenbud 1973).   Except for extreme cases of monazite sand beach in Brazil and a uranium mine, the terrestrial dose rate is generally 0.3 ~ 0.6 mGy/a, often indicated by mSv/a (1 Sv = 1 Gy for γ–rays) at 1 m above the ground.   The dose rate from cosmic rays is 0.2 ~ 0.3 mGy/a.   Hence, the external γ–ray dose rate of 0.5 ~ 1.0 mGy/a is relatively constant in an ordinary environment at the ground level.

The dose rate measured by a TLD is 0.1 ~ 0.3 mGy/a in a limestone cave covered by pure carbonate cave deposits but reaches close to 0.5 ~ 1 mGy/a at the site of flow–in soils (Ikeya 1976).   Since the annual dose rate for age determination in ESR dating involves the internal dose rate of both α– and β–rays, the average apparent dose rate is as low as 0.5 mGy/a or as high as 4 mGy/a.   However, it is rarely lower than 0.5 mGy/a or higher than 5 mGy/a at an ordinary place.   If an apparent dose rate obtained from the *ED* for a sample of known age is outside this range, the *ED* or the known age must be re–examined considering the above dose rate.   The order of magnitude of the age can be deduced from the obtained *ED* and information on the environment.   Some examples of the dose rate are shown in Table 4.8 to indicate an approximate value.

The radiation dose from atmospheric radioactivities, mostly ^{222}Rn and its daughters in aerosols has to be considered in a special place such as a dry cave where ^{222}Rn activity is extraordinarily high (see Section 5.5 of Chapter 5).

Table 4.8   Examples of the annual dose rate for various samples.

Sample	D	D_{in}	D_{ex}	(mGy/a)
Speleothems (limestone cave)	0.2~1.5 (~1.0)	0.1~0.5	0.1~1.0	
Shells in sands	0.6~1.0	0.1	0.5~1.0	
in clays	~1.5	0.1	0.5~1.5	
Coral reef	0.6	~0.1	~0.5	
Quartz in sediments	~2~	~0	~2~	

4.7 Summary

Radiation assessment including the absolute analysis of the uranium and thorium concentration is necessary for the determination of the ESR age. Units, conversion factors and numerical data for additive irradiation and assessment of the annual dose rates for α-, ß- and g-rays are given in tables together with some theoretical formula. The method of radiation assessment for uranium-series disequilibrium or delayed build-up of 230Th is given in detail since the clock time of most materials for ESR dating starts when ^{238}U is included in the materials. Assessment of the early-uptake, linear-uptake saturation uptake, as well as leaching of radioactive elements were made for age determination. Details for practical applications of ESR dating are given in each chapter. For the annual dose evaluation and radioactivity data, TL dating works should be consulted (Aitken 1985).

References

Aitken M. J. (1985): *Thermoluminescence Dating* (Academic Press, London).

Aitken M.J., Clark P.A., Gaffney C.G. and Lovborg L. (1985): Beta and gamma gradients in sample and soil. Nuc. Tracks 10, 647-653.

Apers D., Debuyst R., DeCanniere P., Dejehet F. and Lombard E. (1981): A criticism of the dating by ESR of the stalagmitic floors of the Caun de l'Arago at Tautavel. *Absolute Dating and Isotope Analysis in Prehistory- Methods and Limits* ed. by DeLumley H. and Labeyrie J., 533-550.

Attix F. H. (1986): *Introduction to Radiological Physics and Radiation Dosimetry* (John Wiley and Sons, New York).

Bell W. T. (1979): Thermoluminescence dating: radiation dose rate data. *Archaeometry* 21, 243-245.

Bell W.T. (1980): Alpha dose attenuation in quartz grains for thermoluminescence dating. *Ancient TL* 12, 4-8.

Blackwell B.A. and Schwarcz H.P. (1993): ESR isochron dating for teeth: a brief demonstration in solving the external dose calculation problem. *Appl. Rad. Isot.* 44, 243-252.

Blackwell B.A., Schwarcz H.P., Schick K. and Toth N. (1993): ESR dating tooth enamel from the Paleolithic site at Longola, Zambia. *Appl. Rad. Isot.* 44, 253-260.

Blackwell BAB, Leung HY, Skinner AR, et al. (2000): External dose rate determinations for ESR dating at Bau de l'Aubesier, Provence, France. *Quat. Intern.* **68**, 345-361.

Blackwell BAB, Skinner AR, Blickstein JIB (2001): ESR isochron exercises: how accurately do modern dose rate measurements reflect paleodose rates? *Quat. Sci. Rev.* **20**,1031-1039.

Bowman S.G.E. (1976): Thermoluminescence dating: the evaluation of radiation dosage. Unpublished Ph.D. thesis, University of Oxford, Oxford.

Brennan B.J., Lyons R.G. and Phillips S.W. (1991): Attenuation of alpha particle track dose for spherical grains. *Nucl. Tracks* **18**, 249-253.

Brennan B.J. (2000): Systematic underestimation of the age of samples with saturating exponential behaviour and inhomogeneous dose distribution. *Radiat. Meas.* **32**, 731-734.

Brodsky A. B. ed. (1978): *CRC Handbook of Radiation Measurement and Protection* Vol. 1, Physical Science and Engineering Data, 229-265.

DeCanniere P., Debuyst R., Dejehet F., Apers D. and Grün R. (1986): ESR dating: a study of ^{210}Po-coated geological and synthetic samples. *Nucl. Tracks* **11**, 211-220

Dolo J. M., Lecerf N., Mihajlovic V., Falgueres C. and Bahain J.J. (1996): Contribution of ESR dosimetry for irradiation of geological and archarological samples with a 60Co panoramic source. *Appl. Radiat. Isot.* **47**, 1419-1421.

Eisenbud M. (1973): *Environmental Radioactivity* (Academic Press).

Fain J, Soumana S, Montret M, et al. (1999): Luminescence and ESR dating α- beta-dose attenuation for various grain shapes calculated by a Monte-Carlo method. *Quat. Sci. Rev.* **18**, 231-234.

Fleming S. J. (1979): *Thermoluminescence Techniques in Archaeology* (Claredon Press, Oxford).

Goslar T. and Hercman H. (1988): TL and ESR dating of speleothems and radioactive disequilibrium in the uranium-series. *Quat. Sci. Rev.* **7**, 423-427.

Grün R., Schwarcz H. P. and Chadam J. (1988): ESR dating of tooth enamel: coupled correction factor for U-uptake and U-series disequilibrium. *Nucl. Tracks* **14**, 237-241.

Grün R. (1989): Electron spin resonance (ESR) dating. Quat. Intern. **1**, 65-109.

Grün R., Radtke U. and Omura A. (1992): ESR and U-series analyses on corals from Huon peninsula, New Guinea. *Quat. Sci. Rev* **11**, 197-202.

Guibert P. and Schvoerer M. (1991): TL dating: low background gamma spectroscopy as a tool for the determination of the annual dose. *Nucl. Tracks* **18**, 231-238.

Hennig G. J. and Grün R. (1983): ESR dating in Quaternary geology. *Quat. Sci. Rev.* **2**, 157-238.

ICRP (1983): Annales of the ICRP Radionuclide Transformation: Energy and Intensity of Emission. Vol. 1,11-13.

Hubbell J.H. (1982): Photon mass attennation and energy absorption coefficient from 1 keV to 20 MeV. *J. Appl. Radiat. Isot.* **33**, 1269-1290.

Ikeya M. (1976): Natural radiation dose in Akiyoshi cavern and on karst plateau. *Health Phys.* **31**, 76-78.

Ikeya M. (1978): Electron spin resonance as a method of dating. *Archaeometry* **20**, 147-158.

Ikeya M. (1982): A model of linear uranium accumulation for ESR age of Heidelberg (Mauer) and Tautavel bones. *Jpn. J. Appl. Phys.* **21**, L690-L692.

Ikeya M. and Ohmura K. (1983): Comparison of ESR ages of corals at marine terraces with ^{14}C and ^{230}Th/^{234}U ages. *Earth Planet. Sci. Letters* **65**, 34-38.

Ikeya M. and Ohmura K. (1984): ESR age of Pleistocene shells by radiation assessment. *Geochem. J.* **18**, 11-17.

Ikeya M. (1986): *Dating and Age Determination of Biological Materials.* ed. Zimmerman M.R. and Angel S.L. (Croom Helm, London). Chapt. 3 Electron spin resonance.

Ikeya M., Tani A. and Yamanaka C. (1995): ESR isochrone dating of fracture age utilizing the grain size dependency of the doe correction factor. *Jpn. J. Appl. Phys.***34**, 334-337.

Ivanovich M. and Harmon R. S. (1982): Uranium-Series Disequilibrium Application to Environmental Problems in the Earth Science (Oxford Univ. Press, Oxford).

Karakostanoglou I. and Schwarcz H.P. (1983): ESR isochron dating. *PACT.* **19**, 391-398.

Kigoshi K. (1971): Alpha recoil thorium-234 into water and the uranium-234/uranium-238 disequilibrium in nature. *Science* **173**, 47-48.

Komura K. and Sakanoue M. (1985): Gamma-ray spectroscopy for ESR dating. ESR Dating and Dosimetry (Ionics, Tokyo), 9-17.

Lederer C. M. and Shirley V. S. (1978): Table of Isotopes (Wiley & Sons, New York).

Liritzis Y. and Kokkoris M. (1992): Revised dose-rate data for thermoluminescence/ESR dating. *Nucl. Geophys.* **6**, 423-443.

Lyons R. G. (1988): Determination of alpha effectiveness in the ESR dating using nuclear accelerator techniques: methods and energy dependence. *Nucl. Tracks* **14**, 275-280.

Lyons R. G. and Brennan B. J. (1989): Alpha-particle effectiveness in ESR dating: energy dependence and implications for dose-rate calculations. *Appl. Radiat. Isot.* **40**, 1063-1070.

Lyons R.G. and Brennan B. J. (1991): Alpha/gamma effectiveness ratio of calcite speleothems. *Nucl. Tracks* **18**, 223-227.

Lyons R.G., Brennan B.J. and Readhead M. (1993): Games with gammas: problems in environmental gamma dose determination. *Appl. Radiat. Isot.* **44**, 131-137.

Mudelsee M., Barabas M. and Mangini A. (1992): ESR dating of Quaternary deep-sea sediment core RC-17-177. *Quat. Sci. Rev.* **11**, 181-189.

Mejdahl V. (1979): Thermoluminescence dating: beta-dose attenuation in quartz grains. *Archaeometry* **21**, 61-72.

Nambi K.S. (1979): On ESR dating of minerals. *Jpn. J. Appl. Phys.* **18**, 2319-2320.

Nambi K.S.V. and Aitken M.J. (1986): Annual dose conversion factors for TL and ESR dating. *Archaeometry* **28**, 202-205.

Ogoh K., Ikeda S. and Ikeya M. (1993): Confirmation and recalculation of dose-rate data for TL and ESR dating. *Advances in ESR Appl.* **9**, 22-28.

Olley JM, Roberts RG, Murray AS (1997): Disequilibria in the uranium decay series in sedimentary deposits at Allen's Cave, Nullarbor Plain, Australia: Implications for dose rate determinations. *Radiat. Meas.* **27**, 433-443.

Rambaud X, Fain J, Miallier D, et al. (2000): Annual dose assessment for luminescence and ESR dating: evaluation of nuclides mobility. *Radiat. Meas.* **32**, 741-746.

Prescott J. R. and Stephan L. G. (1982): The contribution of cosmic radiation to the environmental dose for thermoluminescence dating - latitude, altitude and depth dependencies. *PACT* **6**, 17-25.

Radtke U. and Grün R. (1988): ESR dating of corals. *Quat. Sci. Rev* **7**, 465-470.

Seitz M.G. and Taylor R.E. (1974): Uranium variation in dated fossil bone series from Olduvasi Gorge, Tanzania. *Archaeometry* **16**, 129-135.

Wintle A. G. (1978): A thermoluminescence dating study of some Quaternary calcite. *Canad. J. Earth Sci.* **15**, 1977-1986.

Yang Q, Rink WJ, Brennan BJ (1998): Experimental determinations of beta attenuation in planar dose geometry and application to ESR dating of tooth enamel. *Radiat. Meas.* **29**, 663-671.

Yokoyama Y., Nguyen H.V., Quaegebeur J.P. and Poupeau G. (1982): Some problems encountered in the estimation of annual dose-rate in the electron spin resonance dating of fossil bones. *PACT* **6**, 103-115.

Chapter 5

CaCO$_3$

– Cave Deposits –

Practical ESR dating started from dating carbonate stalactites and stalagmites in caves. Speleology physics, the study of caves using physics, is the field where ESR dating was born and bred. Caves were cradles for Quaternary ESR dating. The paleoclimates might be clarified from the ESR dating of CaCO$_3$, cave deposits.

5.1   Introduction

Carbonates are common minerals in nature. The basic constituent unit in all carbonate minerals is the CO_3^{2-} molecular ion. Calcium carbonate ($CaCO_3$) has two main crystal structures of calcite and aragonite. **Calcite** with trigonal (rhombohedral) symmetry is the only thermodynamically stable form of pure $CaCO_3$ at room temperature and one atmosphere pressure. It constitutes most limestones, marble (metamorphic carbonate rock), chalk (soft foraminiferal limestone), tufa and cave deposits (stalactite, stalagmite and travertine). Elementary defects induced by ionizing radiation are an electron center, CO_3^{3-} and a hole center, CO_3^-. An oxygen vacancy with an electron, actually the CO_2^- molecular ion, is also formed by radiation.

Aragonite with an orthorhombic structure is less common and less stable than calcite. It appears in special environments characterized by low temperature and high pressure. Some shells, pearls, corals and marine muds of biological origin are aragonitic. The crystal structure and ESR dating of aragonitic fossils are described in Chapter 6. **Vaterite** (μ–$CaCO_3$) is another polymorph of $CaCO_3$. It is a metastable phase observed only immediately after deposition and soon changes into calcite. **Dolomite** is also a carbonate mineral described as $CaMg(CO_3)_2$, but a considerable amount of Mn^{2+} is involved due to the similar ionic radius of Mn^{2+} and Mg^{2+}. As the ESR signal of Mn^{2+} is intense and therefore interferes with the radiation–induced signal, carbonates with a high concentration of Mn^{2+} are not suitable for ESR dating.

Secondary mineral deposits formed by precipitation and recrystallization contain relatively small amounts of impurities because the impurities remain in the surrounding water rather than in the crystalline solid. The *distribution coefficient (repulsion coefficient)*, $K = C_s/C_1$, the ratio of the impurity content in the solid (C_s) to that in the liquid (C_1), or to that in the melt ($K = C_s/C_m$) is generally smaller than 1.0. Hence, secondary minerals are relatively pure and can thus be better samples than primary minerals with a large amount of impurities. This is the reason why secondary mineral deposits were successfully dated with ESR (Ikeya 1975). Nature has purified minerals by crystallization making them good objects of ESR dating.

"How old is the stalactite ?" and *"How fast does it grow ?"* are questions answered incorrectly by cave guides in most tourist show–caves. The growth rate of a stalactite may be determined by accurately dating carbonates at several positions within a stalactite. The age and the growth rate of stalactites and the radiation environment in limestone caves are described in this chapter together with studies of the paleoenvironment. These underground

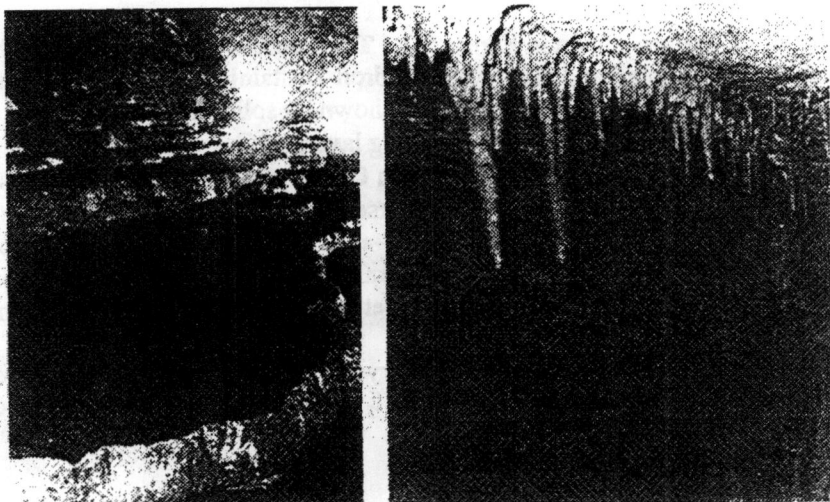

Figure 5.1 (Left) Rimstone formation called *"Hundred Dishes"* in Akiyoshi Cave in Japan. (Right) Arrays of stalactites from the ceiling fissure in Mammoth Cave in Kentucky, USA. Relatively pure secondary deposits are ideal objects of ESR dating.

wonderlands untouched by human beings reveal the history of our past environment (Hudson *et al.* 1976) with a time scale derived from the ESR measurement of radiation effects. Typical carbonate formation in the limestone cave at Akiyoshi Cave in Japan and Mammoth Cave in Kentucky is shown in Figure 5.1.

In this chapter, ESR dating of secondary deposits of CaCO$_3$ is described with the emphasis on limestone cave deposits related to speleology, the study of caves. Models of *radiation-induced defects* in CaCO$_3$ are described mostly from physics and chemistry studies using single crystals or synthetic powders, both pure and doped with impurities. ESR dating of a *stalactite*, a secondary deposit in a limestone cave, is described in detail because Quaternary ESR dating started from cave deposits.

5.2   Crystal Structure and Defects in Calcite (CaCO$_3$)

5.2.1   *Crystal Structure*

Calcite is a rhombohedral form of calcium carbonate (CaCO$_3$) which consists of the divalent cation, Ca^{2+}, and the divalent triangle planar molecular anion, CO$_3^{2-}$ with sp^2 hybrid molecular orbitals. Three different unit cells of calcite are shown in Figure 5.2. The *true unit cell* shown by dashed lines in (a) is a tall and thin rhombohedron containing two formula units. The crystal cleaves into a *flatter cell* shown by solid lines in (a) and the *hexagonal unit cell* shows that alternating layers of Ca^{2+} and CO$_3$$^{2-}$ ions lie along the c–axis as shown in (b). Each CO$_3$$^{2-}$ group within a layer has a common orientation, which is 180° reversed in each adjacent layer. The C–O bonds are coincident with the three a–axis. In a CO$_3$$^{2-}$ ion, the carbon atom occupies the center of an equilateral triangle of oxygens. The bond angle of O–C–O is 120° and the length between the carbon and the oxygen is 0.128 nm.

Although the growth of a single crystal is difficult, synthetic powder calcite can be prepared by mixing CaCl$_2$ and Na$_2$CO$_3$ solutions at room temperature.

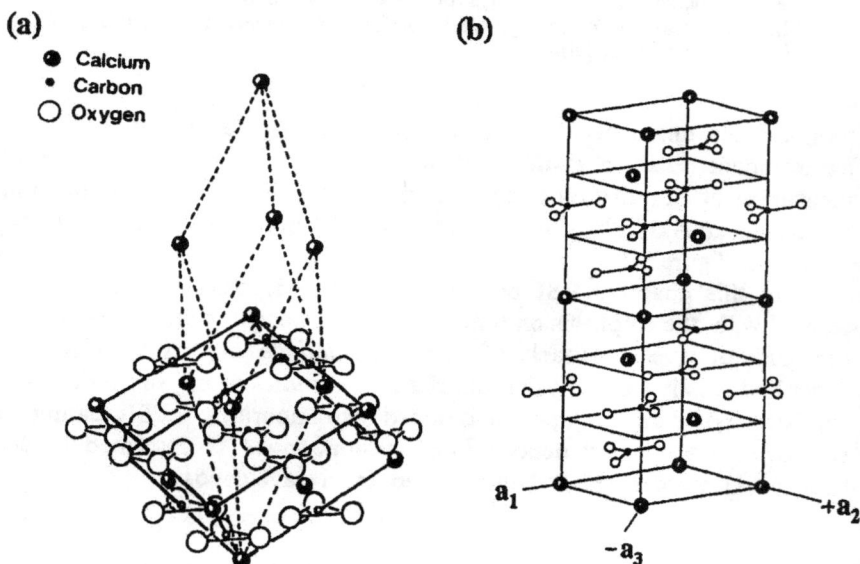

Figure 5.2 The crystal structure of calcite. (a) The true rhombohedral unit cell is shown by dashed lines and the cleaved cell by solid lines. (b) The hexagonal unit cell; $a_1 = a_2 = a_3 = 0.499$ nm and $c = 0.171$ nm (Tucker and Wright 1990).

5.2.2 Radiation–Induced Intrinsic Defects

An ESR spectrum of pure synthetic calcite powder after γ–irradiation is shown in Figure 5.3 in which three molecular ions, CO_3^{3-}, CO_3^- and CO_2^- are observed. The g factors expected from a single crystal are indicated.

The triangle planar (D_{3h} symmetry) CO_3^{2-} molecular ion in $CaCO_3$ is easily ionized by radiation. When an ionized electron is trapped by CO_3^{2-} with 24 electrons, an *electron–excess type defect*, CO_3^{3-} with 25 electrons is formed, and when an electron is deleted, an *electron–deficient defect*, CO_3^- with 23 electrons is formed (Serway and Marshall 1967a, b). These defects are called an "*electron center*" and a "*hole center*", respectively. A planar CO_3^{2-} traps an electron and distorts into a pyramidal CO_3^{3-} with C_{3v} symmetry, while the hole–type CO_3^- still keeps a planar molecular form with D_{3h} symmetry (Walsh 1953). Molecular orbital schemes of CO_3^{3-}, CO_3^{2-} and CO_3^- based on the general scheme of AB_3 molecules with 25, 24 and 23 electrons for atoms A (A = B, C, Si, N, P, As and S) and B (B = O) are shown in Appendix A2.

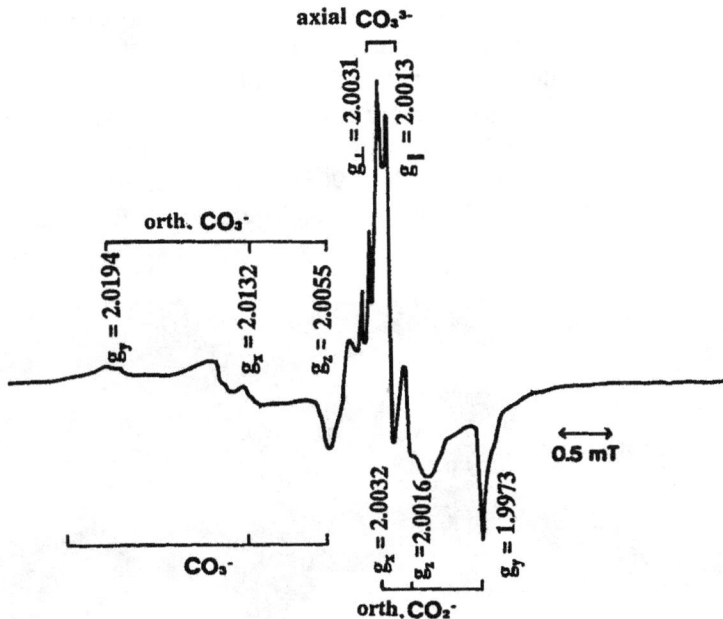

Figure 5.3 ESR spectrum of pure synthetic CaCO₃ powders after γ–irradiation at room temperature. The signal positions of CO_3^{3-}, CO_3^- and CO_2^- centers are indicated with their g factors.

The shift of g factors from the free electron value $g_e = 2.0023$ and the hyperfine (hf) constant A can be estimated theoretically if the molecular orbital calculation is made. The obtained results are $g_\parallel \geq g_\perp \geq g_e$ and a large A for CO$_3^{3-}$, while $g_\perp > g_e \geq g_\parallel$ with a small A for CO$_3^-$.

Molecular orbitals of CO$_3^{2-}$, CO$_3^{3-}$ and CO$_3^-$ in vacuum have recently been calculated with a supercomputer and their graphic presentation of *highest occupied molecular orbital* (HOMO) is shown in Figure 5.4. The calculation must be made in an ionic crystal under Madelung potentials (electrostatic Coulomb potentials due to cations and other anions), but the simplified *ab–initio* calculation in vacuum allows us to visualize molecular orbitals.

Another electron center, CO$_2^-$ molecular ion, is a triatomic species expressed as a AB$_2$ type molecule having 17 electrons and shows a bent configuration (C_{2v} symmetry) with a bond angle of approximately 134° (Marshall *et al.* 1964). Molecular orbital schemes of the AB$_2$ type molecules are shown in Appendix A2.

The CO$_3^{3-}$ and CO$_3^-$ centers are stable only at low temperatures. However, these centers associate with some impurities and are electrically stabilized by Coulomb attraction, for example, in the form of Y^{3+}–CO$_3^{3-}$ by associating with a trivalent Y^{3+} ion (Marshall *et al.* 1968). The CO$_2^-$ ion is an electron center similar to but more stable than CO$_3^{3-}$ (Marshall *et al.* 1964). It also associates with impurities, for example, in the form of CO$_2^-$ – –F$^-$ complex (Marshall and McMillan 1968). Lattice distortion around an impurity or dislocation also stabilizes these molecular ions. Electrons at the

Figure 5.4 A graphic presentation of highest occupied molecular orbitals of (a) CO$_3^{3-}$ with 25 electrons, (b) CO$_3^{2-}$ with 24 electrons and (c) CO$_3^-$ with 23 electrons (courtesy by T. Takada, NEC Fundamental Research Lab.).

CO$_3^{3-}$ center move to some impurity sites and are temporarily trapped and then released to recombine with the CO$_3^-$ center. The long lasting *phospho-rescence* of CaCO$_3$ after the termination of radiation exposure at room temperature is due to recombination between these two centers. ESR param-eters of the carbonate radicals (CO$_3^{3-}$, CO$_3^-$ and CO$_2^-$) in CaCO$_3$ are summarized in Table 5.1.

5.2.3   *Impurity-Related Defects*

Impurities are involved in the carbonate lattice. Divalent molecular anions such as the planer SO$_3^{2-}$ rather than the tetrahedral SO$_4^{2-}$ substitute for CO$_3^{2-}$ in CaCO$_3$. Impurity-related paramagnetic defects, especially those associated with anions are formed by radiation and observed in both synthetic and natural carbonates. Ionizing radiation produces electrons and holes which form CO$_3^{3-}$ and CO$_3^-$, self-trapped electron and holes and impurity-related SO$_3^-$ as shown in Figure 5.5 (Kai and Miki 1991). Quartet hf splitting signals by ^{33}S ($I = 3/2$, abundance 0.75%) has been observed, confirming the association of S.

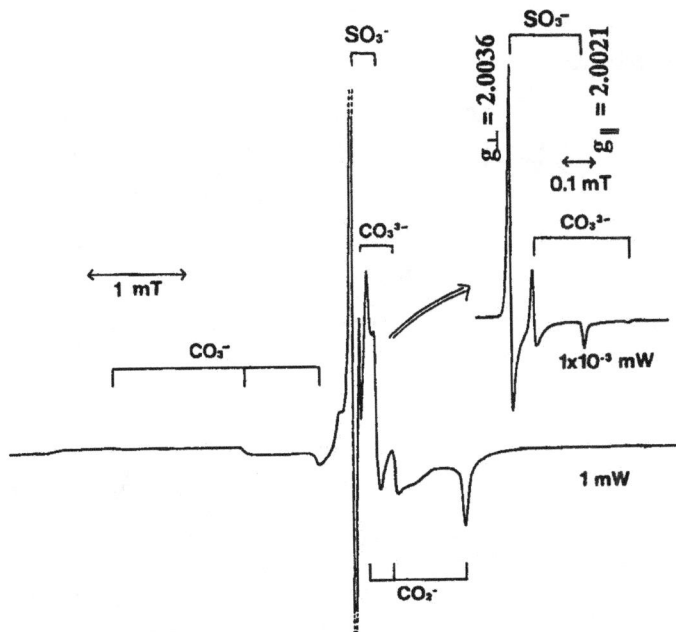

Figure 5.5   ESR spectrum of synthetic calcite doped with SO$_3^{2-}$ and irradiat-ed at room temperature. SO$_3^-$ radicals are formed (Kai and Miki 1991).

Table 5.1 ESR parameters of carbonate radicals in irradiated carbonate minerals.

Species	Materials [a]	g factors [b]				A-tensor (mT)			Ref.
		g_{zz}	g_{xx}	g_{yy}	g_{av} [c]	A_{zz}	A_{xx}	A_{yy}	
CO_3^{3-} axial	N-calcite	$g_∥$=2.0013	$g_⊥$=2.0031		2.0025	17.12	11.13 (^{13}C)		1
	S-calcite	$g_∥$=2.0016	$g_⊥$=2.0032		2.0027				2
	S-calcite		$g_⊥$=2.0034				$A_⊥$=10.9		3
CO_3^{3-}-Y^{3+}	N-calcite	2.0012	2.0024	2.0038	2.0025	18.83	12.61	12.61 (^{13}C)	4
						0.31	0.35	0.29 (^{89}Y)	
HCO_3^{2-}	N-calcite	2.00197	2.00387	2.00502	2.00362	2.78	3.55	4.48 (^{1}H)	5
CO_3^{3-}-Li^+	S-calcite	$g_∥$=2.0012	$g_⊥$=2.0031		2.0025	0.40	0.26 (^{7}Li)		6
CO_3^- orth.	N-calcite	2.0055	2.0132	2.0194	2.0127				7
	N-calcite	2.0164	2.0142	2.0126	2.0144				8
		2.0163	2.0143	2.0128	2.0145				8
	S-calcite	2.0056	2.0100	2.0180	2.0112	1.37	1.08	1.03 (^{13}C)	3
	S-calcite	2.0056	2.0100	2.0210	2.0122	1.37	1.08	1.05 (^{13}C)	3
	S-calcite	2.0055	2.0092	2.0222	2.0123	1.39	1.08	0.95 (^{13}C)	3
	KHCO₃	2.0066	2.0086	2.0184	2.0112	1.4	1.0	1.0 (^{13}C)	9
CO_3^- axial	N-calcite	$g_∥$=2.0051	$g_⊥$=2.0162		2.0125	1.31	0.94 (^{13}C)		1
CO_3^- iso.	S-calcite·H₂O				g_{iso}=2.0115		A_{iso}=1.14 (^{13}C)		10
CO_2^- orth.	N-calcite	2.0016	2.0032	1.9973	2.0007	17.73	13.46	13.17 (^{13}C)	11
	N-calcite	2.0016	2.0032	1.9971	2.0006				12
		2.0026	2.0018	1.9972	2.0005				8
	S-calcite	2.0015	2.0032	1.9974	2.0007	18.9	15.8	15.6 (^{13}C)	3
CO_2^- axial	N-calcite	$g_∥$=2.0032	$g_⊥$=1.9994		2.0007	13.46	15.59 (^{13}C)		13
	N-calcite	$g_∥$=2.0028	$g_⊥$=1.9991		2.0003				8
	N-calcite	$g_∥$=2.0031	$g_⊥$=1.9994		2.0006				8
	S-calcite	$g_∥$=2.0032	$g_⊥$=1.9995		2.0007	$α$-irradiation			14
CO_2^- iso.	S-calcite				g_{iso}=2.0008		A_{iso}=14.9 (^{13}C)		3
	S-calcite·H₂O				g_{iso}=2.0006		A_{iso}=14.8 (^{13}C)		10
	aragonite				g_{iso}=2.0007				15
CO_2^--F^-	N-calcite	2.0022	2.0035	1.9980	2.0012	4.39	9.31	4.39 (^{19}F)	16
CO_2H	KHCO₃	2.0012	2.0031	1.9971	2.0005	14.2	17.9	13.3 (^{13}C)	9
CO_2^- orth.	HCOONa	2.0014	2.0032	1.9975	2.0007	19.5	15.5	15.1 (^{13}C)	17

a) N: natural, S: synthetic.
b) For orthorhombic symmetry g_{zz}, g_{xx} and g_{yy} and for axial symmetry $g_∥$ and $g_⊥$ are given.
c) Calculated using $g_{av} = [(g_{zz}^2 + g_{xx}^2 + g_{yy}^2)/3]^{1/2}$ or $g_{av} = [(g_∥^2 + 2g_⊥^2)/3]^{1/2}$ for convenience (see Section 2.4.3).

Reference: 1) Serway and Marshall 1967a; 2) Eachus and Symons 1968; 3) Debuyst *et al.* 1991; 4) Marshall *et al.* 1968; 5) Cass *et al.* 1974; 6) Bacquet *et al.* 1975; 7) Serway and Marshall 1967b; 8) Rossi *et al.* 1985; 9) Chantry *et al.* 1962; 10) Debuyst *et al.* 1993; 11) Marshall *et al.* 1964; 12) Rossi and Poupeau 1989; 13) McMillan and Marshall 1968; 14) Debuyst *et al.* 1990a; 15) Miki and Kai 1990; 16) Marshall and McMillan 1968; 17) Overnall and whiffen 1961.

Nuclear spin and abundance: see Table 5.2.

Table 5.2 ESR parameters of anion impurity radicals in irradiated carbonate minerals.

Species	Materials [a]	g factors [b]				A–tensor (mT)			Ref.
		g_z	g_{xx}	g_{yy}	g_{av} [c]	A_z	A_{xx}	A_{yy}	
SO_3^- orth.	aragonite [d]	2.0042	2.0038	2.0025	2.0035	11.4 (^{33}S)			1
SO_3^- axial	S-calcite [d]	$g_\parallel$=2.0024	$g_\perp$=2.0038		2.0033	13.2 (^{33}S)			1
	S-$CaCO_3$	$g_\parallel$=2.0021	$g_\perp$=2.0036		2.0031	no hf line of ^{13}C			2
SO_3^- iso.	S-calcite·H_2O				g_{iso}=2.0032	A_{iso}=2.12 (^{1}H)			3
	S-$CaCO_3$				g_{iso}=2.0031				2
	aragonite [d]				g_{iso}=2.0034				1
SO_2^- iso.	aragonite [d]				g_{iso}=2.0060				1
	S-$CaCO_3$				g_{iso}=2.0057				2
CSO_2^{3-}	N-calcite	2.001	2.063	2.008	2.024				4
NO_3^{2-}	S-calcite	$g_\parallel$=2.0027	$g_\perp$=2.0066		2.0053	6.68	3.42 (^{14}N)		5
		$g_\parallel$=2.0017	$g_\perp$=2.0060		2.0046	6.80	3.44 (^{14}N)		6
NO_3	S-calcite	$g_\parallel$=2.0039	$g_\perp$=2.0225		2.0163				5
NO_2^{2-}	S-calcite	2.0087	2.0029	2.0039	2.0052	3.64	0.45 (^{14}N)		6
PO_4^{2-}	N-calcite	2.0122	2.0072	2.0033	2.0076	1.85	1.87	2.00 (^{31}P)	7
PO_3^{2-}	N-calcite	$g_\parallel$=2.00165	$g_\perp$=2.00162		2.00163	81.8	63.9	(^{31}P)	7
PO_2^{2-}	N-calcite	1.9960	2.0036	2.0007	2.0001	4.5	8.8	3.8 (^{31}P)	8
PO_2	N-calcite	2.0011	2.0045	2.0036	2.0031	0.94	0.66	0.49 (^{31}P)	8
$CO_3·PO_2^{3-}$	N-calcite	2.0024	2.0382	2.0090	2.0166	0.62	0.52	0.62 (^{31}P)	9
AsO_3^{2-}	N-calcite	$g_\parallel$=2.00162	$g_\perp$=2.00195		2.00184	93.76	75.25	(^{75}As)	10
AsO_2^{2-}	N-calcite	2.0150	1.9991	1.9910	2.0017	5.39	21.93	6.20 (^{75}As)	11

a), b) and c) see Table 5.1.
d) The *g*–values by these authors were systematically large by 0.0003 or so.
Reference: 1) Kai and Miki 1992; 2) Barabas 1992; 3) Debuyst *et al.* 1993; 4) Marshall *et al.* 1974; 5) Eachus and Symons 1968; 6) DeCanniere *et al.* 1988; 7) Serway and Marshall 1966b; 8) Bershov *et al.* 1968; 9) Serway *et al.* 1969; 10) Serway and Marshall 1966a; 11) Marshall and Serway 1969.

Nuclear spin and abundance:
^{1}H (*I*=1/2, 99.985%); ^{7}Li (*I*=3/2, 92.5%); ^{13}C (*I*=1/2, 1.10);
^{14}N (*I*=1, 99.634%); ^{19}F (*I*=1/2, 100%); ^{31}P (*I*=1/2, 100%);
^{33}S (*I*=3/2, 0.75%); ^{75}As (*I*=3/2, 100%); ^{89}Y (*I*=1/2, 100%).

ESR signals of defects associated with impurity ions can be used to monitor the content of impurities at the sensitivity level that cannot be reached by ordinary analytical techniques. The origin and meaning of these signals in relatively impure natural carbonates have been discussed (Apers *et al*. 1981) and models of defects have been established using synthetic carbonates (Kai and Miki 1991, 1992, Debuyst *et al*. 1990a, b, 1991, Barabas 1992). ESR parameters of radiation–induced defects related to anion impurities in CaCO$_3$ are summarized in Table 5.2.

5.2.4   *Characterization of ESR Signals Relevant to Dating*

An ESR spectrum of natural calcite (*speleothem*) is shown in Figure 5.6. Among many radiation–induced defects listed in Tables 5.1 and 5.2, the following four species have been observed during dating studies and their

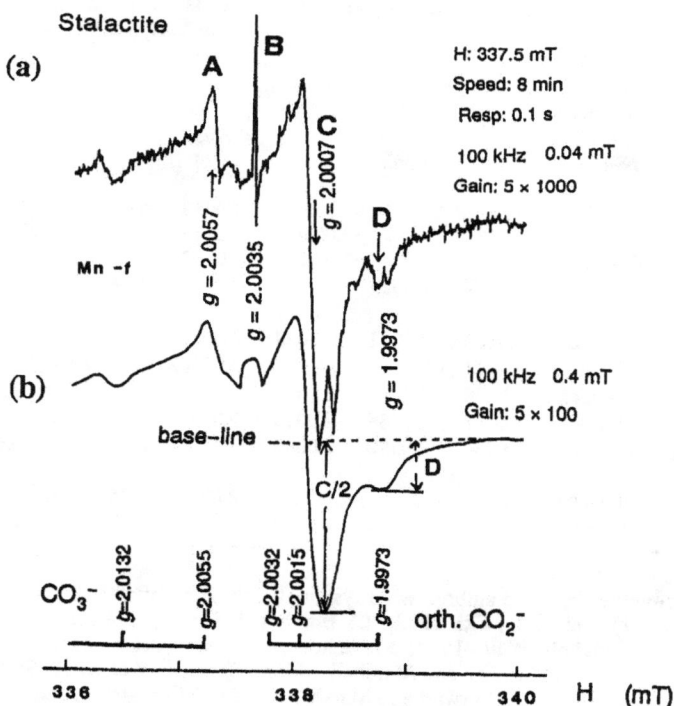

Figure 5.6 ESR spectrum of a stalactite from Akiyoshi, Japan with a 100 kHz field modulation of (a) 0.04 mT and (b) 0.4 mT. The radiation–induced signals are designated by A, B, C and D. The amplitude for the high field side of signals C and D indicated in (b) is used for the signal intensity.

origins in calcitic and aragonitic carbonates have been clarified through the systematic studies on defects in the CaCO$_3$ lattice by the Belgian, German and Japanese groups. These have been termed **signals A, B, C and D** (Ikeya and Ohmura 1981) which correspond to h_1, h_2, h_3 and h_4, respectively (Yokoyama *et al.* 1981). The terms, A, B, C and D are used in this book and are explained as follows. ESR parameters and characteristics of these defects are summarized in Table 5.3, and microwave power dependence and thermal behavior of these signals are shown in Figure 5.7.

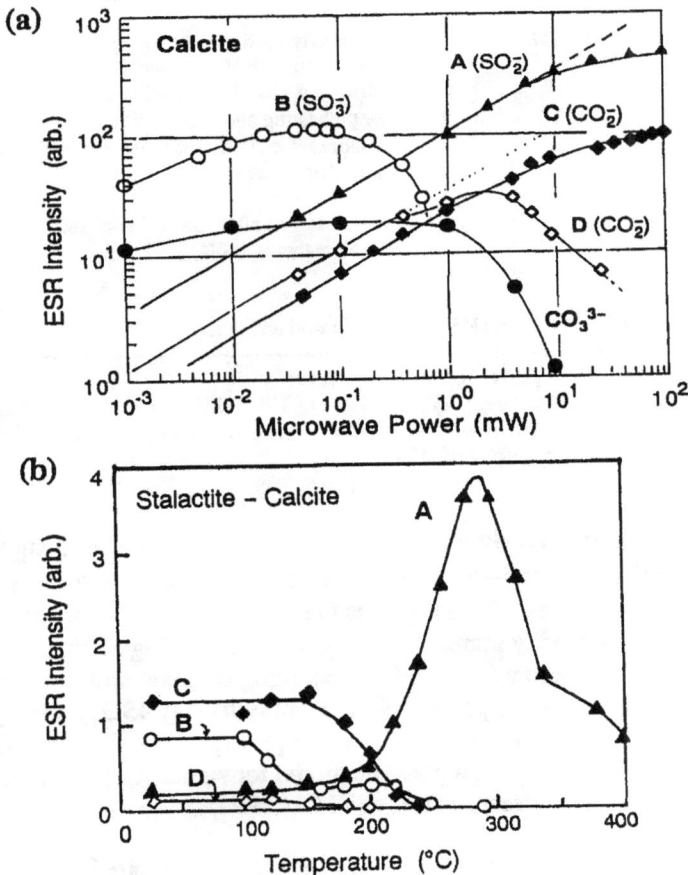

Figure 5.7 (a) Microwave power dependence and (b) isochronal annealing curves of signal A, B, C and D in calcite.

Table 5.3 Models and characteristics of ESR signals relevant to dating.

Models	Notation	g factor	Microwave power (mW) [a]	Characteristics	Reference
SO_2^- (free rotation)	A	$g_A = 2.0057$	>20	thermally formed at the expense of signal B. hardly enhanced by simple γ–irradiation but enhanced by heating before or after irradiation. disappears <~200K and annealed out at 300°C.	1
SO_3^- (axial)	B	$g_B = g_\perp = 2.0036$ ($g_{\parallel} = 2.0021$)	<0.1	less visible in old samples. disappears < 230K and annealed out at ~200°C.	2
CO_2^- (free rotation)	C	$g_C = 2.0007$	>20	practically used as a dating signal. associated with H_2O molecule. formation yield by γ–irradiation is reduced by preheating and dehydration. disappears < 200K, changing into orthorhombic. annealed out at ~200°C.	3
CO_2^- (orthorhombic)	D	$g_D = g_{yy} = 1.9973$ ($g_{xx} = 2.0016$) ($g_{yy} = 2.0032$)	2	still observed after irradiation at 180°C and remains ≤ 150K.	4
Organic radicals		$g = 2.0040 - 2.0045$		humic acid and others	5

a) Conventional microwave power for measurement.
Reference: 1) Kai and Miki 1992; 2) Kai and Miki 1992, Miki *et al.* 1993, Barabas 1992; 3) Debuyst *et al.* 1990a, 1991, 1993; 4) Debuyst *et al.* 1990a, b, 1991; 5) DeCanniere *et al.* 1985.

Signal A: $g_A = 2.0057$ (freely rotating SO_2^-) : The intensity of signal A is not clearly enhanced by γ–irradiation or even is decreased in some cases, especially for old samples. This signal is thermally the most stable of those discussed, and the intensity increases after thermal annealing at the expense of signal B. It is isotropic with no ^{13}C hf coupling constant and is not saturated up to high microwave power. SO_2^- in anhydrite ($CaSO_4$) has $g_{zz} = 2.0058$, $g_{xx} = 2.0022$ and $g_{yy} = 2.0092$, which gives $g_{av} = 2.0057$ (see Table 7.1). A freely rotating SO_2^- is a possible model for signal A, though no hf line due to ^{33}S has been observed (Kai and Miki 1992).

Signal B: $g_B = 2.0036 = g_\perp$ (axial SO_3^-) : Signal B in calcite is axial SO_3^- at $g_{\parallel} = 2.0021$ and $g_\perp = 2.0036$, giving $g_{av} = 2.0031$, while signal B in aragonite is isotropic SO_3^- at $g_{iso} = 2.0031$ (see Chapter 6). Synthetic

$CaCO_3$ doped with SO_3^{2-} clearly shows an enhancement of this signal after γ–irradiation (Miki *et al.* 1993). Signal A is produced by heating at the expense of this signal (Kai and Miki 1992). The intensity of signal B is enhanced by γ–irradiation but the growth shows a tendency to saturation as the radiation dose increases for Holocene samples. Since this signal has a tendency to early saturation with increased microwave power, the micro-wave power level has to be low (0.01 mW).

Signal C: $g_C = 2.0007$ (freely rotating CO_2^-) : This signal has practically been used as a *dating signal* because (1) the absolute intensity per unit weight of the sample is proportional to the known age and (2) γ–ray irradia-tion enhances the intensity. This isotropic signal with $g_{av} = 2.0007$ is associ-ated with the *freely rotating* CO_2^- and it becomes *orthorhombic* at 77K (Hughes and Soo 1970, Debuyst *et al.* 1990a, b). Since signal C is actually a composite signal consisting of several overlapping signals, especially at the low field side, the height of the amplitude at the high field side from the baseline is recommended to be taken as a half of the intensity of signal C (C/2) as shown in Figure 5.6 (b). A high microwave power (50 ~ 100 mW) is preferable to measure this signal (Lyons *et al.* 1993).

The linewidth of CO_2^- is relatively broad (about 1 mT), indicating the presence of some additional element. Carbonates are known to contain water (Gaffey 1988). The decay of this signal by annealing accompanies the release of moisture and hence a model of a rapidly tumbling CO_2^- in a hin-dered state, presumably *hydrated* CO_2^-, is suggested (Murata *et al.* 1993, Debuyst *et al.* 1993). Doping with Al^{3+}, Sc^{3+} and Y^{3+} increases the forma-tion efficiency of signal C by X–irradiation. The increase is considered to be related with abundant water molecules incorporated together with these ions. On the other hand, doping with Fe^{3+} reduces the formation efficiency of signals B and C by γ–irradiation. The suppression effect is explained by the *valance change* of iron (Kai and Miki 1993).

Signal D: $g_D = 1.9973 = g_{yy}$ (orthorhombic CO_2^-) The orthorhombic and axial CO_2^- molecular ions show this signal. A powder derivative signal shows a negative peak. This signal is detected in $CaCO_3$ and other carbon-ates such as $KaHCO_3$ (see Table 7.4) as well as in hydroxyapatite (tooth enamel) (see Tables 8.1 and 8.3). This signal includes a component of unstable broad signals, which gives an *ED* smaller than that of signal C by an additive dose method (Takano and Fukao 1993). A method to anneal out overlapping unstable signals has been proposed as described in Chapter 6.

5.2.5 Alpha–Ray–Induced Defects : Axial CO$_2^-$

The ionizing radiation with different radiation quality of $\alpha-$, $\beta-$ and $\gamma-$ rays produces different types of defects. Figure 5.8 shows the ESR spectra of synthetic CaCO$_3$ irradiated by (a) γ–rays, (b) external α–rays by ^{210}Po–coating and (c) internal α–rays by ^{210}Po doping (DeCanniere et al. 1982, 1986, Debuyst et al. 1990a, b). The obtained g factors indicated in the figure allow identification of the species of defects using Tables 5.1 and 5.2.

The characteristic feature is that CO$_2^-$ with axial g factors of $g_{\parallel} = 2.0032$ and $g_{\perp} = 1.9995$ (rotating along the x–axis perpendicular to the O–C–O plane; see Figure 8.4 and Table 8.2) is observed by α–irradiation.

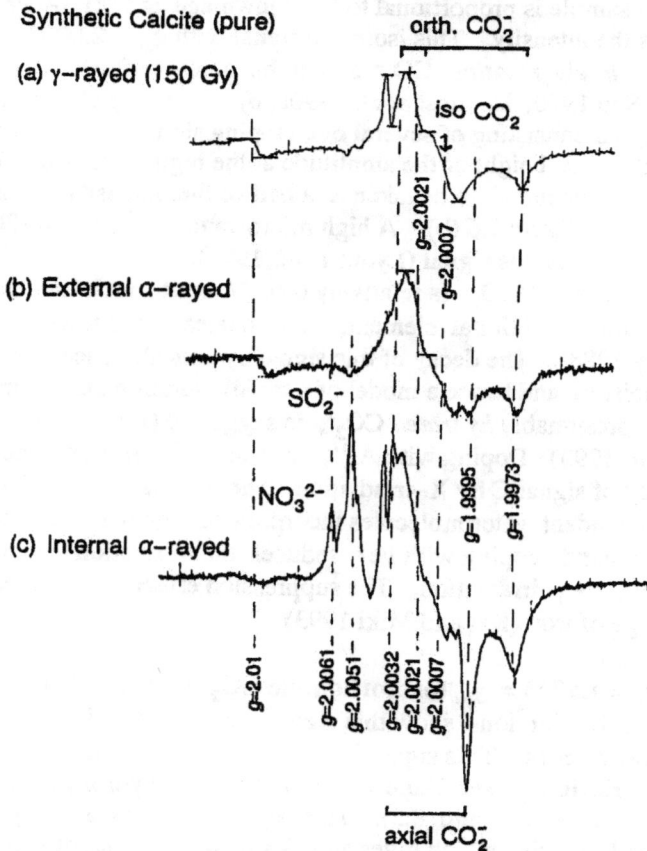

Figure 5.8 ESR spectra of synthetic CaCO$_3$. (a) γ–irradiation, (b) external α–ray irradiation by ^{210}Po–coating and (c) internal α–ray irradiation by ^{210}Po doping (after DeCanniere et al. 1986)

The signal is especially intense after internal α–irradiation. This is presumably because of radiation effects by an α–recoil atom which creates a track
with a high local concentration of defects. The production of isotropic CO_2^-
by α–irradiation is very small. Alpha–ray efficiency (k–value) for signal C
(isotropic CO_2^-) is thus very low ($k = 0.05$; see Table 4.6).

5.3   Thermoluminescence (TL)

If calcite is irradiated and heated, TL peaks are observed at $50 \sim$
$120°C$, $230 \sim 300°C$ and $330 \sim 400°C$ as shown in Figure 5.9 (a). The peak
temperature depends on the heating rate and so differs from one report to
another. If the heating rate is about $5 \sim 6°C/s$, the latter two peaks will be
around $285°C$ and $360°C$ (Meldin 1964, Ikeya 1978a, Miki and Ikeya
1978). The first TL peak at $50 \sim 120°C$ is associated with the migration of
the CO_3^{3-} electron center, which has a decay time of a few hours at room
temperature and is only observed immediately after γ–irradiation. The
second peak may be due to electrons released from freely rotating CO_2^-
(signal C at $g = 2.0007$) or from CO_3^{3-} associated with a trivalent impurity.
The third peak might be associated with the decay of signal A at $g = 2.0057$
(SO_2^-) judging from annealing characteristics. The lifetime of defects

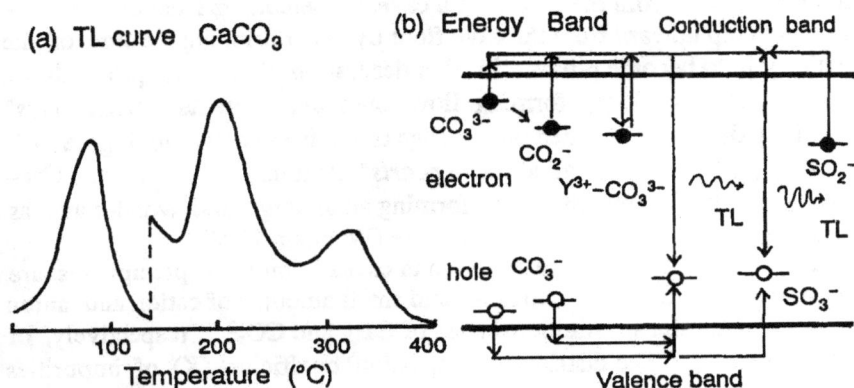

Figure 5.9 (a) Thermoluminescence (TL) curve of a stalactite. (b) The
energy levels of both hole and electron type defects indicate TL processes
from low temperature.

responsible for the second and third peaks are ~ 0.5 Ma and ~ 2 Ma, respectively, at 15°C in the estimation using Eq. (3.18).

An *energy band scheme* of simple defects in $CaCO_3$ and the processes of TL are shown in Figure 5.9 (b). A hole center observed at room temperature after CO_3^- decay at low temperature would be associated with some impurity. The most stable form is SO_3^- (signal B) in most $CaCO_3$.

5.4   Speleothems
5.4.1 *Mechanism of Carbonate Growth*

Karst areas have been developed by water dissolution of limestone rocks. *Limestone* caves have been formed underground along the path of groundwater. Slightly acidic water containing carbon dioxide dissolves the limestone and comes into a cave from the ceiling, where gaseous carbon dioxide is released from the water, depositing $CaCO_3$. The reaction of carbonate dissolution and the back reaction of its deposition are written as

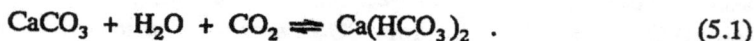

$$CaCO_3 + H_2O + CO_2 \rightleftharpoons Ca(HCO_3)_2 \ . \tag{5.1}$$

Naturally, this reaction is temperature dependent.

A thin hollow carbonate tube called *"straw"*, *"macaroni"* or *"tubular stalactite"*, is first formed by waterdrops and has a diameter of the drop. Next, conical cylindrical deposits of carbonate called *"stalactites"* are formed and hang from the ceiling of a cave. Speleothems called *"stalagmites"* develop upward from the cave floor by water dripping. Some of the deposits join to become pillars. Further decoration of the cave proceeds on the cave wall and floor, forming flowstones and curtains. *"Rimstones"* resembling dishes are formed on the edge of pools as shown in Figure 5.1. Round egg-like pebbles called *"cave pearls"* are found in the pools. Carbonate deposits decorate the cave, forming an underground wonderland as shown in a picture book on limestone caves (Waltham 1974).

Detrital speleothems are common in caves. Impurity precipitates are incorporated into calcite as inclusions and small amounts of cation and anion impurities enter the substitutional sites of Ca^{2+} and CO_3^{2-}, respectively, of the calcite lattice. The distribution (repulsion) coefficient (K) of impurities has extensively been studied in crystal growth. Impurities such as SO_3^{2-}, PO_4^{3-} and NO_3^- as well as organic molecules such as *humic acid* are included in calcite. The source of organic molecules or anions in limestone cave deposits could be *"cave guano"*, droppings from bats or other biological species living in a cave (Moore and Nicholas 1964). The content of these ions reflects the environment of the cave at the time of crystallization. Hence, the *paleoenvironment* can be studied from the variation of impurity ions in cave deposits, while the age can be determined with ESR using signal C due to freely rotating CO_2^-. The saturation levels of the impurity-related centers by heavy irradiation will reflect the content of impurities.

5.4.2 *ESR Spectra*

(a) *Mn^{2+} signals* : Figure 5.10 shows ESR derivative spectra of (a) a limestone and (b) a stalactite from the Akiyoshi karst plateau (Carboniferous to Permian). A prominent feature is the *hf sextet* and the *forbidden transitions* ($\Delta m = \pm 1$) associated with Mn^{2+} nuclear spin ($I = 5/2$). The intensity of Mn^{2+} signals is much stronger in a limestone than in a stalactite because of the repulsion of Mn^{2+} in the recrystallization of $CaCO_3$ during stalactite growth. The ESR intensity of radiation-induced radicals in a natural limestone was relatively weak although it had been exposed to natural radiation for about 200 Ma. It is difficult to create defects in the limestone with a high impurity content as the inactive volume, b (see Chapter 3) around Mn^{2+}, Fe^{2+} and Al^{3+} in the calcite lattice absorbs radiation energy (i.e., creates electrons and holes) without forming stable defects. Experimentally, the spectrum

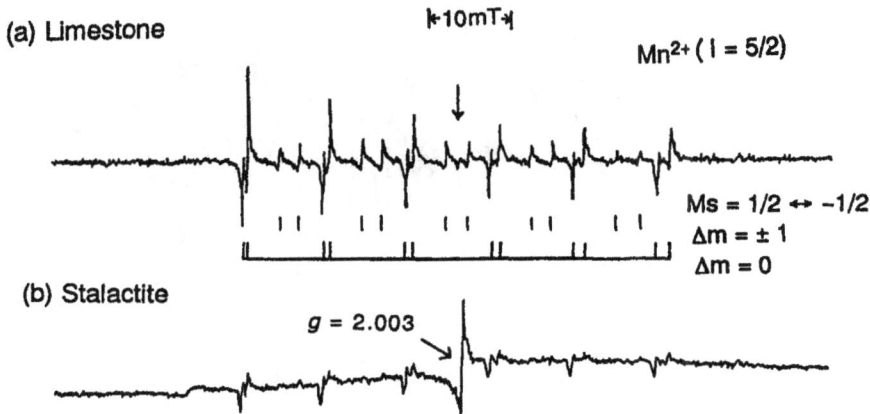

Figure 5.10 ESR spectra of (a) a limestone and (b) stalactite from Akiyoshi, Japan with a 100 kHz field modulation of 0.4 mT. The hf lines of Mn^{2+} associated with the $M = 1/2 \leftrightarrow M = -1/2$ transition are observed with relatively weak forbidden transition lines.

has to be measured with a reduced amplifier gain. The concentration has already reached saturation due to lifetimes of the order of 1 Ma.

(b)   *Humic acid and organic radicals* :   Humic acid and *fluvic acid* with stable radicals show a signal at $g = 2.0045$. This makes the ESR dating of impure carbonate difficult.   Artificial doping of humic acid into synthetic CaCO$_3$ gives a broad signal at $g = 2.0043$ and the enhanced signal of SO$_3^-$ by γ–irradiation (Hennig and Grün 1984, DeCanniere *et al.* 1985).   Periodic variation of organic molecules was detected with *laser–induced lumines–cence* photographed under a microscope giving an *annual growth rate* of 1.58 μm/a and a cycle of the *solar activity change* of 11 years (Shopov *et al.* 1991).

   Organic radicals are sometimes observed in the speleothem irradiated with γ–rays and thermally annealed at 600K.   Figure 5.11 (a) shows a typical organic radical observed in cave deposits after γ–irradiation of 1 kGy (Chong and Fujitani 1984).   The typical hf spectrum of protons shows separation into seven lines (*septet*) due to the interaction of six equivalent protons with further splitting into four lines (*quartet*) with three protons.   The model of such organic molecules is suggested to be *tert–butyl radicals* $[(CH_3)_3C \cdot]$ or a *trimer of methyl radicals* $[(\cdot CH_3)_3]$ stabilized by hydrogen bonding with

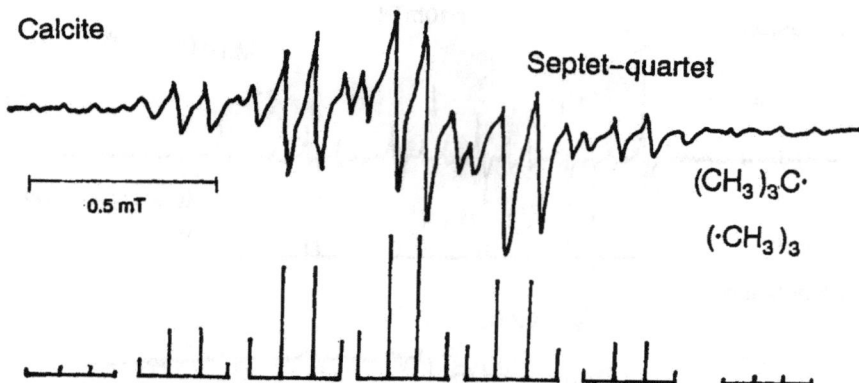

Figure 5.11   ESR spectrum of organic radicals in natural calcite after anneal–ing at 600K following γ–irradiation (Chong and Fujitani 1984). The hf sep–tet–quartet due to interaction with six and three equivalent protons is observed. Models of *tert*–butyl radicals $[(CH_3)_3C \cdot]$ and a trimer of methyl radicals $[(\cdot CH_3)_3]$ were proposed (Ikeya 1984).

oxygens of CO_3^{2-} in $CaCO_3$ (Ikeya 1984). Two of three CH_3- or $CH_3\cdot$ are equivalent and the third one may form hydrogen bonds with three oxygens of CO_3^{2-}. Both wing lines are sometimes too weak to be detected, especially when Mn^{2+} signals interfere. The septet–quartet signal has been observed in petroleum–rich mudstone and carbonates (Ikeya and Furusawa 1989). A similar signal observed in flints has been assigned to stable *perinaphthenyl radicals* (Chandra *et al.* 1988; see Chapter 10). It is not clear whether such organic radicals are actually involved in carbonates.

5.4.3 TD of a Stalactite

The additive dose method described in Chapter 3 (Section 3.2.1) can be used to estimate the *ED* of a stalactite. Figure 5.12 shows the enhancement of the signal intensity at $g_C = 2.0007$ (freely rotating CO_2^-) from several parts of a stalactite from Akiyoshi Cave, Japan by exposure to γ–rays. The signal intensity grows almost linear up to 1 kGy and the *ED*'s are obtained by extrapolation to the zero ordinate.

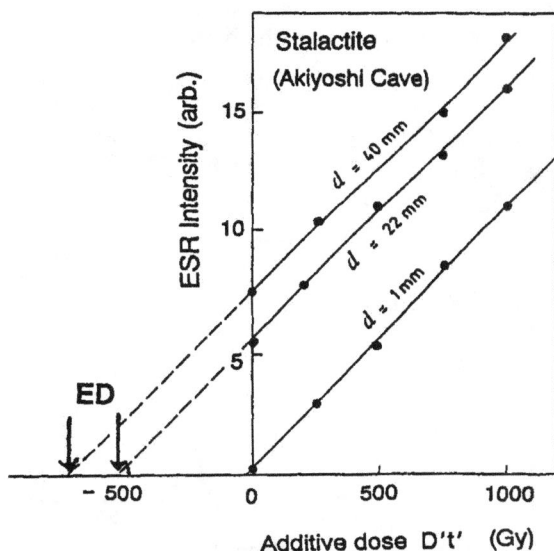

Figure 5.12   The enhancement of the ESR signal intensity as a function of γ–ray dose from a ^{60}Co source. Samples are taken at a distance of d mm from the surface of a stalactite. The *ED* is obtained by extrapolation of the linear enhancement.

Figure 5.13 shows the *ED* as a function of radial distance from the surface of the stalactite. The signal intensity and the *ED* at the surface are very small but increase linearly toward the inside of the stalactite. Since the surface of a stalactite is recent and the calcite becomes older toward the inside, this stratigraphic concordance indicates that the ESR signal intensity can be used for determining at least the relative age. The signal intensity at the center "*straw*" is about 2/3 of that at a position close to the center. This is because the transparent calcite crystal at the straw position was formed later than other positions. A polarization micrograph of a polished thin slice of a stalactite indicates that the c–axis of the calcite crystal is almost parallel to the radial direction except at the center "straw" position, where it is along the longitudinal. The result suggests that the central part inside the straw was crystallized later, which is in concordance with both the *ED* obtained and the model of stalactite growth in speleology (Moore and Nicholas 1964).

The stalactite contains *brown rings* due to inclusion of clay minerals. The age of the brown ring formation can be determined by interpolating the *ED* (or age T_{ESR}) to the position in the stalactite. The ages are concordant with those of other deposits from caves located at coastal areas, which suggests global climate changes causing the brown ring formation.

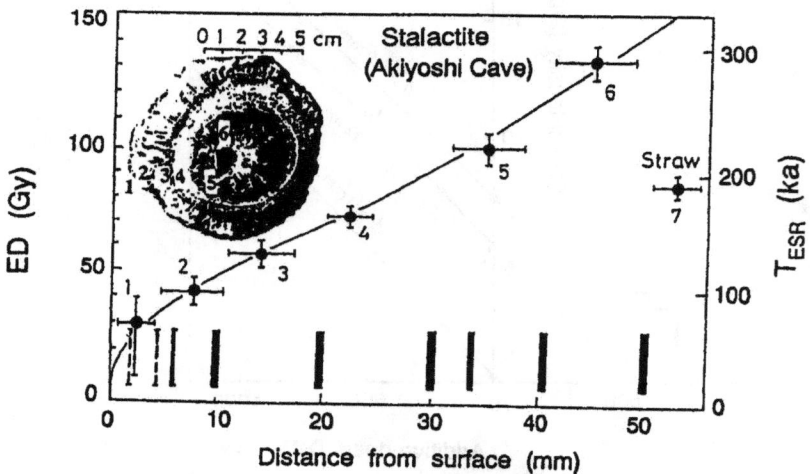

Figure 5.13 The *ED* for a radially sliced stalactite as a function of distance from the surface. The age deduced from the *ED* and the measured external γ–ray dose rate of 0.2 mGy/a is shown on the right ordinate. The positions of brown rings consisting of clay particles are indicated by bars, whose width corresponds to the amount of brown clay inclusion (Ikeya 1985).

5.4.4 Age Deduction from TD using U–Series Disequilibrium

The *TD* obtained from the ESR intensity is next used for the age estimation. The simplest assumption is an *infinite medium model* for cave deposits. The contents of ^{238}U, ^{232}Th and ^{40}K in calcite are 0.43 ppm, 0.39 ppm and less than 10^{-3} %, respectively. Since the dose rate from ^{232}Th–series disintegration is much smaller than that from the ^{238}U–series, the contribution of ^{232}Th and ^{40}K is usually negligible. So, at the early stage of ESR dating, the annual dose rate was simply obtained using Eq. (4.9) assuming *radioactive equilibrium* in ^{238}U–series disintegration without considering *delayed buildup* of ^{230}Th described in Eq. (4.28).

If a *finite medium model* is assumed, the *TD* is expressed as

$$TD = [C_U(kTD_\alpha + TD_\beta) + D_{Th,K}T_{ESR}] + D_{ex}T_{ESR} , \qquad (5.2)$$

where C_U is the ^{238}U content, $D_{Th,K}$ is the dose rate due to ^{232}Th and K_2O, and k is the α–ray efficiency [$k = 0.052 \pm 0.006$ is reported for speleothems by Lyons and Brennan (1991)]. D_{ex} is the external γ–ray dose including the contribution of cosmic rays, and TD_α and TD_β are doses of internal α– and β–rays from ^{238}U, respectively. kTD_α and TD_β are calculated using Eq. (4.28) for the case of ^{238}U–series disequilibrium, assuming the rate of radon (^{222}Ra) loss is negligible. The external radiation dose D_{ex} is measured by TLD ($CaSO_4$: Tm) phosphors hung from a stalactite. Figure 5.14 shows an

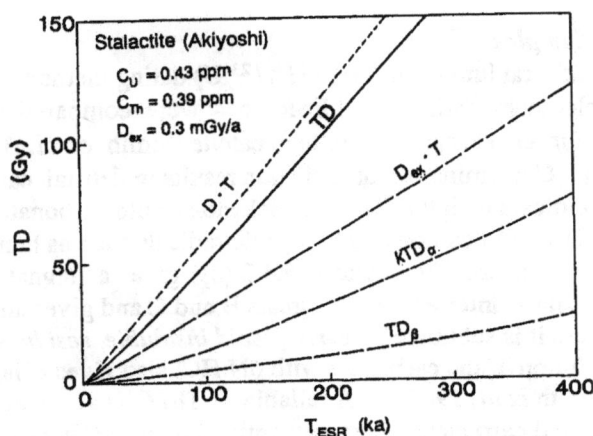

Figure 5.14 An example of *TD* calculation of a stalactite from Akiyoshi using Eq. (5.2). The relation of D versus T_{ESR} is shown using calculated values of kTD_α and TD_β from Eq. (4.28) and the measured $D_{ex}T_{ESR}$.

example of *TD* calculation of a stalactite from Akiyoshi. In measuring the external dose by TL, the inside carbonate is shielded by the outside new formation and so the dose rate is attenuated by the surface carbonate. In an application of *ESR isochron dating* (see Chapter 4) to speleothems, the C_U is not varied but the D_{ex} is different from place to place. Homogeneity of the D_{ex} both in space and time (the D_{ex} in the past) is a major problem. Although the proposal is commendable, the first application of this method was not successful for cave deposits (Karakostangolou and Schwarcz 1983).

5.4.5   Charge Transfer Method : Old Sample Dating with Signal A ?

A suggestion has been made to heat the sample and convert signals B, C and D to a more stable signal A using anthropologically important speleothems from Arago Cave, France (Yokoyama *et al.* 1983a, b). Although the "*charge transfer method*" to convert the unstable signal to the more stable one is good in principle, it was asserted that laboratory heating leads to *overestimation* of ages (Skinner 1983, Debenham 1983). Actually, the *TD* obtained by this method is twice larger than that using signal C.

The integrated intensity of signal A is much smaller than that of signals C and D, though the height of signal A is large. Signal A is also created by heating some aragonitic coral samples. Thus, it is too naive to consider that the charge is simply transferred from a freely rotating CO_2^- to form signal A (SO_2^-). Dating old carbonates over a million years old is still controversial.

5.4.6   Suitable Samples

Specialists of uranium–thorium ($^{230}Th/^{234}U$) dating method began to use ESR for speleothem dating, and ESR ages were compared with the $^{230}Th/^{234}U$ ages for Quaternary speleothem calcite (Smith *et al.* 1985a, b, Smart *et al.* 1988). Contamination or inclusion results in detrital carbonates whose age might indicate a high *ED* due to inclusion of old carbonate microcrystals compared to clean carbonates. Organic radicals such as humic acid and fluvic acid described in Section 5.4.2 (b) give a signal around $g = 2.0045$. This signal interferes with signals B and C and gives an erroneous large *TD* unless it is subtracted clearly. *Acid insoluble residues* (AIR) obtained by dissolution of the carbonate with 6N HCl also gives a large *ED* by about 20% (Smith *et al.* 1985a). A reliable $^{230}Th/^{234}U$ age can not be obtained for a detrital carbonate where contamination of ^{230}Th is estimated from the presence of $^{230}Th/^{232}Th$: usually Th is not included in $CaCO_3$ (Schwarcz 1980). *Uranium leaching* is also noted when comparing the ESR data with the U–series ages of speleothems (Grün 1985).

5.4.7 Examples in Paleo–Anthropology

Systematic ESR studies of caves related with *paleo–anthropology* have been made at Petralona (Greece), Arago (France) and other places (Ikeya 1977, 1978a, Ikeya and Poulianos 1979, Yokoyama *et al.* 1981, 1983a, b, Falgueres *et al.* 1990, Wieser *et al.* 1993). The following possibility and limitation of ESR dating in paleo–anthropology have been clarified:

(1) The content of radioactive elements and environmental radiation are not uniform in a cave like Arago. The tentative dose rate of 1 mGy/a used at an early stage of ESR dating is no longer applicable though it is usually less than 2 mGy/a and larger than 0.3 mGy/a. The approximate age can be estimated without an accurate radiation assessment.

(2) The content of uranium is low in the recrystallized carbonate in the cave. Hence, external radiation, mainly due to the inflow of soils or terrarossa, plays a major role.

Some examples of travertine ages are shown in Figure 5.15.

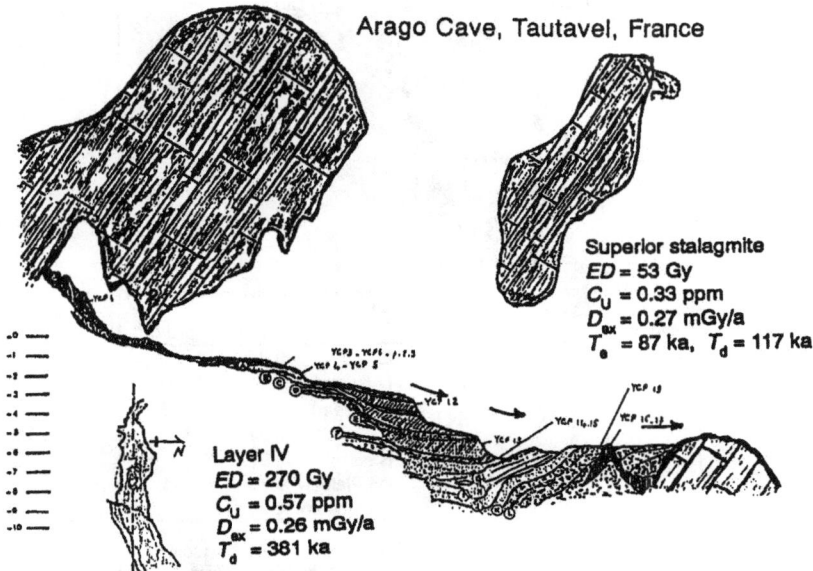

Arago Cave, Tautavel, France

Superior stalagmite
ED = 53 Gy
C_U = 0.33 ppm
D_{ex} = 0.27 mGy/a
T_e = 87 ka, T_d = 117 ka

Layer IV
ED = 270 Gy
C_U = 0.57 ppm
D_{ex} = 0.26 mGy/a
T_d = 381 ka

Figure 5.15 A drawing of the cave at Arago, Tautavel, France, where a paleo–anthropological cranium was excavated. The ED, the content of ^{238}U, D_{ex} and the ESR ages T_e (for radioactive equilibrium) and T_d (for disequilibrium) are shown for carbonate travertines (after DeLumley *et al.* 1981).

5.5 Speleology (Study of Caves) and Paleoenvironment

5.5.1 *Radiation Environment in a Cave*

(a) *Gamma–ray dose rate* : The contents of ^{238}U and ^{232}Th in cave deposits are small. This is also true for a stalactite at Akiyoshi Cave and for a
stalagmite at Petralona Cave. The *external radiation dose rate* measured by
TLD in Akiyoshi Cave is shown in the map in Figure 5.16 together with the
^{222}Rn content in the atmosphere during the summer season. External γ–ray
dose rate in the cave is 0.1 ~ 0.4 mGy/a while it is 0.75 mGy/a on the

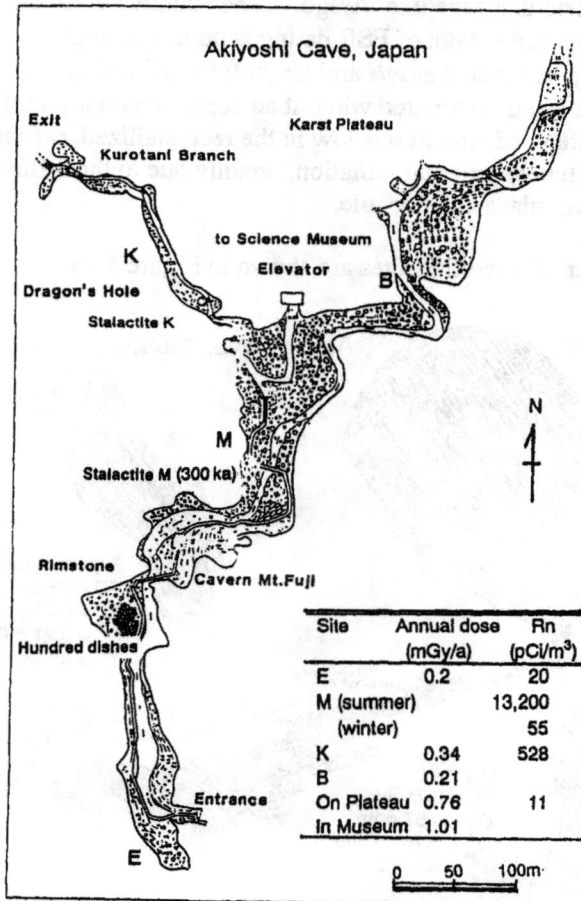

Site	Annual dose (mGy/a)	Rn (pCi/m³)
E	0.2	20
M (summer)		13,200
(winter)		55
K	0.34	528
B	0.21	
On Plateau	0.76	11
In Museum	1.01	

Figure 5.16 Guide map of the Akiyoshi limestone cave, showing the external
γ–ray dose rate, D_{ex} and atmospheric radon (^{222}Rn) activity.

ground about 90 m above the cave. The dose rate in the cave is high at the site where inflow soil or terrarossa is observable (Ikeya 1976, Miki and Ikeya 1980). Assessment of the external dose rate in the past is difficult since the cave environment may have been changed. A dose rate must be obtained by emplacing TLD or using a portable dose rate meter at the excavation site before the site is disturbed by sampling or excavating.

(b)   *Atmospheric radon* (^{222}Rn) *activity* : Atmospheric radon is extremely high in many caves, especially in summer (Wilkening and Watkins 1976, Miki and Ikeya 1980, Quindos *et al.* 1988). Radon emanation from fissures and the lack of ventilation in caves allow the accumulation of atmospheric radon. In some cases, it reaches the *"Working Level"* (1 WL = 100 pCi/l = 3.7 Bq/l), which is the *recommended danger level* in uranium mines. All workers have to immediately withdraw from the mine at 3 WL. The incidence of lung cancer may be doubled, especially for smokers, if they work 40 hours a week for 20 years at 1 WL.

The lack of ventilation allows accumulation of radon in a limestone cave as well as in a closed room, tunnel, coal or metal ore mine. An artificial ventilation is not allowed to protect natural growth of cave deposits in a limestone cave. The park patrol cave guides at the Mammoth Cave National Park, USA, wear radiation dosimeters and restrict their working hours in the cave to protect their health from radiation. Radon accumulation easily reaches 1 WL in such a cave, the longest (500 km) in the world. In spite of the severe restriction and radiation protection for personnel in uranium and other mines, we pay no attention to atmospheric radiation in natural limestone caves. Cave guides at show-caves are usually not protected from the radiation hazard caused by atmospheric radon in most countries.

The effect of atmospheric radon activity on the age of a stalactite is considered negligible in wet caves (Ikeya 1978a), but this seems not to hold for cave deposits in dry caves (Lyons *et al.* 1989). The annual dose rate at the carbonate surface due to atmospheric radon and its daughters has been calculated (Toyoda and Ikeya 1990).

5.5.2   Rainfall

(a) *Brown rings* ; *flooding* : The brown rings in a stalactite and a stalagmite or layers of clay sediments in the cave floor travertine may be markers of the same age. The origin of such brown rings in cave deposits may be considered due to flooding or to an extraordinary high precipitation that brought a large amount of small clay particles as inclusions into the stalactite

(Ikeya 1980). Brown rings can be used as an indicator of the *paleoclimate* like an age indicator in dendrochronology (tree ring counting) at the local level.

These brown rings or clay sediments might have occurred at the time of global change in the climate or changes in the groundwater level related to sea level changes. The ages of the brown rings obtained by interpolation have already been indicated in Figure 5.13 for a stalactite at Akiyoshi. There are five or six main brown rings and the approximate ages from the outside are about 120, 140, 200, 220, 250 and 320 ka. The age might shift about ± 20% considering the overall uncertainty in the *ED* and the average annual dose rate. Similar ages of brown rings have been obtained for stalagmites, travertines and cave pearls (Ikeya 1978b).

Three or four faint brown rings were formed in the last 100 ka as is seen in the stalactite and recent spelenthems. However, innumerable faint rings of clay particles are observable under a microscope, suggesting that the dominant ring formation is due only to an extremely severe condition of an ordinary phenomenon. The inclusions suggest heavy rainfalls in the past as clay particles come into the cave from fissures at a time of high precipitation on the ground. This was confirmed by Arakawa and Hori (1989).

(b) *Growth steps ; variation of precipitation ?* : The growth steps of stalactites can be observed under a microscope after chemical etching of a small calcite crystal from a stalactite. The interval of about 10 μm in growth steps corresponds to a cyclic change of 30 ~ 40 years using the growth velocity estimated by ESR dating, which may correlate with fluctuations in the rainfall. Inclusion of clay particles prevents the growth of stalactites. Luminescence produced by nitrogen laser excitation reveals variation in the concentration of *humic acid* and *fulvic acid* in cave *flowstones* (Shapov *et al.* 1991), which gives a similar growth rate of 1.7 μm/a; the growth of flowstones is generally rapid. Although the growth steps may arise from different reasons in crystal growth, a systematic investigation will clarify the climate change recorded in cave deposits.

(c) *Impurity distribution ; acid rainfall from SO_3^- ?* : The concentration of any SO_3^{2-} impurity in carbonates can easily be detected by irradiating the sample with γ-rays and measuring the ESR signal intensity at $g_\perp = 2.0036$ due to axial SO_3^- (signal B). *Spatial distribution* of some impurity ions such as SO_3^{2-} as revealed from SO_3^- of irradiated cave deposits would indicate changes in the *cave environment* in the past. The ESR microscope described in Chapter 14 would help to visualize the distribution of impurity ions in a slice of a stalactite or other cave deposits. Such a project using aragonite corals is described in Chapter 6 (see Figure 6.17).

5.5.3. Growth Rate and Temperature

ESR dating tells the age of a particular local portion of a cave deposit. The growth rate of a stalactite related to the reaction in Eq. (5.1) is temperature dependent. Recent growth rates along the radial direction for stalactites at several caves were determined and plotted against the reciprocal temperature in an Arrhenius plot. The *average growth velocity* may depend on the amount of underground water or the precipitation and on the location in the cave. However, it is the growth along the *longitudinal direction* that is directly affected by the precipitation. The *radial direction* would not be affected greatly since the growth occurs when it is wet with dripping water.

The reaction in Eq. (5.1) depends on the temperature. The amount of the initial supply of CO_2 for dissolving the limestone would also depend on the vegetation and temperature on the ground above the cave. Hence, the overall temperature dependence is obtained. If the average temperature in a cave is below 10°C, the growth velocity is extremely small. The average temperature change in the past may be recorded since the growth velocity changes by one order of magnitude with a temperature variation of 4 ~ 5°C (Ikeya *et al.* 1984). The fact that relatively transparent pure stalactites with few brown rings are observed in the caves in warm areas indicates the high growth rate in such caves. The growth rate of speleothems in glacial periods would be different from that during ordinary climate periods.

5.5.4. Cave Pearls, Flowstones and Travertines

Speleothems such as cave pearls, travertines and flowstones are some—times porous and contain *brines*. Growth lines due to colored inclusions tell the site where *recrystallization* occurred at later times. Such portions give young ages and should be avoided in ESR dating (Ikeya 1978b).

The distribution of ^{238}U in a cave floor or a stalactite may fluctuate to some extent, which results in a similar fluctuation in the *TD*. Also, external radiation may be shielded by subsequent deposition and the growth may have been interrupted by temperature variation and by chemical change in the dripping cave water. These result in a direct change in the *TD*. Hence, the stratigraphic order of the *TD*'s may be obtained for an old stalactite as shown in Figure 5.13 but may not be for a cave floor that has grown in a short time. A detailed calculation of *radiation shielding* is necessary to estimate the *TD*.

The history of growth is needed for such a calculation. Thus, an early study to correlate ESR ages of young cave travertines with the detailed *geomagnetic reversal* was unfortunately not successful (Morinaga *et al.* 1985).

5.6 Other Calcium Carbonates
5.6.1 *Spring Deposits*

Secondary calcite of *spring travertines* has been dated with ESR (Grün *et al.* 1988). A broad line at $g = 2.0045$ due to organic radicals was observed in addition to the signals of allowed and forbidden transitions of Mn^{2+}. Only 15% of the samples examined showed the dating signal at $g_C = 2.0007$ (freely rotating CO_2^-). They contain a considerable amount of acid insoluble residue (AIR) composed of clay minerals with variable amounts of U, Th and K and of organic substances. Modern samples contain excess initial ^{230}Th which must be included in radiation assessment as is done for deep–sea sediment foraminifera in the next chapter. Some travertines may constitute an open system where uranium and other radioactive elements enter or leach and thus are not appropriate for dating. However, well crystallized crusts with less AIR may be an object of ESR dating.

5.6.2 *Calcrete*

Calcrete is a conglomerate consisting of surficial sand and gravel cemented into a hard mass by precipitation of calcium carbonate (Bates and Jackson 1980). The genesis of calcretes is of particular interest. Broad ESR signals at $g = 2.0045$ and 2.0008 and a weak signal at $g = 1.9972$ have been observed. The intensity of these signals are enhanced after γ–irradiation. ESR and $^{230}Th/^{238}U$ dating of Spanish calcrete have been attempted for the former two signals considering ^{238}U–series disequilibrium (Radtke *et al.* 1988, Chen *et al.* 1989). The signal at $g = 2.0008$ (2.0007, signal C) was observed only in a few samples and the *ED* from this signal was 122 Gy, while it was 226 Gy for $g = 2.0045$ due to organic radicals (humic acids). Organic radicals may have entered carbonate at the time of crystallization (Megro and Ikeya 1992) or they may have been created by thermal decomposition. In the latter case, an apparent *ED* or ESR age obtained from the signal at $g = 2.0045$ for samples at the same area may be related to the age in a certain time interval since the signal intensity is related to the age and temperature in a chemical reaction (see Chapter 12). Subtraction of this signal is needed to obtain an accurate age using signal C.

5.6.3   *Stromatolite*

A stromatolite is an organosedimentary structure produced by the sediment trapping, binding, and/or precipitation activity of micro–organisms, primarily by cyanobacteria (Awramik 1971). Figure 5.17 shows the ESR spectra of Chinese stromatolite of the late Precambrian (900 Ma). In addition to the intense Mn^{2+} sextet signals and weak Fe^{3+} signals ($g = 4.28$ and $g = 9$), radical signals associated with SO_3^-, SO_2^- and CO_2^- are observed between the third and fourth hf lines of Mn^{2+}. It is interesting to study the distribution of these paramagnetic species in a *bandage pattern* of the stromatolite using ESR microscope described in Chapter 14.

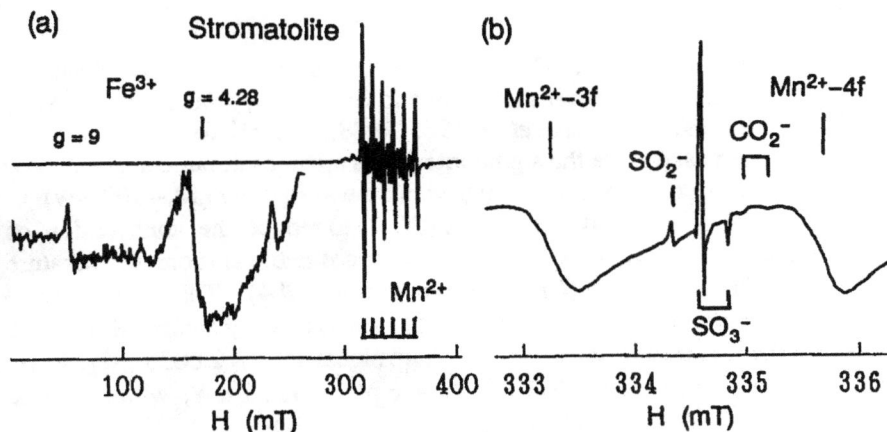

Figure 5.17 (a) A full ESR spectrum of Precambrian stromatolite and (b) its central portion (333 ~ 336 mT).

5.7   Technical Notes

(a)   *Sample preparation* :   Cutting of samples is usually carried out by using a saw with sufficient water to prevent heating. The *recrystallized porous portion* should be avoided since the *ED* sometimes shows a small value at the site of brine recently recrystallized. This portion can be distinguished if the sample is polished and the growth line is carefully observed (Ikeya *et al.* 1984). A *powder sample* (grains of 100 ~ 300 μm in diameter) is prepared by soft crushing, grinding and sieving. Washing with weak acid (0.1N acetic acid) and distilled water, sometimes in an ultrasonic bath, is the routine procedure for chemical etching of the surface. Regions containing defects along dislocations created by crushing and grinding as well as adherent fine particles are dissolved.

An interfering ESR signal at $g = 2.0001$ produced by *crushing* causes an underestimation of the *ED* (Hennig *et al*. 1983, Smith *et al*. 1989). *Etching* with absolute ethanol and HCl reduces as much as 35% in weight and 2 μm of the surface irradiated by external α–rays. Use of absolute ethanol is recommended in order to prevent the movement of Mn^{2+} ions (Lyons *et al*. 1985, 1988). Generally, grains smaller than 100 μm in diameter are not used in our laboratory as they stick to the inner wall of the sample tube. It is proposed to anneal the sample around 100°C for about 15 min to eliminate interference by unstable ESR signals because the presence of such signals gives a low *ED* at low microwave power.

(b) *ESR measurements* : Conditions for ESR measurements are discussed mostly for carbonate deposits (Hennig and Grün 1984, Liang *et al*. 1989, Özar *et al*. 1989, Yokoyama *et al*. 1985, 1988). A microwave power of 1 mW has been used since the signal intensities are not saturated at such a low microwave power. However, a higher microwave power (10 ~ 100 mW) is preferable for signals C (CO_2^-) and A (SO_2^-) to reduce the contribution of broad background signals, some of which are not stable at room temperature after artificial irradiation (Grün and DeCanniere 1984). This was demonstrated for a stalagmite (Lyons *et al*. 1988, 1993) and egg shells (Kai *et al*. 1988). As shown in Figure 5.7, the sharp signal at $g = 2.0036$ (signal B: SO_3^-) is saturated even at a low microwave power of 0.1 mW, while signals A and C do not saturate up to a high microwave power. The *ED* at the low microwave power (1 mW) is slightly smaller (10 ~ 20%) than that at the high microwave power (50 mW), depending on the contribution of the interfering broad signal.

5.8 Summary

Radiation–induced defects in carbonates have been well identified by recent investigations using synthetic carbonates doped with impurities and ^{13}C–isotope. These are the CO_3^- hole center, CO_3^{3-} and CO_2^- electron centers as well as the impurity–related SO_3^- hole center and SO_2^- electron center. The ESR age is deduced from the *ED* and the contents of ^{238}U, ^{232}Th and ^{40}K using the method of radiation assessment. Although there are some problems and limitations, ESR began to give overall reliable ages of crystallization of calcite using signals due to CO_2^- and SO_3^-. It is interesting to see how we have clarified the models of defects for the establishment of ESR dating and reached the present level of understanding since the first work in 1975.

Dating of secondary deposits of carbonate is not limited to the study of caves (speleology). A paleo-anthropological study of caves inhabited by human ancestors needs information on the age as to when carbonate speleothems were deposited. Studies of carbonates at K-T boundary related with the mrteorite impact indicated the anomalous increase of SO_3^- and Mn^{2+} (Miura et al., 1985; Griscom, 2001).Therefore, ESR dating is not restricted to the dark underground world but emerges into a bright world of widespread interest. The physics of lattice defects in minerals has thus developed and contributed to Quaternary geochronology.

References

Abeyratne M., Spooner N.A., Grun R., et al. (1997): Multidating studies of Batadomba cave, Sri Lanka. *Quat. Sci. Rev.* **16**, 243-255.

Apers D.J., Debuyst R., DeCanniere P., Dejehet F. and Lombard E. (1981): A criticism of dating by electron paramagnetic resonance (ESR) of the stalagmitic floors of the Caune de l'Arago at Tautavel. *Proc. Absolute Dating Isotope Anal. Prehistory Meth. Limits* 533-550.

Arakawa T. and Hori N. (1989): ESR dating of carbonate speleothem rings and late Quaternary climatic changes in the Ryukyu Islands, Japan. *Appl. Radiat. Isot.* **40**, 1143-1146.

Awramik S.M. (1971): Precambrian columnar stromatolite diversity: reflection of metazoan appearance. *Science* **174**, 825-826.

Bacquet G, Dudas J., Escribe C., Youdri L. and Belin C. (1975): ESR of CO_3^{3-}-Li^+ center in irradiated synthetic single crystal calcite. *Le J. de Physique* **36**, 427-429.

Bassiakos Y (2001): Assessment of the lower ESR dating range in Greek speleothems. *J. Radioanaly. Nucl. Chem.* **247**, 629-633.

Bahain J.J., Yokoyama Y., Masaoudi H., Falgueres C. and Laurent M. (1994): Thermal behaviour of ESR signals observed in various natural carbonates. *Quantern. Geochr.* **13**, 671-674.

Bates R.L. and Jackson J.A. ed (1980): *Glossary of Geology* (AGI, Virginia).

Barabas M. (1992): The nature of paramagnetic centers at $g = 2.0057$ and $g = 2.0031$ in marine carbonates. *Nucl. Tracks* **20**, 453-464.

Bartoll J., Stosser R., Nofz M. (2000): Generation and conversion of electronic defects in calcium carbonates by UV/Vis light. *Appl. Radiat. Isot.* **52** ,1099-1105.

Bershov L.V., Tarashchan A.N., Samoilovich M.I. and Lushnikov M.I. (1968): Electron-hole centers in natural calcites with phosphors impurities. *Zh. Strukt.Chim* **5**, 309-311.

Blackwell B.A.B, Spalding C.N, Blickstein J.I.B., et al. (2001): ESR dating the hominid-bearing breccias at the Makapansgat Limeworks Cave, South Africa. *J. Hum. Evol.* **40**, A3-A4.

Cass J., Kent S., Marshall S.A. and Zager S.A. (1974): Electron spin resonance absorption spectrum of HCO_3^{2-} molecular ions in irradiated single crystal calcite. *J. Magn. Reson.* **14**, 170-181.

Chandra H., Symons M.C.R. and Griffiths D.R. (1988): Stable perinaphthenyl radicals in flints. *Nature* **332**, 526-527.

Chantry G.W., Horsefield A., Morton J.W. and Whiffen D.H. (1962): Electronic absorption spectra of CO_2^- trapped in γ-irradiated crystalline sodium formate. *Molecular Phys.* **5**, 189-194.

Chen Y., Arakel A.V. and Lu J. (1989): Investigation of sensitive signals due to g- ray irradiation of chemical precipitates. A feasibility study for ESR dating of gypsum, phosphate and calcrete deposits. *Appl. Radiat. Isot.* **40**, 1163-1170.

Chen Y., Brumby S. and Gao J. (1994): Observations on the suitability of the signal ATg=2.0040 for ESR dating of secondary carbonates. *Quanter. Geochr.* **13**, 675-678.

Chong T.S. and Fujitani Y. (1984): Organic radicals in a stalactite. *Chikyuu* 58, 237-239.

Debenham N.C. (1983): Reliability of thermoluminescence dating of stalagmite calcite. *Nature* **304**, 154-156.

Debuyst R., Bidiamambu M. and Dejehet F. (1990a): Diverse CO_2^- radicals in γ- and α-irradiated synthetic calcite. *Bull. Soc. Chim. Belg.* **999**, 535-541.

Debuyst R., DeCanniere P. and Dejehet F. (1990b): Axial CO_2^- in α-particle irradiated calcite : potential use in ESR dating. *Nucl. Tracks* **17**, 525-530.

Debuyst R., Bidiamambu M. and Dejehet F. (1991): An EPR study of γ- and α irradiated synthetic powdered calcite labeled with ^{13}C. *Nucl. Tracks* **18**, 193-201.

Debuyst R., Dejehet F. and Idrissi S. (1993): Paramagnetic centers in γ-irradiated synthetic monohydrocalcite. *Appl. Radiat. Isot.* **44**, 293-297

DeCanniere P., Debuyst R., Dejehet F. and Apers D. (1982): ESR study of radiation effects in ^{210}Po-doped calcium carbonate: contribution to the estimation of the paleodose in a stalagmite floor. *Radiochem. Radioanal. Lett.* **50**, 345-353.

DeCanniere P., Joppart T., Debuyst R., Dejehet F. and Apers D. (1985): ESR dating: a study of humic acids incorporated synthetic calcite. *Nucl. Tracks* **10**, 853-863.

DeCanniere P., Debuyst R., Dejehet F. and Apers D. (1986): ESR dating: A study of ^{210}Po-coated geological and synthetic samples. *Nucl. Tracks* **11**, 211-220.

DeCanniere P., Debuyst R., Dejehet F. and Apers D. (1988): ESR study of internally α-irradiated (^{210}Po nitrate doped) calcite single crystal. *Nucl. Tracks* **14**, 267-273.

DeLumley H., Park Y., Camara A., Geleijnse V., Beiner M., Fournier A., Miskovsky J., Hoffer M. and Schaaf O. (1981): Characterisques sedimentologiques et mineralogiques du remplissage Quaternaire de la Caun de l'Arago a Tautavel origine et mise en place des sediments. *Datations absolues et analyses Isotopiques in Prehistoire Methodes et Limites* ed. by DeLumley H. and Labeyrie J. 43-73.

Eachus R. S. and Symons M. C. R. (1968): Unstable Intermediate, the NO_3^{2-} impurity center in irradiated calcium carbonate. *J. Chem. Soc.* (A) **45**, 790-793.

Engin B., Guven O., Koksal F. (1999): Electron spin resonance age determination of a travertine sample from the southwestern part of Turkey. *Appl. Radiat. Isot.* **51**, 689-699.

Falgueres C., Yokoyama Y. and Bibron R. (1990): Electron spin resonance (ESR) dating of hominid-bearing deposits in the Caverna delle Fate, Italy. *Quat. Res.* **34**, 121-128.

Gaffey S. J. (1988): Water in skeletal carbonates. *J. Sed. Petrology* **58**, 397-414.

Griscom D. (2001): Electron spin resonance investigation of KT spheroids and diamictites from Quitana Roo, Mexico and Belize. *Proc. 2001-ESRDD-Osaka* in press.

Grün R. and DeCanniere P. (1984): ESR dating; problems encountered in the evaluation of the naturally accumulated dose (AD) of secondary carbonates. *J. Radioanal Nucl. Chem. Lett.* **85**, 213-226 .

Grün R. (1985): ESR dating speleothem: limits of the method. *ESR Dating and Dosimetry*, 61-72.

Grün R., Schwarcz H.P., Ford D.C. and Hentzsch B. (1988): ESR dating of spring deposited travertines. *Quat. Sci. Rev.* **7**, 429-432.

Grün R. and Beaumont P. (2001): Border Cave revisited: a revised ESR chronology. J. Hum. Evol. 40, 467-482.

Hennig G.J., Grün R. and Brunnacker, K. (1983): Interlaboratory comparison project of ESR dating, phase I 1982. *PACT* **9**, 447-452.

Hennig GJ. and Grün R. (1984): ESR dating in Quaternary geology. *Quat. Sci. Rev.* **2**, 157-238.

Hoffmann D., Woda C., Strobl C., et al. (2001): ESR-dating of the Arctic sediment core PS1535 dose-response and thermal behaviour of the CO_2--signal in foraminifera. Quat. Sci . Rev. 20, 1009-1014.

Hudson J. H., Shinn E. A., Halley R.B. and Lidz B. (1976): Speleochronology: a tool for interpreting past environments. *Geology* **4**, 361-364.

Hughes R.C. and Soo Z.G. (1970): EPR of CO_2^- defects in calcite: motional and nonsecular contribution. *J. Chem. Phys.* **52**, 6302-6310.

Ikeya M. (1975): Dating a stalactite by electron paramagnetic resonance. *Nature* **255**, 48-50.

Ikeya M. (1976): Natural radiation dose in Akiyoshi Cavern and on Karst Plateau. *Health Physics* **31**, 76-78.

Ikeya M. (1977): Electron spin resonance dating and fission track detection of the Petralona stalagmite. *Anthropos* **4**, 152-168.

Ikeya M. (1978a): Electron spin resonance as a method of dating. *Archaeometry* **20**, 147-158.

Ikeya M. (1978b): Spin resonance ages of brown rings in cave deposits. *Naturwissen.* **65**, 489.

Ikeya M. and Poulianos A. (1979): ESR age of the trace of fire at Petralona. *Anthropos* **6**, 44-47.

Ikeya M. (1980): ESR dating of carbonates at Petralona Cave. *Anthropos* **7**, 143-151.

Ikeya M. and Ohmura K. (1981): Dating of fossil shells with electron spin resonance. *J. Geology* **89**, 247-250.

Ikeya M., Baffa F. O. and Mascarenhas S. (1984): ESR dating of cave deposits from Akiyoshi Cave in Japan and Diabo Cavern in Brazil. *J. Speleol. Soc. Japan* **9**, 58-67.

Ikeya M. (1985): *Dating Method of Pleistocene Deposits* ed. by Rutter N. W., Geoscience Canada Reprint Series 2, Chapt. XI Electron Spin Resonance.

Ikeya M. and Furusawa M. (1989): A portable spectrometer for ESR microscopy, dosimetry and dating. *Appl. Radiat. Isot.* **40**, 845-850.

Jacobs C., DeCanniere P., Debuyst R., Dejehet F. and Apers D. (1989): ESR study of g-ray irradiated synthetic calcium carbonates. *Appl. Radiat. Isot.* **40**, 1147-1152.

Kai A,. Miki T. and Ikeya M. (1988): ESR dating of teeth, bone and egg shells excavated at a paleolithic site of Douara cave, Syria. *Quat. Sci. Rev.* **7**, 503-507.

Kai A. and Miki T. (1991): Sulfite radicals in irradiated calcite. *Jpn. J. Appl. Phys.* **30**, 1109-1110.

Kai A. and Miki T. (1992): Electron spin resonance of sulfite radicals in irradiated calcite and aragonite. *Radiat. Phys. Chem.* **40**, 469-476.

Kai A. and Miki T. (1993): Effect of trivalent metal impurities on radiation-induced radicals in calcite. *Appl. Radiat. Isot.* **44**, 311-314.

Karakostangolou I. and Schwarcz H.P. (1983): ESR isochron dating. *PACT J.* **9**, 391-398.

Kohno H., Yamanaka C. Ikeya M., Ikeda S. aand Horino Y. (1994): An ESR study of radicals in $CaCO_3$ produced by 1.6 MeV He^+-and γ-irradiation. *Nucl. Instr. Method*

Physs. Res. B **91** 366-369.

Kohno H., Yamanaka C. and Ikeya M. (1994): Pulsed annd cw-ESR study of CO2- in CaCO$_3$ produced by 1.6 MeV He$^+$-and γ-irradiation. *Jpn. J. Appl. Phys.* **33**, 5743-5746.

Liang R., Peng Z., Jin S. and Huang P. (1989): The estimation of the influence of experimental conditions of ESR results. *Appl. Radiat. Isot.* **40**, 1071-1075.

Lyons R.G., Williams P.W. and Wood W.B. (1985): Determination of α-efficiency in speleothem calcite by nuclear accelerator techniques. *ESR Dating and Dosimetry*, 3948.

Lyons R. G., Bowmaker G. A. and O'Connor C. J. (1988): Dependence of accumulat ed dose in ESR dating on microwave power: a contra-indication to the routine use of low power levels. *Nucl. Tracks* **14**, 243-251.

Lyons R.G., Crossley P.C., Ditchburn R.G., McCabe W.J. and Whitehead N. (1989): Radon escape from New Zealand speleothems. *Appl. Radiat. Isot.* **40**, 1153-1158.

Lyons R. G. and Brennan B. J. (1991): Alpha/gamma effectiveness ratios of calcite speleothems. *Nucl. Tracks* **18**, 223-230.

Lyons R.G., Bowmaker G.A. and O'Conner C.J. (1993): High or low ? the optional microwave power for determination of accumulated dose in ESR dating of speleothem calcite. *Appl. Radiat. Isot.* **44**, 825-827.

Lyons R. G. (1996): Back to basics: qualitative spectral analysis as an investigatory tool, using calcite as a case study. *Appl. Radiat. Isot.* **47**, 1385-1391.

Marshall S.A., Reinberg A.R., Serway R.A. and Hodges J.A. (1964): Electron spin resonance spectrum of CO$_2^-$ molecule-ions in single crystal calcite. *Mol. Phys* **8**, 225-231.

Marshall S.A. and McMillan J.A. (1968): Electron spin resonance spectrum of CO$_2^-$ molecule ions associated with F$^-$ ions in single crystal calcite. *J. Chem. Phys.* **49**, 4887-4890.

Marshall S.A., McMillan J.A. and Serway R.A. (1968): Electron spin resonance absorption spectrum of Y^{3+}-stabilized CO$_3^{3-}$ molecule-ion in single crystal calcite. *J. Chem. Phys.* **48**, 5131-5137.

Marshall S.A. and Serway R.A. (1969): Electron spin resonance absorption spectrum of the AsO22- molecule-ion in g-irradiated single crystal calcite. *J. Chem. Phys.* **50**, 435-439.

Marshall R.L., Ohlsen W.D., Marshall S.A. and Serway R.A. (1974): Tentative identification by ESR of CSO$_2^{3-}$ impurity radical in irradiated single crystal of calcite. *J. Magn. Reson.* **15**, 89-97.

McMillan J.A. and Marshall S.A. (1968): Motional effects in the electron spin resonance absorption spectrum of CO$_2^-$ molecule-ion in single crystal calcite. *J. Chem. Phys.* **48**, 467-471.

Meguro K. and Ikeya M. (1992): A new spin trapping method with inorganic microcrystals. *Chem. Express* **6**, 399-402.

Meldin W. L. (1964): Trapping centers in thermoluminescent calcite. *Phys. Rev.* **135**, 1770-1779.

Miki T. and Ikeya M. (1978): Thermoluminescence and ESR dating of Akiyoshi stalactite. *Jpn. J. Appl. Phys.* **17**, 1703-1704.

Miki T. and Ikeya M. (1980): Accumulation of atmospheric radon in calcite caves. *Health Physics* **39**, 351-354.

Miki T. and Kai A. (1990): Rotating CO$_2^-$ centers in corals and related materials. *Jpn. J. Appl. Phys.* **29**, 2191-2192.

Miki T., Kai A. and Murata T. (1993): Radiation-induced radicals in sulfite-doped CaCO$_3$. *Appl. Radiat. Isot.* **44**, 315-319.

Miura Y., Ohkura Y., Ikeya M. Miki T., Ruckridge J., Takaoka N., Nielson TFD (1985):

ESR data of Danish calcite at Cretaceous and Tertiary boundary. *ESR Dating and Dosimetry*, 469-476.

Moore B.W. and Nicholas B.G. (1964): *Speleology: The Study of Caves* (Heath & Company, Mass).

Morinaga H., Inokuchi H., Yasukawa K., Ikeya M., Miki T. and Kusakabe M. (1985): Paleomagnetism, paleoclimatology and ESR dating of stalagmite deposits. *ESR Dating and Dosimetry*, 31-37.

Murata T., Kai A. and Miki T. (1993): Hydration effects on CO_2^- radicals in calcium carbonates and hydroxyapatite. *Appl. Radiat. Isot.* **44**, 305-309.

Overnall D. W. and Whiffen D. H. (1961): Electron spin resonance and structure of the CO2- radical ion. *Molec. Phys.* **4**, 135-144.

Özer A. M., Wieser A., Göksu H. Y., Müller P., Regulla D. F. and Erol O. (1989): ESR and TL age determination of caliche nodules. *Appl. Radiat. Isot.* **40**, 1159-1161.

Quindos L.S., Fernandes P.L., Soto J., Villar E. and Miki T. (1988): Evolution of radon concentration in Altamila Cave. *J. Speleol. Soc. Jpn.* **13**, 45-51.

Pike A.W.G. and Hedges R.E.M. (2001): Sample geometry and U uptake in archaeological teeth: implications for U-series and ESR dating. *Quat. Sci. Rev.* **20**,1021-1025.

Porat N., Amit R. and Zilberman E. (1994). Electron Spin Resonance dating of carbonate nodules from Shivta, western Negev - preliminary results. Geological Survey of Israel Current Research 9, 41-46.

Radtke U., Brückner H., Mangini A. and Hausmann R. (1988): Problems encountered with absolute dating (U-series, ESR) of Spanish calcrete. *Quat. Sci.* **7**, 439-445.

Rossi A., Poupeau G. and Danon J. (1985): On some paramagnetic species induced in natural calcites by ß- and g-ray irradiations. *ESR Dating and Dosimetry*,77-85.

Rossi A. and Poupeau G. (1989): Radiation-induced paramagnetic species in natural calcite speleothems. *Appl. Radiat. Isot.* **40**, 1133-1138.

Schramm D.U., Rossi A.M. (1997): ESR studies on the linewidth temperature variation of CO_2^- radicals in carbonated apatite and calcium carbonates. *Jpn. J Appl. Phys.* **36** (2B), L202-L205.

Schramm D.U., Rossi A.M. (2000): Electron spin resonance (ESR) studies of CO_2^- radicals in irradiated A and B-type carbonate-containing apatites. *Appl. Radiat. Isot.* **52**, 085-1091.

Schwarcz H.P. (1980): Absolute age determination of archaeological sites by uranium-series dating of travertines. *Archaeometry* **22**, 3-24.

Schwarcz H.P, Rink W.J. (2001): ESR dating of the Die Kelders Cave 1 site, South Africa. *J. Human. Evol.* **38**,121-128.

Serway R.A. and Marshall S.A. (1966a): Electron spin resonance absorption spectrum of the AsO_3^{2-} molecule-ion in γ-irradiated single crystal calcite. *J. Chem. Phys.* **45**, 2309-2314.

Serway R.A. and Marshall S.A. (1966b): ESR absorption spectra of two phosphorous defect centers in irradiated single crystal calcite. *J. Chem. Phys.* **45**, 4098-4104.

Serway R.A. and Marshall S.A. (1967a): Electron spin resonance absorption spectra of CO_3^- and CO_3^{3-} molecule ions in irradiated single crystal calcite. *J.Chem. Phys.* **46**, 1949-1952.

Serway R.A. and Marshall S.A. (1967b): Electron spin resonance absorption spectrum of orthorhombic CO_3^- molecule-ions in irradiated single crystal calcite. *J. Chem. Phys.* **47**, 868-869.

Serway R.A., Marshall S.A., McMillan J.A., Marshall R.L. and Ohlsen W.D. (1969): Electron spin resonance absorption spectra of hole-trap centers associated with phosphorous in irradiated single crystal calcite. *J. Chem. Phys.* **51**, 4978-4981.

Shopov Y., Dermendjiev V. and Buyukliev G. (1991): A new method for dating of natural materials with periodical macrostructure by autocalibration and its application for study of the solar activity in the past. *Proc. Intern. Conf. Environmental Methods* 17-22.

Skinner A. F. (1983): Overestimation of stalagmite calcite ESR dates due to laboratory heating. *Nature* **304**, 152-154.

Smart P.L., Smith B.W., Chandra H., Andrew J.N. and Symons M.C.R. (1988): An intercomparison of ESR and uranium series ages for Quaternary speleothem cal cites. *Quat. Sci. Rev.* **7**, 411-416.

Smith B.W., Smart P.L., Symons M.C.R. and Andrews J.N. (1985a): ESR dating of detritally contaminated calcites. *ESR Dating and Dosimetry*, 49-59.

Smith B.W., Smart P.L. and Symons M.C.R. (1985b): ESR signal in a variety of speleothem calcites and their suitability for dating. *Nucl. Tracks* **10**, 837-874.

Smith B.W., Smart P.L., Fox P.J., Prescott J.R. and Symons M.C.R. (1989): An investigation of ESR signals and their related TL emission in speleothem calcite. *Appl. Radiat. Isot.* **40**, 1095-1104.

Takano M. and Fukao Y. (1993): ESR dating of Pleistocene fossil shells of the Atsumi Group, Central Honshu, Japan: on the discrepancy in TD value among different ESR peaks. *Appl. Rad. Isot.* **44**, *in press*.

Toyoda S. and Ikeya M. (1990): Annual dose rate from atmospheric radon activity. *Anthropos* **12**, 36-49.

Tucker M.E. and Wright V.P. (1990): *Carbonate Sedimentology* (Blackwell Scientific, London) pp.284.

Uesugi A. and Ikeya M. (2001): Electron spin resonance measurement of organic radicals in petroleum source rock containing transition metal ions. Jpn. J. Appl. Phys. 40, 2251-2254.

Waltham T. (1974): *Caves* (McMillian, London).

Walsh A. D. (1953): The electron orbitals, shapes and spectra of polyatomic molecules. Part V. Tetraatomic, non-hydride molecules, AB$_3$. J. *Chem. Soc.* **47**, 2301-2306.

Wieser A., Göksu H.Y., Regulla D.F., Fritz P., Vogenauer A. and Clark I.D. (1993): ESR and TL dating of travertine from Jordan: Complications in paleodose assessment. *Appl. Radiat. Isot.* **44**, 149-152.

Wilkening M. H. and Watkins D. E. (1976): Air exchange and Rn-222 concentrations in Carlsbad Caverns. *Health Phys.* **31**, 139-145.

Ye Y.G., Diao S.B., He J., et al. (1998): ESR dating studies of palaeo-debris-flow deposits in Dongchuan, Yunnan Province, China. Quat. Sci. Rev.17 (11): 1073-1076.

Yokoyama Y., Quaegebeur J. P., Bibron R., Leger C., Nguyen H. V. and Poupeau G. (1981): Electron spin resonance (ESR) dating of stalagmites of the Caune de l'Arago at Tautavel. Datations absolues et analyses Isotopiques in Prehistoire Methodes et Limites ed. by DeLumley H. and Labeyrie J. 507-532.

Yokoyama Y., Quaegebeur J.P., Bibron R. and Leger C. (1983a): ESR dating of Paleolithic calcite; thermal annealing experiment & trapped electron lifetime. *PACT* **9**, 372-379.

Yokoyama Y., Quaegebeur J.P., Bibron R.and Leger C. (1983b): ESR dating of stalagmites of the Caune de l'Arago, the Grotte du Lazaret, the Grotte du Vallonnet and the Abri Pielombard : comparison with the U-Th method. *PACT J.* **9**, 381-389.

Yokoyama Y., Bibron R., Leger C. and Quaegebeur J.P. (1985): ESR dating of paleolithic calcite: fundamental studies. *Nucl. Tracks* **10**, 929-936.

Yokoyama Y., Bibron R. and Leger C. (1988): ESR dating of paleolithic calcite: a comparison between powder and monocrystal spectra with thermal annealing. *Quat. Sci. Rev.* **7**, 433-438.

Chapter 6

CaCO$_3$

– Biocarbonates (Fossils) –

" A scientist is like a child collecting shining pebbles on a beach. "

--- Newton ---

Nowadays, scientists use machines to collect precious pebbles, but must find a beach where they can enjoy science, child–like. Shining speleothems in muddy caves are the first shining pebbles. Then follow aragonitic shells and corals, all of which yield their ages.

6.1 Introduction

Fossils are the remains of biological species and are indices of geological age. Detailed classification of shell species has been done in paleontology to estimate the age of sediments. *Shells* and *corals* are composed of the mineral part, $CaCO_3$, mostly microcrystalline *aragonite* or *calcite*, and the organic part of the protein called *conchiolin*. Biological carbonate fossils with organic remains can be the object of ESR dating as secondary carbonate deposits. Radicals or defects produced by natural radiation in the mineral part have been used for dating with ESR.

Thermoluminescence (TL) dating of fossils has often been tried, but chemical luminescence (CL) due to the combustion of organic remains by heating made it difficult to use TL for dating fossils. ESR dating of fossil shells was done first in a relative way and then with radiation assessment (Ikeya and Ohmura 1981, 1983, 1984, Radtke *et al.* 1981, 1985, Hütt *et al.* 1983). ESR dating of corals from coral reef islands (Ikeya and Ohmura 1983) and carbonate microfossils, *planktonic foraminifera* in deep-sea cores (Sato 1982, Mangini *et al.* 1983), can contribute to oceanography or marine geography and geology. *Egg shell* and *shell ornaments* as well as shells from *shell mounds* at archeological sites are also carbonate and datable.

In this chapter, dating procedures for carbonate fossils, shells, corals and foraminifera are described with geological application to age determination of marine terraces. Comparison of ESR ages with $^{14}C-$ and $^{230}Th/^{234}U$ ages indicates that ESR dating of carbonate fossils is reliable in the time range from a few hundred to a million years, beyond the upper limit of ^{14}C dating (50 ka).

Figure 6.1   A photograph of shell samples used for ESR dating.

6.2 Crystal Structure and Defects in Aragonite (CaCO₃)

6.2.1 Crystal Structure

Aragonite is one of the trimorphous forms of $CaCO_3$ different from calcite and vaterite. Aragonite is less common and less stable than calcite and is readily converted into calcite at normal pressure and temperature. It is widely observed as a metamorphic mineral in schists and in volcanic rocks. Many organisms (mollusks and corals) build their skeletons of aragonite.

The structure of aragonite is orthorhombic with higher density and hardness than calcite. The unit cell of aragonite has Ca^{2+} ions in pseudo-hexagonal layers parallel to the [001] plane separated by two distinct layers of CO_3^{2-} planar molecular ions, as shown in Figure 6.2.

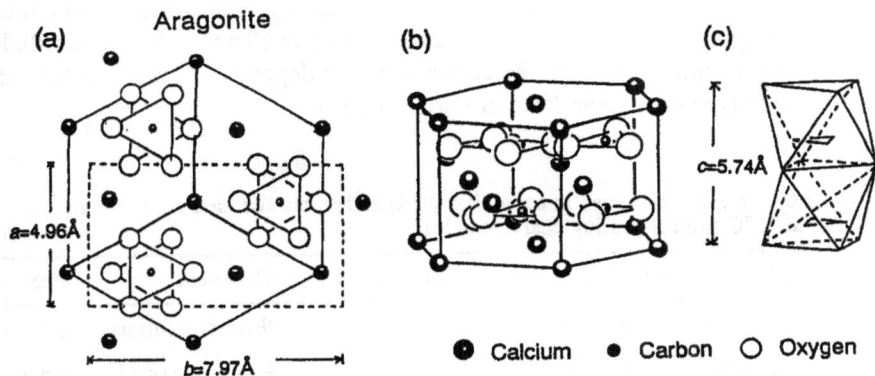

Figure 6.2 Crystal structure of aragonite with orthorhombic symmetry. (a) and (b) CO_3^{2-} ions lie within an octahedral arrangement formed by Ca^{2+} ions at 1/3 and 2/3 of the height of the octahedron measured along the c–axis. (c) The octahedra are joined face to face in columns parallel to the c–axis, and edge to edge with octahedra of neighboring columns (after Batty 1981).

6.2.2 Radiation–Induced Defects

Single crystals of natural aragonite or aragonitic shells and corals show signals almost identical to those in calcite. Among four signals relevant to dating of calcite, signals **A, C,** and **D** are the identical radical species to those in aragonite, SO_2^-, freely rotating CO_2^- and orthorhombic CO_2^-, respectively. The g factor of signal **B** in aragonite ($g_B = g_{av} = 2.0031$) is different from that of the similar signal in calcite ($g_B = g_\perp = 2.0036$ and $g_\parallel = 2.0024$) due to SO_3^-. It is noted that the former is **isotropic** SO_3^- and the latter is **axial** SO_3^-. Since the g factor is determined by the spin–orbit interaction and so by the molecular orbitals, it is not appreciably affected by the surrounding lattice of the molecular species while the linewidth and the mode of the hindered rotation are affected. The saturation level of $g_B = 2.0036$ decreases as the concentration of Mg (Mg/Ca ratio) increases above 0.1 mol/mol, while that of $g = 2.0031$ increases (Barabas *et al.* 1989). Hence, the signal at $g = 2.0031$ is considered to be the g_{av} of axial SO_3^- (see Table 5.2). Mg^{2+} ions might be accompanied with H_2O molecules leading to rapidly tumbling hydrated radicals. In fact, the saturation level of isotropic CO_2^- (signal C) is enhanced as the Mg/Ca ratio is increased.

In addition to isotropic SO_3^-, **orthorhombic SO_3^-** has been identified in aragonite (Kai and Miki 1992). The observed spectrum in fossil carbonates is likely to be an overlap of several signals, especially in the region around $g = 2.001$ (Katzenberger and Willems 1988). Table 6.1 shows the assigned signals in aragonitic shells and corals (see Chapter 5 for detailed description of these signals). Microwave power dependency and isochronal annealing curves are shown later in Figure 6.20.

Table 6.1 Signals observed in aragonitic shells and corals and their approximate life-times at 15°C from annealing characteristics.

Models	Notation	g factor (A tensor)	Comment	Ref.
SO_2^- (free rotation)	A	$g_A = 2.0057$	thermally formed	1,2,3,4
SO_3^- (isotropic)	B	$g_B = g_{av} = 2.0031$ [a]	$\tau \approx 3$ Ma (15°C)	1,2,3
SO_3^- (orthorhombic)		$g_{zz} = 2.0038$ $g_{xx} = 2.0021$ $g_{yy} = 2.0034$		1
CO_2^- (free rotation)	C	$g_C = 2.0007$	$\tau \approx 1$ Ma (15°C)	2,5,6
CO_2^- (orthorhombic)	D	$g_D = g_{yy} = 1.9973$	$\tau \approx 1$ Ma (15°C)	2
$(CH_3)_2C\cdot-R$ [b]	septet	$g = 2.0039$ ($A = 2.17$ mT: ^{1}H)	thermally formed	7,8
?		$g = 2.0018 \sim 2.0020$	observed in shells	2,9,10

a: Signal B due to axial SO_3^- in calcite is at $g_\parallel = 2.0021$ and $g_\perp = 2.0036$.
b: The septet signal due to proton hf splittings is ascribed to isopropyl radicals in valine-doped aragonitic CaCO₃ thermally annealed following γ-irradiation.
Reference: (1) Kai and Miki 1992; (2) Katzenberger *et al.* 1989; (3) Barabas 1992; (4) Walther *et al* 1992; (5) Miki and Kai 1990, 1991; (6) Barabas *et al.* 1992a, b; (7) Kai and Miki 1989; (8) Murata *et al.* 1993; (9) Skinner and Weicker 1992; (10) Molodkov 1988a.

6.3   Fossil Shells in Sediments

6.3.1   *ESR Spectra*

(a) *Aragonitic shell* : A small broken piece of a shell in a sediment regard–
less of the paleontological species is sufficient for absolute dating with ESR.
Carbonate in a shell is not randomly oriented but the crystal directions are
oriented to harden the shell. Consequently, ESR spectra of a piece of shell
depend strongly on the direction of the magnetic field and grains larger than
500 μm show an angular dependence. This is the reason why grains of 100 ~
250 μm are used.

Figure 6.3 shows an ESR spectrum of aragonitic shell (*Polinices*)
whose age was determined by $^{230}Th/^{234}U$ dating as 65 ka. The g factors for
signals A and B are $g_A = 2.0057$ and $g_B = 2.0031$, respectively.   Signal B
overlaps a peak of the central hf line of *isopropyl radicals* with hf splitting of
2.17 mT.   Two more lines were observed in the quintet signal originally
ascribed to alanine radicals (Ikeya 1981) and the septet was ascribed to
isopropyl radicals (Kai and Miki 1989).

ESR spectra of shells with geologically different ages show distinctly
different signal intensities. The relative intensities of signals A, C and D

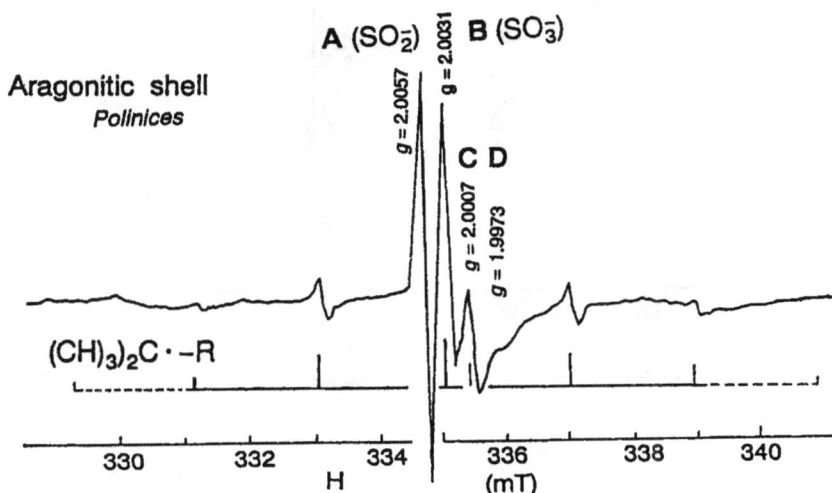

Figure 6.3   ESR spectrum of an aragonitic shell (*Polinices*) of about 65,000
years BP. The septet signal due to the hf interaction associated with isopropyl
radicals is observed in addition to four signals indicated by A, B, C and D (g_A
= 2.0057, g_B=2.0031, g_C = 2.0007 and g_D = 1.9973).

increase as the age increases from about 60 ka to 500 ~ 600 ka, while the intensity of signal B due to SO_3^- is saturated. One can tell the approximate age of the sea shells at least qualitatively from the absolute magnitude of the intensity per 300 mg sample. The result indicates that the natural dose rate is more or less the same. The spectral pattern is different for samples of different ages due to different stability of the signals and so the relative intensity serves as a good indication of the age (Ikeya and Ohmura 1981).

(b)   *Calcitic shell* : The ESR spectrum of calcitic shells shows signals at $g = 2.0057$ (signal A) and $g_\perp = 2.0036$ and $g_\parallel = 2.0021$ (signal B in calcite) in addition to Mn^{2+} signals as shown in Figure 6.4. Calcitic shells contain a large amount of Mn^{2+} which masks radiation–induced signals and thus makes the dating difficult as also for other calcite such as stromatolite (see Figure 5.17).

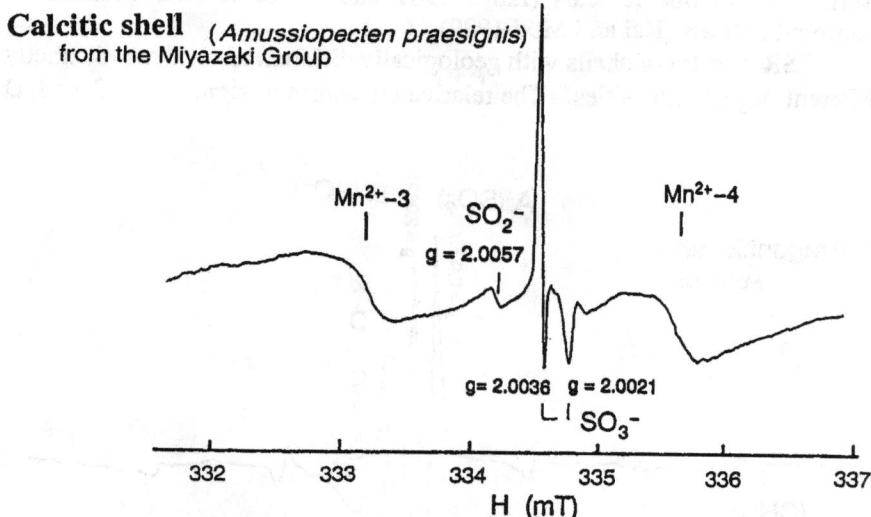

Calcitic shell   (*Amussiopecten praesignis*)
   from the Miyazaki Group

Mn²⁺–3   SO₂⁻   Mn²⁺–4
        g = 2.0057

g= 2.0036   g = 2.0021
              SO₃⁻

332   333   334   335   336   337
                  H (mT)

Figure 6.4   ESR spectrum of fossil calcitic shell.

6.3.2   *Relative Signal Intensity and Age*

A proposal has been made to deduce an absolute age from the relative intensities taking into account that the thermal stabilities are different for different signals (Debuyst *et al.* 1984). The basic philosophy is similar to $^{230}Th/^{234}U$ dating in which the activity ratio is used. The signal intensity, I, increases as a function of time before the saturation tendency appears;

$$I(t) = a D\tau (1 - e^{-t/\tau}) \;, \tag{6.1}$$

where τ is the lifetime, a the defect production efficiency (Gy^{-1}) and D the annual dose rate (Gy/a). The relative intensity for two signals is described by

$$I_1(t)/I_2(t) = (a_1/a_2)(\tau_1/\tau_2)(1 - e^{-t/\tau_1})/(1 - e^{-t/\tau_2}) \;. \tag{6.2}$$

Hence, it is possible in principle to determine the age without knowing the dose rate D. Figure 6.5 shows a plot of the *intensity ratio* I_1/I_2 as a function of the age normalized by a lifetime τ. One can determine the age if the intensity ratio and the lifetime ratio are known for some signals.

The situation is not so simple, however, since the defect production efficiency, a, may also be different for two signals. Neither the *charge trans-fer process* nor the *interaction volume*, b, described in Chapter 3 was considered in this simple model. Hence the applicability is for young samples in which the signal intensity increases linearly with the radiation dose. Since the lifetime of defects is temperature dependent with different activation energies, this method may have limitations in applicability very similar to

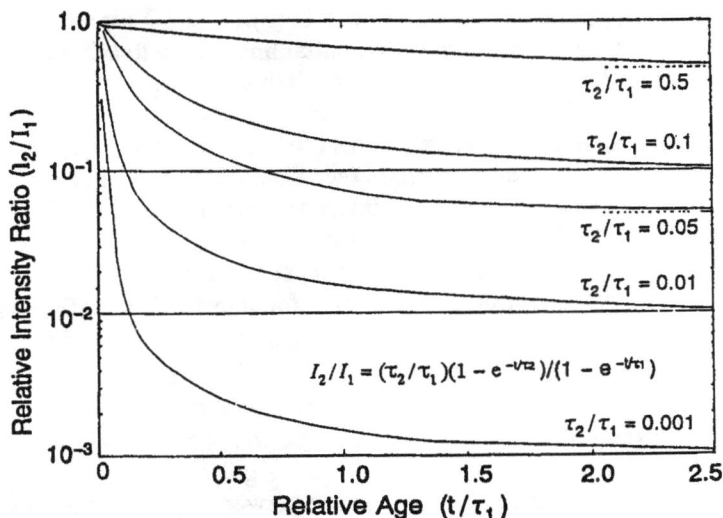

Figure 6.5   Relative signal intensity (I_2/I_1) normalized by the ratio of the lifetime (τ_2/τ_1) as a function of the age t normalized by the lifetime τ_1. ESR dating from the intensity ratio is difficult as the lifetimes of defects are temperature dependent. Paleotemperature may be assessed from the ratio.

that in chemical ESR dating. An attempt at a dating application was made on Italian mollusks using the signals at $g = 2.0018$ and 2.0014, but was not successful because of the uncertainty of the lifetimes (Grün 1985).

The effect of temperature variation could be used to get a more precise ESR age from the intensity ratio using the record of $\delta^{18}O$ variation in deep–sea cores. Since we now know models for signals A and B (SO$_2^-$ and SO$_3^-$, respectively), the conversion mechanism from SO$_3^-$ to SO$_2^-$ must be considered in the above arguments. Nevertheless, the relative intensity ratio may be used for the estimation of *paleotemperature* for samples of known ages and known paleoenvironment from the presence of sulfite radicals. Preliminary study of Holocene corals indicates a high content of SO$_2^-$ of about 3 ka suggesting nearby volcanic action at that time.

6.3.3   ED of Shells

The *ED* for shells has been obtained by the conventional additive dose method in which the extrapolation of the growth to the zero ordinate is made, as described in Chapter 3. Figure 6.6 (a) shows ESR derivative lines of a shell before and after artificial γ–ray irradiation. The intensity of signal A (SO$_2^-$) was not enhanced by γ–irradiation, while that of signals B, C and D clearly grew with radiation dose as shown in (b). The *ED* for signals B and C is obtained by a least–square fit of the initial linear growth according to Eq. (3.3). Usually the full amplitude of the derivative signal is taken as the signal intensity. However, the amplitude of signal C from the baseline at the high field side (indicated by C/2 as the half amplitude of signal C) is preferred. The "*ED plateau*" method (see Chapter 3) was applied to shells using the ESR absorption spectrum instead of a derivative one (Hütt *et al.* 1985, Molodkov 1988). Fitting to a saturation curve rather than least–square fitting to a linear growth line was used in these studies.

The *ED*–values differ considerably for signals C and D. Thermal annealing at 180°C reduces the unstable signal which overlaps signal D and produces a concordant *ED* for signals C and D as shown in Figure 6.7 (Takano and Fukao 1993). When the broad signal at $g = 2.0045$ overlaps, we might possibly overestimate the *ED*. Caution is needed since heating at higher temperature produces interfering Mn^{2+} signals.

Using a logarithmic scale a relation between the *ED* obtained using signal C of aragonitic shells and the geologic ages of the sediments or the radiometric ages is observed (Ikeya 1981, 1986). An overall correlation is observed even though the age shows variation by a factor of nearly three for shells from different environments. The apparent annual dose rate is

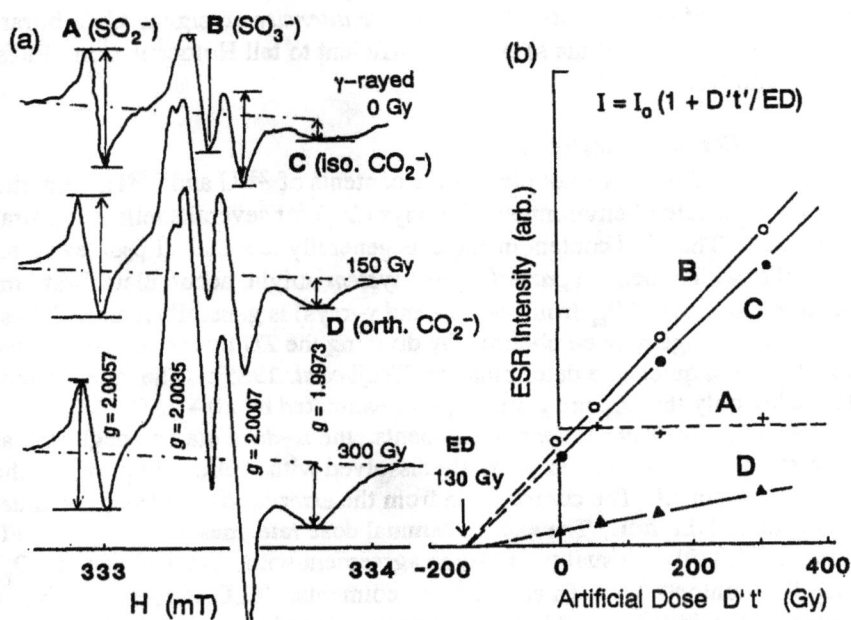

Figure 6.6 (a) ESR spectra of a shell before and after γ–irradiation. (b) Growth of the signal intensity as a function of artificial dose. The *ED* is obtained according to Eq. (3.3).

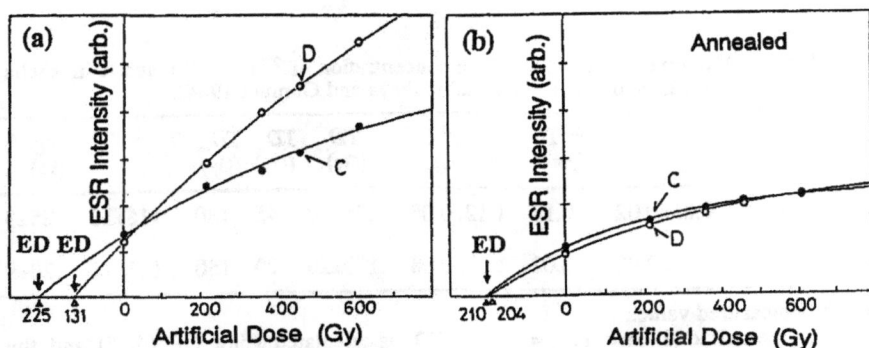

Figure 6.7 Growth curve of signals C and D as a function of artificial dose (a) before and (b) after annealing for 16 min at 180°C (Takano and Fukao 1993).

~ 1 mGy/a, which enables an approximate age estimation without radiation assessment. Measurements of the *absolute intensity* of signal C (arbitrary units per unit weight of the sample) is sufficient to tell Holocene from Pleistocene samples.

6.3.4 Radiation Assessment

Table 6.2 shows examples on the contents of ^{238}U and ^{232}Th and the external dose rate of environmental γ–rays (D_{ex}) for several shells in natural shell beds. The ^{238}U content in shells is generally less than 1 ppm even for old shells which being a *partial open–system* might accumulate uranium. The internal dose (TD_{in} from α–, β– and γ–rays) is generally small. Thus, the minimum age may be obtained by dividing the *TD* by the external dose rate D_{ex} for a quick age determination (Tsuji *et al.* 1985). The ages calculated using only the D_{ex} are generally overestimated by 10 ~ 15%.

Among the *external dose* components, the α–dose may be neglected as the surface of the sample is normally dissolved with acetic acid prior to the ESR measurement. The contribution from the *external β– and γ–rays* must be considered in full. The external annual dose rate measured by the TL dosimeter (TLD) is usually grossly in agreement with that calculated ($D_γ$) using the contents of U, Th and K$_2$O in sediments. TLD elements within a polyvinyl chloride pipe or thin (2 mm) copper (Cu) pipe are generally inserted 50 cm below the ground surface. The dose rate of cosmic rays (0.15 ~ 0.2 mGy/a) must be added to the calculated $D_γ$ from the sediment.

The internal dose rate of α– and β–rays is taken into account for age determination. The *internal dose* (TD_{in}) is calculated using Eq. (4.28) for

Table 6.2   Examples of the *TD* and the concentration of ^{238}U, ^{232}Th and K in shells and sediments for ESR dating (revised after Ikeya and Ohmura 1984).

Shells	^{238}U (ppm)	^{232}Th (ppm)	D_{ex} [a] (mGy/a)	TD (Gy)	TD_{in} (Gy)	TD_{ex} (Gy)	T_{ESR} [b] (ka)	Th/U age [c] (ka)
Arca boncardi	0.85±0.02	0.12	1.12±0.08	175±20	45	130	115±15	125±9
Glycymeris	0.43±0.01	<0.02	1.12±0.08	170±20	20	150	133±15	120±9

a) TLD measured value.
b) The internal total dose $TD_{in} = kTD_α + TD_β$ is calculated using Eq. (4.28) and the external total dose is $TD_{ex} = D_{ex} T_{ESR}$.
   The age T_{ESR} is calculated using $r_0 = {}^{234}U/{}^{238}U = 1.15$ and $k = 0.2$.
c) Th/U ages by A. Omura.

^{238}U–series disequilibrium for α– and β–rays. An α–ray efficiency (k–value) of 0.15 was used at an early stage but recently a small value of $k = 0.05$ has been proposed (Lyons and Brennan 1991). The TD_{in} is about $10 \sim 20\%$ of the TD and therefore k–value reduction from 0.15 to 0.05 does not affect the ESR age greatly. Hence, accurate assessment of the average external β– and γ–ray dose rates during the burial period is vital to the age determination.

The β–rays from the environment produce defects at the surface 1 mm of shells. Dissolving the surface with acetic acid reduces the contribution of β–rays as well as the signal intensity of Mn^{2+} for *colored shells*. The effect of shell thickness and surface removal rate by acid treatment on the ED has been studied (Nakazato *et al.* 1993). The results can be analyzed using the β–attenuation factor of a thin plate shown in Figure 4.3 (c). If the external β–rays come from both sides of the shell, twice of the attenuation factor should be used. The results indicate a larger dependence on the etching thickness than was expected in early studies (Ikeya and Ohmura 1984). Much attention should be paid to the *average thickness* of the sample and the high dose rate at the surface due to β–rays. The *shape of shells* in surrounding soil sediment must be taken into account for a precise age determination. Assessment of the D_{ex} is non–trivial for *spiral shells* because the dose depends on the shape or local radiation dose.

There is an argument in ^{230}Th/^{234}U dating studies that shells are almost an open system. The initial uranium concentration in shells is usually very low. Uranium may be accumulated in old fossil shells but its content is still generally very low. The present content of ^{238}U may be used for assessment of the current internal dose. If uranium has accumulated at a constant rate up to the present level, the internal dose can be calculated as for the case of an open–system fossil bone in Chapter 8. However, this effect is negligible since it is the external dose that constitutes most of the TD for shells because of their low ^{238}U content.

The ^{230}Th/^{234}U dating is directly affected by the incorporation of uranium but in contrast the effect is small for ESR ages which strongly depend on the environmental radioactivity and on the TD rather than the specific radioactivities of incorporated radionuclides. *Movement of radioactive elements* within sediments affects the ESR age. Some shells in clays have given reasonable ages, while others in sand sediment have given much younger ages. The *water content* and the *compression* of sediments due to sediment overburden may result in variation of the dose rate.

ESR dating of fossil shells has been made for Pleistocene samples from

Japan: Ikeya and Ohmura 1981, 1984, Ninagawa *et al.* 1985, Tsuji *et al.* 1985,
     Shimokawa *et al.* 1992, Imai and Shimokawa 1993, Nakazato *et al.* 1993
Italy: Radtke *et al.* 1981, 1985, Radtke 1986
Estonia: Hütt *et al.* 1983, Molodkov 1986, 1988, 1989, 1993,
     Molodkov and Raukas 1987
Western Australia: Hewgill *et al.* 1983
Arctic Sea: Katzenberger and Grün 1985
South Africa: Goede and Hitchman 1987
Chile: Radtke 1987
Argentina: Huang *et al.* 1989
China: Peng *et al.* 1989
USA: Skinner 1989, Skinner and Weicker 1992, Skinner and Mirecki 1993

The ages of Holocene shells in Hamana Lake bore–hole cores in Japan have been obtained using signal C at a high microwave power of 50 mW (Kai and Ikeya 1989). Holocene samples from Estonia have been studied (Hütt *et al.* 1985, Molodkov and Hütt 1985).

Problems encountered in ESR dating still exist in determining the age of shells; *how to subtract the component of overlapping signals* and *how accurately one can determine the radiation dose* including radioactive disequilibrium. Dating shells older than Pleistocene using signal C poses a problem as the stability may not be sufficient depending on the ambient temperature. Signal B due to SO_3^- gave the same *ED* as signal C but the growth curve shows a tendency to saturation. Hence, it was considered that signal B may be used for young (~ 100 ka) shells (Ikeya and Ohmura 1981).

6.4 Corals on Marine Terraces
6.4.1 *Samples and Localities*

Marine terraces of coral reef islands have been dated to know the *sea level changes* and *uplift of the islands* due to Quaternary tectonic movements with various dating techniques (Bender *et al.* 1979, Bloom *et al.* 1974, Chapell 1974, Ku *et al.* 1990). Corals contain nearly the same amount of uranium as seawater (about 3 ppm) in contrast to shells which contain a much reduced level of uranium. Since the internal dose predominates at the inside of a large coral, an *infinite medium model* for radiation assessment may be used if the sample is taken from the core of a massive coral. Aragonitic corals constitute almost a *closed system* where no radioactive element leaches out or enters the coral since its formation. Thus, reliable ages have been obtained for coral samples by several radiometric dating methods.

Aragonitic coral samples so far studied using signal C are from the coral reefs of:

Ryukyu Islands, Japan: Ikeya and Ohmura 1983, Koba *et al*. 1985,
                                 Ikeda *et al*. 1992a
Barbados in the Caribbean Sea: Ikeya 1984, Radtke and Grün 1988,
                                 Radtke *et al*. 1988, Skinner 1988,
                                 Walther *et al*. 1992
Haiti: Skinner 1985
Hainan Island, China: Peng *et al*. 1989
Sumba Island, Indonesia: Pirazzoli *et al*. 1991
Huon Peninsula, New Guinea: Grün *et al*. 1992
Hawaii: Jones *et al*. 1993

Initial work on corals was done using samples from Kikai–jima (28° 19'N, 129° 40'E) which is the island closest to the Ryukyu Trench. The Ryukyu Group unconformably lies over the Shimajiri Group of Late Miocene to Early Pleistocene as shown in the geologic section of Kikai–jima in Figure 6.8.   The Shimajiri Group began to fold in the latest Pliocene to Early Pleistocene, and an erosional flat surface was developed along the anticlinal axis. The Lower Subgroup of the Ryukyu group was deposited on the erosional surface.   The last emergence occurred at 80 ka and the island has been uplifting since then at a rate of about 2.5 mm/a. Terrace ages are considered to be 120 ~ 80 ka old (Koba *et al*. 1985).

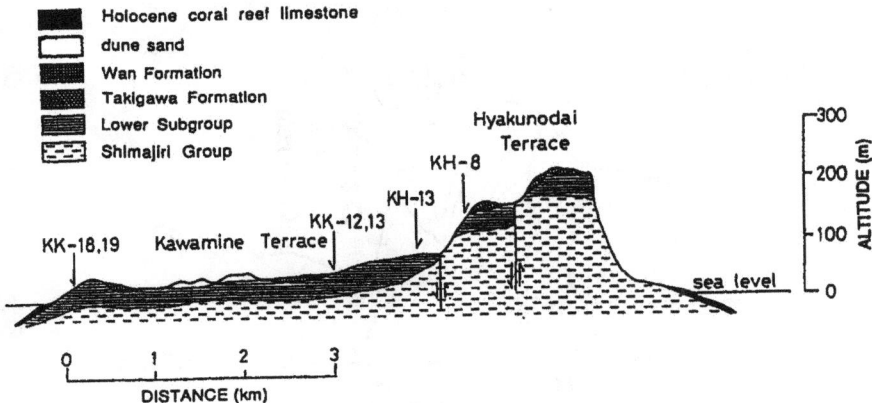

Figure 6.8   Schematic illustration of the geological cross section of Kikai–jima, Ryukyu Islands, Japan for sampling aragonitic corals from marine terraces of the coral reef island.

6.4.2 *ESR Spectra and Age Determination*

Figure 6.9 shows typical ESR spectra of both modern and ^{14}C–dated 2,400 year–old corals from the Ryukyu Islands, Japan. The signals are indicated by A, B, C and D as assigned in the dating of shells. Only a broad signal around $g = 2.0045$ is observed in modern corals. The intensity of signal C is enhanced by γ–irradiation and 1/10 of the signal intensity for the sample K–16 may still be measurable, suggesting that the lower limit of age detectable without computer signal averaging is possibly one or two hundred years. The *ED* for Pleistocene corals has been obtained by linear fitting and the obtained ages were concordant with those by other methods (Ikeya and Ohmura 1983, Ikeya 1985).

Figure 6.10 shows the *ED*–plateau method for a Pleistocene coral from Barbados in the Caribbean Sea (see Section 3.2.5 in Chapter 3). The *ED*'s obtained from the enhancement of the signal intensity after artificial irradiation were plotted against the magnetic field. A plateau was obtained in the region around signals B and C (Miki and Ikeya 1985).

The *average uranium contents* (C_U) of 25 coral samples from the Ryukyu Islands and Barbados are 3.1 and 2.85 ppm, respectively. Hence, an approximate estimation of the ESR age is possible from the obtained *ED*'s

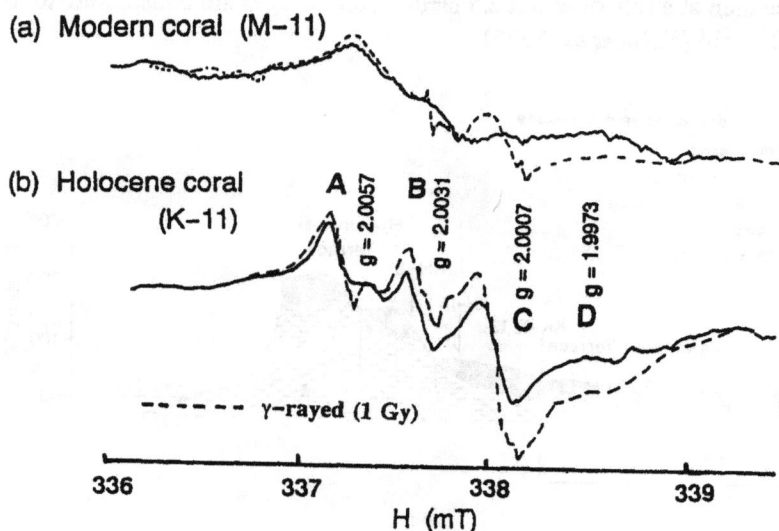

Figure 6.9 ESR spectra of (a) modern (M–11) and (b) ^{14}C–dated Holocene corals (K–16) from the Ryukyu Islands, Japan before (solid curves) and after (dashed ones) additive γ–ray irradiation of 1 Gy.

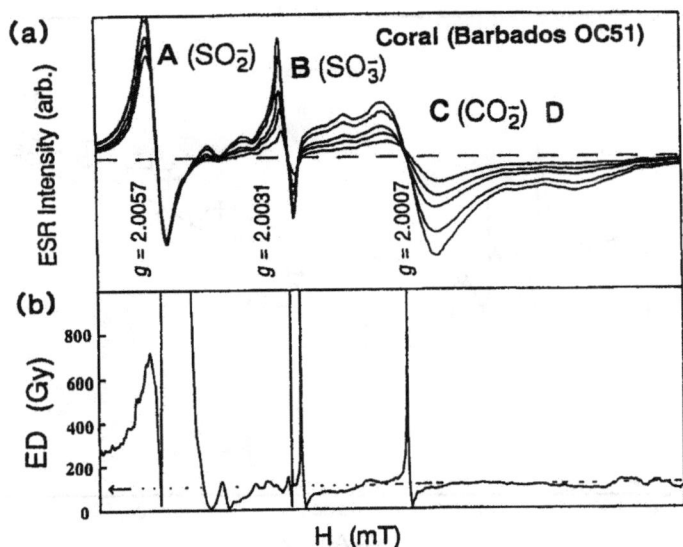

Figure 6.10 A plateau method of ESR dating for a coral from Barbados (terrace II). (a) ESR spectra before and after γ–irradiation and (b) the obtained *ED*'s are plotted against the magnetic field.

assuming $C_U = 3$ ppm.   The variation of the C_U in fossil corals at Barbados terraces I, II and III is presumably due to a *diagenetic change* of the hard coralline carbonate skeleton.

The enhancement of the signal intensity by γ–irradiation is almost the same for Holocene corals. Hence, the intensity of signal C before γ–irradiation is almost proportional to the *ED*.   A linear relation was obtained between the signal intensity and the additive dose from which the intensity was converted to *ED*. The signal intensities and the estimated *ED* are plotted against ^{14}C–ages in Figure 6.11, from which an apparent dose rate of 700 ± 150 μGy/a was obtained. The dashed curve is a theoretical one using Eq. (4.28) empirically best fitted for an apparent $k = 0.15$ and an average $C_U = 3$ ppm for Holocene corals. Thus, relative ESR dating is possible by comparing the signal intensities of corals with those of standard corals dated by ^{14}C– or other methods (Ikeya and Ohmura 1983).   This simple method was used for *screening* Holocene and Pleistocene carbonates before analyzing with ^{14}C dating to avoid needless cost (Radtke 1988).

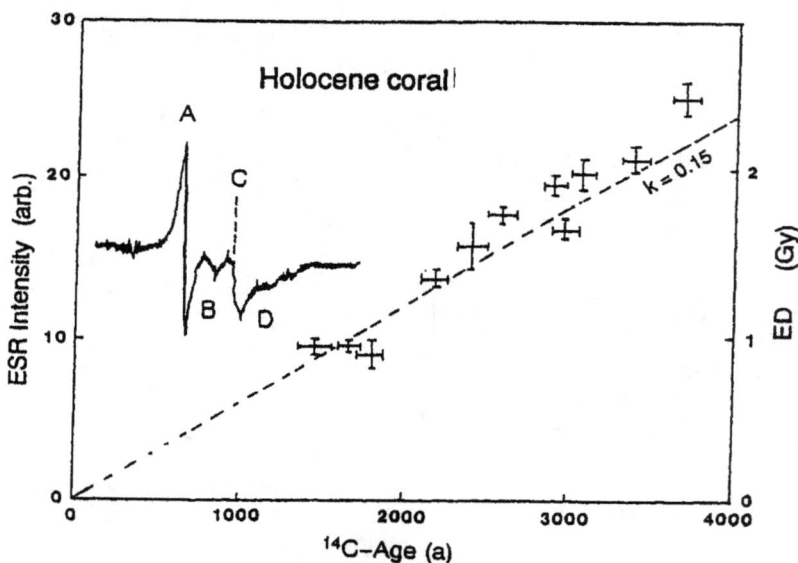

Figure 6.11   Relative ESR dating of corals from a calibration curve using ¹⁴C-age. The *TD* for *k*-value of 0.15 is shown for an average C_U =3 ppm assuming an infinite medium model.

The absolute ESR age can be determined by evaluating the natural radiation. The initial activity ratio of $^{234}U/^{238}U$ (r_o = 1.14) and $^{226}Ra/^{238}U$ (p_o = 0.027 ± 0.005) for modern corals were obtained by γ-ray spectroscopy (Komura 1982), which indicates that ^{238}U enters corals as UO_2^{2+} more easily than $^{226}Ra^{2+}$ does. The *TD* due to initial ^{226}Ra incorporation and to its daughters is described by

$$TD_{26} = C_U p_o (D_{26}/\lambda_{26})(1 - e^{-\lambda_{26} t}) \; , \qquad (6.3)$$

where D_{26} is the dose rate of ^{226}Ra and its daughters corresponding to 1 ppm of ^{238}U in secular equilibrium, and λ_{26} is the decay constant. The shift of the ESR age for p_o = 0.03 is about 100 years at most, which does not affect the coral age much compared with the case in which ^{226}Ra is not incorporated (p_o = 0).   The effect is shown by the dashed line in Figure 6.12 (a). The effect of ^{222}Rn loss (N_{loss} %) on the *TD* versus age relation for N_{loss} = 0% and 10% is shown in (b). The *rate of radon loss* is less than 10% (Yokoyama

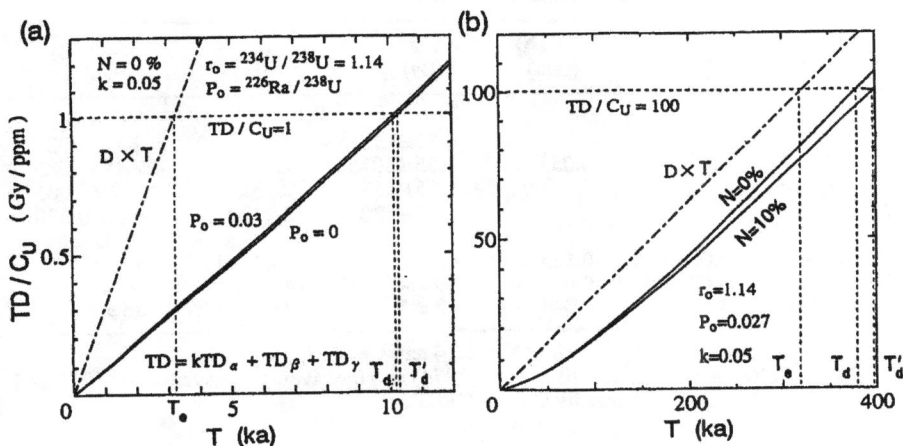

Figure 6.12 The TD (normalized by the C_U) versus age relation for (a) a young coral of $TD/C_U = 1$ Gy/ppm for $p_o = 0.03$ (T_d) and $p_o = 0$ (T'_d), and (b) for Pleistocene corals of $TD/C_U = 100$ Gy/ppm for $N_{loss} = 0\%$ and 10%. The age determination for uranium–series equilibrium (T_e) and for disequilibrium (T_d) is given graphically. The initial activity ratios (r_o and p_o), the rate of radon loss (N_{loss}) and k-value are indicated in the figure.

and Nugyuen 1980) though radon diffuses more easily in other media such as soil (Nozaki *et al.* 1978). Hence the effect of the rate of radon loss on the age can be neglected.

Table 6.3 shows the contents of ^{238}U and ^{232}Th, the ED and the age of corals from the Ryukyu Islands. The age is obtained by assuming ^{238}U–series disequilibrium and initial ^{226}Ra incorporation without ^{222}Rn loss in an infinite medium.

Various ESR ages have been determined assuming *uranium–series disequilibrium in an infinite medium*. The obtained ESR ages are plotted against other radiometric ages in Figure 6.13. A reasonable agreement was obtained with the ^{14}C– and $^{230}Th/^{234}U$ ages within the time range of the latter two methods. The upper limitations of ^{14}C– and $^{230}Th/^{234}U$ age are about 50 ka and 400 ka, respectively. ESR ages can cover from a few hundred years to well above 1 Ma.

Table 6.3 Some examples of *ED* and the contents of ^{238}U and ^{232}Th for corals and the age obtained by assuming ^{238}U–series disequilibrium and initial ^{226}Ra incorporation without ^{222}Rn loss in an infinite medium.

Corals Localities	^{238}U (ppm)	^{232}Th (ppm)	*ED* (Gy)	T_{ESR} [a] (ka)	Ref–age [b] (ka)
Kikai Island					^{14}C–age
76–4–1			1.35±0.15	2.30±0.20	2.17±0.087
75–3–27–12	3.0 [b]	0.05 [b]	1.55±0.015	2.61±0.20	2.39±0.085
75–3–27–4			1.75±0.17	2.95±0.20	2.57±0.085
75–3–27–7			1.95±0.20	3.27±0.29	3.06±0.090
					Th/U–age
CK19 A–Lm.	3.00±0.11	0.125	45 ± 3	49.3 ± 4	41 ± 2
CK21 Y–Lm.	3.42±0.06	0.02	49 ± 4	49.5 ± 4	54 ± 2
CK24 M–Lm.	2.30±0.10	0.044	62 ± 5	76.7 ± 7	86 ± 4

a: Age calculated using $r_o = {}^{234}$U/^{238}U = 1.14 and k = 0.15.
b: Radiometric ages by Konishi *et al.* (1974) and the average contents of ^{238}U and ^{232}Th for corals in this area by Omura (1983).

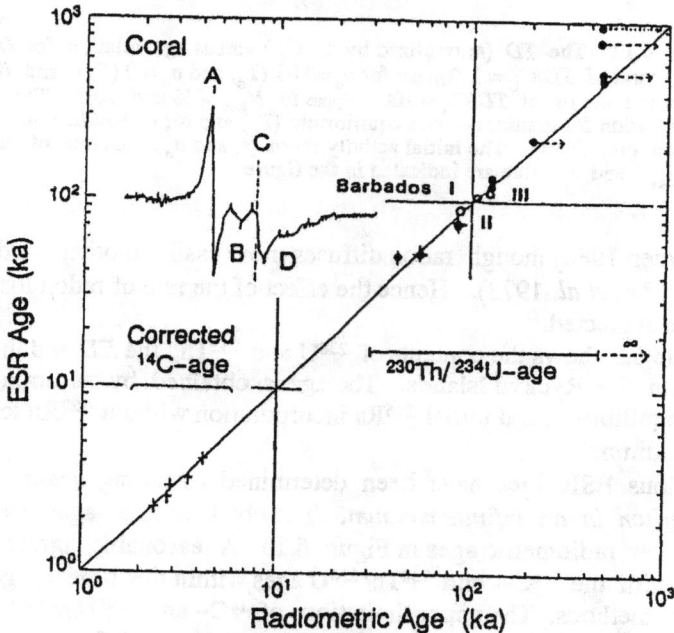

Figure 6.13   Concordance of ESR ages of corals using signal C at g_C = 2.0007 with the results of ^{14}C and ^{230}Th/^{234}U dating. The average ESR ages for Barbados terraces I, II and III are shown by open circles (Ikeya 1986).

6.4.3   *Spread of ESR Ages in Barbados Terraces*

Figure 6.14 shows the spread of ESR ages for the corals at Barbados terraces I (Worthing), II (Ventnor) and III (Rendezvous). Twenty samples were supplied with their ^{238}U contents. The average ESR ages for Barbados terraces I, II and III are 87 ka, 100 ka and 120 ka, in concordance with the reported $^{230}Th/^{234}U$ ages of 82 ± 5 ka, 105 ± 5 ka and 118 ± 9 ka, respectively (Bender *et al.* 1979). According to geologists, terraces of coral reef islands are formed in 5 ~ 10 ka. Hence the ESR ages for corals within the same terrace should be within this time range. However, the spread of the ESR ages within the same terrace is appreciably larger than this range, suggesting the possibility of similar diagenetic alteration at an outcrop. Although corals are considered to constitute a closed system, unlike shells, some *diagenetic change* may occur due to fresh–water flow after the coral rises above the sea level. The local radiation environment may also contribute to the spread as the AFS samples in terrace III are located close to the soil environment. The calculation of ESR ages was made assuming the infinite medium of a massive coral reef and thus external radiation from the nearby soil environment might give high ESR ages. Work on Barbados corals (Ikeya 1986) was repeated, including corals from higher terraces of interest (Radtke and Grün 1988, Skinner 1988).

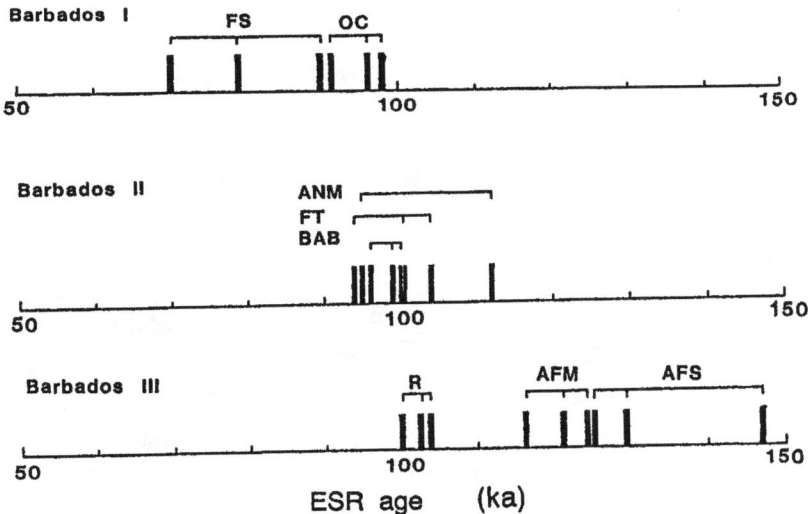

Figure 6.14   The spread of ESR ages for corals from Barbados terraces I, II, and III at different localities indicated by FS, OC, ANM, etc.

6.4.4   Correction Factors: Variation of Trap Concentration

(a)   *Preheating effects* : Growth of the intensities of signals A, B and C by artificial γ–irradiation for samples preheated at 120, 160 and 180°C are shown in Figure 6.15 (Ikeda *et al.* 1992a). *Gamma–ray sensitivities* of signals A, B and C first increase and then decrease with preheating time. This implies that the concentration of the impurity causing each signal is not constant but varies as a function of time and temperature; an aspect as yet overlooked in ESR dating. The SO$_2^-$ radical (signal A) is produced by the thermal decomposition of organics without γ–irradiation and irradiation enhances the intensity to the saturation level. No SO$_3^-$ radical (signal B) is produced by heating before irradiation but subsequent γ–irradiation produced this signal. It is interesting to note that the production rate and the saturation level of signal B are higher at 120°C than at 160°C in contrast to signal A which shows a higher saturation level at 160°C. This suggests that SO$_2^-$ may arise from diamagnetic SO$_3^{2-}$ rather than SO$_3^-$. This is supported by the fact that SO$_2^-$ is an electron center while SO$_3^-$ is a hole center. The preheating effect on signal C (freely rotating CO$_2^-$) may be interpreted as loss of water molecules on heating above 160°C (Katzenberger *et al.* 1989, Walther *et al.* 1992).

Formation and decay of the traps for signal B or C may be explained by the following rate equations,

$$dN_1/dt = -K_1 N_1 , \tag{6.5}$$

$$dN_2/dt = K_1 N_1 - K_2 (N_2)^i , \tag{6.6}$$

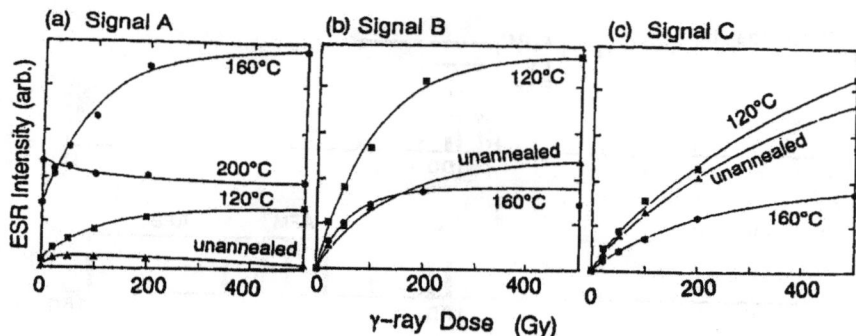

Figure 6.15   Growth of the ESR intensities by γ–irradiation for a non-heated sample and samples preheated at 120, 160, and 200°C for 1 hour. (a) signal A (*g* = 2.0057) (b) signal B (*g* = 2.0031) and (c) signal C (*g* = 2.0007).

and

$$dN_3/dt = K_2(N_2)^i , \qquad (6.7)$$

where N_1 is the number of the main source from which N_2 is formed, N_2 is the number of the sources for signal B or C and N_3 is the number of the products formed by the decay of N_2. K_1 and K_2 are the rate constants of the two reactions and i is the order of the reaction.

The formation and decay of the traps at several temperatures were explained well by assuming third–order decay. For example, some impurity centers in NaCl decay according to third–order kinetics and aggregate to form stable trimers (Ikeya and Crawford 1973). Although rigorous solution of the rate equations described above is difficult, an approximation to Eq. (6.6) by changing K_1, K_2 and N_1 finds their appropriate values. An Arrhenius plot of the rate constants K_1 and K_2 of each annealing temperature gave the extrapolated constants at 20°C as shown in Figure 6.16 (a). The variation in the number of sources (N_2) as a function of geologic time was obtained using K_1 and K_2. The number of traps for signal C drastically decreases in

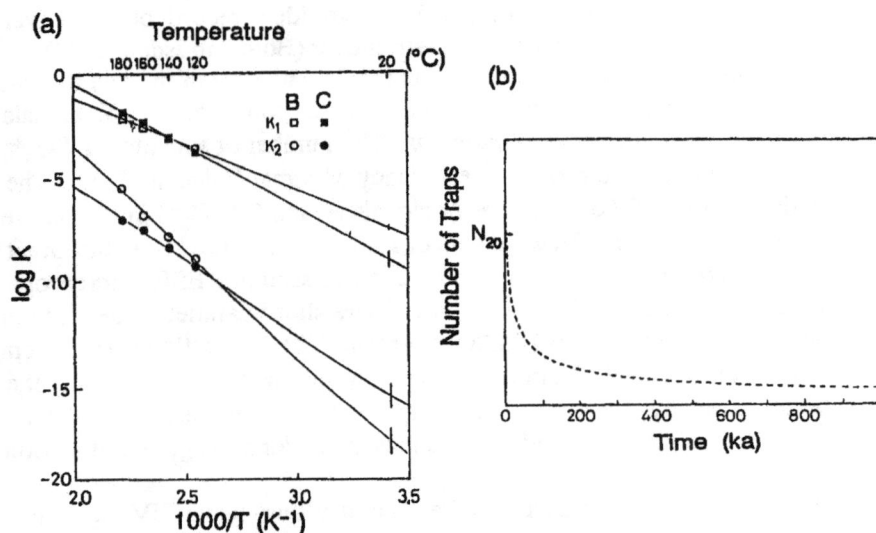

Figure 6.16 (a) An Arrhenius plot of the rate constants K_1 and K_2 against reciprocal annealing temperature. (b) Variation in the number of traps at 20°C as a function of time using K_1 and K_2 of signal C. N_{2o} means the initial value of N_2. The number of traps for signal C decreases rapidly in the first 100 ka. Correction is necessary for samples younger than 100 ka.

the first 100 ka as shown in (b). Some correction of ESR ages is necessary for precise dating.

The number of traps at a certain geologic age is in proportion to the sensitivity to artificial γ–rays when we estimate the ED. Hence, the variation in the sensitivity can cause underestimation of ESR ages using signal C for Holocene and late Pleistocene corals and overestimation for Early and Middle Pleistocene corals. In early work, the difference was attributed to an empirical k–value. The change in the defect formation efficiency during geologic time may be related to decomposition of organic materials or release of crystal water in a coral, which may lead to the increased formation efficiency of signals A, B and C.

6.4.5 Distribution of SO₃⁻ Radicals (signal B) in a Coral

Massive coral colonies grow at a rate of $4 \sim 20$ mm/a, recording changes of ocean surface conditions in their calcium carbonate skeleton. *Annual skeletal density bands* record changes in sunlight and water temperature (Knutson *et al.* 1972). *Stable oxygen isotope ratios* ($\delta^{18}O$) provide a record of water temperature and salinity (Fairbanks and Dodge 1979). Fluorescent humic compounds in coral skeleton provide a record of past river discharge, rainfall and shallow water environment (Boto and Isdale 1985).

A variety of radicals are observed in corals. In view of the fact that the relative impurity content depends on environmental conditions, ESR signals could serve as an *environmental indicator*. The number of substituted SO_4^{2-} ions in coral affects the formation efficiency of signal B due to SO_3^-. The spatial distribution of the intensities of signals B and C (CO_2^-) in a massive coral *Porites sp.* from the leeward side of Hudson Island (18° S) in the Great Barrier Reef was investigated using a microwave scanning ESR microscope described in Chapter 14. The sample coral was sliced parallel to the growth direction and was exposed to 10 kGy of γ–rays. The flat section was set on an aperture on the wall of a microwave cavity and was scanned along the growth direction by 1 mm step over an aperture 3 mm in diameter. Thus, ESR intensities of signals B and C were measured along the growth direction of the coral (Ikeda *et al.* 1992b).

Figure 6.17 shows fluorescent bands under long–wave UV light and the measured ESR intensities in 15 years of growth. The yellow–green fluorescent bands appear as the high density regions at the same periods or slightly earlier. Because the *fluorescence* is caused by terrestrial *humic compounds*, it is considered that the fluorescent bands correlate with summer, a season of monsoonal rainfall in the Great Barrier Reef (Klein *et al.* 1990).

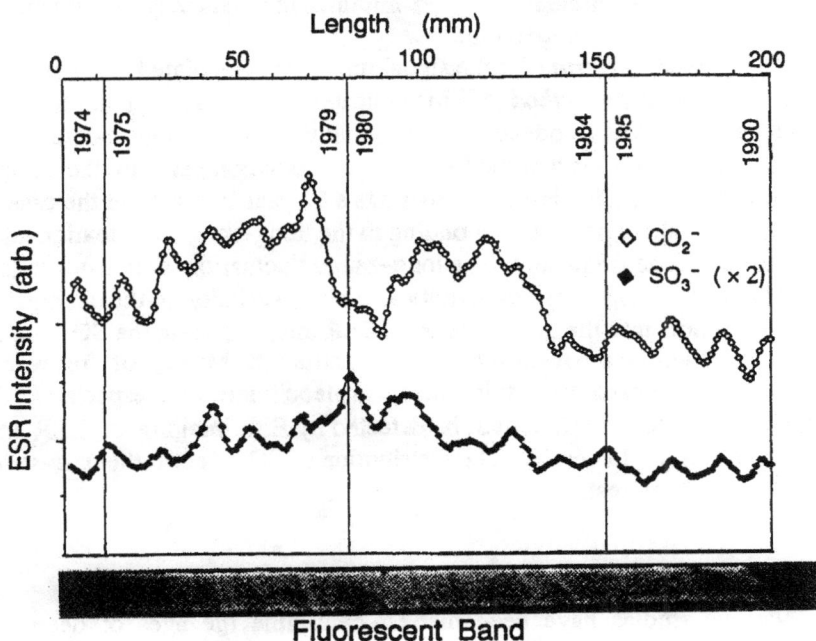

Figure 6.17   Spatial distribution of fluorescent bands under long-wave UV light and the ESR intensities of CO_2^- and SO_3^- radicals in the 15 years' growth of a massive coral (Ikeda *et al.* 1992b).

The intensities of ESR signals also vary periodically and thus, periodic structures in ESR signal intensities consist of *annual fluctuations*. High intensity regions of signal C correspond to the annual skeletal high density regions, and those of signal B correlate with the fluorescent bands. This indicates that the fluctuation in the intensity of signal C is attributable to variation in the density, while that of signal B may originate from variation in the amount of SO_3^{2-} ions incorporated in coral probably due to changes in the SO_3^{2-} concentration in seawater. The SO_3^{2-} ions dissolved from terrestrial humic compounds or those from meteoric water may be responsible for the increased incorporation of SO_3^{2-} in summer. The synchrony of fluorescent bands with high intensity regions of signal B supports this possibility. Another possible cause for increased SO_3^{2-} incorporation is the biologically induced *mineralization* of the coral. The growth rate of corals varies periodically, mainly by water temperature and light level. The *distribution coeffi-*

cient or repulsion coefficient for SO_3^{2-} impurity in corals may depend on the growth rate and so the temperature.

The *power spectrum* for the signal intensities calculated by means of the *maximum entropy method* (MEM) indicates the annual fluctuation of an intense peak at a low frequency region in addition to the long–range variation. The frequency ratio of the long–range fluctuation peak to the annual one is about 1/12, indicating the former has a 12 year interval. In the case of the CO_2^- signal, a peak corresponding to the long–range fluctuation also arises at the same frequency. The long–range fluctuation seems to reflect global climatic change or solar activity as well as variation in the amount of freshwater runoff into the nearshore sea. ESR imaging using the SO_3^- radical in corals will be a powerful tool to reconstruct the history of freshwater runoff into the nearshore environment. Paleoclimate and especially the atmospheric content of SO_2 will be detected by ESR imaging or ESR microscopy of SO_3^- in corals. The distribution of SO_3^- after the *Industrial Revolution* is of interest.

6.5   Deep–Sea Sediment : Planktonic Foraminifera
Various studies have been made to determine the ages of deep–sea sediments. One is by TL dating which is based on the hypothesis that the defects in minerals are optically bleached by sunlight while they are floating in air or in the water. Another is by variation in paleomagnetism and oxygen isotopes in deep–sea sediment cores. The ESR age at each site is of much interest as we can determine the deposition rate and check the ESR age against the paleomagnetic age. More importantly, an independent absolute core age by the non–destructive ESR method would contribute to paleoclimatology.

A systematic study on *calcite foraminifera* has been made by the Heidelberg group (Mangini *et al.* 1983, Barabas *et al.* 1988) following the work by Sato (1981, 1982). ESR spectra of foraminifera show the following signals:

$$g_A = 2.0057 \text{ (freely rotating } SO_2^-)$$
$$g_B = g_\perp = 2.0036 \text{ and } g_{\parallel} = 2.0021 \text{ (axial } SO_3^-)$$
$$g_C = 2.0007 \text{ (freely rotating } CO_2^-)$$

The signal at $g = 2.0036$ ($g_\perp$ of SO_3^-; signal B in calcite) has been used for dating foraminifera. The relation between the depth of the core and the *ED* was shown to be in stratigraphic order. In Sato's first calculation, the *ED* at the site of *geomagnetic reversal* was tuned to give 780 ka and the annual dose rate was treated as a constant. However, it was pointed out that uranium–series disequilibrium (Wintle and Huntley 1983, Sato 1983) and the initial incorporation of ^{230}Th in the deep–sea sediment (Ikeya 1985) must be taken into account to calculate the *TD*. Agreement between the ESR ages and $\delta^{18}O$ stratigraphy was observed up to 800 ka. It was suggested that signal C (CO_2^-) could not be used for samples over 100 ka from the lifetime obtained by annealing experiments (Mudelsee *et al.* 1992, Barabas *et al.* 1992b). The lifetime of such experiments has a large ambiguity, especially for the signal associated with water molecules.

Figure 6.18 shows the *TD* versus age relation for the reported content of ^{238}U (0.26 ppm) and initial incorporation of ^{230}Th estimated from the surface content of the daughter ^{226}Ra which corresponds to 13.5 ppm of the parent ^{238}U. Using the contents of ^{232}Th (1.67 ppm) and ^{40}K (0.56%), one can calculate the *TD* as a function of time assuming an infinite medium. The rapid increase of the *TD* from the surface with depth is due to the defects produced by the decay of the initial ^{230}Th. The more gradual increase of the

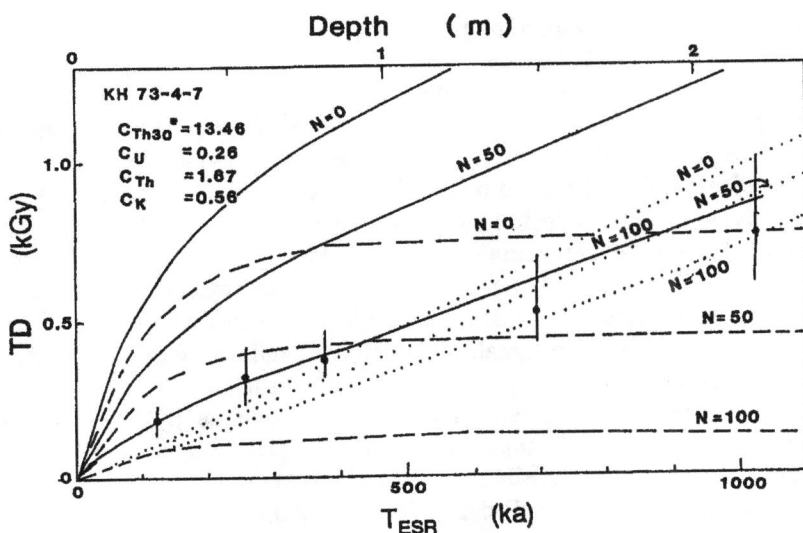

Figure 6.18 The *TD* versus *T* relation for planktonic carbonate fossil of foraminifera taking the content of ^{238}U into account. The TD_{230} due to the initial incorporation of ^{230}Th is shown as dashed curves.

TD subsequently arises from a nearly constant dose rate due to disequilibrium in decay of parent uranium. Reasonable agreement is obtained for a rate of radon loss (N_{loss}) of 30 ~ 50%. If the low content of ^{226}Ra is due to diffusion or dissolution from the sediment, the effect must be included in the calculation. The rate of ^{222}Rn loss must be studied for deep–sea sediments.

When the intensity of the signal at $g = 2.0036$ was plotted as a function of depth, an overall increase in the intensity up to the last 2 Ma was observed (Mangini *et al.* 1983, Siegele and Mangini 1985, Takeuchi and Saeki 1985). The saturation of the signal growth above 2 Ma has been attributed to the lifetime of the defect.

An interesting observation is that the intensity of the Mn^{2+} signal increases as a function of the depth and so the age up to 4 Ma, while signal B increases up to 2 Ma. Presumably Mn^{2+} diffuses into the $CaCO_3$ lattice with the passage of time even at low temperature in deep–sea sediments. Thus, relative dating is possible using the *thermal diffusion* of Mn^{2+} (Ikeya 1985). Foraminifera can be used as a pilot sample in ESR dating (Barabas *et al.* 1988, 1992b) since the age can be estimated by other methods of dating, oxygen isotopes and the rate of sedimentation (Mudelsee *et al.* 1992).

6.6 Application in Archaeology
6.6.1 *Shells from Shell Mounds*

Shells and other archaeological remains can help to solve a variety of archaeological problems. Shells about a few thousand years old show only a very weak signal of radiation–induced radicals in addition to signals due to organic radicals. The signal to noise (*S/N*) ratio for young shells is poor unless a signal averaging technique is employed. Some shells, especially those in fresh water, contain small amounts of Mn^{2+} and can be dated with ESR if they are older than a thousand years. However, marine shells, especially layered shells like oysters contain a high concentration of Mn^{2+} which masks the radiation–induced signals. Therefore, shell dating of archaeological remains is more difficult than for old geological shells.

A broken shell piece in cemented soil at the Palenque pyramid (1,300 BP) in the Maya zone, Yucatan peninsula, Mexico gave an age of 40 ~ 50 ka. Naturally, this is the age of shells in cemented soil and not the date when the pyramid was constructed. ESR dating of shells at the Japanese Jomon and Yayoi shell mounds (Nakajima *et al.* 1993), shells in the pre–Inca era (2 ka) and Sambaqui shell mounds in Brazil (Baffa and Mascarenhas 1985) indicates that ESR dating of shells is possible using signal C at $g = 2.0007$ as well as signal B at $g = 2.0031$. Figure 6.19 (a) shows the typical ESR spectra of a shell from a shell mound in Australia together with the enhancement of the signal intensity by artificial irradiation. The *ED* of a few Gy is obtained. The *ED* is compared with ^{14}C–age in (b) and an apparent annual dose rate of 1.2 mGy/a is obtained by tuning to the ^{14}C–ages. Also, organic radicals are produced in some cases by artificial irradiation, which makes the evaluation of the *ED* using signal C difficult. Organic remains have not yet greatly decomposed in shells of a few thousand years old, and therefore create a broad signal, not a septet spectrum, by irradiation.

6.6.2 Egg Shell

Egg shell is also calcium carbonate of biological origin. A septet due to organic radicals was first analyzed in chicken egg irradiated by γ–rays and subsequently heated (Kai *et al.* 1988). The detectable radiation dose for a modern egg shell was about 0.5 ~ 0.1 Gy with a commercial ESR spectrometer (sensitivity: 10^{10} spins/0.1 mT). This suggests that dating with ESR is possible for shells older than a few thousand years depending on the environmental dose rate. The *ED* of ostrich egg shells from the anthropological Douara cave, Syria is 42 Gy and the ^{14}C–age is > 50 ka (see Chapter 8).

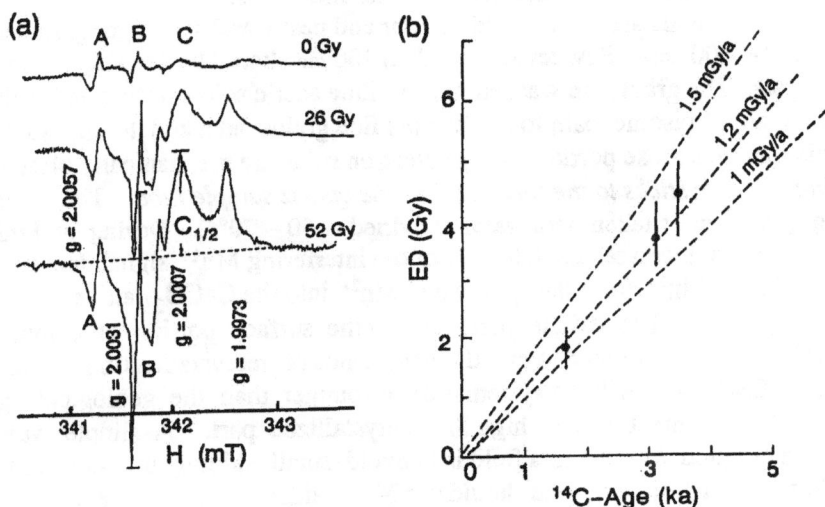

Figure 6.19 (a) ESR spectra of shells at shell mound in Australia. (b) The *ED* versus ^{14}C–age plot gives an appropriate dose rate of 1.2 mGy/a mostly from the external radiation by β– and γ–rays.

6.7   Technical Notes

(a) *Sampling in fields* : For calcitic shells, avoid colored shells and layered shells such as oysters as they contain paramagnetic Mn^{2+} which masks radiation–induced radical signals. Fresh–water shells are good in many cases. Aragonitic shells contain almost no Mn^{2+}. For corals, take a sample from the center of a large coral (> 30 cm) so that the sample could be regarded as an infinite medium. This makes radiation assessment easy.

Samples that seem to have been in a homogeneous radiation environment should be selected. In other words, samples close to a big rock or deposited in soils should be avoided since assessment of the radiation dose rate is difficult. Unless the *in situ* measurement of the external dose rate D_{ex} is made by embedding TLD, collect sands or soils in the vicinity of the sample.

(b) *Sample preparation* :

<u>Shell</u> : An important practical point in dating shells with ESR is to remove the surface colored region by dipping shells in 0.1 N acetic acid. The surface part is in some cases a calcite deposit with a high content of Mn^{2+} (Hütt *et al.* 1983). The effect of external α–rays from surrounding sediments can thus be removed. Etching with acetic acid is better than with HCl for calcite speleothems and possibly for shells and corals.

The shells are ground with mortar and pestle and sieved to grain sizes of 100 ~ 300 μm.   Powder smaller than 100 μm should be used for uranium analysis. The grains are washed with a dilute acetic acid solution and agitated in an ultrasonic bath to remove the fine grains attached to the surface. This removes those portions whose electron spins are mechanically bleached. *Fine powder sticks to the inner wall of the quartz sample tube.* The remaining grains are shaken with water and dried at 50 ~ 70°C. Drying at higher temperature increases the intensity of the interfering Mn^{2+} signal due to the diffusion of Mn^{2+} from the aggregated Mn^{2+} into the CaCO$_3$ lattice.

<u>Coral</u> : The calcite part, usually the surface portion of a massive coral, must be removed to avoid the influence of recrystallization. Otherwise, ESR ages will be systematically younger than the geological ages. The Mn^{2+} content is also high in a recrystallized part. A simple way to obtain a good sample in a field is to avoid small shining microcrystals of calcite. In laboratory, one should use X–ray diffraction to check a ratio of calcite/aragonite (Davies and Hooper 1963). Sample pretreatment such as grinding and washing with acid is identical to that for shells and speleothems.

(c) *ESR dating* : The conditions of ESR measurement, additive γ–ray irradiation and *ED* determination from the dating signal component (g_C =
2.0007) are essentially the same as those for speleothems in Chapter 5.
Figure 6.20 shows (a) *microwave power dependence* and (b) *isochronal
annealing* of signals A, B, C and D as well as of the signal of orthorhombic
SO_3^- for corals. Appropriate microwave power should be chosen to measure
the signal. Unstable signals must be excluded by heating at 60°C for a day or

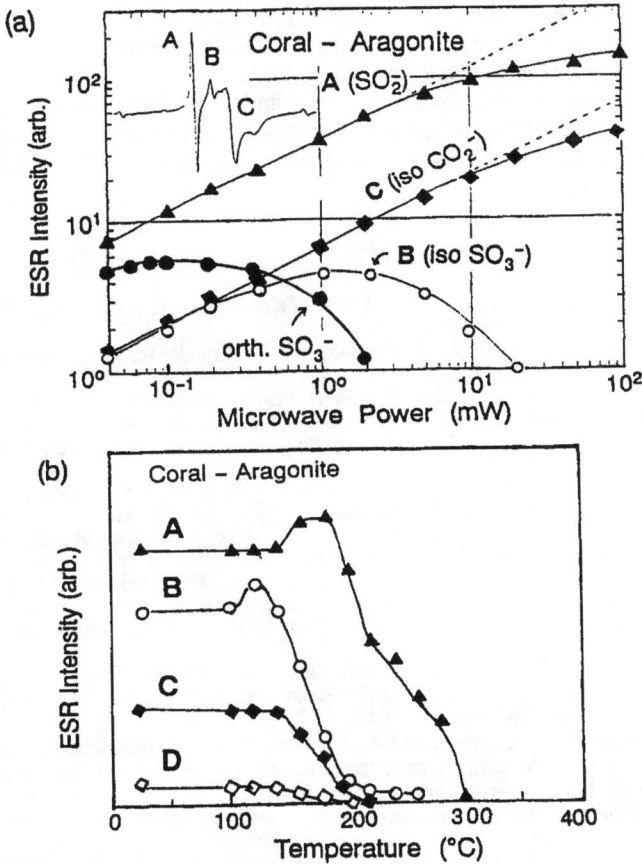

Figure 6.20 (a) Microwave power dependence and (b) isochronal annealing
curves of the signal intensity for an aragonitic coral sample (after Kai and Miki
1991).

100 °C for 15 min, or by keeping the specimen for a few weeks after γ-irradiation. Efforts are going on to establish practical applications for determination of terrace ages were made in 20-th century (Radtke, 2001).

6.8 Summary

Aragonitic shells are the most suitable samples for ESR dating as a partial recrystallization can be recognized by seeing the shining crystal phase of calcite. ESR ages of these fossils concord well with ^{14}C- and ^{230}Th/^{234}U ages within their respective time spans. ESR can cover the time range of entire Quaternary, which is not possible by other methods. The impurity-associated signals such as SO$_2^-$, which are not usable for dating, are useful for assessing the historical environmental change. Annual variation in the SO$_3^-$ signal intensity, for example, might indicate paleo-environmental changes and climate fluctuations. As the required time and the cost for ESR dating are considerably less than for other methods of dating, this method should become a reliable chronological method in oceanography or marine geology.

References

Baffa O. and Mascarenhas S. (1985): ESR dating of shells from Sambaqui. ESR *Dating and Dosimetry* (Ionics, Tokyo) 139 -143.

Barabas M., Bach A. and Mangini A. (1988): An analytical model for the growth of ESR signals. *Nucl. Tracks* **14**, 231-236

Barabas M., Bach A., Mudelsee M. and Mangini A. (1989): Influence of the Mg-content on ESR signals in synthetic calcium carbonate. *Appl. Rad. Isot.* **40**, 1105-1111.

Barabas M. (1992): The nature of the paramagnetic centers at g = 2.0057 and g = 2.0031 in marine carbonates. *Nucl. Tracks* **20**, 453-464.

Barabas M., Bach A., Mudelsee R. and Mangini A. (1992a): General properties of the paramagnetic center at g = 2.0006 in carbonates. *Quat. Sci. Rev.* **11**, 165-171.

Barabas M., Mudelsee R., Walther R. and Mangini A. (1992b): Dose response and thermal behavior of ESR signal at g = 2.0006 in carbonates. *Quat. Sci. Rev.* **11**, 173-179.

Barabas, M., Walther R., Wieser A., Radtke U. and Grun R. (1993): Second interlaboratory-comparison project on ESR dating. *Appl. Radiat. Isot.* **44**, 119-129.

Batty M. H. (1981): *Carbonates. In Mineralogy for Students* (Longman, London and New York) 219.

Bender M.L., Fairbanks R.C., Tailor F.W., Matthew R.K., Goddard J.G. and Broecker W.S. (1979): Uranium-series dating of the Pleistocene reef tracts of Barbados, West Indies. *Geol. Soc. Am. Bull.* **90**, 577-594.

Bloom A.L., Broecken W.S., Chappell J.M.A., Matthews R.K. and Mesolella K.J. (1974): Quaternary sea level fluctuations on a tectonic cast: new ^{230}Th/^{234}U dates from the Huon Peninsula, New Guinea. *Quat. Res.* **4**, 185-205.

Boto K. G. and Isdale P. (1985): Fluorescent bands in massive corals result from terrestrial fulvic acid inputs to nearshore zone. *Nature* **315**, 396-397.

Brumby S. and Yoshida H. (1994): ESR dating of mollusc shell: investigations with modern shells of four speccies. *Quatern. Geochr.* **13**, 157-162.

Brumby S. and Yoshida H. (1994): The annealing kinetics of ESR signals due to paramagnetic centers in mollusk shell. *Rad.at. Meas.* **24**, 255-263.

Chappell J. (1974): Geology of coral terrace, Huon Peninsula, New Guinea: a study of Quaternary tectonic movements and sea-level changes. *Geol. Soc. Am. Bull.* **85**, 553-570.

Davies T.T. and Hooper P.R. (1963): The determination of the calcite: aragonite ratio in mollusk shells by X-ray diffraction. *Mineral. Magazine* **33**, 608-612.

Debuyst R., Dejehet F., Grun R., Apers D. and DeCanniere P. (1984): Possibility of ESR dating without determination of the annual dose. *J. Radioanal. Nucl. Chem. Lett.* **86**, 399-410.

Fairbanks R.G. and Dodge R.E. (1979): Annual periodicity of the $18O/16O$ and $^{13}C/^{12}C$ ratios in the coral Montastrea annularis. *Geochim. Cosmochim. Acta.* **43**, 1009-1020.

Goede A. and Hitchman M. A. (1987): Electron spin resonance analysis of marine gastropods from coastal archaeological sites in South Africa. *Archaeometry* **29**, 163-174.

Grun R. (1985): ESR-dating without determination of annual dose: a first application on dating mollusk shells. *ESR Dating and Dosimetry*, 115-123.

Grun R., Radtke U. and Omura A. (1992): ESR and U-series analyses on corals from Huon Peninsula, New Guiana. *Quat. Sci. Rev.* **11**, 197-202.

Hantoro W.S. , P. Pirazzoli, C. Jouannic,et al., (1994): Quaternary uplifted coral reef terraces on Alor Island, East Indonesia. *Coral Reefs,* **13**, 215-223.

Hewgill F.R., Kendrick G.W., Webb R.J. and Wyrwoll K.H. (1983): Routine ESR dating of emergent Pleistocene marine units in Western Australia. *Search* **14**, 215-217.

Huang P., Liang R., Jing S., Peng Z. and Rutter N.W. (1989): Study on accumulated dose in littoral shells of Argentina. *Appl. Radiat. Isot.* **40**, 1119-1122.

H utt G., Molodkov A., Punning J.M. and Pung L. (1983): The first experience in ESR dating of fossil shells in Tallin. *PACT J.* **9**, 433-438.

H utt G., Molodkov A., Kessel H. and Rankas A. (1985): ESR dating of subfossil Holocene shells Estonia. *Nucl. Tracks* **10**, 891-898.

Ikeda S., Kasuya M. and Ikeya M. (1992a): ESR dating of corals and pre-annealing effects on ESR signals. *Quat. Sci. Rev.* **11**, 203-207.

Ikeda S., Neil D., Ikeya M., Kai A. and Miki T. (1992b): Spatial variation of CO_2^- and SO_3^- radicals in massive coral as environmental indicator. *Jpn. J. Appl. Phys.* **31**, L1644-L1646.

Ikeda S., Furusawa M. and. Ikeya M. (1994): Spatial variation of CO_2^- and SO_3 in a massive coral from Ishigaki Island, Japan and its implication. *Proc. IGC,* Part B, 225-228. (VSP).

Ikeya M. and Crawford J.H. (1973): EPR study of impurity-vacancy aggregate in NaCl containing both Mn^{2+} and other divalent cations. *Phys. Stat. Solidi* (b) **58**, 643-654.

Ikeya M. (1981): Paramagnetic alanine molecular radicals in fossil shells and bones. *Naturwissenschaften* **68**, 474-475.

Ikeya, M. and Ohmura K. (1981): Dating of fossil shells with electron spin resonance. *J. Geology* **89**, 247-250.

Ikeya M. and Ohmura K. (1983): Comparison of ESR ages of corals at marine terraces with ^{14}C and $^{230}Th/^{234}U$ ages. *Earth Planet. Sci. Letters* **65**, 34-38.

Ikeya M. (1984): Limitation of ESR age for carbonate fossils. *Naturwissenschaften* **71**, 421-423.

Ikeya M. and Ohmura K. (1984): ESR age of Pleistocene shells by radiation assessment. *Geochem. J.* **18**, 11-17.

Ikeya M. (1985): Electron Spin Resonance. *Dating Methods of Pleistocene Deposits and Their Problems* ed. by Rutter N. W., Geoscience Canada, 73-87.

Ikeya M. (1986): *Dating and Age Determination of Biological Materials* ed. by Zimmerman M. R. and Angel L. Chapt. 3. Electron Spin Resonance.

Imai N. and Shimokawa K. (1993): ESR ages trace elements in fossil mollusk shell. *Appl. Radiat. Isot.* **44**, 161-165.

Jones A., Blackwell B. A. and Schwarcz H. P. (1993): Annealing and etching of corals for ESR dating. Appl. Radiat. Isot. **44**, 153-156.

Kai A., Miki T. and Ikeya M. (1988): ESR dating of teeth, bones and eggshells excavated at paleolithic site of Douara Cave, Syria. *Quat. Sci. Rev.* **7**, 503-507.

Kai A. and Ikeya M. (1989): ESR study of fossil shells in sediments at Hamana Lake. *Appl. Radiat. Isot.* **40**, 1139-1142.

Kai A. and Miki T. (1989): Electron spin resonance of organic radicals derived from amino acids in calcified fossils. *Jpn. J. Appl. Phys.* **28**, 2277-2282.

Kai A. and Miki T. (1992): Electron spin resonance of sulfite radicals in irradiated calcite and aragonite. *Rad. Phys. Chem.* **40**, 469-476.

Katzenberger O. and Gr n R. (1985): ESR dating of circumarctic mollusk. *Nucl. Tracks* **10**, 885-890.

Kai A. (1996): ESR study of radiation-induced organic radicals in CaCO$_3$. *Appl. Radiat. Isot.* **47**, 1483-1487.

Katzenberger O. and Willems N. (1988): Interferences encountered in the determination of AD of mollusk samples. *Quat. Sci. Rev.* **7**, 485-489.

Katzenberger O., Debuyst R., DeCanniere P., Dejehet F., Apers D. and Barabas M. (1989): Temperature experiments on mollusk samples: an approach to ESR signal identification. *Appl. Radiat. Isotop.* **40**, 1113-1118.

Klein R., Loya Y., Gvirtzman G., Isdale P.J. and Susic M. (1990): Seasonal rainfall in the Sinai Desert during the Late Quaternary: evidence from fluorescent bands in fossil corals. *Nature* **345**, 145-147.

Knutson D.W., Buddemeier R.W. and Smith S.V. (1972): Coral chronometers: seasonal growth bands in reef coral. *Science* **177**, 270-272.

Koba M., Ikeya M., Miki T. and Nakata T. (1985): ESR ages of the Pleistocene coral reef limestones in the Ryukyu Island, Japan. *ESR Dating and Dosimetry* 93-104.

Koba M., Ikeya M., Miki T. Kaigara T., Nakashima H. and Kan H. (1987): Quaternary shorelines and crustal movements on Minamidaito-Jima, Northwestern Pacific. in eds. Qin Y. and Zhao S., *Late Quaternary Sea-Level Changes*, Chinese Ocean Press, 188-198.

Komura K. (1982): Dating by non-destructive γ-ray spectroscopy. *Archaeometry* and Natural *Science* **14**, 1-24.

Konishi K., Omura A. and Nakamichi O. (1974): Radiometric coral ages and sea level records from the Late Quaternary reef complexes of the Ryukyu Islands. *2nd Intern. Coral Reef Symp.* vol.2, 595-613.

Ku T.L., Ivanovich M. and Luo S. (1990): U-series dating of last interglacial high sea stand: Barbados revised. *Quat. Res.* **33**, 129-147.

Lyons R. and Brennan B. J. (1991): Alpha/gamma effectiveness ratios of calcite speleothems. Nucl. Trucks 18, 223-227.

Malmberg M. andRadtke U. (2000): The α-efficiency of corals and its importance for ESR dating. *Radiat. Meas.*, **32**, 747-750.

Mangini A., Segl M. and Schmitz W. (1983): ESR studies of CaCO$_3$ in deep-sea sediments. *PACT J.* **9**, 439-446.

Miki T. and Ikeya M. (1985): A plateau method for total dose evaluation in ESR dating with a digital data processing. *Nucl. Trucks* **10**, 913-919.

Miki T. and Kai A. (1990): Rotating CO$_2^-$ centers in coral and related materials. *Jpn. J. Appl. Phys.* **29**, 2191-2192.

Miki T. and Kai A. (1991): Thermal annealing of radicals in aragonitic CaCO3 and CaPO$_4$_2H$_2$O. *Jpn. J. Appl. Phys.* **30**, 404-410.

Molodkov A. and H tt G. (1985): The ESR dating of subfossil shells: some refinement. *ESR Dating and Dosimetry*,145-155.

Molodkov A. (1986): Application of ESR to the dating of subfossil shells from marine deposits. *Ancient TL* **4**, 49-55.

Molodkov A. and Raukas A. (1987): The age of upper Pleistocene marine deposits of the Boreal transgression on the basis of electron spin resonance (ESR) dating of subfossil mollusk shells. *Boreas* **17**, 267-272.

Molodkov A. (1988): ESR dating of Quaternary shells: recent advances. *Quat. Sci. Rev.* **7**, 477-484.

Molodkov A. (1989): The problem of long-term fading of absorbed palaeodose on ESR dating of Quaternary mollusk shells. *Appl. Radiat. Isot.* **40**, 1087-1093.

Molodkov A. (1993): ESR dating of non-marine mollusc shells. *Appl. Radiat. Isot.* **44**, 145-148.

Molodkov A. (1996): ESR dating of *Lymnaea baltica and Cerastoderma glaucum* from low Ancylus level and transgressive Litorina Sea deposits. *Appl. Radiat. Isot.* **47**, 1427-1432.

Molodkov A. (2001): ESR dating evidence for early man at a Lower Palaeolithic cave-site in the Northern Caucasus as derived from terrestrial mollusc shells. *Quat. Sci.Rev.* **20**, 1051-1055.

Mudelsee M., Barabas M. and Mangini A. (1992): ESR dating of the Quaternary deep- sea sediment core RC17-177. *Quat. Sci. Rev.* **11**, 181-189.

Murata T., Kai A. and Miki T. (1993): Hydration effects on CO_2^- radicals in calcium carbonates and hydroxyapatite. *Appl. Radiat. Isot.* **44**, 305-309.

Nakajima T., Otsuki T., Aonuma M., Satou J. and Shouji K. (1993): Dating of shell mound at Chiba-city, Japan. *Appl. Radiat. Isot.* **44**, 157-159.

Nakazato H., Shimokawa K. and Imai N. (1993): ESR dating for Pleistocene shell fossils and value of annual dose. *Appl. Radiat. Isot.* **44**, 167-173.

Ninagawa K., Yamamoto I., Yamashita Y., Wada T., Sakai H. and Fujii S. (1985): Comparison ESR with TL for fossils calcite shells. *ESR Dating and Dosimetry*, 105-114.

Nozaki Y., DeMasters D.M., Lewis D.M. and Turekian K.K. (1978): Atmospheric 210Pb fluxes determined from soil profile. *J. Geophys. Res.* **83**, 4047-4051.

Omura A. (1983): Uranium-series ages of some solitary corals from the Rikiu limestone on the Kikai-jima, Ryukyu Islands. *Trans. Proc. Palaeont. Soc. Japan* no. 130, 117-122.

Ochiai H., Ikeya M., Morsy M.A. (1999): Geologic aspect and ESR dating of corals from Pleistocene terraces in Ras Mohammed, Egypt. *Carbonate Evaporite* **14**, 138-145.

Peng Z., Jing S., Liang R., Huang H., Yucai Q. and Ikeya M. (1989): Study on comparison of ESR dating of coral and shells with 230Th/234U and 14C methods. *Appl. Radiat. Isot.* **40**, 1127-1132.

Pirazzoli P.A., Radtke U., Hantoro W.S., Jouannic C., Hoang C.T., Causse C. and Borel-Best M. (1991): Quaternary raised coral-reef terraces on Sumba Island, Indonesia. *Science* **252**, 1834-1836.

Radtke U., Hennig G.J., Linke G.J. and M ngersdorf J. (1981): ^{230}Th/^{234}U and ESR dating problems of fossil shells in Pleistocene marine terraces (Northern Latium, Central Italy). *Quaternaria* **23**, 37-50.

Radtke U., Mangini A. and Grun R. (1985): ESR dating of fossil marine shells. *Nucl. Tracks* **10**, 879-884.

Radtke U. (1986): Value and risks of radiometric dating of shore lines - geomorphological and geochronological investigations in Central Italy, Eolian Islands and Ustica (Sicily). *Z. Geomorph. Supll.* **62**, 167-181.

Radtke U. (1987): Paleo sea levels and discrimination of the last penultimate interglacial fossiliferous deposits by absolute dating methods and geomorphological investigations. *Berliner Geogra. Studien* **25**, 313-342.

Radtke U. (1988b): How to avoid "useless" radiocarbon dating. *Nature* **333**, 307-308.

Radtke U. and Grun R. (1988): ESR dating of corals. *Quat. Sci. Rev.* **7**, 465-470.

Radtke U., Grun R. and Schwarcz H.P. (1988): Electron spin resonance dating of the Pleistocene coral reef tract of Barbados (W.I.). *Quat. Res.* **29**, 197-215.

Radtke U. and Schellmann G. : Comment on: Aguirre, M.L. and Whatley, R.C. (1996): Late Quaternary marginal marine deposits and palaeoenvironments from Northeastern Buenos Aires Province, Argentina: A review. *Quat. Sci. Rev.*, **15**, 1064-1065.

Radtke U., Grün R. and Mangini A. (1997): The Pleistocene coral reefs of Kikai-jima (Ryukyu Island, Japan) and their significance for palaeo sea-level reconstructions. *Ocean Research*, **12**, 245-256.

RadtkeU. (1998): Upper and Middle Quaternary coral reefs as a tool in palaeo sea-level and neotectonic research - with examples from Barbados (W.I.), Papua New Guinea, Sumba Island (Indonesia), Ryukyu Islands (Japan)and Cook Islands. In: D. Kelletat

German Geographical Coastal Research. The Last Decade. IGU-Sonderband, 259-288.

Sato T. (1981): Electron spin resonance dating of calcareous Microfossils in deep-sea sediment. *Rock Magn. and Palaeogeophys.* **8**, 85-88.

Sato T. (1982): ESR dating of planktonic foraminifera. *Nature* **300**, 518-521.

Sato T. (1983): ESR studies of planktonic foraminifera. *Nature* **305**, 161-162.

Schellmann G. and Radtke U.(1997): Electron spin resonance (ESR) techniques applied to mollusc shells from South America (Chile, Argentina). *Quat. Geol.* **16**, 257-264.

Schellmann G. and Radtke U. (1997): Electron spin resonance (ESR) techniques applied to mollusc shells from South America and implications for the palaeo sea-level curve. *Quat. Sci. Rev.*, **16**, 465-476.

Schellmann G. and Radtke U.(1999): The status of ESR dating of mollusc shells - illustrated on samples from Patagonia (Argentina). *Quatern. Sci Rev.*, **18**, 1515-1527

Schellmann G. and Radtke U. (2000): ESR dating of stratigraphically well constrained marine terraces along thePatagonian Atlantic coast (Argentina) - a question of correspondence. *Quaternary International*, **68/71**, 261-274.

Schellmann G. and Radtke U. (2001): Progress in ESR dating of young and middle Quaternary corals. *Quat. Sci. Rev.*, **20**, 1015-1020.

Schramm D. U. and Rossi A. M. (1996): Electron spin resonance (ESR), electron nuclear double resonance (ENDOR) and general triple resonance of irradiated biocarbonates. *Appl. Radiat. Isot.* **47**, 1443-1455.

Shimokawa K., Imai N., Nakazato H. and Mizuno K. (1992): ESR dating of fossil shells in the Middle to Upper Pleistocene strata in Japan. *Quat. Sci. Rev.* **11**, 219-224.

Siegele R. and Mangini A. (1985): Progress of ESR studies on CaCO₃ of deep-sea sediments. *Nucl. Tracks* **10**, 937-943.

Skinner A.F. (1985): Comparison of ESR and ²³⁰Th/²³⁴U ages in fossil aragonitic corals. *ESR Dating and Dosimetry*, 135-138.

Skinner A.F. (1988): ESR dating of marine aragonite by electron spin resonance. *Quat. Sci. Rev.* **7**, 461-464.

Skinner A.F. (1989): ESR dosimetry and dating in aragonitic mollusks. *Appl. Radiat. Isot.* **40**, 1081-1086.

Skinner A.F. and Weicker, N. (1992): ESR dating of Chione cancellata and Chama sinuosa. *Quat. Sci. Rev.* **11**, 225-229.

Skinner A.F. and Mirecki J. (1993): ESR dating of molluscs: is it only a "shell game"? *Appl. Radiat. Isot.* **44**, 139-143.

Skinner A.R. and Shawl C.E. (1994): ESR dating of terrestrial quaternary shells. *Quater. Geochr.* **13**, 679-684.

Takano M. and Fukao Y. (1993): ESR dating of Pleistocene fossil shells of the Atsumi Group, Central Honshu, Japan: on the discrepancy in TD value among different ESR peaks. *Appl. Rad. Isot.* **in press.**

Takeuchi A. and Saeki R. (1985): Electron spin resonance (ESR) signals of pelagic sediments in the Southern Pacific. *ESR Dating and Dosimetry* 125-133.

Tanaka K., Hataya R., Spooner N.A., Questiaux D G., Saito Y. and Hashimoto T.Åi1997Åj: Dating of marine terrace sediments by ESR, TL and OSL methods and their applicabilities. *Quatern. Geol.* **16**,257-264

Tsuji Y., Sakuramoto E., Iwasaki E., Ishiguchi M. and Ohmura K. (1985): Ages of pelecoypod shells of the last Shimosueyoshi stage by electron spin resonance. *ESR Dating and Dosimetry*, 87-92.

Tsukamoto S. and Heikoop J. M. (1996): Sulfite radicals in banded coral. *Appl. Radiat. Isot.* **47**, 1437-1441.

Walther R., Barabas M. and Mangini A. (1992): Basic ESR studies on recent corals. *Quatern. Sci. Rev.* **11**, 191-196.

Wintle A.G. and Huntley D.J. (1983): ESR studies of planktonic foraminifera. *Nature* **305**, 161-162.

Chapter 7

Evaporites

– Sulfates and Other Minerals –

Crystallization by evaporation of aqueous solutions in nature is the start of the geochronological clock for evaporites such as sulfates (gypsum and anhydrite), halite (NaCl) and nahcolite ($NaHCO_3$). Development of deserts can be studied by ESR dating of these evaporite minerals (e.g., desert rose). A dream of traveling with a portable ESR for field survey.

7.1   Introduction

Sulfates are minerals containing sulfate ions SO_4^{2-}. Anhydrous calcium sulfate ($CaSO_4$) called **"anhydrite"** changes readily into hydrous calcium sulfate, called **"gypsum"** ($CaSO_4 \cdot 2H_2O$). Gypsum is the most common sulfate mineral associated with halite (NaCl) and anhydrite ($CaSO_4$) in sedimentary salts precipitated from saline solution. It is found in the surface of cave walls as cave flowers (*gypsum flowers*) or as gypsite–containing dirts and sands. Gypsum cemented rocks (*gypcretes*) are found in the playa–lake environment in an arid climate.

Secondary mineral deposits precipitated or crystallized by evaporation of aqueous solution are called **"evaporites"**. Recent concepts and discoveries in the field of evaporite sedimentology are reviewed by Melvin (1990). The precipitation age and impurity centers of evaporites are important for the study of sedimentary history and paleo–environment. $BaSO_4$ **(barite)**, $SrSO_4$ **(celestite)**, $NaHCO_3$ **(nahcolite)**, NaCl **(halite)**, KNO_3 **(niter)** and $Na_2B_4O_7 \cdot 10H_2O$ **(borax)** are known as evaporites in addition to $CaSO_4$ (anhydrite) and $CaSO_4 \cdot 2H_2O$ (gypsum). However, ESR dating of evaporites has been done so far only for gypsum, nahcolite and halite.

The ESR dating of gypsum deposits at Mammoth Cave in Kentucky was first noted as a possibility (Ikeya 1978). Stratigraphic marine gypsum crystals have been dated in India (Nambi 1982), followed by gypsum deposits in playa lakes (Ikeya 1985, Chen *et al.* 1988, 1989), in caves (Ikeda *et al.* 1989) and at a surface of the San Andreas fault in California (Ikeda and Ikeya 1992). Nahcolite and halite have been dated using saline lake sediments at Searles Lake in California (Ikeya and Kai 1988). Meanwhile lattice defects were studied with ESR for gypsum (Albuquerque and Isotani 1982, Kasuya *et al.* 1991a), anhydrite (Bershov *et al.* 1975), barite and celestite (Ryabov *et al.* 1983, 1989). Still, only a few investigations on evaporites have been carried out by the early 1990's in spite of the importance for studying the paleo–environment in conjunction with the development of deserts.

7.2   Anhydrite ($CaSO_4$), Barite ($BaSO_4$) and Celestite ($SrSO_4$)

7.2.1   *Crystal Structures*

Sulfate minerals consist of divalent cations such as Ca^{2+}, Sr^{2+} and Ba^{2+} and a divalent tetrahedral molecular anion, SO_4^{2-}. Anhydrite ($CaSO_4$), barite ($BaSO_4$) and celestite ($SrSO_4$) have an orthorhombic crystal form different from monoclinic gypsum ($CaSO_4 \cdot 2H_2O$) described later. The crystal structure of anhydrite and barite is shown in Figure 7.1. The unit cell of anhydrite has two edges. Sulfur and Ca atoms lie in an approximately even space on the lines of intersection of mirror planes. In barite, the SO_4^{2-} and Ba^{2+} ions lie on the mirror planes and Ba^{2+} ions link the SO_4^{2-} ions in such a way that each Ba^{2+} is coordinated by twelve oxygens. The structure of celestite is similar to that of barite, with Sr^{2+} taking the place of Ba^{2+}; its lattice parameter is shown in the figure.

Anhydrite ($CaSO_4$) is easily formed from gypsum by dehydration and is also hydrated into gypsum in nature and is sometimes found with gypsum and halite in evaporites. Sometimes they are fibrous, granular or scaly. Synthetic anhydrite doped with rare earth ions such as Tm^{3+}, Dy^{3+} or Tm^{3+} is a commercial TL dosimeter (TLD) element material used in radiation dosimetry as described in Chapter 4.

Barite ($BaSO_4$) is a common mineral that often occurs in tabular crystals, in granular form or in compact masses. *Barite rose* (rosette) is an aggregate of tabular sand–filled crystals in sandstone. The small disk–shaped barite found in sandy shale is called a *"barite dollar"*. A cluster of crystals called *"desert rose"* found in the sand of Oklahoma is usually calcite and less commonly barite, gypsum or celestite. Celestite ($SrSO_4$) and barite are principal ores of strontium and barium (Bates and Jackson 1980).

(a) Anhydrite ($CaSO_4$)   (b) Barite ($BaSO_4$)

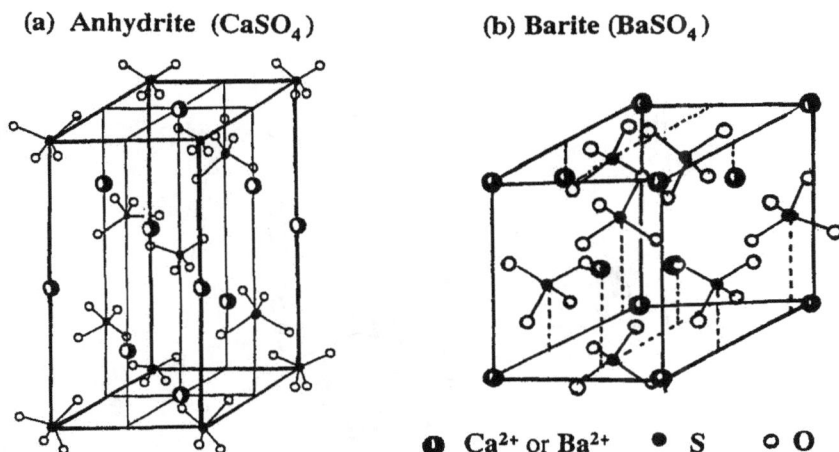

$\bullet$ Ca^{2+} or Ba^{2+} $\bullet$ S $\circ$ O

Figure 7.1 The structure of the unit cell of (a) anhydrite and (b) barite (after Deer *et al.* 1966 and Jong 1959).

Anhydrite ($CaSO_4$):	$a = 0.6991$	$b = 0.6996$	$c = 0.6230$ nm
Barite ($BaSO_4$):	$a = 0.8878$	$b = 0.5450$	$c = 0.7152$ nm
Celestite ($SrSO_4$):	$a = 0.8359$	$b = 0.5352$	$c = 0.6866$ nm

7.2.2   *ESR Spectra and Models of Defects*

Radiation–induced defects in anhydrite, barite and celestite have been studied extensively by Russian groups who found about 40 different types of defects as described in a book by Marfunin (1979). Identification of TL centers in $CaSO_4$ has been done with ESR (Danby *et al.* 1982, Sankaran *et al.* 1983, Nambi 1985, 1993). ESR parameters of defects assigned in these minerals are summarized in Table 7.1 although no ESR dating has been done so far. ESR spectra and some defects in anhydrite and barite are described below.

(a) *Anhydrite* ($CaSO_4$) :   Figure 7.2 shows ESR spectra of γ–irradiated anhydrite powder containing (a) 0.2% and (b) 0.1% alkali. Four types of defects, SO_4^-, SO_3^-(I) and SO_3^-(II) and SO_2^-, are observed.

SO_4^- is a distorted tetrahedral radical with 31 electrons. It is a hole–type center. The production of this center is enhanced by the presence of alkali, indicating that the hole center is stabilized by alkali ions substituted at Ca^{2+} sites.

SO_3^-(I) and SO_3^-(II) are electron–type centers in the $CaSO_4$ lattice though SO_3^- formed from SO_3^{2-} in $CaCO_3$ is a hole–type center. They are 25–electron pyramidal molecules of AB_3–type isoelectronic with CO_3^{3-}.

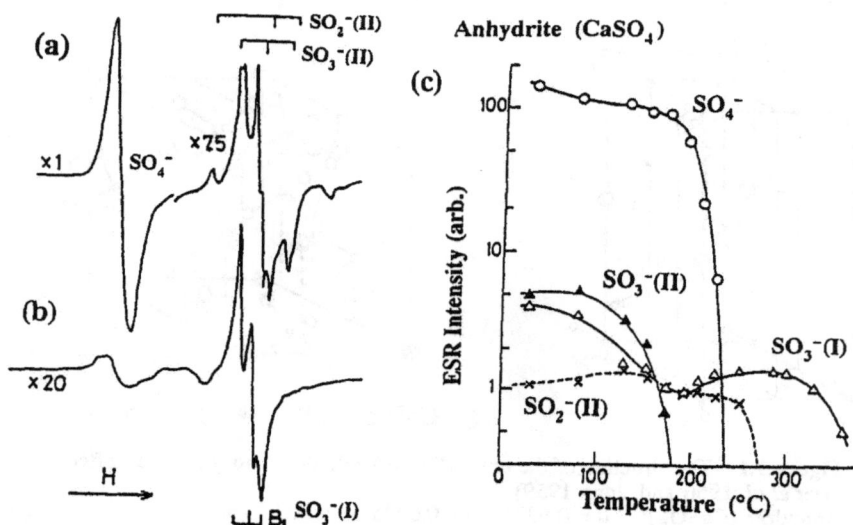

Figure 7.2 ESR spectra of γ–irradiated anhydrite powder containing (a) 0.2% and (b) 0.01% alkali. (c) Isochronal annealing of the signals.

SO_2^- is an electron–type center. It is a bent radical with 19 electrons. Molecular orbitals of AB_4, AB_3 and AB_2 type molecules are described in Appendix A2. Isochronal annealing curves of these centers are shown in Figure 7.2 (b). The hole center SO_3^--(I) can be used for dating anhydrite.

(b) *Barite* ($BrSO_4$) : Figure 7.3 shows ESR spectra of (a) barite powder and (b) a single crystal. In addition to SO_3^- signals at $g = 2.0028$, 2.0001 and 2.0036, a hole–type center with g factors of 2.0191, 2.0127 and 2.0103 is observed (Kasuya *et al.* 1991b). The signal intensity of SO_3^- saturates at 0.4 mW, while that of the hole center does so at 25 mW.

The identification of SO_3^- was confirmed by quartet hf splitting of ^{33}S ($I = 3/2$, 0.76%) (Ryabov *et al.* 1983). The hole center shows no hf lines of ^{33}S. A possible model arising from analogous hole centers in anhydrite ($CaSO_4$) (Bershov *et al* 1971) is O_2^{3-} stabilized by an M^{3+} ion such as Y^{3+}. The result of isochronal annealing for 20 min indicates that SO_3^- and O_2^{3-} are stable up to 300°C and 200°C, respectively, as shown in Figure 7.3 (c).

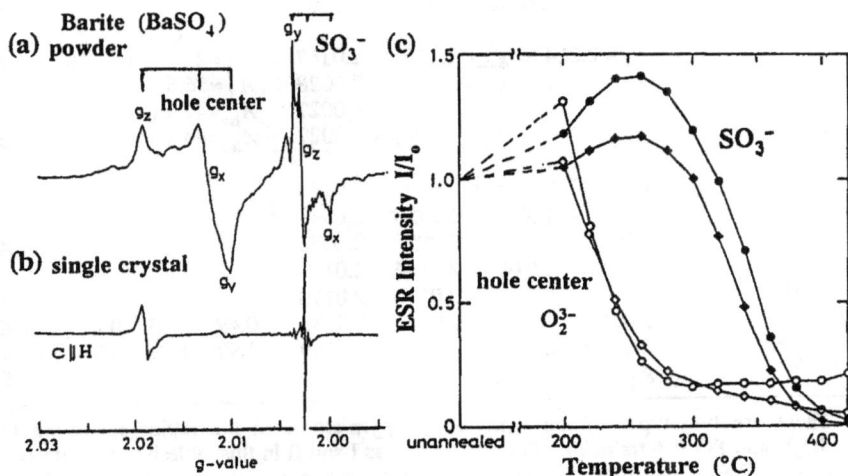

Figure 7.3   ESR spectra of (a) natural barite powder (Colorado) and (b) a single crystal for the direction of the magnetic field along the *c*-axis. (c) Isochronal annealing of the signals for 20 min (Kasuya *et al.* 1991b).

Table 7.1 ESR parameters of defects assigned in irradiated anhydrite ($CaSO_4$), barite ($BaSO_4$) and celestite ($SrSO_4$).

Defects *)	g factors				A tensor (mT)			Ref.
	g_{zz}	g_{xx}	g_{yy}	g_{av} **)	A_{zz}	A_{xx}	A_{yy}	
Anhydrite ($CaSO_4$)								
SO_4^-	$g_{\parallel}=2.011$	$g_{\perp}=2.012$		2.0117	$A_{iso}=1.44$ (³³S)			1
$SO_4 \cdot V_{Ca}(I)$ a)	2.0395	2.0006	2.0091	2.0165				2
$SO_4 \cdot V_{Ca}(II)$ a)	2.0256	2.0011	2.0084	2.0117				2
$SO_3^-(I)$	2.0048	2.0038	2.0029	2.0038	no hf line			1
$SO_3^-(I)$	2.0025	2.0031	2.0041	2.0032	12.1	10.7	9.6 (³³S)	3
$SO_3^-(I)$	2.0020	2.0042	2.0036	2.0033				4
$SO_3^-(II)$	2.0050	2.0035	2.0018	2.0034				1
$SO_3^-(II)$	2.0022	2.0039	2.0041	2.0034	13.1	9.3	9.4 (³³S)	3
$SO_3^-(II)$	2.0012	2.0040	2.0022	2.0025				4
$SO_2^-(I)$	2.0058	2.0022	2.0092	2.0057				4
$SO_2^-(II)$	2.0058	2.0020	2.0098	2.0059				4
$SO_2^-(II)$	2.0066	2.0031	2.0015	2.0037	no hf line			1
SO_2^- b)	2.005	2.003	2.008	2.0053				5
SSO^- b)	2.009	2.003	2.016	2.0093				5
SSO_3^- b)	2.003	2.028	2.023	2.018				5
SSO_3^-	2.0114	2.0270	2.0218	2.0201				4
PO_4^{2-} b)	$g_{\parallel}=2.014$	$g_{\perp}=2.018$		2.0167	$A_{\parallel}=2.9$	$A_{\perp}=2.8$		5
PO_3^{2-} b)	$g_{\parallel}=2.0012$	$g_{\perp}=2.0036$		2.0028	$A_{\parallel}=56.6$	$A_{\perp}=49.9$		5
$PO_3^{2-}(I)$				$g_{iso}=2.0027$	$A_{iso}=47.4$ (³¹P)			3
$PO_3^{2-}(II)$				$g_{iso}=2.0027$	$A_{iso}=44.8$ (³¹P)			3
O^- b)	$g_{\parallel}=2.011$	$g_{\perp}=2.019$		2.0163				5
$O_2^-(I)$	2.0238	1.9940	1.9950	2.0043				4
$O_2^-(II)$	2.0176	1.9940	1.9950	2.0022				4
$O_3^-(I)$	2.0056	2.0140	2.0144	2.0113				4
$O_3^-(III)$	2.0102	2.0112	2.0116	2.0110				4
$O_2^{3-}-Y^{3+}$	2.0122	2.0029	2.0262	2.0138	0.49	0.63	0.49 (⁸⁹Y)	6
$O_2^{3-}-B^{3+}$	2.0120	2.0086	2.0122	2.0109	0.92	1.17	0.90 (¹¹B)	4
S_2^- b)	$g_{\parallel}=2.74$	$g_{\perp}$ (not measured)						5

*　There are two types of non–equivalent oxygen atoms in the lattice involving the SO_4^{2-} and PO_4^{3-} tetrahedra. Defects termed as I and II in this table involve oxygens at type I and type II lattice sites, respectively. For example, $SO_3^-(I)$ and $SO_3^-(II)$ are the SO_3^- centers trapped at the vacancy of type I and type II oxygens in the $CaSO_4$ lattice, respectively, as indicated in Figure 7.4.

**　g_{av} is calculated using Eq. (2.10) or Eq. (2.11).

Table 7.1 (continued)

Defects [*]	g factors				A tensor (mT)			Ref.
	g_{zz}	g_{xx}	g_{yy}	g_{av} [**]	A_{zz}	A_{xx}	A_{yy}	
Barite (BaSO$_4$)								
SO$_4^-$ [b]	2.003	2.071	2.016	2.030	1.57	0.56 (^{135}Ba)		5
SO$_3^-$(I)	1.9995	2.0024	2.0031	2.0017	13.5	9.6	9.6 (^{33}S)	3
SO$_3^-$ [b]	$g_\parallel = 2.0036$ $g_\perp = 2.0050$			2.0045				5
SO$_3^-$	2.0028	2.0001	2.0036	2.0022				7
SO$_2^-$	2.0032	2.0104	2.0128	2.0088	5.38	0.82	0.71 (^{33}S)	8
SO$_2^-$ [b]	2.010	2.004	2.014	2.0093				5
SO$_2^-$ [b]	2.010	2.007	2.030	2.016				5
PO$_2^{2-}$	1.9992	2.0020	2.0037	2.0016	30.0	9.1	9.2	8
PO$_3^{2-}$(I)				$g_{iso} = 1.9999$	$A_{iso} = 50.4$ (^{31}P)			3
PO$_3^{2-}$(II)				$g_{iso} = 2.0000$	$A_{iso} = 49.9$ (^{31}P)			3
O$_2^{3-}$	2.0191	2.0127	2.0103	2.0140				7
O$^-$ [b]	$g_\parallel = 2.011$ $g_\perp = 2.022$			2.0183				5
S$_2^-$ [b]	$g_\parallel = 3.972$ $g_\perp = 0.42$			2.319				5
Celestite (SrSO$_4$)								
SO$_3^-$(I)	2.0019	2.0038	2.0040	2.0032	13.3	9.5	9.5 (^{33}S)	3
PO$_3^{2-}$(I)				$g_{iso} = 2.0025$	$A_{iso} = 50.8$ (^{31}P)			3

ESR measurement at 290 K except for a) 77 K and b) not reported.
Reference: 1) Huzimura 1979; 2) Danby *et al.* 1982; 3) Ryabov *et al.* 1983; 4) Bershov
  et al. 1975; 5) Samoilovich and Tsinober 1970; 6) Bershov *et al.* 1971;
  7) Kasuya *et al.* 1991b; 8) Ryabov *et al.* 1989.
Nuclear spin and abundance: ^{11}B ($I = 3/2$, 80.1%); ^{31}P ($I = 1/2$, 100%);
  ^{33}S ($I = 3/2$, 0.75%); ^{89}Y ($I = 1/2$, 100%); ^{135}Ba ($I = 3/2$, 6.59%).

Figure 7.4 Projection showing the arrangement of two types of oxygen atoms around the Ca^{2+} ion in CaSO$_4$ (after Danby *et al.* 1982).

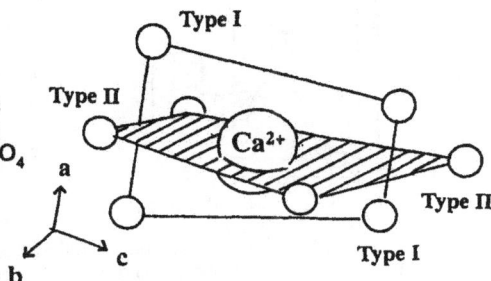

7.3   Gypsum ($CaSO_4 \cdot 2H_2O$)

7.3.1   *Crystal Structure*

Gypsum ($CaSO_4 \cdot 2H_2O$) is a common evaporite mineral that has precipitated or crystallized from aqueous solution under dry or arid conditions. White Sands National Monument in New Mexico is a white landscape, the world's largest gypsum dunefield. Crystalline gypsum or *selenite* occurs in caves and limestone cavities as well as in clays, shales and some sands. Gypsum is also produced by action of sulfuric acid solution on the rocks containing calcium. The action of sulfurous vapors on calcium bearing minerals produces gypsum in volcanic areas.

Gypsum is a monoclinic crystal containing four molecules in the unit cell. There are six different ways of defining the unit cell. Figure 7.5 shows the structure of gypsum projected to the planes perpendicular to the c– and a–axes of the I unit cell (Deer *et al.* 1966). There are pairs of adjacent layers parallel to [010] that contain Ca^{2+} and SO_4^{2-} ions. The water molecules with their hydrogen atoms bonded to oxygen atoms of the SO_4^{2-} group are located between successive pairs of layers.

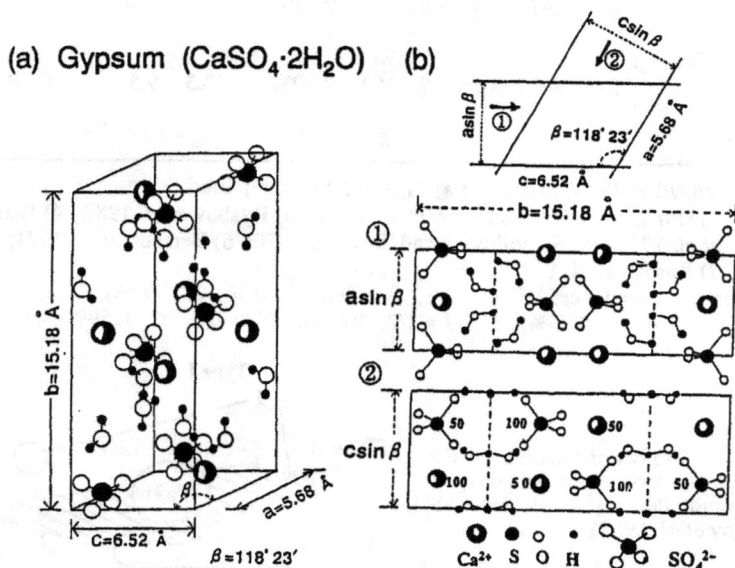

Figure 7.5   (a) Monoclinic crystal structure of gypsum ($CaSO_4 \cdot 2H_2O$) and (b) the structures projected to the planes perpendicular to (1) the c–axis and (2) the a–axis of the I unit cell. The directions of projection in the unit cell are shown above (revised after Deer *et al.* 1966, Atoji and Rundle 1958).

7.3.2 ESR Spectra and Models of Defects

The ESR study of radiation–induced paramagnetic centers in gypsum was first reported by Wigen and Cowen (1960). X–ray irradiated natural gypsum crystal gave four types of ESR signals designated as A, B, C and D centers (Albuquerque and Isotani 1982). Natural gypsum from deserts shows four types of ESR signals termed as G1, G2, G3 and G4, among which two signals G1 and G4 correspond to the B and A centers, respectively (Kasuya et al. 1991a). Figure 7.6 (a) shows the ESR spectra of the irradiated single crystal of gypsum for the magnetic field perpendicular to the b–c plane and parallel to the b– and c–axes. Microwave power dependence of these centers are shown in (b). The anisotropic g factors were determined from the studies of angular dependence as summarized later.

Models of radiation–induced defects have been studied using synthetic gypsum powder doped with cation and anion impurities, Na$^+$, K$^+$, Ba^{2+} and CO$_3^{2-}$, using Na$_2$SO$_4$, K$_2$SO$_4$, Ba(OH)$_2$·8H$_2$O and CaCO$_3$ (Ikeda and Ikeya 1992). Figure 7.7 shows the ESR spectra of these synthetic gypsum irradiated by γ–rays at room temperature. The hole–type center (G2) and electron–type center (G3) observed in natural gypsum are observed as indicated in spectrum (a) for nominally pure (undoped) gypsum. The isotropic signal (G1 = B) at g = 2.004 is also clear.

The intensity of signal G3 is enhanced by doping with K$^+$. It is tempting to assume that the anisotropic g–factor for signal G3 is due to CO$_2^-$ at

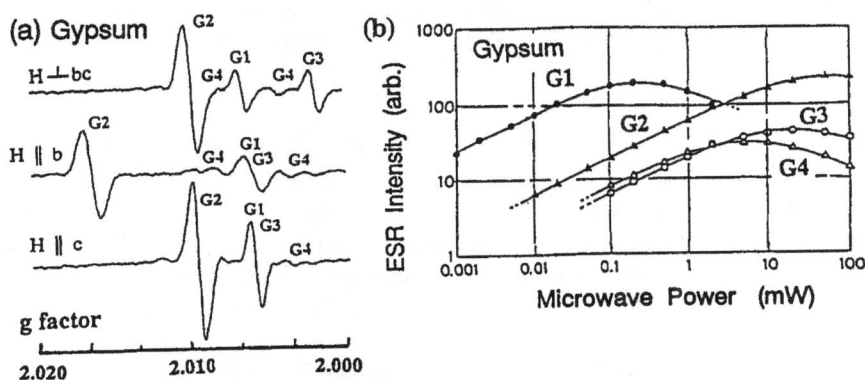

Figure 7.6 (a) ESR spectra of the irradiated single crystal of gypsum from Australia for the direction of the magnetic field perpendicular to the bc–plane, along the b and c–axes. (b) Microwave power dependence of the signal intensity (Kasuya et al. 1991a).

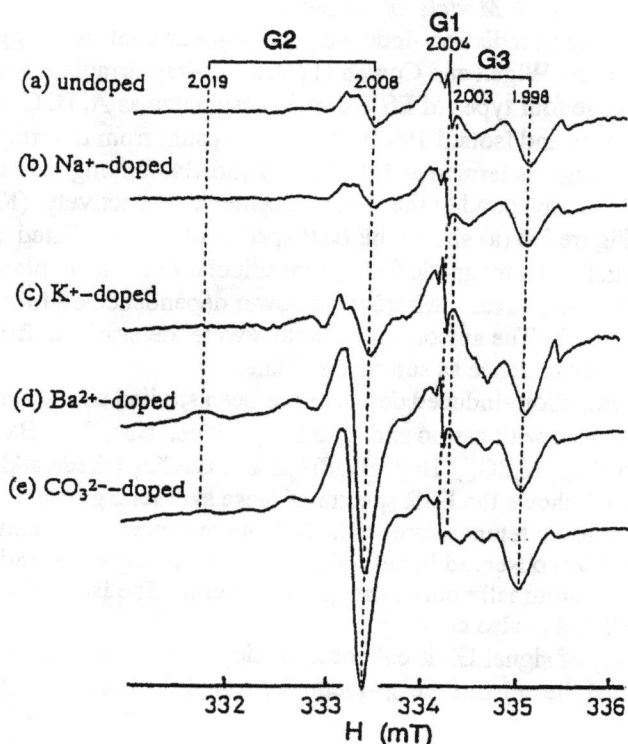

Figure 7.7   ESR spectra of impurity–doped synthetic gypsum powder after γ–irradiation: (a) undoped gypsum and (b) – (e) gypsum doped with Na⁺, K⁺, Ba²⁺ and CO_3^{2-} (Ikeda and Ikeya 1992).

the SO_4^{2-} site associated with K⁺ at the Ca²⁺ site in an axial symmetry where the axis of hindered rotation is in the O–O direction. The g–factor for signal G3 agrees fairly well with that for the CO_2^- center in $CaCO_3$. The hindered rotation of CO_2^- along the c–axis may be related to hydrogen bonding of oxygen atoms by nearby hydrogens of water molecules. Stabilization of the CO_2^- center may be made by charge–compensating K⁺ at the Ca²⁺ site.

The ESR parameters for defects in gypsum are summarized in Table 7.2 together with speculative models based on defects in natural and synthetic gypsum. Thermal stability is not clear since dehydration occurs by heating gypsum in isothermal annealing experiments.

Table 7.2 ESR parameters of the radicals in irradiated natural and synthetic gypsum.

Center[*]	Model	Sample	g factor (A–tensor in mT))	Ref.
G1 (=B)	SO$_3^-$	natural crystal	$g = 2.0028 - 2.0042$	1
		natural powder	$g = 2.004$	1
		natural crystal	$g = 2.004$	2
		synthetic powder	$g = 2.004$	3
G2	CO$_3^-$ or O$_2^{3-}$	natural crystal	$g_{zz} = 2.0192$ $g_{xx} = 2.0084$ $g_{yy} = 2.0088$	1
		natural powder	$g = 2.009$	1
		synthetic powder	$g_\parallel = 2.019$ $g_\perp = 2.008$	3
G3	CO$_2^-$	natural crystal	$g_{zz} = 1.9973$ $g_{xx} = 2.0029$ $g_{yy} = 2.0027$	1
		synthetic powder	$g_\parallel = 1.998$ $g_\perp = 2.003$	3
G4 (=A)	O$_2$H·	natural crystal	$g = 1.999 - 2.008$ ($A = 0.2 - 0.5$)	1
		natural crystal	$g_{zz} = 2.003$ $g_{xx} = 1.999$ $g_{yy} = 2.088$ ($A_{zz} = 0.52$ $A_{xx} = 0.21$ $A_{yy} = 0.33$)	2
C	SO$_4^-$?	natural crystal	$g = 2.002$ $g = 2.004$	2
D	OH·	natural crystal	$g_{zz} = 2.037$ $g_{xx} = 2.000$ $g_{yy} = 2.002$ ($A_{zz} = 0.4$ $A_{xx} = 1.4$ $A_{yy} = 1.2$)	2
	OH·	natural crystal	$g_\parallel = 2.1108$ $g_\perp = 2.028$ ($A_{zz} = 0.33$ $A_{xx} = -4.3$ $A_{yy} = -3.25$)	4
	OH·	natural crystal	$g_\parallel = 2.051$ $g_\perp = 2.003$ ($A_{zz} = 0.2$ $A_{xx} = -1.7$ $A_{yy} = -1.7$: ^{1}H)	5

[*] The G1, G2, G3 and G4 centers are designated by Kasuya et al. (1991a) and the A, B, C and D centers by Albuquerque and Isotani (1982).

Reference: 1) Kasuya et al. 1991a; 2) Albuquerque and Isotani 1982; 3) Ikeda and Ikeya 1992; 4) Gunter 1967; 5) Wigen and Cowen 1960.

G1 (=B) center (SO$_3^-$) : $g = 2.0028 \sim 2.0042$

A "*G1 center*" with an isotropic g–factor was identified as SO$_3^-$ from the hf structure of ^{33}S ($I = 3/2$, abundance 0.76%). This center may correspond to the B center which has been assigned to a trapped electron at a vacancy or lattice defect (Albuquerque and Isotani 1982). SO$_3^-$ is observed in other sulfates such as anhydrite (CaSO$_4$), barite (BaSO$_4$) and celestite (SrSO$_4$) as well as in calcite and aragonite (CaCO$_3$).

G2 center (CO_3^- or O_2^{3-} ?) : $g_{zz} = 2.0192$, $g_{xx} = 2.0084$, $g_{yy} = 2.0088$

A "*G2 center* consists of two magnetically non–equivalent sites (the *ac*–plane being the mirror plane), both having a *g*–tensor of rhombic symmetry with principal *g* factors:

$g_{zz} = 2.0192$ 90° from the *c*–axis and 10° from the *b* axis
$g_{xx} = 2.0084$ parallel to the *c*–axis
$g_{yy} = 2.0088$ 90° from the *c*–axis and 80° from the *b* axis

Two G2 signals with the principal axes tilted by ± 10° from the *b*–axis seem to have nearly axial symmetry. The signal intensity is enhanced in synthetic gypsum doped with CO_3^{2-} impurity. A possible candidate is CO_3^- but no hf splitting was observed for labeling with ^{13}C (Ikeda and Ikeya 1993). A small value of $A \sim 1$ mT is observed for CO_3^- in $CaCO_3$ (see Table 5.1). An alternative model is O_2^{3-} from analogous models in barite (Kasuya *et al.* 1991b) and anhydrite (Bershov *et al.* 1971).

G3 center (CO_2^-) : $g_{zz} = 1.9973$, $g_{xx} = 2.0029$, $g_{yy} = 2.0027$

A "*G3 center*" has orthorhombic symmetry with principal *g* factors:

$g_{zz} = 1.9973$ 90° from the *b*–axis and 90° from the *c* axis
$g_{xx} = 2.0029$ parallel to the *c*–axis
$g_{yy} = 2.0027$ parallel to the *b* axis

From the doping experiment mentioned previously and resemblance of the *g* factor, a model of the axial CO_2^- rotating around the *y*–axis (along the O–O direction) of the molecule has been proposed. The splitting might be due to hf splitting for a proton in nearby hydrogen atom that bonds with the CO_2^-. Sulfate and sulfide contain appreciable amount of CO_3^{2-} anion impurity.

G4 (=A) center ($O_2H\cdot$): $g = 1.999 \sim 2.008$, $A = 0.2 \sim 0.5$ mT

A "*G4 center*" consists of two magnetically non–equivalent centers, each having two lines of hf structure. This center was previously reported as an A center by Albuquerque and Isotani (1982), who identified it as $O_2H\cdot$.

C center (SO_4^- or SO_2^- ?) : $g = 2.002$ and 2.006

A "*C center*" has two magnetically nonequivalent sites. A model of SO_4^- or models of SO_3^- for $g = 2.002$ and SO_2^- for $g = 2.006$ were suggested (Albuquerque and Isotani 1982). SO_2^- is a possible candidate from the similarity of *g* factor in anhydrite ($g_{zz} = g_{\parallel} = 2.0058$, $g_{xx} = 2.0022$ and $g_{yy} = 2.0092$) (Bershov *et al.* 1975).

D center (OH·) : g_{zz} = 2.037, g_{xx} = 2.000, g_{yy} = 2.002
   A "*D center*" with an anisotropic *g* factor has also two nonequivalent sites. A model of the OH· radical has been proposed.

The molecular species such as SO$_3^-$ (G1), CO$_3^-$ (G2 ?), CO$_2^-$ (G3) and SO$_2^-$ (C ?) have already been observed in calcite and aragonite. The microwave power saturation behavior of these centers shown in Figure 7.6(b) is consistent with the above identification. For example, SO$_3^-$ (G1) saturates at 0.1 mW, while CO$_2^-$ (G3) at 100 mW (see Figures 5.7 and 6.20).

7.3.3 ESR Dating of Natural Gypsum

(a)   *Gypsum in caves* : Gypsum precipitate is found in dry caves (Moore and Nicholas 1964). This is due to the reaction of Ca^{2+} with SO$_4^{2-}$ ions in percolating water. Gypsum in volcanic caves may be formed by the reaction of Ca^{2+} in volcanic rocks and SO$_4^{2-}$ of volcanic gas origin.   Recrystalliza-tion of primary small gypsum deposits is common in caves. Thus, gypsum speleothems form after the cave formation and the age of the oldest one indicates the minimum age of the cave formation.
   Intense ESR signals (G1 and G2) associated with defects in gypsum were observed during ESR dating of cave deposits at Mammoth Cave Na-tional Park, Kentucky (Ikeya 1978, 1985) as shown in Figure 7.8. Dating

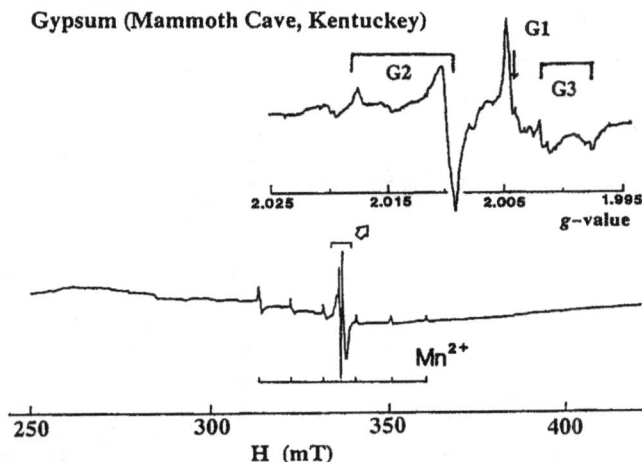

Figure 7.8   ESR spectrum of gypsum deposits at Mammoth Cave in Ken-tucky, USA.   Mn^{2+} signals are observed together with signal G1 (SO$_3$), G2 (CO$_3^-$?) and G3 (CO$_2$).

was done using signal G2 at $g = 2.008$.

ESR spectra of volcanic cave deposits show a signal at $g = 2.004$ (G1). The dating signal (G2) at $g = 2.008$ and also G3 were not observed presumably due to the absence of CO_3^{2-} impurities in gypsum from the volcanic cave. The *ED* was obtained using signal G1 by an additive dose method (Ikeda *et al.* 1989).

(b)   *Gypsum in playa lakes* :   Figure 7.9 shows an ESR spectrum of gypsum in a bore–hole core sample at Lake Eyre in Australia. Signals at $g = 2.001$ (G3), 2.003 (G1), 2.006 and 2.008 (G2) were observed. Mn^{2+} signals observed in the sample as received was removed by washing with dilute HCl. Presumably, Mn^{2+} derives from the surface–coating carbonates.   The signal at $g = 2.006$ may be due to SO_2^{-} ($g = 2.0057$) in $CaCO_3$. Gypsum sands from the lacustrine terraces at a paleolithic site, Douara in Syria gave concordant ESR ages as estimated from the stratigraphy (Ikeya 1985).

(c)   *Gypsum at a fault surface* (*fault displacement*) :   Gypsum crystals precipitate at the fault surface after the displacement of faults and thus the minimum age of fault displacement is determined by dating mineral deposits at the fault plane. The San Andreas fault, a boundary between the Pacific

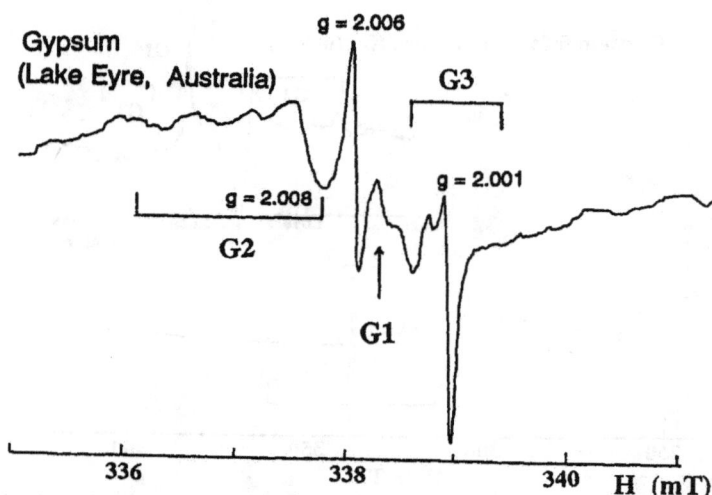

Figure 7.9   ESR spectrum of gypsum in a bore–hole core of lake sediments at Lake Eyre in Australia.

and North American plates in plate tectonics, is a gigantic strike slip fault
that stretches almost the entire length of the state of California as shown in
Figure 7.10 (a). Very small pieces of gypsum precipitates (1 – 3 mm in
diameter) were collected from the fault surface at a trench site on the Carrizo
Plain in the central California. The last fault break with surface displace-
ments at the trench site occurred during the 1857 Fort Tijon earthquake.

Figure 7.10 (b) shows the ESR spectra of gypsum precipitates from the
fault plane at the trench site. The gypsum before additive irradiation shows
signals associated with Mn^{2+}, two lines at $g = 2.012$ and $g = 1.996$ due to
forbidden transitions, together with weak signals. Additive irradiation
enhances the intensity of the signals designated as G1, G2 and G3. Growth
of the signal intensity of G2 at $g = 2.008$ was used to determine the ED of
0.54 Gy, and the minimum ESR age of the fault formation of 260 (or 300)
years for 0% (or 100% Rn loss) was obtained. The discrepancy between the
ESR age and the historical record of the displacement may be due to the
influence of internal α–ray dose and/or the contamination of older gypsum
particles. Systematic sampling of gypsum and precise assessment of the
environmental dose rate will be needed for the practical nse of this method.

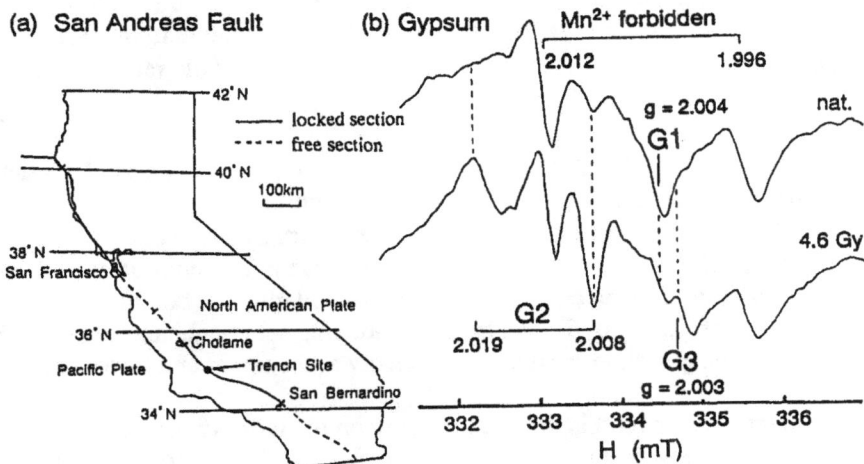

Figure 7.10   (a) Map of the San Andreas Fault and the trench site in 1989.
(b) ESR spectra of natural and γ–irradiated gypsum samples from the trench
site. Two Mn^{2+} lines interfere with the radical signals designated G1, G2 and
G3. The signal G2 at $g_\perp = 2.008$ was used to determine the ED.

(d) *Marine gypsum* : Marine gypsum crystals of Mediterranean (Italy) and Arabian (India) sea origins have been dated with ESR using signal G1 at $g = 2.0040$ (Nambi 1982). Marine gypsum in India gave consistent ages. In this study, the height from the peak to peak of the derivative line seems to have been taken as the intensity. The contribution of a broad signal in the spectrum should be subtracted for a precise age determination. The possibility of an overlapping organic signal at $g = 2.0045$ must be considered.

(e)   *Gypcrete* : In an arid climate, gypcrete, a gypsum–cemented crust or rock is found in some playa–lake beachrock environment. Gypcrete cores from Amadeus Basin, Northern Australia have been studied with ESR (using the signal at $g = 2.0090$; G2) and ^{14}C dating (Chen *et al.* 1988, 1989).

7.3.4 *Distribution of Radicals in a Gypsum Crystal*

Gypsum precipitates obtained from pluvial lake sediments excavated at Turkey have been dated with ESR (Omura *et al.* in preparation). Pieces of small crystals are obtained among gypsum crystals. A slice of crystal was imaged using a microwave scanning ESR microscope as shown in Figure 7.11 (see Chapter 14). The central position gives a high intensity for the signal at $g = 2.006$ (SO_2^-) in agreement with the expected growth of the crystal. The intensities of both upper and lower surfaces are high for the dating signal G2 at $g = 2.008$. It is interesting to note that the absolute signal intensity within a crystal is not uniform. The possible reasons are as follows:

1) Non–uniform distribution of radioactive elements. If this is the case, the distribution of SO_2^- in Figure 7.11 (a) will indicate the relative age not related to radiation effects. On the other hand, the distribution of signal G2 used for dating might be either due to the non–uniform distribution within the crystal or due to different climatic conditions.
2) Range of external β–rays from surrounding sediments for signal G2 may be responsible for the high intensity at the upper and lower edges.

A systematic work on this material is going on to clarify the reason and the paleo–climatic environment that caused such a distribution of the defects. Additive irradiation must be done to obtain the formation efficiency and the *ED* as an image.

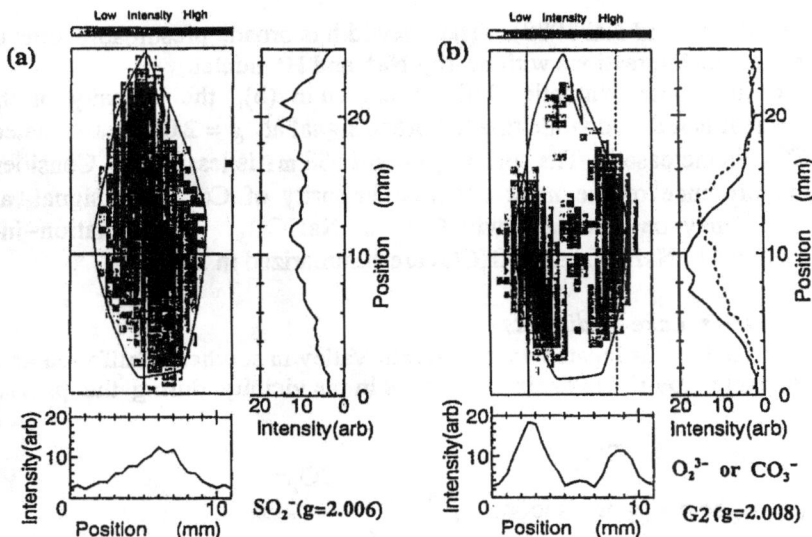

Figure 7.11   Distribution of the signal intensity of (a) SO_2^- ($g = 2.006$) and (b) G2 ($g = 2.008$) in a gypsum crystal taken from the bore–hole core at Turkey (after Omura *et al.* in preparation).

7.4   Nahcolite (NaHCO₃) and Halite (NaCl)

7.4.1   *ESR Spectra and Defects in Synthetic NaHCO₃*

The Great Salt Lake in Utah, USA and the Dead Sea in Israel have high concentration of salts in an arid or semiarid environment.   Nahcolite (NaHCO₃) with a monoclinic structure and **halite (NaCl)** with face centered cubic (fcc) structures are also evaporites.   Defects called *"color centers"* have been extensively studied in NaCl, historically using a rock salt and then using a synthetic crystal.   F centers in NaCl show a broad linewidth (~ 5 mT) and is therefore not a dominant signal for Quaternary halite.

Figure 7.12 (a) shows the ESR spectra of special reagent grade synthetic NaHCO₃ after γ–irradiation at several different doses.   Comparing with the reported g factors for radiation–induced CO_3^- and $CO_2(H)$ radicals in KHCO₃ (Chantry *et al.* 1962), the signal at $g_{\parallel} = 2.0058$ and $g_{\perp} = 2.0183$ was assigned to CO_3^- split into hf doublets ($A_{av} = 1.1$ mT) by a nearby proton nucleus.   The low magnitude of the derivative line at a higher magnetic field may be due to broadening by anisotropic hf interaction.   The broad signal with the minimum at $g = 1.9975$ and the maximum slope at $g = 2.0025$ may be associated with CO_2^- rather than CO_3^{3-} as previously assigned (Ikeya and Kai 1988). The linewidth is broad, presumably due to superhyperfine interactions with nearby Na^+ and H^+ nuclei.

For less pure synthetic $NaHCO_3$ shown in (b), the intensity of the CO_3^- signal is reduced while that the broad signal at $g = 2.0025$ associated with CO_2^- is increased. The splitting around 338 mT is less clear. Considering the presence of the most common impurity of Ca^{2+}, the signal at $g = 2.0025$ may be associated with Ca^{2+} in $NaHCO_3$. The radiation–induced defects in $NaHCO_3$ and $KHCO_3$ are summarized in Table 7.3.

7.4.2 Searles Lake Sediments

Searles Lake is located west of Death Valley in southern California and was connected by rivers to six other lakes in the vicinity during the pluvial

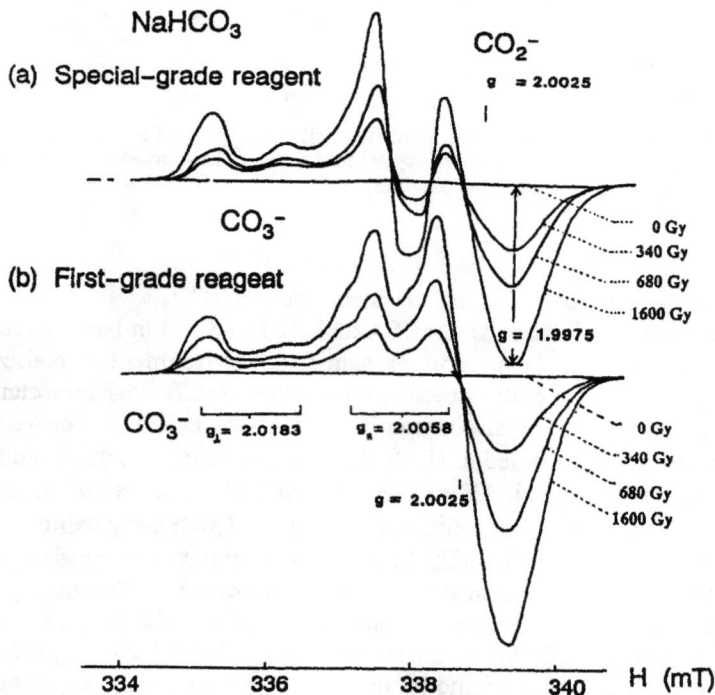

Figure 7.12   ESR spectra of (a) special reagent–grade and (b) first–grade $NaHCO_3$ irradiated by γ–rays. The signal of a hole–type center, CO_3^-, is split by hf coupling ($A_{av} = 1.1$ mT) with a nearby proton. The electron–type center, indicated as CO_2^-, shows the signal at $g = 2.0025$ and 1.9973.

Table 7.3 ESR parameters of defects assigned in irradiated NaHCO$_3$ and KHCO$_3$.

Material	Model	g factors (A–tensor in mT)			Reference
NaHCO$_3$	CO$_3^-$ (HCO$_3$)	$g_∥$ = 2.0183	$g_⊥$ = 2.0058	(A_{av} = 1.1: ^{13}C)	1
	CO$_2^-$ (HCO$_2$)	$g_∥$ = 1.9975	$g_⊥$ = 2.0025	(A_{av} = 0.4: ^{13}C)	1
KHCO$_3$	CO$_3^-$	g_{zz} = 2.0066	g_{xx} = 2.0086	g_{yy} = 2.0184	2
		(A_{zz} = 1.4	A_{xx} = 1.0	A_{yy} = 1.0: ^{13}C)	
	CO$_2$(H)	g_{zz} = 2.0012	g_{xx} = 2.0031	g_{yy} = 1.9971	2
		(A_{zz} = 17.9	A_{xx} = 14.2	A_{yy} = 13.3: ^{13}C)	
	(HCO$_3$)$_2^-$ [a]	g_1 = 2.0261	g_2 = 2.0079	g_3 = 2.0063	3
		(A_1 = 0.28	A_2 = –0.40	A_3 = –0.46: ^{13}C)	
	HCO$_3$ [a]	g_1 = 2.0182	g_2 = 2.0089	g_3 = 2.0048	3
	HCO$_3$ [a]	g_1 = 2.0189	g_2 = 2.0076	g_3 = 2.0060	3
	HCO$_3$	g_1 = 2.0184	g_2 = 2.0087	g_3 = 2.0059	3
	CO$_3^{3-}$ [a]	$g_∥$ = 2.0013	$g_⊥$ = 2.0032		3

ESR measurement at room temperature except for a) at 77 K.
Reference: 1) Ikeya and Kai 1988; 2) Chantry *et al*. 1962; 3) Holmberg 1971.
Nuclear spin and abundance: ^{13}C (1/2, 1.10%)

period. The Pleistocene river connection between these lakes was interrupted during dry periods and salts accumulated in the lake. The paleomagnetic stratigraphy of the lake sediments dates back to 3 Ma: the first datum at 185 m is the Brunhes–Matsuyama reversal (730 ka). The sediments down to 40 m have been dated by ^{14}C and down to 150 m by U–series (Smith 1984).

The size and the salinity of Searles Lake have fluctuated in response to climatic changes. The lake sediments are composed of clay, marl and salt layers deposited during the last 3 Ma. Samples of nahcolite (NaHCO$_3$) and halite (NaCl) collected from core LDW–6 in the north–central part of Searles Lake were dated with uranium–series isotopes up to its limit of about 330 ka (Bischoff *et al*. 1985). Figure 7.13 shows the ESR spectra of natural and γ–ray irradiated NaHCO$_3$ and NaCl samples for these samples (Ikeya and Kai 1988). The spectra of two NaCl samples shown in (b) and (c) are completely different and one of them shows almost the same spectrum as NaHCO$_3$. The NaHCO$_3$ sample and one of NaCl samples (83–66–19) show a weak signal of CO$_3^-$ and intense signal of CO$_3^{3-}$, while another NaCl sample

(83–63–24) shows an intense signal of CO_3^- and a weak signal of CO_2^-. Among eight samples shown in Table 7.4, only NaCl (83–63–24) from the deepest core (153 m) shows a completely different spectrum. The different spectra of NaCl may arise from different precipitation conditions at the lake.

    The intensities of these signals are increased by γ−irradiation. Table 7.4 shows the ESR age of the $NaHCO_3$ and NaCl samples using the CO_2^-

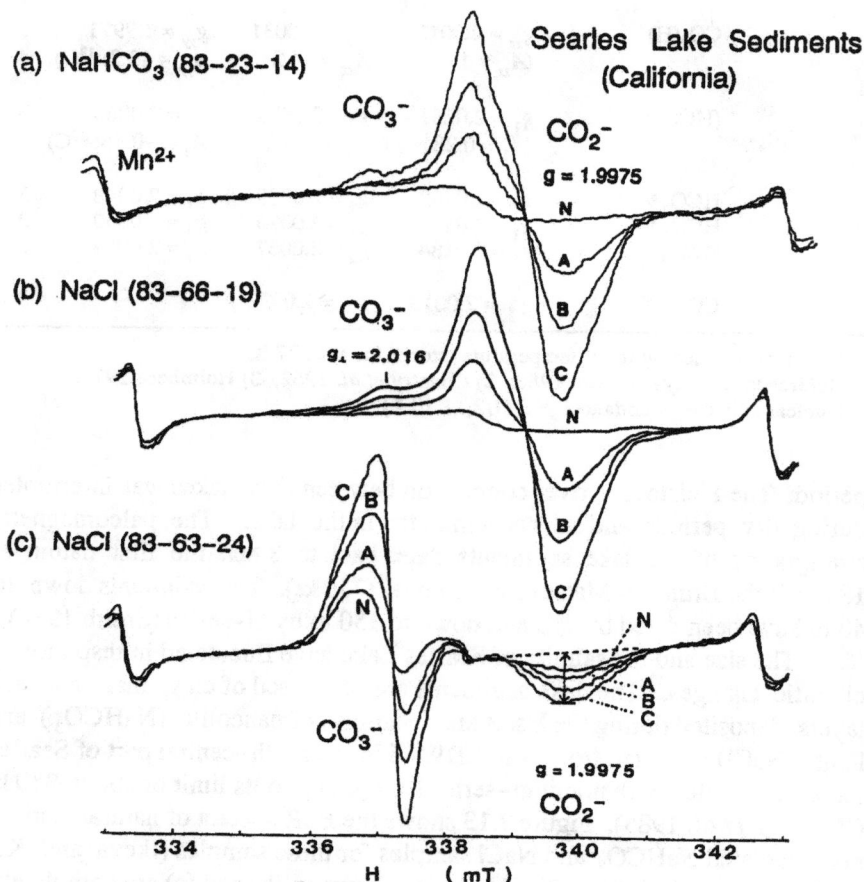

Figure 7.13   ESR spectra of (a) natural $NaHCO_3$ (83–23–14), (b) NaCl (83–66–19) and (c) NaCl(83–63–24) out of a bore–hole core from Searles Lake before and after γ−irradiation. The radiation dose are N (0 Gy), A (150 Gy), B (350 Gy) and C (680 Gy).

Table 7.4   The contents of ^{238}U, ^{232}Th and K as well as the *ED* for ESR dating of NaHCO$_3$ and NaCl in the bore–hole core of Searles Lake Sediments near Death Valley, southern California.

Sample	Depth (m)	C_U (ppm)	C_{Th} (ppm)	K$_2$O (%)	*ED* (Gy)	$N=0\%$ $k=0$	$N=0\%$ 0.2	$N=100\%$ 0	$N=100\%$ 0.2	Th/U age (ka)
NaHCO$_3$										
82–16–1	35.6	0.23	<0.004	0.24	**18.5**	76	63	80	70	35.3
84–6–9	79.6	3.50	0.34	0.6	**127**	120	70	165	96	192
83–22–13	99.6	0.68	0.08	0.24	**56.1**	169	109	215	148	>330
83–23–14	110.9	0.36	0.095	0.24	**57.4**	202	143	235	185	>330
NaCl			all above 330 ka	(upper limit of ^{230}Th/^{238}U dating)						
84–4–17	122	1.09	0.12	0.12	**50.6**	172	92	281	139	>330
83–21–18	126.5	2.67	0.70	0.05	**178**	295	131	518	233	>330
83–66–19	130	0.08	0.02	0.06	**30.9**	428	294	729	402	>330
83–63–24 *	153	0.22	<0.004	0.12	**370**	2275	1357	3050	2163	

* The age was determined using the CO$_3^-$ signal, while all other samples were dated using the CO$_2^-$ signal assuming an infinite medium.

signal assuming an infinite medium for the α–ray efficiency $k=0$ and 0.2 and the rate of radon loss $N_{loss}=0\%$ and 100%. Since the initial incorporation of ^{226}Ra and ^{230}Th upon precipitation is neglected in this calculation, the absolute ESR ages may not be accurate without a detailed radiation assessment. The ESR ages are generally younger than Th/U ages, even for complete radon loss ($N_{loss}=100\%$) and $k=0$. However, overall ESR ages generally increase with stratigraphic depth. The age uncertainty is largest for NaCl with a low uranium content. The external radiation D_{ex} from the sediments above and below the deposited layer may play a major role.

The fact that the α–ray track or α–recoil track is dissolved by ground–water (Kigoshi 1971) may indicate that the effective k–value is zero or even negative for such microcrystalline materials with brines and also for those surrounded by groundwater. Constant dissolution and recrystallization along α–tracks might be a cause of young ESR ages. The result that the ESR ages go up close to 700 ka indicates a sufficient stability of the CO$_2^-$–type center in both NaHCO$_3$ and NaCl. The age obtained by using the CO$_3^-$ center as indicated by sample 83–63–24 is too old to be accepted.

Some NaCl, especially rock salt, contains Mn^{2+} and colloid signals. Irradiation produces Mn^0 ($3g^5\,4s^2$ ^{6}S) with $A=2$ mT (Ikeya and Itoh 1975).

7.5 Nitrates, Borates and Other Evaporites
7.5.1 *Nitrates*

Nitrates are evaporite minerals containing the NO_3^- anion. **Niter** (KNO_3) and **soda niter** ($NaNO_3$) occur in some natural caves and in arable soil in dry and hot regions. The famous Chilean saltpeter deposits consist mostly of soda niter ($NaNO_3$, 18%), salt ($NaCl$, 16%) and sulfates (Na_2SO_4, $CaSO_4$, $MgSO_4$, 16%), in addition to sands and clay minerals.

The crystal structure of KNO_3 is orthorhombic as aragonite ($CaCO_3$) (see Chapter 6). The molecular orbital schemes of NO_3^- with 24 electrons are the same as CO_3^{2-}. Paramagnetic NO_3^{2-} and NO_3 are 25 and 23 electron systems, similar to CO_3^{3-} and CO_3^-, respectively (see Appendix A2). ESR parameters and suggested models of defects so far reported for nitrates are summarized in Table 7.5 though no ESR dating work has been reported. Some of these defects are not stable at room temperature.

7.5.2 *Borates*

Borates are mineral compounds characterized by fundamental structures of planar BO_3^{3-} and tetrahedral BO_4^{5-}. Borax ($Na_2B_4O_7 \cdot 10H_2O$), ulexite ($NaCaB_5O_9 \cdot 8H_2O$) and colemanite ($Ca_2B_6O_{11} \cdot 5H_2O$) are important minerals as resources for boron (Melvin 1990). Boracites are halogeborates which involve monovalent Li^+ ($Li_4B_7O_{12}X$; $X = Cl$, Br) or divalent cations M^{2+} ($M_3B_7O_{13}X$; $X = Cl$, Br, I).

Radiation–induced defects have been studied in boracites, alkali borates and borosilicates (Griscom *et al.* 1976, Haddad *et al.* 1985, Vignaud *et al.* 1986). BO_3^{2-}, B^{2+}, O^-, boron–oxygen hole center (BOHC), boron electron center (BEC), E' center, etc., have been reported and their ESR parameters are summarized in Table 7.5. BO_3^{2-} is isoelectronic with CO_3^- and NO_3 (see Appendix A2). Both ^{10}B and ^{11}B show hf splitting with the ratio $A(^{11}B)/A(^{10}B) = \mu_1(^{11}B)/\mu_1(^{10}B) = 2.99$. Synthetic $Li_2B_4O_7$ and $Na_2B_4O_7$ have been studied as TLD materials, which suggests the possibility of ESR dating or dosimetry.

7.5.3 *Other Minerals*

Carbonated evaporites such as Na_2CO_3 (soda ash) show similar signals as those in $CaCO_3$, i.e., signals A (SO_2^-), B (SO_3^-) and C (isotropic CO_2^-). Since most evaporites contain CO_3^{2-}, SO_3^{2-} or SO_4^{2-} as impurities, similar signals may be observed. Dolomite $[Ca,Mg(CO_3)_2]$ contains iron as well as Mn^{2+} and their signals predominate over the radical signals.

Table 7.5 ESR parameters of defects assigned in irradiated nitrates and borates.

Defect	Material	g factor			A tensor (mT)			Ref.
		$g_\parallel$ (g_1)	(g_2)	$g_\perp$ (g_3)	$A_\parallel$ (A_1)	(A_2)	$A_\perp$ (A_3)	
Nitrates								
NO_3	$NaNO_3$	2.0060		2.0237	no hf line			1
	$NaNO_3$	2.0047		2.0231	no hf line for ^{14}N			2
					3.92		0.3 (^{17}O)	
	KNO_3	2.0031		2.0232	0.431		0.346 (^{14}N)	3
	NH_4NO_3	$g = 2.0195$			$A = 0.363$ (^{14}N)			4
	urea nitrate	2.0066	2.0114	2.0203	no hf line			5
NO_3^{2-}	KNO_3	2.0015		2.0057	6.34		3.18 (^{14}N)	6
	KNO_3	2.025		2.008	$A = 0.9$ (^{14}N)			7
	NH_4NO_3	2.0018		2.0062	6.7		3.08 (^{14}N)	4
	$Ba(NO_3)_2$	1.9997		2.0053	6.63		3.47 (^{14}N)	8
NO_2	$NaNO_3$	2.0060		1.9980	5.2		5.8 (^{14}N)	9
NO_2^{2-}	KNO_3	2.002		2.006	6.1		3.2 (^{14}N)	7
NO	KNO_3	2.008	2.010	2.005	0.35		3.1 (^{14}N)	7
O_2^-	NH_4NO_3	$g = 2.0053$			no hf line			4
Borates								
BEC[a]	B_2O_3–K_2O	$g_{iso} = 2.0018$			$A_{iso} = 10.8$			10
E'	$B_2O_3 \cdot 3SiO_2$	2.0020		1.9996	23.4		20.7 (^{11}B)	11
BOHC[a]	$B_2O_3 \cdot 3SiO_2$	2.0025	2.0115	2.0355	1.36	1.53	0.87 (^{11}B)	11
B^{2+}	Zn–Cl boracite	$g = 2.0030$			$A = 9.0$ (^{11}B)			12
	Zn–Br boracite	$g = 2.0016$			$A = 7.2$ (^{11}B)			12
	Mg–Cl boracite	$g = 2.0048$			$A = 10.6$ (^{11}B)			12
BO_3^{2-}	Zn–Cl boracite	2.0079	2.0119	2.0126	1.01	1.1	1.11 (^{11}B)	12
	Zn–Br boracite	2.0094	2.0119	2.0148	1.23	1.0	0.80 (^{11}B)	12
	$K_2B_2O_4$	2.0091	2.0128	2.0143	1.14	0.85	0.85 (^{11}B)	13
O^-	Zn–Cl boracite	2.0042		2.0287				12
	Zn–Br boracite	2.003		2.030				12
Cl_2^-	$LiCl$–B_2O_3	2.0042		2.035	10.0		0.8 (^{35}Cl)	14

a) BEC = boron–electron center; BOHC = boron–oxygen hole center.

Reference: 1) Gesi and Kazumata 1964; 2) Reuveni and Luz 1976; 3) Livingston and Zeldes 1964; 4) Jarke and Ashford 1975; 5) Chantry *et al.* 1962; 6) Zeldes 1963; 7) Cunningham 1962; 8) Zdansky and Sroubek 1965; 9) Silver *et al.* 1971; 10) Griscom 1971; 11) Griscom *et al.* 1976; 12) Haddad *et al.* 1985; 13) Taylor *et al.* 1971; 14) Griscom *et al.* 1969.

Nuclear spin and abundance:
^{11}B $(I = 3/2, 80.1\%)$; ^{14}N $(I = 1, 99.63\%)$; ^{17}O $(I = 5/2, 0.038\%)$.

TL studies have been done on miscellaneous minerals and results are summarized (Sankaran et al. 1983). Although a TL peak does not necessarily correspond to the decay of a paramagnetic species, one can roughly estimate the stability of defects from TL peaks and therefore their lifetimes.

Antarctic carbonates showed intense signals of NO_3^2 (Whitehead *et al.*, 2001). The NO_3^2 centers are also formed artificially by electric discharges and UV-irradiation of calcite powder which was found during the study of electroluminescence (EL) (Sato , 2001). It must be noted that UV irradiation of ordinary carbonate did not show this signal. Antarctic carbonate may contain NO_3^- by some reason and intense UV must have generated them.

7.6 Summary

Paramagnetic defects and ESR dating of evaporites, mostly sulfate minerals and some carbonates, are described in this chapter. Stable defects common to all are SO_2^-, SO_3^-, CO_2^- and oxygen interstitial (peroxy) type O_2^{3-} (or O_3^-?) which are created from molecular anions of SO_4^{2-} or CO_3^{2-}. Examples of ESR dating of a fault (the San Andreas Fault) and arid lake sediments are described using precipitates of gypsum and nahcolite. The use of impurity signals for studying environmental changes is also discussed. Few ESR dating studies of evaporites have been made so far but, considering the importance of desert evolution, the techniques have to be refined for accurate age determination.

References

Albuquerque A.R.P.L. and Isotani S. (1982): The ESR spectra of X-ray irradiated gypsum. *J. Phys. Soc. Jpn.* **51**, 1111-1118.

Atoji M. and Rundle R.E. (1958): Neutron diffraction study of gypsum, $CaSO_4\_2H_2O$. *J. Chem. Phys.* **29**, 1306-1311.

Bates R.L. and Jackson J.A. ed. (1980): *Glossary of Geolog* (AGI, Virginia).

Bershov L.V., Martirosyan V.O., Marfunin A.S. and Speranskii A.V. (1971): The yttrium-stabilized electron-hole center in anhydrite. *Phy. Status Solid* (b) 44, 505-512.

Bershov L.V., Martirosyan V.O., Marfunin A.S. and Speranskii A.V. (1975): EPR and structure models for radicals in anhydrite crystals. *Fortschr. Mineral.* 52, 591-604.

Bischoff J.L., Rosenbauer J. and Smith G.I. (1985): Uranium-series of sediments from Searles Lake: difference between continental and marine climate record. *Science* 227, 1222-1224.

Chantry G.W., Horsfield A., Morton J.R. and Whiffen D.H. (1962): Structure, electron resonance and optical spectra of trapped CO_3^- and NO_3. *Mol. Phys.* **5**, 589-599.

Chen Y., Lu J., Head J., Arakel A.V. and Jacobson G. (1988): ^{14}C and ESR dating of calcrete and gypcrete cores from the Amadeus Basin, Northern Territory, Australia. *Quatern. Sci. Rev.* **7**, 447-453.

Chen Y., Arakel A. V., and Lu J. (1989): Investigation of sensitive signals due to g- rays irradiation of chemical precipitates: a feasibility study for ESR dating of gypsum, phosphate and calcrete deposits. *Appl. Radiat. Isot.* **40**, 1163-1170.

Cunningham J. (1962): Radiation chemistry of ionic solids II. Free radicals detected in irradiated potassium nitrate by EPR. *J. Phys. Chem.* **66**, 779-786.

Danby R.J., Boas J.F., Calvert R.L. and Polbrow J.R. (1982): ESR of thermoluminescent centers in $CaSO_4$ single crystals. *J. Phys. C: Solid State Phys.* **15**, 2483-2493

Deer W.A., Howie R.A. and Zussman J. (1966): *An Introduction to Rock Forming Minerals* (Longman, Essex).

Gesi K. and Kazumata Y. (1964): Electron spin resonance of g-ray irradiated $NaNO_3$. *J. Phys. Soc. Japan* **19**, 1981.

Griscom D.L., Taylor P.C. and Bray P.J. (1969): Paramagnetic resonance of room- temperature-stable V-type centers in g-irradiated alkali halide-boron oxide glasses. *J. Chem. Phys.* **50**, 977-983.

Griscom D.L. (1971): ESR studies of an intrinsic trapped-electron center in X-irradiated alkali borate glasses. J. Chem Phys. 55, 1113-1122.

Griscom D.L., Sigel Jr. G.H. and Ginther R.J. (1976): Defect centers in a pure-silica-core borosilicate-clad optical fiber: ESR studies. *J. Appl. Phys.* 47, 960-967.

Günter T.E. (1967): Electron paramagnetic resonance studies of the radiolysis of H2O in the solid state. *J. Chem. Phys.* **6**, 3818-3829.

Haddad M., Vignaud G., Berger R. and Levasseur A. (1985): ESR of X-ray irradiated boracites. *J. Phys. Chem. Solids* **46**, 997-1005.

Hofstaetter A., Scharmann A. and Scheib S. (1996): EPR and ENDOR investigations on the X-ray storage phosphors BaSO4 and SrSO4. *Appl. Radiat. Isot.* **47**, 1579-1587.

Holmberg R.W. (1971): ESR study of g-irradiated single crystal of potassium bicarbonate at 77 K. *J. Chem. Phys.* **55** 1730-1735.

Huzimura R. (1979): ESR studies of radical ion centers in irradiated $CaSO_4$. *Jap. J. Appl. Phys.* **18**, 2031-2032.

Ikeda S., Ikeya M. and Kashima N. (1989): ESR dating of gypsum speleothems. *J. Speleo. Soc. Japan* **14**, 68-73.

Ikeda S. and Ikeya M. (1992): ESR signals in natural and synthetic gypsum: An application of ESR to the age estimation of gypsum precipitates from San Andreas Fault. *Jpn. J. Appl. Phys.* **31**, L136-L138.

Ikeda S. and Ikeya M. (1993): ESR signals in synthetic gypsum doped with $^{13}CO_3^{2-}$ ions. *Appl. Radiat. Isot.* **44**, 321-323.

Ikeya M. and Itoh N. (1971): Optical and electron paramagnetic resonance of atomic manganese in NaCl. *J. Phys. Soc. Japan* **29**,1295-1300.

Ikeya M. (1978): ESR as a method of dating. *Archaeometry* **20**, 147-158.

Ikeya M. (1985): *Dating Method of Pleistocene Deposits and their Problems* ed. Rutter, N.W., Geoscience Canada Reprint Series 2, Chapt. XI.

Ikeya M. and Kai A. (1988): ESR dating of saline sediments using $NaHCO_3$ and NaCl. *Quatern. Sci. Rev.* **7**, 471-475.

Ikeya M., Oka T., Omura T., Okawa M. and Takeno S. (1997): Evaluation of environment using electron spin resonance (ESR) microscope image of gypsum ($CaSO_4 2H_2O$) nicrocrystalss in borehole cores at Konya Basin, Turkey. *Japan Review* **8**, 193-208.

Jarke F.H. and Ashford N.A. (1975): Electron paramagnetic resonance of NO_3, NO_3^{2-} and O_2^- in irradiated $NaNO_3$. *J. Chem. Phys.* **62**, 2923-2924.

Jong W. F. de (1959): *Kompendium der Kristallkunde* (Springer-Verlag, Wien)

Kasuya M., Brumby S. and Chapell J. (1991a): ESR signals in gypsum single crystals: implications for ESR dating. *Nucl. Tracks* **18**, 329-333.

Kasuya M., Kato M. and Ikeya M. (1991b): ESR signals of natural barite ($BaSO_4$) crystals: possible application to geochronology. *Essays in Geology*, Prof. H. Nakagawa Commemorative Volume 95-98.

Kigoshi K. (1971): Alpha recoil thorium-234 into water and uranium-234/uranium- 238 disequilibrium in nature. *Science* **173**, 47-48.

Kohno H., Yamanaka C. and Ikeya M. (1996): Effects of alpha-irradiation and pulsed ESR measurements of evaporites. *Appl. Radiat. Isot.* **47**, 1459-1463.

Livingston R. and Zeldes H. (1964): Paramagnetic resonance study of NO_3 in irradiated KNO_3. *J. Chem Phys.* **41**, 4011-4012.

Melvin J.L. ed. (1990): *Evaporites, Petroleum and Mineral Resources* (Elsevier, Amsterdam).

Moore B.W. and Nicholas B.G. (1964): *Speleology: The Study of Caves* (Heath & Company, Mass.).

Nambi K. S. V. (1982): ESR and TL dating studies on some marine gypsum crystals. *PACT* **6**, 314-321.

Nambi K.S.V. (1985): Scope of electron spin resonance in thermally stimulated luminescence studies and in chronological applications. *Nucl. Tracks* **10**, 113-131.

Nambi K. S. V. (1993): On the sulphoxy radicals in $CaSO_4$:Dy,Na thermoluminescent phosphor : electron paramagnetic resonance studies. *J. Phys.: Condens. Matter* **5**, 1791-1800.

Ochiai H., Ikeya M. and Morsy MA (1999): Geologic aspect and ESR dating of corals from Pleistocene terraces in Ras Mohammed, Egypt. *Carbonates and Evapotites* **14**, 138-145.

Omura T. and Ikeya M. (1995): Evaluation of the ambient environment of mineral (gypsum($CaSO_4$ $2H_2O$)) growth by ESR microscope. *Geochem. J.* **29**, 317-324.

Reuveni A and Luz Z (1976):ESR of NO_3 in sodium nitrate. *J. Magn. Reson.***23**, 271-274.

Ryabov I. D., Bershov L. V., Speranskiy A. V. and Ganeev I. G. (1983): Electron paramagnetic resonance of PO_3^{2-} and SO_3^- radicals in anhydrite, celestite and barite: the hyperfine structure and dynamics. *Phys. Chem. Minerals* **10**, 21-26.

Ryabov I. D., Bershov L. V. and Ganeev I. G. (1989): Electron paramagnetic resonance of PO_{22} and SO_2^- radicals in barite. *Phys. Chem. Minerals* **16**, 374-377.

Samoilovich M. I. and Tsinober L. I. (1970): Characteristics of radiation color centers and microisomorphism in crystals. *Sov. Phys. Crystallogr.* 14, 656-666.

Sankaran A. V., Nambi K. S. V. and Sunta C. M. (1983): Progress of thermoluminescence research on geological materials. *Proc. Indian Nat. Sci. Acad.* **49**, 18-112.

Silver B.L., Koresh J., Schlick S. and Luz Z (1974): ESR study of NO_2 in X-irradiated sodium nitrate 17O. *Mol. Phys.* **45**, 225-231.

Smith G.I. (1984): Paleohydrologic regimes in the southwestern Great Basin, 0-3.2 Ma ago, compared with other long records of "Global" climate. *Quatern. Res.* **22**, 1-17.

Taylor P.C., Griscom D.L. and Bray P.J. (1971): ESR studies of BO_3^{2-} ions in potassium borate ceramics. *J. Chem. Phys.* **54**, 748-760.

Vignaud G., Berger R., Haddad M. and Levasseur A. (1986): Review of ESR centers induced by X-ray irradiation in both vitreous and crystalline borates or halogenoborates. *J. Non-Cryst. Solids* **86**, 6-12.

Whitehead N., Lyon GL, Claridge G.C., Sato H. and Ikeya M. (2001): ESR studies of Antarctic carbonates and sulfates. *ESR Appl. Metrol.* **17** - *Proc. 2001-ESRDD-Osaka* in press.

Wigen P.E. and Cowen J.A. (1960): Paramagnetic resonance absorption in several electron-irradiated molecular crystals. *J. Phys. Solids* 17, 26-33.

Zdansky K. and Sroubek Z. (1965): Electron spin resonance of NO_3^{2-} in the irradiated $Ba(NO_3)_2$ single crystal. *Phys, Stat. Sol.* **10**, 571-574.

Zeldes H. (1963): *Proc. Int.. Conf. Paramagnetic Resonance 1st Jerusalem*, p.764.

Chapter 8

Phosphates

– Bioapatite for Anthropology –

Dear to me is sleep
still more, being made of stone –––––
––––– then do not wake me. Keep your voices low.
Michaelangelo Sonnet 78

8.1   Introduction

The phosphate ion (PO_4^{3-}) is an important constituent of phosphate minerals. Bones and teeth consist of an organic part, **collagen,** and an inorganic part, **hydroxyapatite** $[Ca_{10}(PO_4)_6(OH)_2]$, a mineral that belongs to the phosphate group. Bones and teeth are often the only materials that are adequately preserved in archaeological sites. Mineral samples of $CaHPO_4$ and apatite $[Ca_{10}(PO_4)_6 X_2 ; X = OH, F, Cl, Br, I]$ have been studied extensively. The apatite changes in structure due to replacement of host ions by both cationic and anionic impurities coming from the environment. It constitutes an *"open system"* to the environment in contrast to a "closed system" such as coral in Chapter 6.

Man, *where did he come from* and *where does he go* ? This is a fundamental question that interests us as all human beings, the *Homo sapiens sapiens.* In the glacial era, man lived in caves to avoid cold weather. ESR dating of a stalactite at Akiyoshi cave attracted an anthropologist in Greece and led to ESR dating of bones and teeth excavated at Petralona cave (Poulianos 1971, Ikeya 1977, 1978, Hennig *et al.* 1981). Work on Arago, Tautavel, France (DeLumley and Labeyrie 1981) and Choukoutien, China followed with participation of European scientists in ESR dating. The absence of an appropriate dating method in the Quaternary beyond the upper limit of ^{14}C-dating was the main reason for the rapid development of this method. Petralona cave thus can be called as the cradle for the growth of ESR dating in paleo–anthropology.

Radiocarbon–dating of bones using organic collagen has been done within the technique's limits of 50 ka. The relative ages of bones have also been estimated indirectly from the content of *fluorine* or *uranium* as these elements accumulate from the environment with the passage of time. Nitrogen, on the other hand, is gradually leached from bones. *Amino acid racemization* has also been used for dating fossils. However, only relative dating is possible with these chemical methods as the environment affects the chemical composition (Oakley 1980). TL dating of fossil bones is very difficult due to the oxidation or burning of organic remains.

Dentists have studied defects and radicals with unpaired electrons produced by radiation both in the inorganic apatite part and in the organic part of bones and teeth (Ostrowski *et al.* 1980). The possibility of ESR dating (Ikeya 1978) and relative ESR dating for bones and teeth were demonstrated using anthropologically known fossils (Ikeya and Miki 1980). ESR dating gives only an equivalent dose *(ED)* related to the age. The dose rate may changing through burial time consequent to uranium accumulation from the

the environment. Sensitivity of ESR induction with radiation may also depend on structural changes of bones. Hence, these earlier works tuned the ED to the known age for bones and tooth enamel and obtained an apparent dose rate, D. Such a study gave only the order of magnitude of the ages or the age within an accuracy of a factor 2 or 3. Subsequent work aimed to refine the age by dealing with assessment of radiation dose and structural changes (Ikeya 1982a, 1985, Yokoyama *et al.* 1982, Grün and Invernati 1985, Grün *et al.* 1987, 1988, Blackwell and Schwarcz 1993).

Figure 8.1 shows a photograph of the cranium of *Homo erectus petra-lonasis* which was once considered *Homo sapiens neanderthalensis*. A controversy on its age exists among anthropologists (Day 1982). Two ESR ages of 200 ka (Hennig *et al.* 1981, 1982) and 450 ka (Ikeya 1982a) were presented in addition to faunal evidence of 700 ~ 500 ka (Poulianos 1982). Radiation assessment at the site is considered as necessary for dating bones.

The first part of this chapter describes radiation–induced defects in synthetic apatites and bioapatites such as bone and tooth enamel. Typical ESR spectra of anthropologically important bones are given together with the obtained ED. Lastly, the ages are deduced based on *open–system models of uranium accumulation* and compared with the accepted ages of standard fossils. ESR dosimetry for accident and atomic–bomb radiation exposure using tooth enamel is described in Chapter 13.

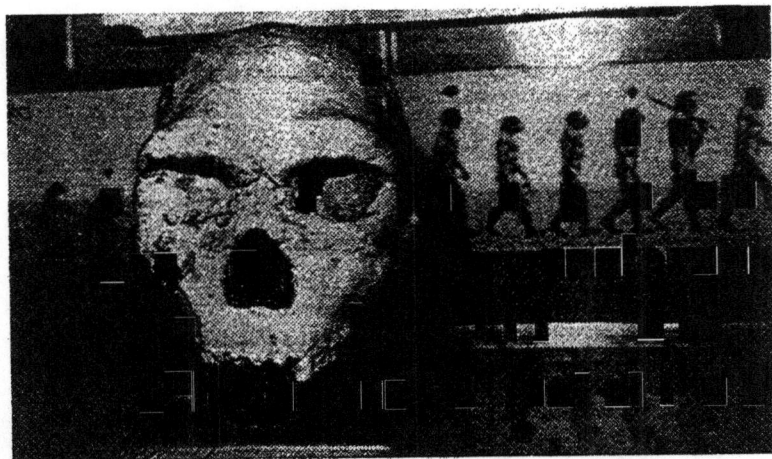

Figure 8.1 The cranium *Homo erectus petralonasis* recovered from the Petralona Cave, northern Greece. The photograph was taken at the Geology Institute, University of Thesaloniki, Greece. An ESR age of 450 ka was presented in addition to the faunal assignment of 700 ka.

8.2   Structure and Defects in Apatites

8.2.1   *Structure*

Bone and tooth consist of an inorganic part called **"hydroxyapatite"** and an organic part called **"collagen"**.   Teeth consist of three main tissues; dentine, enamel and cement.   **Enamel** contains 95 ~ 98% hydroxyapatite and a few percent organic materials.

Hydroxyapatite, $Ca_{10}(PO_4)_6 (OH)_2$, is a microcrystalline hexagonal mineral.   The crystal structure of apatites, [$Ca_{10}(PO_4)_6 X_2$: X = OH, F, Cl, Br, I], is shown in Figure 8.2.   The structure of apatites changes by the replacement of each constituent cation and anion by other impurity ions.

(1) Ca^{2+} can be replaced by monovalent (Na^+, K^+ and H^+), divalent (Mg^{2+}, Sr^{2+}, Ba^{2+}, Fe^{2+}, etc.) and trivalent cations (Al^{3+}, Y^{3+}, Ce^{3+}, etc.).

(2) OH^- can be replaced by monovalent anions (F^- and Cl^-), divalent anions (CO_3^{2-}) and neutral H_2O.

(3) PO_4^{3-}, a tetrahedral phosphate ion, can be replaced by divalent (SO_4^{2-}, CO_3^{2-}, etc.) and trivalent molecular anions (BO_3^{3-}, As_4O^{3-}, etc.).

The physics of defects in apatite is rather complicated due to a large unit cell structure and symmetry involved (Brown and Chow 1976).

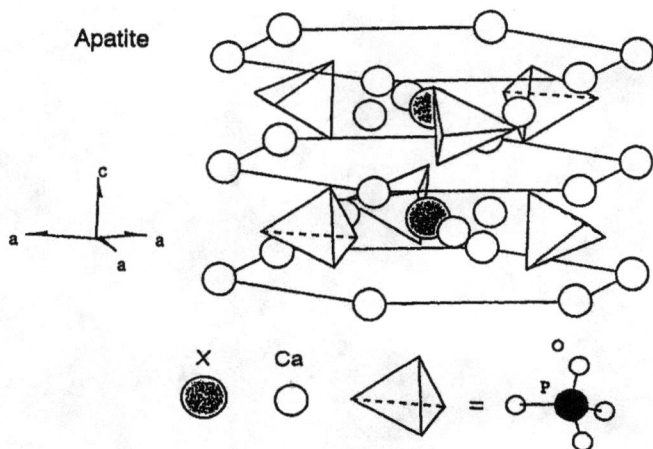

Figure 8.2   Crystal structure of apatites $[Ca_{10}(PO_4)_6 (X)_2$; X = OH, F, Cl]. Lattice parameters:

Hydroxyapatite (X = OH) : $a = 0.941$ $c = 0.687$ nm
Fluorapatite   (X = F) : $a = 0.935$ $c = 0.687$ nm (Deer *et al.* 1966).

8.2.2 *ESR Spectra and Defects in Synthetic Apatites*

Radiation–induced defects in synthetic apatites have been studied by various investigators, as summarized in Table 8.1. A study on models of radiation–induced defects in synthetic apatites has been done using *hydroxy– apatite* and *fluorapatite* powder in a wet state and *sintered fluorapatite* in a dry state (Ishii and Ikeya 1993). Sintering is done using a molding press of 1 MPa (1 t/cm^2) and a furnace in a nitrogen atmosphere at 1000°C. ESR spectra of these apatites after γ–irradiation are shown in Figure 8.3. A prominent observation is that these apatites have different types of CO_2^- radicals, that is,

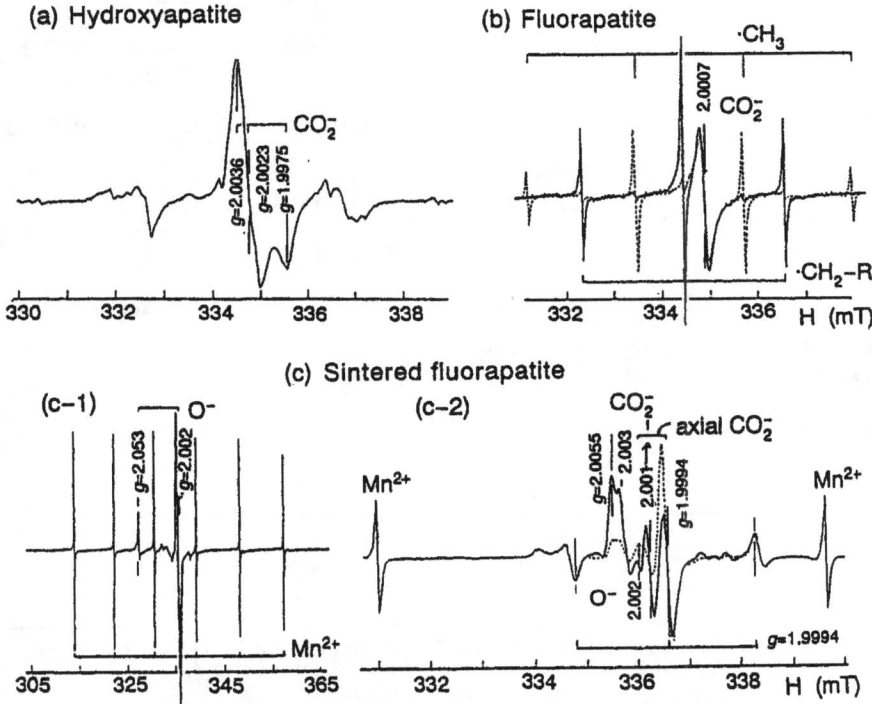

Figure 8.3   ESR spectra of γ–irradiated apatites. (a) Hydroxyapatite [$Ca_{10}(PO_4)_6(OH)_2$]. (b) Fluorapatite [$Ca_{10}(PO_4)_6F_2$]; the dotted curve shows the enhancement of the quartet signal and disappearance of the triplet after heating at 200°C for 15 min. (c) Sintered fluorapatite; sweep width of mag– netic field is (c–1) 100 mT and (c–2) 10 mT; the dotted curve shows the spec– trum when synthesized by doping with 0.1 mol% CO_3^{2-}.

Table 8.1   ESR parameters of defects in irradiated synthetic apatites.

Model	$g_1 (g_\parallel)$	g_2	$g_3 (g_\perp)$	A_1	A_2	A_3	Material	Ref.
CO_2^-	2.0023	2.0036	1.9975				OH–apatite	1
orthorhombic	2.0015	2.0030	1.9970	15.9	16.4	19.9 (^{13}C)	C–apatite (B) [a]	2
	2.0017	2.0031	1.9972				C–apatite (B) [a]	3
	2.0018	2.0034	1.9971				C–apatite (A) [a]	4
	2.0024	2.0035	1.998	20.5	17.5	16.4 (^{13}C)	C–apatite (A) [a]	5
CO_2^- axial	2.003		1.999				sin–F–apatite [b]	1
CO_2^- isotropic	2.0007			A_{iso} =14.8 (^{13}C)			F–apatite	1
	2.0007			A_{iso} =14.7 (^{13}C)			C–apatite	2,3
CO_3^- axial	2.0060	2.0170	2.0084	A_{iso} =1.3 (^{13}C)			C–apatite (B) [a]	2
	2.0066	2.0178	2.0087				C–apatite (B) [a]	3
CO_3^- rotating	2.0115			A_{iso} =1.12 (^{13}C)			C–apatite (B) [a]	2,3
O^-	2.0276	2.0406	2.0330				C–apatite (B) [a]	2
	2.0018		2.0683	0.56	0.59	(^{1}H)	OH–apatite	6
O^- at an F^- site	2.002		2.053				sin–F–apatite	1
	2.0018		2.0522	0.70	0.03	(^{19}F)	F–apatite	7
O_2^-	2.06	2.001	2.001				OH–apatite	8
O^{3-} at three F^-	1.9983		1.9994	3.2	2.1	(^{19}F)	F–apatite	7
vacancies	1.9995		2.0000	0.41	0.27	(^{19}F)	F–apatite	7
Hole center [c]	2.0068	2.0032	2.0148				OH–apatite	9
SO_2^-	2.0055						sin–F–apatite	1
$\cdot CH_3$	2.0026 quartet			2.3 (^{13}C)			F–apatite	1
$\cdot CH_2$–R	2.0033 triplet			2.1 (^{13}C)			F–apatite	1

a) Carbonated apatite (A–type is CO_3^{2-} substitution at OH^- and B–type at PO_4^{3-} sites).
b) Sintered fluorapatite
c) Hole center trapped at both OH^- and neighboring PO_4^{3-}.
Reference: 1) Ishii and Ikeya 1993; 2) Callens *et al.* 1989; 3) Meons *et al.* 1993;
            4) Bacquet *et al.* 1981; 5) Geoffroy and Tochon–Danguy 1982; 6) Mengeot
            et al. 1975; 7) Piper *et al.* 1965; 8) Dugas and Rey 1976; 9) Close *et al.*
            1981.
Nuclear spin and abundance: ^{1}H (I = 1/2, 99.99%); ^{13}C (I = 1/2, 1.10%);
                             ^{19}F (I = 1/2, 100%).

Hydroxyapatite : orthorhombic CO_2^- at $g = 2.0023$, 2.0036, 1.9975
Fluorapatite : freely rotating CO_2^- at $g = 2.0007$
Sintered fluorapatite: axial CO_2^- at $g_{\parallel} = 2.003$, $g_{\perp} = 1.999$

Orthorhombic CO_2^- detected in hydroxyapatite is presumably due to hydrogen bonding. Two oxygen atoms in CO_2^- would form hydrogen bonds with the H atom in OH^- along its crystalline c-axis. Then an orthorhombic spectrum appears. The hindered rotation of the bent molecular ion of CO_2^- around the y-axis (O–O direction)(see Figure 8.4), if it occurs, will lead to an axial signal with anisotropic g factors, $g_{\perp} = 2.0025$ and $g_{\parallel} = 1.9973$.

Freely rotating CO_2^- in fluoroapatite, presumably surrounded by water molecules as in $CaCO_3$, results in the single line at $g = 2.0007$. The signal amplitude is enhanced with a decrease in the linewidth by decreasing the temperature. The line shape becomes broad again below 200K and changes into the spectrum of orthorhombic CO_2^- as in the case of calcite (Ishii and Ikeya 1993, see also Chapter 5).

In addition to freely rotating CO_2^-, the g factors ($g = 2.003$ and $g = 1.999$) of an axial signal detected in sintered fluorapatite are in agreement with those of **axial CO_2^-**, presumably substituting F^- and rotating around the x-axis perpendicular to O–C–O in the c-plane as shown in Figure 8.4 (b); $g_{yy} = 1.9973$ and $g_{zz} = 2.0016$ associated with orthorhombic CO_2^- in calcite

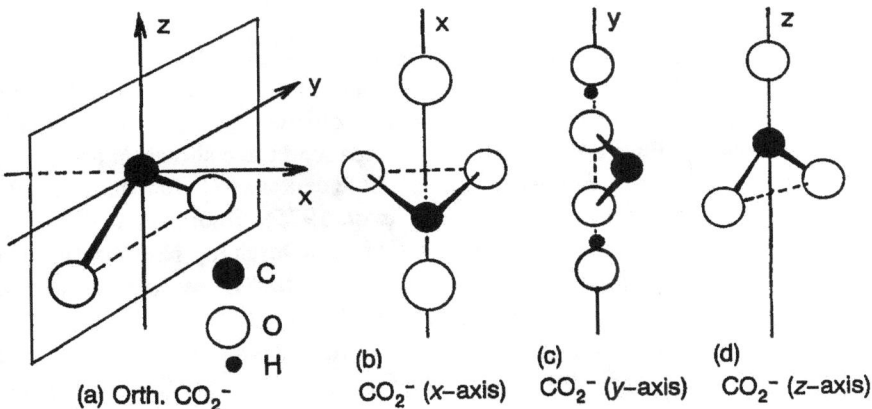

Figure 8.4 (a) Principal axes of orthorhombic CO_2^- and three modes of *hindered rotation of CO_2^-* around the (b) x-axis in fluorapatite (c-axis), (c) y-axis in hydroxyapatite due to hydrogen bonding and (d) z-axis for adsorbed CO_2^-. A mode of free rotation is also observed in fluorapatite.

(see Table 5.1) are averaged out to give $g_\perp = 1.9995$. This defect was neither observed in hydroxyapatite nor in fluorapatite in a wet state. The formation yield of this defect by radiation is increased by doping with $CO_3{}^{2-}$ and thus H_2O molecules may be no longer involved in the sintered apatite.

The g factors of $CO_2{}^-$ in hindered rotation around the z–axis give a somewhat isotropic signal at $g = 2.0016$. The *coordinate system* of $CO_2{}^-$ and its *axes of hindered rotation* are summarized in Table 8.2 and Figure 8.4.

Table 8.2 Various $CO_2{}^-$ radicals detected in irradiated apatite, bone, tooth enamel and $CaCO_3$.

Species	g factor [*)	Material
$CO_2{}^-$		
orthorhombic:		
	$g_{zz} = 2.0016$ $g_{xx} = 2.0032$ $g_{yy} = 1.9973$	OH–apatite, tooth enamel CaCO_3
axial:		
z–rotation:	$g_{\parallel} = g_{zz} = 2.0016$ $g_\perp = 2.0016 \approx g_{iso}$	not observed
x–rotation:	$g_{\parallel} = g_{xx} = 2.0032$ $g_\perp = 1.9995$	sintered F–apatite CaCO_3 (α–irradiated)
y–rotation:	$g_{\parallel} = g_{yy} = 1.9973$ $g_\perp = 2.0025$	tooth enamel, bone
isotropic: free rotation: $g_{av} = 2.0007$		F–apatite, tooth enamel, CaCO_3

* See Chapter 2 (Section 2.4.3) for calculation of g factors in hindered rotation.

In fluorapatite, *triplet* and *quartet* signals presumably due to $\cdot CH_2R$ ($R \neq H$) and $\cdot CH_3$, respectively, are observed. In sintered fluorapatite, the sextet due to hf splitting of Mn^{2+} is observed together with O^- and probably $SO_2{}^-$ as shown in Figure 8.3 (c–1) and its expanded trace shown in (c–2).

O^- at $g = 2.053$ and $g = 2.002$ is a hole center of O^- substituting for F^- in synthetic hydroxyapatite (Mengeot et al. 1975), fluorapatite (Piper et al. 1965) and also in natural apatite crystals from Durango, Mexico (Ishii et al. 1991). Substitutional O^- is created by radiation in samples heated at high temperature, presumably particularly in dry apatite.

The signal at $g = 2.0055$ grows with microwave power up to above 10 mW. The g factor is close to that of $SO_2{}^-$ in $CaCO_3$ (see Table 5.2) and $CaSO_4$ (see Table 7.1). Since the formation yield of this signal is decreased by doping with $CO_3{}^{2-}$, a model of $SO_2{}^-$ seems to be appropriate. The above models are summarized in Table 8.1.

8.2.3 Defects in Bioapatites : Bone and Tooth

Extensive physics and chemistry studies of radiation–induced defects in bones and tooth enamel or in bioapatite (hydroxyapatite) have been made with ESR from low to high temperature in the field of biophysics related to clinical and dental radiation measurements as summarized in Table 8.3.

The ESR spectrum of bones is angle dependent because of preferential orientation of the c–axis of microcrystals along the growth direction (Pane-pucci and Farach 1977). Studies on radiation–induced defects with high frequency ESR spectrometry have also been done for tooth enamel, in which overlapping signals are avoided by orienting the sample (Bouchez et al. 1988, Rossi and Poupeau 1990). Figure 8.5 shows ESR spectra of fossil tooth enamel at X–band (9.5 GHz) and Q–band (35 GHz) frequencies.

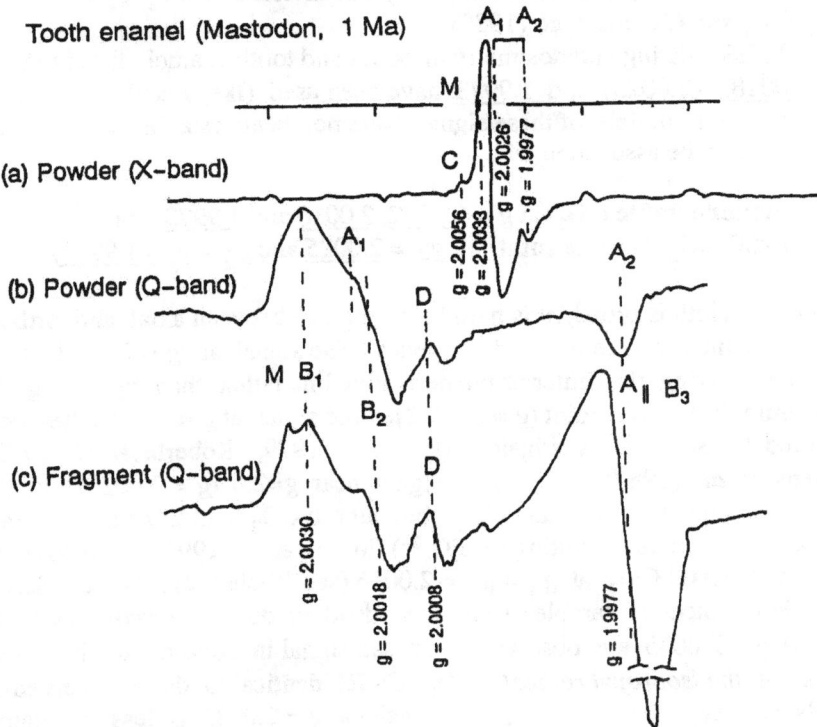

Figure 8.5 ESR spectra of powdered tooth enamel (1 Ma old Mastodon) at (a) X–band (9.5 GHz) and (b) Q–band (35 GHz), and (c) an enamel fragment at Q–band (35 GHz) (after Rossi and Poupeau 1990). Models for these signals are summarized in Table 8.3.

The following signals are observed:

Axial signals (A) at $g_\perp = 2.0026$ and $g_\parallel = 1.9977$
Orthorhombic signals (B) at $g_1 = 2.0030$, $g_2 = 2.0018$ and $g_3 = 1.9977$
Two isotropic signals (C) at $g_{iso} = 2.0056$ and (D) at $g_{iso} = 2.0007$
A weak septet signal (M) centered at $g = 2.0033$ ($A = 2.16$ mT)

These signals and their suggested models are summarized in Table 8.3 together with some models of radiation–induced defects reported in the literature. The models asterisked (*) and underlined in the table are those suggested by the author based on the considerations in Section 8.2.2 (see also Table 8.2). Signal A might correspond to signal B broadened due to shf by nearby protons. CO_3^{3-} suggested for the main signal around $g = 2.0$ by Cevc *et al.* (1972) and Doi *et al.* (1979) was ascribed to CO_2^- by Bacquet *et al.* (1981) and Callens *et al.* (1989).

In ESR dating and dosimetry of bones and tooth enamel, the signals at $g = 2.0018$, (2.0025) and 1.9973 have been used (Ikeya and Miki 1980). Though clear models of these signals have not been established, they are considered to be associated with

orthorhombic CO_2^- at $g = 2.0032$, 2.0018 and 1.9973   or
axial CO_2^- (y–axis rotation: $g_\perp = 2.0025$ and $g_\parallel = g_y = 1.9973$)

If the linewidth is broad, it is hard to distinguish between axial and orthorhombic signals at the X–band frequency. The signal at $g = 2.0018$ is obtained by taking the center of the derivative line rather than by taking the maximum inflection point ($g = 2.0032$). The signal at $g = 2.0018$ has been reported by several investigators (Doi *et al.* 1979, Köberle *et al.* 1973). Callens *et al.* (1987) assigned the signal near $g=2.0$ ($g = 2.0025$?) to the composite signal of CO_2^- and CO_3^{3-} but later to CO_2^- and CO_3^- at a surface site and a B site (substituting for PO_4^{3-}) (Meons *et al.* 1993). Although no one noted, **axial CO_2^-** at $g_\parallel = g_\perp = 2.0016$ (see Table 8.2) might overlap.

For Holocene samples (with low adsorbed doses), *organic radicals* around $g = 2.0045$ are observed as a broad signal in addition to the septet signal of the *isopropyl radical* [$(CH_3)_2C\cdot{-}R$] identical to those observed in corals and egg shells. The signal intensity at $g = 2.0045$ is less in enamel than in dentine. This signal must be subtracted to obtain the contribution of radiation–induced signal. The intensity of this signal is also enhanced by heating the bone (Ikeya 1986). Chemical ESR dating of Holocene bones might be done based on radical formation or oxidation (see Chapter 12).

Table 8.3 ESR parameters of defects in irradiated bone and tooth (enamel).

g factor (A tensor in mT)	Model and comment	Reference
Tooth enamel (see Figure 8.5)		
A : $g_\parallel = 1.9975$ $g_\perp = 2.0026$ (axial symmetry, no hf line)	axial CO_3^{3-} or a trapped electron at an OH^- site * <u>axial CO_2^- (y rotation)</u>	1
B a): $g_1 = 2.0030$ $g_2 = 2.0018$ $g_3 = 1.9977$ (orthorhombic symmetry)	a distorted form of the above center or an electron trapped near an OH^- site * <u>orthorhombic CO_2^-</u>	1
C : $g_{iso} = 2.0056$	appears after heating at 280°C * <u>SO_2^- or lipid peroxy–type signal</u>	1, 2
D : $g_{iso} = 2.0007$	observed at a Q–band frequency * <u>freely rotating CO_2^-</u>	1, 2
M : Septet centered at $g = 2.0033$ ($A = 2.16$: ^{13}C)	isopropyl radical [$(CH_3)_2 C$ ·–R]	1, 3
Tooth enamel in literature		
$g_1 = 2.0032$ $g_2 = 2.0018$ $g_3 = 1.9972$	orthorhombic CO_2^-	4
$g_1 = 2.0032$ $g_2 = 2.0016$ $g_3 = 1.9973$	orthorhombic CO_2^-	5
$g_x = 2.0044$ $g_z = 2.0020$	CO_3^{3-} at B (PO_4^{3-}) site	5
$g_x = 2.0039$ $g_z = 2.0014$	CO_3^{3-}	5
$g_\parallel = 1.9983$ $g_\perp = 2.0036$ ($A = 17$: ^{13}C)	CO_3^{3-} * <u>axial CO_2^- (x or y–rotation)</u>	6
$g_1 = g_2 = 2.0029$ $g_3 = 1.9972$ ($A = 17$: ^{13}C)	axial CO_3^{3-} * <u>axial CO_2^- (y rotation)</u>	7
$g = 2.0029$ ($A = 50$)	H°; unstable at room temperature	8
Bone in literature		
$g_\parallel = 1.9978$ $g_\perp = 2.0036$	h center (an electron trapped at a PO_4^{3-} site)	9
$g_\parallel = 1.997$ $g_\perp = 2.002$	* <u>CO_2^-</u>	10
$g_\parallel = 1.997$ $g_\perp = 2.0018$	* <u>CO_2^-</u>	11

* Models suggested by the author based on Table 8.2.
a) Species B with $g_1 = 2.0032$ $g_2 = 2.0018$ $g_3 = 1.9975$ was also reported.
Reference: 1) Rossi and Poupeau 1990; 2) Bouchez *et al.* 1988; 3) Roufosse *et al.* 1976;
4) Bacquet *et al.* 1981; 5) Callens *et al.* 1987; 6) Cevc *et al.* 1972; 7) Doi *et al.* 1979; 8) Cole and Silver 1963; 9) Ostrowski *et al.* 1980; 10) Panepucci and Farach 1977; 11) Köberle *et al.* 1973.

8.3　ESR Spectra and *ED*

8.3.1　*ESR Spectra of Bones and Teeth*

ESR dating in paleo–anthropology started with *stalagmites* and *bones* from Petralona cave, northern Greece. This was followed by the studies on bones and *teeth* excavated at Java in Indonesia, Choukoutien in China, Heidelberg and Steinheim in Germany, Arago, Tautavel in France and Qafzeh and Kebara in Israel (see Table 8.4). In the following, some examples of ESR spectra of bones and teeth related to anthropology are discussed.

(a)　*Fossil bones from Petralona, Greece* :　A typical ESR spectrum of the stable defect in a fossil bone excavated at Petralona, Greece is shown in Figure 8.6.　Petralona is an anthropologically well–known site, where ESR dating was first carried out using stalagmitic calcite and bones (Ikeya 1977, 1978).　Signals associated with CO_2^- ($g_{\parallel}$ = 1.9973 and $g_{\perp}$ = 2.0025) were observed, which were ascribed to axial CO_2^-.

The *human cranium* in the Petralona cave covered with carbonate encrustation was discovered by a villager in northern Greece.　It was considered at first as Neanderthaloid and later as middle or early Pleistocene *Homo erectus* (Poulianos 1971, Stringer *et al.* 1979).　ESR dating of bones as well as stalagmites excavated by Greek anthropologists was done (Ikeya 1977, 1978).　These results were used for the first relative ESR dating of fossil bones and teeth in anthropology (Ikeya and Miki 1980).　A piece of cranium and carbonate encrustation were ESR dated to be 157 ka (Hennig *et al.* 1981), which led to the Petralona cave dating controversy (Poulianos 1982, Ikeya 1982a, Hennig *et al.* 1982).

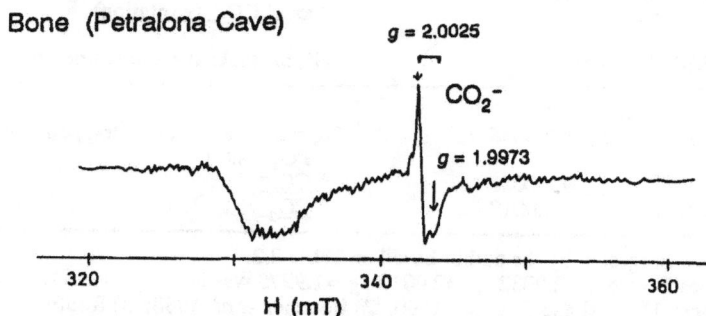

Figure 8.6　ESR spectrum of a bone from Petralona, Greece. Signals associated with CO_2^- are observed (Ikeya 1978).

In this controversy the major issue was the external dose rate (D_{ex}) of 1.9 ± 0.2 mGy/a which is too high to be trusted since the previous work showed it to be less than 0.5 mGy/a in the Petralona cave (Ikeya 1977). The internal dose is small due to the low uranium content in the limestone. The age estimate was 495 ka using the same *ED* and $D_{ex} = 0.5$ mGy/a and 675 ka using the small estimation of $D_{ex} = 0.35$ mGy/a at the site covered with pure carbonate for radon loss of 50%. The D_{ex} later measured by Papastefanou *et al.* (1986) supports the low dose rate of 0.35 mGy/a for the skull. This work on Petralona samples nonetheless stimulated ESR dating in anthropology.

(b) *Fossil ivory and horse molar from Choukoutien, China* : Figure 8.7 shows ESR spectra at the Q–band frequency (35 GHz) of (a) a horse molar and (b) an ivory piece excavated at Choukoutien, China. Although it is still difficult to exclude the presence of axial CO_2^- ($g_{\parallel} = 1.9973$ and $g_{\perp} = 2.0025$), both spectra might be due to orthorhombic CO_2^- ($g_{zz} = 2.0019$, $g_{xx} = 2.0032$ and $g_{yy} = 1.9973$). Septet signals due to $(CH_3)_2C\cdot-R$ radicals overlap in the ivory. The septet was observed in tooth enamel in the middle Pleistocene (Steinheim, Arago, etc.).

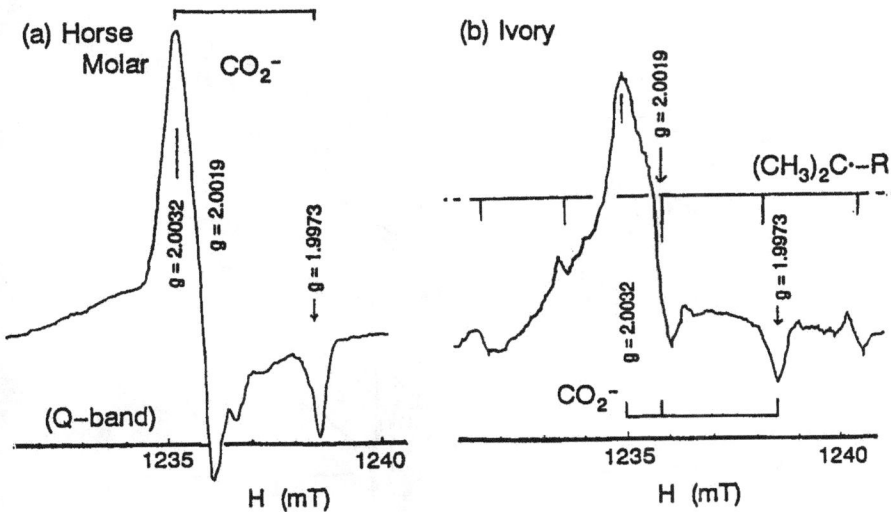

Figure 8.7 Q–band ESR spectra of (a) horse molar and ivory from Choukou-tien, China. The signals of axial and orthorhombic CO_2^- are indicated.

(c) *Fossil bone from Namib, Tanzania* : Figure 8.8 shows the ESR spectrum of bone from the Namib IV site in Tanzania, Africa. The spectrum shows a signal around 160 mT at $g = 4.27$ due to Fe^{3+} and six hf lines due to Mn^{2+}. A broad line is presumably due to Fe^{3+} and Mn^{2+} in the coagulated state and may arise from the adsorbed clay–like particles. It must be noted that one still observes the characteristic CO_2^- signals overlapped with the broad spectrum between the third and fourth lines of the six hf lines of Mn^{2+}. Old bones at Java in both Pucangan and Kabuh layers where Java man was discovered and those at Heidelberg (Mauer), Germany showed nearly the same ESR spectra with signals due to Fe^{3+} and Mn^{2+} (Ikeya 1982b).

(d) *Relatively young bones (Yamashita–cho and Jomon bones, Japan)*: Figure 8.9 (a) shows an ESR spectrum of a deer bone from an anthropological site at Yamashita–cho, Japan considered to be 45 ka old by accelerator ^{14}C dating. The signals associated with CO_2^- are observed in addition to organic radical signal(s) at the low field side of $g_\perp$ as a shoulder (at $g = 2.0045$ by subtracting signals of CO_2^-).

ESR spectra of a piece of human cranium found with old fossil bones in an Akiyoshi limestone fissure in Japan are shown in (b) and (c). The broad signal and signals due to Mn^{2+}, which are associated with detrital clay

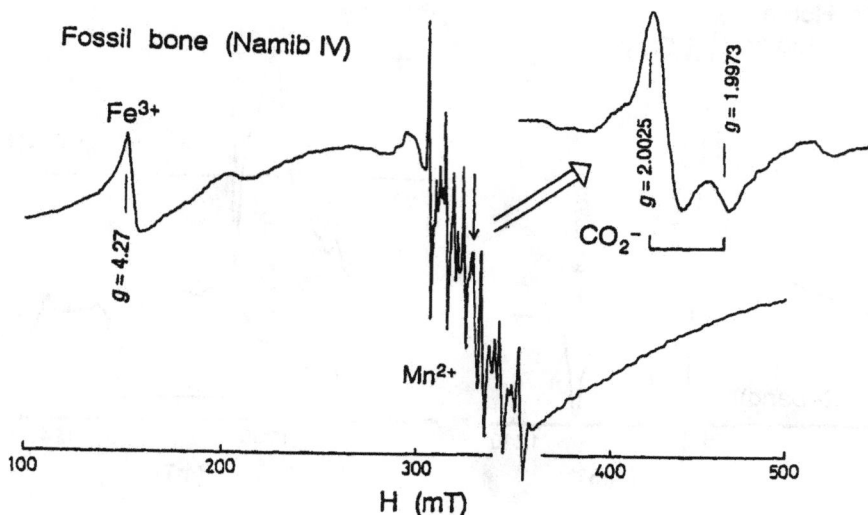

Figure 8.8   X–band ESR spectrum of fossil bone from the Namib IV site, Tanzania. The central signal in–between the hf lines of Mn^{2+} is due to axial CO_2^- as shown in the insertion.

and carbonates, are observed. The signals of CO_2^- are barely seen in the elongated spectrum drawn in (b): the intensity can be estimated to be about 1/12 of that of the deer bone shown in (a). Although the cranium was once considered to be as old as Choukoutien from faunal evidence, ESR clearly indicated that the cranium is considerably younger. Morphologically, the cranium was later assigned to that of the late Jomon era (3 ka) as expected from the absolute ESR intensity of CO_2^- (45 ka/12 = 3.8 ka).

   Holocene bones usually show a relatively intense signal around g = 2.0045. In some cases, this signal is extraordinary intense presumably due to oxidation of organic materials. Figure 8.10 (a) shows an ESR spectrum of a burnt Holocene bone before and after γ–irradiation. The enhanced difference spectrum indicates that subtraction produces the specific signals of CO_2^-, and that ESR dating can be extended to much younger bones in archaeology in which the signal due to organic decomposition dominates. Organic radicals around g = 2.0045 dominate for a burnt black bone, as shown in Figure 8.10 (b); this results in a falsely higher *ED* unless the signal around g = 2.0045 is subtracted. Work to determine the age of rela- tively young bones in New Zealand tried to use this subtraction method (Dennison *et al*. 1985, 1992).

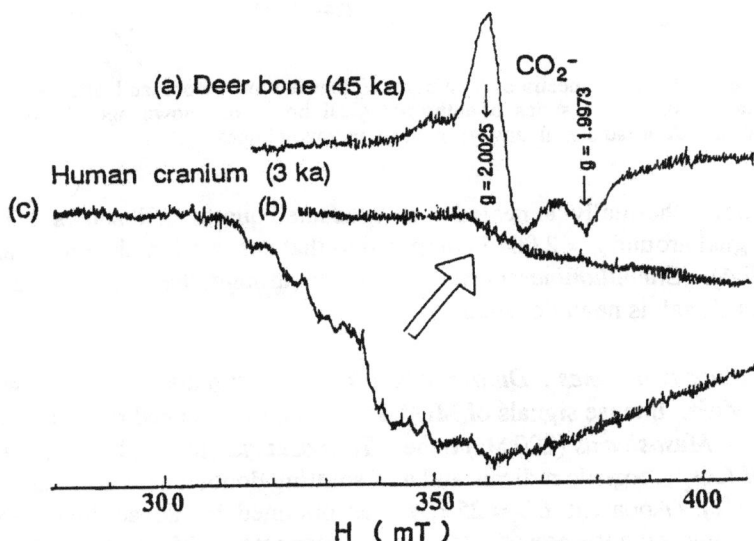

Figure 8.9 ESR spectra of a deer bone (45 ka) at an anthropological site in Yamashita–cho, (b) and (c) human cranium identified as that of the late Jomon era (3 ka). A spectrum of (b) is the central (330– 340 mT) region of (c).

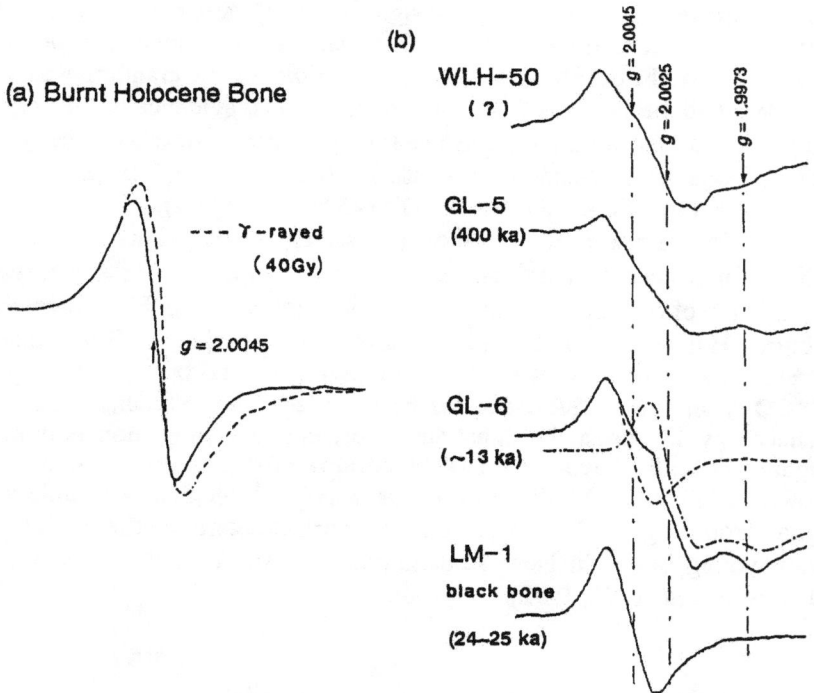

Figure 8.10 ESR spectra of (a) a burnt Holocene bone before and after γ–irradiation and (b) a series of anthropological bones of known ages from Australia. A broad signal always overlap in young bones.

A bone thermally annealed in an evacuated quartz cell shows a less intense signal around $g = 2.0045$ compared to that annealed in air (Ikeya and Miki 1984). *Chemiluminescence* is observed around the temperature at which this signal is newly created.

(e) *Old dinosauer bones* : *Dinosaur bones* showed signals associated with Fe^{3+} and Mn^{2+}. Intense signals of Mn^{2+} and radiation–induced radicals were observed in *Mesosaurus* (200 Ma) bone. *Titanosaurus* (80 Ma) bone showed signals of CO_2^-, organic radicals and a hf splitting doublet ($A = mT$) due to ^{31}P ($I = 1/2$). Apparent $ED = 25$ kGy was obtained by an additive dose method. This gave the age of 1 Ma using a dose rate of 25 mGy/a based on the uranium content of 100 ppm and $k = 0$. ESR dating of such old bones involving recrystallization of apatite has not been successful.

8.3.2   *ED and Age Relation : Relative Dating*

Bones are an *open system* in respect of the radioactivity. The assessment of *ED* is difficult, as uranium accumulates into bones. Soft spongy bones accumulate uranium more easily than solid rigid bones or tooth enamel as indicated by the fission track study (see Figure 8.13). Although the environmental conditions must be considered more carefully, it seems that ^{226}Ra is incorporated into relatively young bones of less than 10 ka, while old bones seem to lose ^{226}Ra produced as a result of ^{230}Th disintegration (Papastefanou and Charalambous 1978, Komura and Sakanoue 1985). The loss is presumably due to dissolution of the α–damaged region which is the cause of a high ^{234}U/^{238}U ratio in the groundwater as discussed in Chapter 4. Therefore, we have to deal with the "*ED in an open system*" by including the effect of leaching and accumulation of radioactive elements.

The *ED* is larger for older bones. The annual dose rate *D*, which must be deduced from the assessment of the radiation environment, varies from sample to sample due to the local variation in the radioactivity. Generally, fossil bones have been excavated by anthropologists without assessing the radiation environment. The soils surrounding the bone or tooth are not kept for the determination of environmental radiation unless ESR dating is taken into consideration at the time of sampling. Hence, the assessment of age simply from ESR determination of the *ED* is a difficult task.

A realistic assessment of the *D* for fossil bones is not straightforward, as the concentration of uranium increases and the uranium–series is in disequilibrium. The *ED*'s of several fossil bones of known origin have been obtained and compared with the anthropological ages for relative dating purposes. The age information from the supplier of the sample is not necessarily be correct as the ages are generally overestimated by the excavators. Admitting the difficulty of the real age estimation for fossil bones, the *ED* is plotted against the best estimate age on a logarithmic scale as shown in Figure 8.11, which gives interesting information.

The apparent *D*'s of 0.1, 1 and 10 mGy/a are shown by dashed lines with a slope of 1.0. A dose rate of nearly 1 mGy/a seems to hold for most samples. Thus, one can get an approximate idea about the apparent *D* of fossil bones for different environment. The apparent *D* varies from about 0.3 mGy/a for a pure carbonate travertine to about 1 or 2 mGy/a in soils at the Petralona and Arago caves.

An unbelievably large apparent *D* of 10 mGy/a is needed to explain the young bones (< 10 ka) excavated at shell mounds or at archaeological sites in earlier studies (Ikeya and Miki 1980, Mascarenhas *et al.* 1982, Caddie

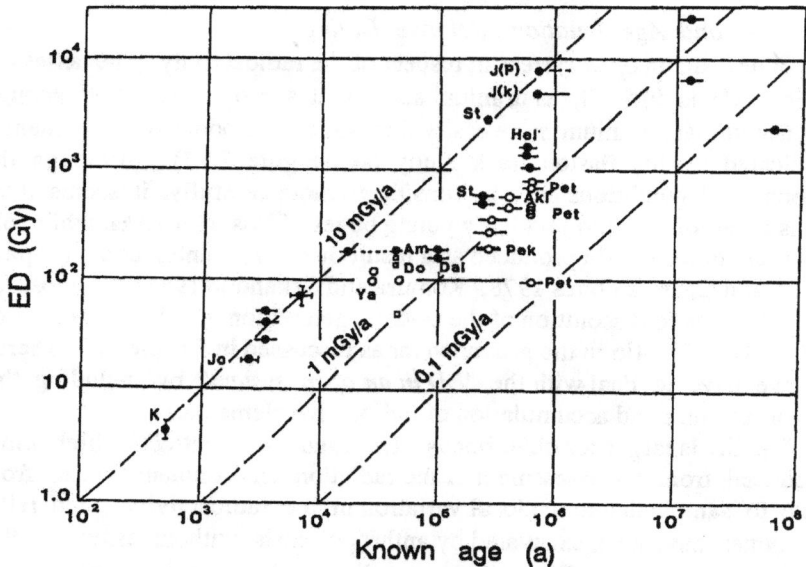

Figure 8.11 *ED* versus age relation for anthropological bones and teeth with apparent dose rates ($D = 0.1$, 1 and 10 mGy/a) indicated by dashed lines. The open circles indicate bones from limestone caves where the environmental radiation is considered low and the closed circles indicate bones from soil sediments. Pet: Petralona, Aki: Akiyoshi, Do: Douara, Pek: Peking, Ya: Yamashita–cho, Am: Amud, J(P): Jawa–Puchangan, J(K): Jawa–Kabuh, Hei: Heidelberg, St: Steinheim, K: Kamakura–era, Jo: Jomon–era.

et al. 1985). If one estimates the internal dose from the ^{238}U content and the external dose by means of TLD, the *ED* should be an order of magnitude smaller. However, the apparent *D* of 10 mGy/a could explain the apparent age of relatively young bones at archaeological sites other than in a limestone cave. This is because the signal intensity around $g = 2.0045$ due to organic radicals was included for the *ED* estimation in these early experiments to detect the weak signal intensity by overmodulation. The tendency of the *ED* to saturate beyond 10 ka may be due to the saturation of organic decomposition (see Chapter 12).

We simply note here that the *ED* versus age relation gives an order of magnitude age as do analytical uranium and fluorine dating. The apparent *D* has an appropriate value in the range of 1 ~ 10 mGy/a. A careful assessment of the radiation environment at the excavation site is cardinal to a precise age determination.

8.4   Uranium Accumulation

It must be noted that present uranium concentration in a bone and tooth enamel depends on the local environment and is not necessarily correlated with the age. Models of uranium accumulation (*early uptake, linear uptake* and *later uptake models*) are described in Chapter 4 (Section 4.5.3). The uranium uptake might be abrupt at an early stage (~ 10 ka) up to a uranium content of a few ppm as organic materials in bone decompose. For a uranium content up to about 100 ppm, it may be assumed to be linear. Such a linear approximation may be appropriate for fossil bones older than 10 ka. A dinosaur bone was found to contain a few hundred ppm of uranium.

8.4.1   *Fossil Bone*

A practical application of these models to fossil bones was first made for the bones from Heidelberg and Java (Ikeya 1982b). Here, a case of fossil bone with a small content of fluorine (less than 0.5%) is presented using bone from a Neolithic site at Yamashita–cho, Ryukyu, Japan, as shown in Figure 8.12   The TD_d for an *early uranium uptake* and the TD_a for a *linear uptake*

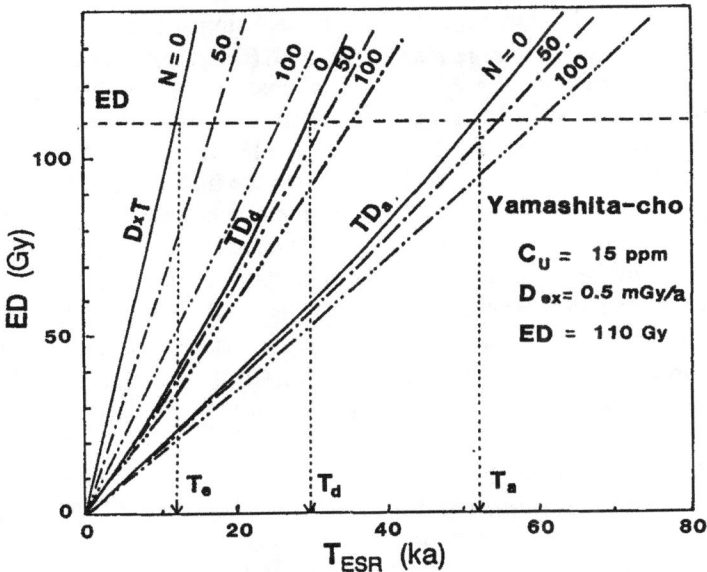

Figure 8.12   *TD* versus age relation for a fossil bone of Neolithic site of Yamashitacho, Okinawa using models of early uranium uptake (TD_d) and linear uptake (TD_a) for rates of radon loss N_{loss} = 0, 50 and 100%.. The ages T_d and T_a are obtained graphically.

are given as a function of age. The TD's for the internal α-rays (TD_α) and β-rays (TD_β) are calculated using Eq. (4.33) to give the internal total dose, TD_{in}. The external total dose, TD_{ex}, is the product of the age (T_{ESR}) and the external γ-ray dose rate (D_{ex}) measured with TLD. The TD is calculated using Eq. (4.11), where $D_{ex} = D_\gamma + D_{cos}$.

The effect of rate of radon loss N_{loss} can be included in the calculation of TD, as described in Chapter 4. If the radon loss for the bone at a burial site is included, the age estimate becomes higher. In one case, the radon loss for a powdered bone was found to be 15 ~ 20%. Generally for old fossil bones with tens of ppm of ^{238}U, the effect of the external dose is negligibly small. The ages T_d for an early uranium uptake and T_a for a linear uptake are obtained graphically using the ED (see Figure 8.12).

8.4.2 *Tooth Enamel*

The uranium accumulation of tooth enamel is negligibly small while it can be appreciable for dentine as is shown in the fission track map for the Choukoutien horse molar in Figure 8.13 (Komura and Sakanoue 1985). Efforts to determine the β-ray dose from the "*external dentine*" to enamel have been made to assess the ED of enamel using a fission track map of uranium distribution in a large mammoth tooth (Grün and Invernati 1985).

Later (recent) uptake of uranium is possible when the groundwater brings in uranium at a late stage after burial or some time after organic decomposition. A functional form of $C_U(t/T)^m$ and the ^{230}Th/^{234}U ratio have been introduced in the equation to assess the history of uranium uptake from the ED (Grün *et al.* 1988). The procedure is essentially a tuning of the ^{230}Th/ ^{234}U-age to the ESR age and attributing the disagreement, for example, a smaller ED to a late uranium uptake. Other factors such as stability of defects or recrystallization along α-tracks also affect the ED. It may thus be too early to attribute the whole discrepancy to a history of late uranium uptake. A functional form of $C_{Us}(1 - e^{-ct})$ (see Eq. (4.35)) would perhaps be a more realistic approximation.

8.4.3 *ESR Isochron Method*

Models of early, linear and recent uranium uptake (EU, LU and RU, respectively) have been applied to obtain the age of tooth enamel in paleo-anthropology using an ESR isochron method (Blackwell and Schwarcz 1993; see Section 4.5.4). The ED's of multiple samples of the same age were plotted as a function of the internal dose (rather than the uranium content) in which the β-dose from dentine to enamel is included based on each model

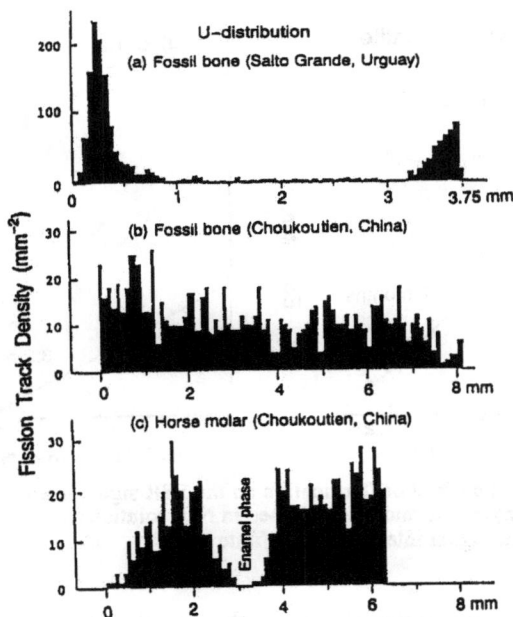

Figure 8.13   Distribution of uranium determined by fission track maps for (a) fossil bone from Salto Grande, Uruguay, (b) fossil bone from Choukoutien, China and (c) horse molar from Choukoutien. The uranium content is high at the surface of a young bone and in the tooth dentine (Komura and Sakanoue 1985).

of uranium uptake. Time–averaged dose rates for enamel and the inner and outer dentines were calculated, and the age and the external dose were obtained from the slope and the coordinate intercept, respectively, in an ESR isochron plot (see Figure 4.10). The internal dose was recalculated with the age and the procedure repeated by iteration. Since most samples in museum or in anthropologist's storage have no information on the D_{ex}, the ESR isochron method allows the age determination of such samples.

8.5   Effect of Fluorination

Hydroxyapatite in bones and teeth is fluorinated. It is preferable to make an effort to locate samples which have suffered less fluorination or those which are close to a closed system. Figure 8.14 shows the effect of *fluorination* and water on the ESR signal intensity of an irradiated bone and

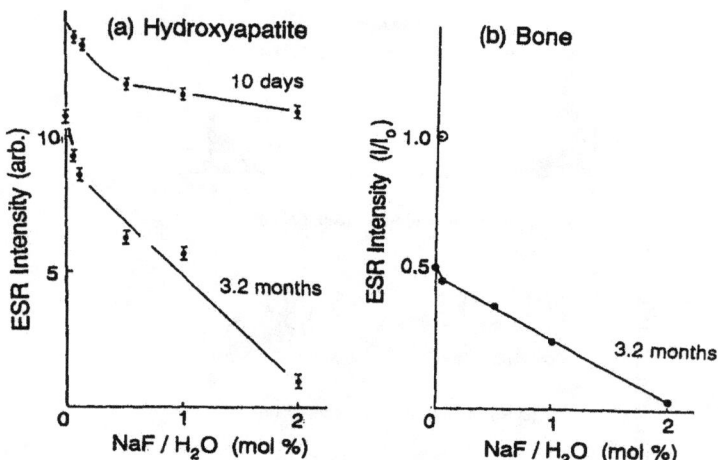

Figure 8.14   The effect of fluorination on the ESR signal intensity of CO_2^- centers in hydroxyapatite and bones dipped in NaF solutions at 50°C.   Fluorination reduces the signal intensity at a definite rate.

hydroxyapatite powder. In the laboratory the signal intensity gets reduced by dipping the sample into water with varying contents of fluoride ions at 50°C, which could presumably be due to a preferentially higher dissolution at the α–damaged part. The effect of fluorination on the TD versus age relation thus must be calculated considering the formation as well as the destruction of the defects created by fluorination.

Numerically, the defect concentration, n, may be written using the contents of fluorine $C_F(t)$ and uranium $C_U(t)$ as

$$dn / dt = a\, C_U(t)\, D(t)[1 - W C_F(t)] - n/\tau - Z n\, [dC_F(t)/dt] , \quad (8.1)$$

Production   Decay   Destruction by fluorination

where W and Z are the interaction parameters of fluorine with the CO_3^{2-} center in hydroxyapatite, a is the defect production efficiency and τ is the lifetime of the ESR signal. The *first term* in the right side of the equation indicates the defect production by the time–dependent dose rate $D(t)$ and the $C_U(t)$, where the decrease in the effective volume of the lattice by the presence of defects is included. The *second term* is due to the thermal decay process.   The lifetime τ of the signal for a bone may be about 10 Ma at room temperature. The *third term* is due to destruction of defects by fluorination.

The C_F is time dependent and may be assumed to be

$$dC_F/dt = c(1 - C_F) ,$$ (8.2)

where c is the fluorination rate depending on the local C_F and temperature. As the maximum weight percent of fluorine for total replacement of the OH⁻ site (fluoroapatite formation) is 3.77%, the present concentration, C_F^*, at age T may be written as

$$C_F^* = 3.77(1 - e^{-cT}) .$$ (8.3)

Therefore,

$$c = -(1/T)\ln(1 - C_F/3.77) .$$ (8.4)

For simplicity we use the uranium accumulation age (T_a) or disequilibrium age (T_d) in Eq. (4.36) as a first step and calculate the age considering the effect of fluorination. Then the obtained age is again used in Eq. (4.36) and the procedures are repeated iteratively.

It must be noted that additive artificial irradiation enhances the signal intensity not by the defect production efficiency without fluorination, a, but by the reduced efficiency due to fluorination. The ED obtained by extrapolating the additive dose growth curve is

$$ED(t) = n(t)/a[1 - WC_F(t)] .$$ (8.5)

In the early work on the fluorination effect (Ikeya 1985), $n(t)/a$ in Eq. (8.5) was simply taken as the value of ED. This approximation however is only valid for the dose estimation in materials with a very low fluorine content.

8.6 ESR Ages of Bones and Teeth in Paleo–anthropology

A generalized computer program is now available to calculate the ESR ages from the ED based on the models of uranium uptake. The obtained ages for anthropologically dated bones are summarized in Table 8.4. De-tailed descriptions on the samples and their place in paleo–anthropology are given in the literature listed in the table. Unfortunately, available data on the external dose rate D_{ex} and the uranium content of the fossil bones are limited. If the content of ^{238}U is high, the contribution is from the internal radiation of α- and β-rays. The variation in D_{ex} from 0.5 mGy/a to 1.5 mGy/a does not affect the ESR age appreciably. The results shown in Table 8.4 are

Table 8.4　Tentative ESR ages of anthropological bones and teeth based on the *ED* and the limited available or estimated environmental radiation dose. ESR ages associated with early uranium uptake (T_d), linear uptake for no radon loss (T_a) and 50% loss of radon (T_{a50}) are shown. Recent results in Europe are added.

Name	C_U (ppm)	C_{Th} (ppm)	C_K (%)	D_{ex} (mGy/a)	ED (Gy)	T_d (ka)	T_a (ka)	T_{a50} (ka)	Ref-age (ka)
Sangiran, Java [a]									
Bone Kabuh	69.5	0	0	1.0	6000	253	456	256	500–700
Puchangan	35.0	0	0	1.0	9000	637	1185	1510	900–1400
Mauer, Heidelberg, Germany [a]									
Bone M3	45	0	0	0.5?	145	124	227	270	400–700
Bone	26	0	0	–	140	144	260	329	
Petralona, Greece [a]									
Cranium–H	0.4	0	0	1.7?	270	150	155	157	127?
–I	–	–	–	0.50	–	404	480	495	200–700
–Ike–Po	–	–	–	0.35	–	545	648	675	
Choukoutien, Beijing, China [a, b]									
Bone Ly6	14	0	0	–	400	105	181	208	200–550
Ly–10	28	0.6	0.16	0.5?	420	65	116	132	
Horse molar Ly10	32	0	0	–	800	100	182	226	
Deer enamel Ly3–4	1.95 (19.85*)			0.86	609	222	342	389	250
Ly6–7	1.67 (24.30*)			0.73	803	283	454	534	350
Ly8–9	0.74 (16.1*)			0.82	726	343	502	566	400
Ly11	0.41 (14.25*)			0.79	905	483	688	773	500
Arago, Tautavel, France [a]									
Bone III–G	26	0	0	0.98	900	121	212	242	200–450
I–K	9.9	0	0	–	940	227	349	434	
Namib IV, Namibia [c]									
Bone	65.1	< 1	0.1	0.5	500			140	$347 \pm {}^{78}_{48}$
Murr, Steinheim, Germany [a]									
Bone	37	0	0	0.5?	600	73	109	150	200–300
Amud, Syria [a]									
Bone	7.8	0	0	1.0?	200	65	98	106	45 ?
Yamashitacho, Japan [a]									
Bone	15	0	0	0.5?	110	35	60	65	34–45
	17	0	0	0.5?	120	35	60	65	(^{14}C–age)
Pech de l'Aze II, France [d]									
Tooth Ly.3	0.33 (18.8*)			0.44	39	44	55		
Ly.8	0.27 (8.9*)			0.47	160	155	189		
Le Moustier, France [e]									
Tooth Ly.H–705A	0.03 (0.49*)			0.52	22	42	43		43
Ly.G–850C	0.39 (10.7*)			0.77	41	43	47		56
Bilzingsleben, Germany [f]									
Tooth 161a	0.32 (26.9*)			0.55	375	309	412		350
Skhul, Israel [g]									
Tooth 521a	0.3 (6.0*)			0.51	72	88	102		92
Longoia, Zambia [h]									
Teeth T70	0.03	0.84	0.02	0.24	23	178	195		

* C_U in dentine from which the β–dose was calculated.

Reference: a) Ikeya 1986 (review); b) Huang *et al.* 1993; c) Shackley *et al.* 1985;
    d) Grün *et al.* 1991c; e) Mellars and Grün 1991; f) Schwarcz *et al.* 1988a;
    g) Stringer *et al.* 1989; h) Blackwell *et al.* 1993.
In addition, the following work has been done on:
  Teeth enamel from
        La Micoque, France: Schwarcz and Grün 1988
        Piégu, Biache Saint–Vaast, Arago and Vallonet, France: Bahain *et al.* 1993
        La Chapelle–aux–Saints, France: Grün and Stringer 1991
        La Chaise–de–Vouthon, France: Blackwell *et al.* 1992
        Siegsdorf, southern Germany: Wieser *et al.* 1988
        Isernia, Italy: Bahain *et al.* 1992
        Monte Circeo, Guattari, Italy: Schwarcz *et al.* 1991
        Jabel Qafzeh, Israel: Schwarcz *et al.* 1988b
        Kebara, Israel: Schwarcz *et al.* 1989
        Tabun, Israel: Grün *et al.* 1991a
        Border Cave, South Africa: Grün *et al.* 1990
        Klasies River, South Africa: Grün *et al.* 1991b
        Jebel Irhoud, Morocco: Grün and Stringer 1991
        Singa, Sudan: Grün and Stringer 1991
        Old Crow River, Canada: Schwarcz 1985
        Alberta, Canada: Schwarcz and Zymela 1985
        Alberta and Saskatchewan, Canada: Zymela *et al.* 1988
  Bones from
        Sambaquis, Brazil: Mascarenhas *et al.* 1982
        Australia: Caddie *et al.* 1985
        New Zealand: Oduwale and Sales 1991; Oduwale *et al.* 1993;
                Dennison *et al.* 1985, Dennison and Peake 1992
        Hamana Lake, Japan: Chong *et al.* 1989

based on an assumed estimate of radiation dose rate which is found to depend on the local environment from studies at Tautavel and Petralona.

Radiation assessment of the excavation site and the analysis of radioactivity were carried out to establish the ESR chronology of a 100 ka archaeological sequence using a series of 27 teeth excavated at the site of Pech de l'Aze II, 2 km north of the Dordogne valley, south west France (Grün *et al.* 1991c). Neutron activation analysis (NAA) of uranium in enamel and dentine was done to assess the β–ray dose from dentin to enamel and an internal α–ray contribution to enamel. ESR age estimates are generally consistent with the results of earlier $^{230}Th/^{234}U$ ages for layer 3 of $103 \pm ^{30}_{25}$ ka (Schwarcz and Blackwell 1983).

ESR ages have been compared with mass–spectrometric $^{230}Th/^{234}U$ ages for some dental fragments from anthropological sites as shown in Figure 8.15. This kind of systematic study with detailed radiation assessment will contribute to more frequent use of ESR dating for the studies of *human evolution* as has been done for a series of *Homo sapiens* bones in Africa

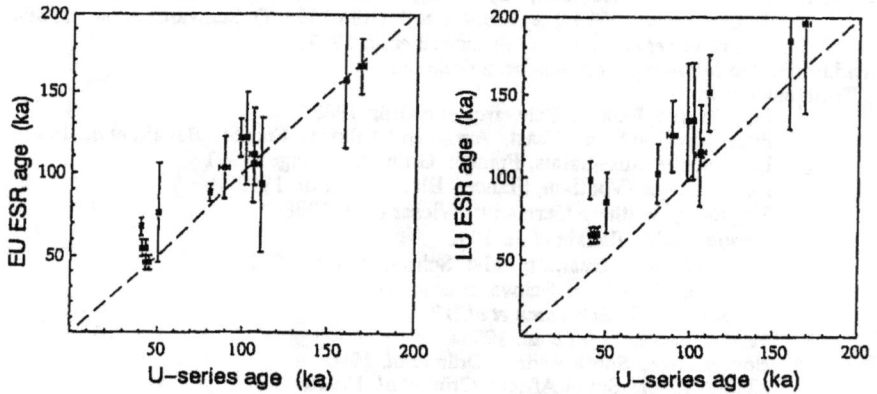

Figure 8.15   ESR ages based on (a) a linear U uptake (LU) and (b) early uptake (EU) are plotted against $^{230}Th/^{234}U$ ages (McDermott *et al.* 1993).

(Grün and Stringer 1991, Blackwell *et al.* 1993) and Israel (Schwarcz *et al.* 1988b, 1989).

ESR dating has contributed to the study of human evolution by dating Petralona, Arago, Choukoutien, Java, etc. An impact of ESR dating on the study of human evolution is the proposal of parallel evolution of Neandertha- loid and early modern human *Homo sapiens sapiens* (Grün and Stringer 1991, McDermott *et al.* 1993). ESR could provide stratigraphically consist- ent ages to samples that could not be dated even with accelerator ^{14}C-dating. It will be later clarified whether the old age is due to parallel evolution. It must be noted that the ESR age obtained using the additive dose method is generally younger than the real age as described in Chapter 3 and Appendix 1 due to natural fading. This indicates that the age of the human evolution must be further shifted to the old side.

8.7   Other Phosphates

8.7.1   *Phosphate Nodules*

Phosphate nodules collected from a deep-sea floor off the coast of New Zealand and Peru have been dated with the $^{230}Th/^{234}U$ method (Burnnet and Veeh 1977). The nodules contain more than 100 ppm of ^{238}U. Geomag- netic reversal associated with the *Matsuyama Boundary* was detected using slices of the phosphate nodules (Morinaga *et al.* 1989). Figure 8.16 (a) shows ESR spectra of the surface, inside and centrals part of a phosphate nodule collected off the coast of New Zealand. The following signals were observed:

(1) A broad signal at $g = 2.0043$ dominant at the surface presumably due to organic radicals.
(2) Two signals at $g = 2.006$ and $g = 2.0010$ clearly observed in the inner portions probably due to isotropic SO_2^- and CO_2^-, respectively.
(3) A weak shoulder at $g = 1.999$ probably due to axial CO_2^-.
(4) An anisotropic hf doublet centered at $g = 2.0035$ ($A_{\parallel} = 4.5$ mT and $A_{\perp} = 3.6$ mT) due to ^{31}P ($I = 1/2$, abundance 100%) presumably associated with phosphate radicals such as PO_2^0 and PO_2^{2-} (Bershov et al. 1968).

The signal at $g = 2.0043$ is intense at the black surface and the signal intensities at $g = 2.006$ and $g = 2.0010$ increase progressively from the surface to the inside as shown in (b). The result indicates the growth of the phosphate

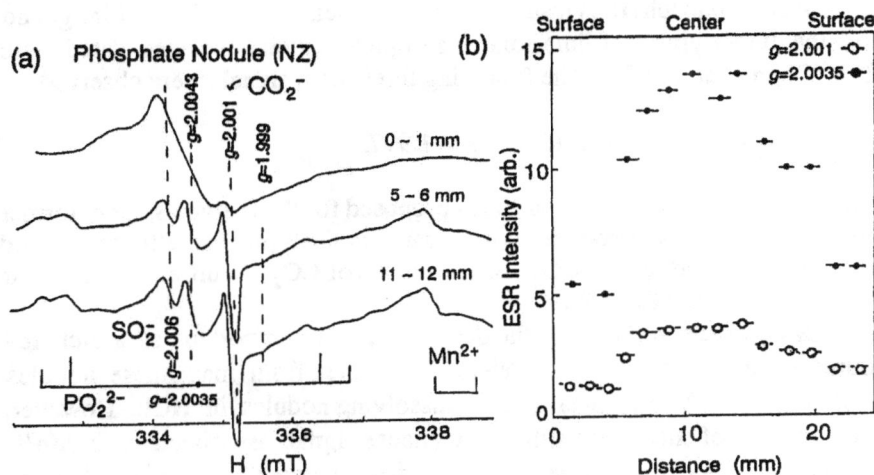

Figure 8.16 (a) ESR spectra for three positions of a phosphate nodule from a deep–sea floor off the coast of New Zealand. The black surface has a broad signal at $g = 2.0045$ in addition to a weak signal at $g = 2.006$ and 2.001. The brown inside portions show spectra with intense signals at $g = 2.006$ and $g = 2.001$. (b) An increase in the signal intensities of $g = 2.006$ and $g = 2.001$ toward the inside indicating the proportionality to the age as in a stalactite.

nodule;   the interior is progressively older as in a stratigraphic sequence or a stalactite.

Gamma–irradiation enhanced the intensity of all signals; the enhanced intensity is broad as the spectrum itself. The ESR ages obtained using the ^{238}U content of 320 ppm are somewhat younger than 400 ka, even if the k-value is assumed as zero. Since the phosphate nodule consists of small particles, presumably amorphous or polycrystalline materials, it is speculated that a natural etching of the α–ray track by seawater occurs; the α–ray damaged region might be dissolved and recrystallized since the phosphate nodule particles are submerged in the seawater. This implies that effectively there is no contribution of α–rays to the ED. Now if a model of linear uranium uptake up to the present level is assumed as is used for bone dating, the ESR age for 100% radon loss and $k = 0$ is close to the age expected from geo-magnetic studies. It may be cautioned that the distribution of ^{238}U might not be uniform in the phosphate nodule. Further research is necessary as in another case of a phosphate nodule from Peru, no detectable signal could be seen.

8.7.2   *Insular Phosphorite*

Insular phosphorites resulting from the chemical reaction of bird guano and the underlying reef limestone in a tropical region have been dated with ESR (Chen *et al.* 1992). The following three main signals were observed:

$$g = 2.0042, \quad g = 2.0015 \text{ and } g = 1.9973$$

Though no concrete model has been proposed for these signals, the former signal appears to be associated with organic radicals and the latter two could be due to $g_\perp$ and $g_\parallel$ of CO_2^- or the signals of CO_2^- in different modes of hindered rotation (Table 8.2).

Manganese nodules in the deep sea are very important as a rich re-source of minerals. A silica skeleton is obtained from manganese nodules after removing heavy metal ions by dissolving nodules in HCl. However, ESR spectra of such silica still show intense signals associated with Mn^{2+} and Fe^{3+}. It may however be possible to date shark teeth incorporated in the manganese nodule during growth.

8.8 Technical Note

In dating bones or teeth, samples must be collected with the sediment attached to them so that the external radiation effects from the sediment by β-rays (surface 2 mm) and by γ-rays (30 cm) can be assessed properly. The contents of U, Th and K in the sediment allow calculation of the external dose rate D_{ex} assuming an infinite medium. The water content is needed to estimate the dose attenuation (see Chapter 4). The external γ-ray dose rate can be measured with TLD or some apparatus. ESR isochron is a method to avoid the sediment dose measurement.

In the laboratory a small piece of solid bone is broken off and the surface soil is removed with a quartz flake and sometimes cleaned with an ultrasonic cleaner before ESR measurement. A shaving blade made of iron or stainless steel sometimes gives a spurious ESR signal that could mask the dating signal for young samples.

Sample preparation is carried out by breaking a bone or tooth into pieces less than 500 μm to as small as 100 μm. Crushing into fine particles produces surface defects (Desrosiers et al. 1989) as in carbonates. Another method is to use a piece of bone as it is and attach it to a marked sample holder so that the direction of the magnetic field as well as geometrical position in a cavity is always the same (Kai et al. 1988). If tooth enamel is separated from dentine, signals due to organic radicals are reduced.

A four-fold improvement of the ESR sensitivity has been made by detecting an out-of-phase second harmonic signal of tooth enamel at 77K (Galtsev et al. 1993). Measurements at 77K and 4.2K increase the sensitivity but saturation occurs at a low microwave power. This method using so-called "rapid passage" at 100 kHz field modulation at 77K may be used to date young bones.

8.9 Summary

Radiation-induced defects in phosphates are described with emphasis on hydroxyapatite and fluorapatite. The defect used for ESR dating of bone and tooth enamel is orthorhombic CO_2^- presumably hydrogen-bonded with protons of OH^- in hydroxyapatite. Considering the broad signal, it may be associated with axial CO_2^- (rotating around the O–O direction i.e., y-axis). ESR dating of bones and tooth enamel is described using models of a rapid uranium uptake and linear uranium uptake for an open system, considering uranium-series disequilibrium. Fluorination affects the age for old bones ~ 1 Ma. ESR ages of fossil bones and teeth in paleo-anthropology are discussed. ESR dating of bones and teeth has a potential application to paleo-anthropology, provided dosimetry aspects are kept in mind while sampling. ESR dating of phosphate nodules in deep-sea sediments is described considering the effect of α-damage dissolution.

References

Baffa O., Brunetti A., Karmann I. and Dias N.C.M. (2000): ESR dating of a toxodon tooth from a Brazilian karstic cave. *Appl. Radiat. Isot.* **52**, 1345-1350.

Bahain J.J., Yokoyama Y., Falgueres C. and Sarcia M.N. (1992): ESR dating of tooth enamel: a comparison with K-Ar dating. Quat. Sci. Rev. **11**, 245-250.

Bahain J.J., Sarcia M.N., Falgueres C. and Yokoyama Y. (1993): Attempt at ESR dating of tooth enamel from French Middle Pleistocene sites. *Appl. Radiat. Isot.* **44**, 267-272

Bacquet G., Quang-Truong Vo, Vignoles M., Trombe J.C. and Bonel G. (1981): ESR of CO_2^- in X-irradiated tooth enamel and A-type carbonate apatite. *Calcified Tissue Intern.* **33**, 105-109.

Bershov L V., Tarashchan A.N., Samoilovich M.I.and Lushnikov V.G. (1968): Electron hole centers in natural calcites with phosphors impurities. *Zh. Strukt. Chim.* **5**, 309-311

Blackwell B., Porat N., Schwarcz H. P. and Debenath A. (1992): ESR dating of tooth enamel: comparison with $^{230}Th/^{234}U$ speleothem dates at La Chaise-de-Vouthon (Charente), France. *Quat. Sci. Rev.* **11**, 231-244.

Blackwell B. and Schwarcz H.P. (1993): ESR isochron dating for teeth: a brief demonstration in solving the external dose calculation problem. *Appl. Radiat. Isot.* **44**, 243-252.

Blackwell B., Schwarcz H.P., Schick K. and Toth K. (1993): ESR dating of tooth enamel from the Paleolithic site at Longola, Zambia. *Appl. Radiat. Isot.* **44**, 253-260.

Blackwell B. A. (1994) : Problems associated with reworked teeth in electron spin resonance (ESR) dating. *Quatern. Geochr.* **13**, 651-660

Blackwell B.A. (1995) Electron spin resonance dating. In: Dating Methods for Quaternary Deposits (eds Rutter N.W. and Catto N.R.), 209-268, Geol. Assn. Canada. Geotext 2.

Bouchez R., Cox R., Herve A., Lopez-Carranza E, Ma J. J., Piboule M., Poupeau G. Blackwell Bonnie A. (1994): Problems associated with reworked teeth in electron spin resonance (ESR) dating. *Quatern. Geochr.* **13**, 651-660.

Brennan B.J., Rink W.J., McGuirl E.L., Schwarcz H.P. and Prestwich W.V. (1997) Beta doses in tooth enamel by "One-Group" theory and the ROSY ESR dating software. *Radiat. Meas.* **27**, 307-314.

Brennan BJ, Prestwich WV, Rink WJ, et al. (2000): Alpha and beta dose gradients in tooth enamel Radiat. Meas. 32, 759-765

Brown W. E. and Chow L. C. (1976): Chemical properties of bone mineral. *Ann. Rev. Mater. Sci.* **6**, 213-235.

Burnnet W.C. and Veeh H.H. (1977): Uranium-series disequilibrium studies in phosphorite nodules from the west coast of South America. *Geochim. Cosmochim. Acta* **41**, 755-764.

Caddie D.A., Hall H.J., Hunter D.S. and Pomery P.J. (1985): ESR considerations in the dating of Holocene and late pleistocene bone material. ESR Dating and Dosimetry (Ionics, Tokyo) 353-361.

Callens F.J., Boesman E.R., Matthys P.F.A., Martens L.C. and Verbeeck R.M.H. (1987): The contribution of CO_3^{3-} and CO_2^- to the ESR spectrum near $g = 2$ of powdered human tooth enamel. *Calcified Tissue Intern.* **41**, 124-129.

Callens F.J., Verbeeck R. M. H., Naessens D.E., Matthys P F.A. and Boesman E.R. (1989): Effect of carbonate content on the ESR spectrum near g=2 of carbonated calciumapatite synthesized from aqueous media. *Calcified Tissue Intern.* **44**, 114-124.

Cetin O. Ozer A.M. and Wieser A. (1994): ESR dating of tooth enamel from karan excavation (Antalya, Turkey). *Quater. Geochr.* **13**, 661-669.

Cevc P. and Schara M. (1972): Electron paramagnetic resonance study of irradiated tooth enamel. *Radiat. Res.* **51**, 581-589.

Charalambous S. and Papastefanou C. (1977): On the radioactivity of fossil bones. *Nucl. Instr. Methods* **142**, 581-588.

Chen Y., Brumby S., Jacobson G., Beckwith A.LJ. and Polach H.A. (1992): A novel application of the ESR method: dating of insular phosphorites and reef limestone. *Quatern. Sci. Rev.* **11**, 209-217.

Chen T., Quan Y., and En, W. (1994): Antiquity of Homo sapiens in China. *Nature* **368**, 55-56.

Chen T.M., Yang Q., Hu Y.Q., Bao W.B. and Li T.Y.(1997): ESR dating of tooth enamel from Yunxian *Homo eretus* Site China, *Quatern. Geol.* **16**, 455-458

Close D. M., Mengeot M. and Gilliam O.R. (1981): Low temperature intrinsic defects in X-irradiated hydroxyapatite synthetic single crystals. *J. Chem. Phys.* **74**, 5497-5503.

Cole T. and Silver A. N. (1963): Production of hydrogen atoms in teeth by X-irradiation. *Nature* **232**, 257-258.

Chong T.S., Ohta H., Nakashima Y., Iida T., Ieda K. and Saisho H. (1989): ESR dating of elephant teeth and radiation dose rate estimation in soil. *Appl. Rad. Isot.* **40**, 1199-1202.

Curnoe D, Grun R, Taylor L, et al. (2001): Direct ESR dating of a pliocene hominin from Swartkrans. *J. Hum. Evol.* **40**, 379-391.

Day M.H. (1982): *Greek fireworks. Nature* **300**, 484.

Debuyst R., Callens F., Frechen M. and Dejehet F. (2000): ESR Study of elephant tooth enamel from the Karilich-Seeufer site in Germany. *Appl. Radiat. Isot.* **52** , 1327-1336.

Deer W.A., Howie R.A. and Zussman J. (1966): *An Introduction of Rock Forming Minerals* (Longman, Essex).

DeLumley H. and Labeyrie J. ed. (1981): *Datations Absolues et Analyses Isotopiques en Prehistoire Methodes et Limites Datations du remplissage de la Caun de l'Arago à Tautavel*. pp.51.

Dennison K. J., Houghton P., Leach B. F. and Peake B. M. (1985): Sample preparation and instrumental aspects of ESR dating of New Zealand human bone. *ESR Dating and Dosimetry*, 341-352.

Dennison K.J. and Peake B.M. (1992): ESR dating of bones in New Zealand. *Quat. Sci. Rev.* **11**, 251-255.

Dennison KJ., Oduwole A.D. and Sales K.D. (1997): The anomalous ESR dating signal intensity observed for human remains from the namu burial sites on the Island of Taumako, Solomon Islands. *Quatern. Geol.* **16**, 459-464.

Desrosiers M.F. and Rey P. (1988): Q-band ESR studies of fossil teeth: consequences for ESR dating. *Quat. Sci. Rev.* **7**, 497-501.

Desrosiers M F., Simic M.G., Eichmiller F.C., Jhonston A.D. and Bowen R.L. (1989): Mechanically induced generation of radicals in tooth enamel. *Appl. Radiat. Isot.* **40**, 1195-1197.

Doi Y., Aoba T., Okazaki M., Takahashi J. and Moriwaki Y. (1979): Analysis of paramagnetic centers in X-irradiated enamel, bone and carbonate containing hydroxyapatite by electron spin resonance spectroscopy. *Calcif. Tissue Intern.* **28**, 107-112.

Dugas J. and Rey C. (1976): Paramagnetic O$_2^-$ in apatites. *J. de Physique* **37**, 449-451.

Falgueres C., Bahain J.J., Yokoyama Y., et al. (1999): Earliest humans in Europe: the age of TD6 Gran Dolina, Atapuerca, Spain. *J. Hum. Evol.* **37**, 343-352.

Galtsev V.E., Grinberg O.Ya., Lebedev Ya.S. and Galtseva E.V. (1993): EPR dosimetry sensitivity enhancement by detection of rapid passage signal of the tooth enamel at low temperature. Appl. Magn. Reson. 4, 331-333.

Geoffroy M. and Tochon-Danguy H.J. (1982): ESR identification of radiation damage in synthetic apatites: a study of the ^{13}C-hyperfine coupling. *Calcif. Tissu Intern.* **34**, S99-S102.

Grün R. and Invernati C. (1985): Uranium accumulation in teeth and its effect on ESR dating: a detailed study of a mammoth tooth. *Nucl. Tracks* **10**, 867-877.

Grün R., Schwarcz H.P. and Zymela S. (1987): ESR dating of tooth enamel. *Can. J. Earth Sci.* **24**, 1022-1037.

Grün R., Schwarcz H. P. and Chadam J. (1988): ESR dating of tooth enamel: coupled correction factor for U-uptake and U-series disequilibrium. *Nucl. Tracks* **14**, 237-241.

Grün R., Beaumont P. and Stringer C.B. (1990): ESR dating evidence for early modern human at Border Cave in South Africa. *Nature* **344**, 537-539.

Grün R. and Stringer C.B. (1991): Electron spin resonance dating and evolution of modern humans. *Archaeometry* **33**, 153-199.

Grün R., Stringer C.B. and Schwarcz H.P. (1991a): ESR dating of teeth from Garrod's Tabun Cave collection. *J. Human Evolution* **20**, 231-248.

Grün R., Shackleton N.J. and Deacon H. (1991b): ESR dating of tooth enamel from Klasies River Mouth Cave. *Current Anthropol.* **31**, 427-432.

Grün R., Mellars P. and Laville H. (1991c): ESR chronology of a 100,000-year archaeological sequence at Pech de l'Aze II, France. *Antiquity* **65**, 544-551.

Grün R. (1992): Electron spin rsonance dating inpaleo-anthropology. *Evolutionary Anthropology* **2**, 174-181.

Grün R, Abeyratne M., Head J. Tuniz C.and Hedges E.M. (1997): AMS ^{14}C analysis of teeth from archaeological site showing anomalous ESR dating results. *Quatern. Geol.* **16**, 437-444

Grün R., Huang P.H., Wu X.Z., et al. (1997): ESR analysis of teeth from the paleoanthropological site of Zhoukoudian, China. *J Hum. Evol.* 32, 83-91.

Grün R., Moriarty K., Wells R. (2001): Electron spin resonance dating of the fossil deposits in the Naracoorte Caves, South Australia. *J. Quat. Sci.* **16**, 49-59.

Hennig G.J., Herr W., Weber E. and Xirotiris N.I. (1981): ESR dating of the fossil hominid cranium from Petralona Cave, Greece. *Nature* **292**, 533-536.

Hennig G.J., Herr W., Weber E. and Xirotiris N.I. (1982): Petralona Cave dating controversy. *Nature* **299**, 282-284.

Huang P.H., Jin S.Z., Peng Z.C., Liang R.Y., Lu Z. J., Wang Z.R., Chen J.B. and Yuan Z.X. (1993): ESR dating of tooth enamel: comparison with U-series, FT and TL dating at the Peking man site. *Appl. Radiat. Isot.* **44**, 239-242.

Ikeya M. (1977): Electron spin resonance dating and fission track detection of Petralona stalagmite. *Anthropos.* **4**, 152-168.

Ikeya M. and Miki T. (1980): Electron spin resonance dating of animal and human bones. *Science* **207** 977-979.

Ikeya M. (1982a): Petralona Cave dating controversy. *Nature* **299**, 280-282.

Ikeya M. (1982b): A model of a linear uranium accumulation for ESR age of Heidelberg (Mauer) and Tautavel bones. *Jpn. J. Appl. Phys.* **21**, L690-L692.

Ikeya M and Miki T (1984): A decade of ESR dating from speleology in Akiyoshi cave. *J. Speleol. Soc. Japan* **9**, 36-57.

Ikeya M. (1985): ESR ages of bones in paleo-anthropology uranium and fluorine accumulation. *ESR Dating and Dosimetry*, 373-379.

Ikeya M. (1986): *Dating and Age Determination of Biological Materials* ed. Zimmerman M.R. and Angel, L. (Croom Helm, London), Chap. 3 Electron Spin Resonance, 59-125.

Ikeya M., Ochiai H., and Ikeya M. (1997): Total dose formula for uranium saturation uptake model for ESR and TL dating. *Radiat. Meas.* **27**, 339-334.

Ishii H., Kasuya M., Furusawa M. and Ikeya M. (1991): ESR, TL and FT measurements of a natural apatite. *Nucl. Tracks* **18**, 189-192.

Ishii H. and Ikeya M. (1993): Defects in synthesized apatite and sintered materials. *Appl. Radiat. Isotop.* **44**, 95-100.

Kai A., Miki T. and Ikeya M. (1988): ESR dating of teeth, bone and eggshells excavated at a paleolithic site of Douara cave, Syria. *Quat. Sci. Rev.* **7**, 503-507.

Köberle D., Terrile C., Panepicci H. C. and Mascarenhas S. (1973): On the paramagnetic resonance of bone irradiated in vivo. *An. Acad. Brasil. Ciénc.* **45**, 157-160.

Komura K. and Sakanoue M. (1985): Gamma-ray spectroscopy for ESR dating. *ESR Dating and Dosimetry*, 9-17.

Marsh R.E. and Rink W.J. (2000): Beta-gradient isochron dating of thin tooth enamel layers using ESR. *Radiat. Meas.* **32**, 567-570.

Mascarenhas S., Baffa O. and Ikeya M. (1982): ESR dating of human bones from Brazilian shell mound Sambaquis. *Amer. J. Anthropol.* **59**, 413-417.

McDermott F., Grün R., Stringer C.B. and Hawkesworth (1993): Mass-spectrometric U-series dates for Israeli Neanderthal/early modern homonid sites. *Nature* **363**, 252-256.

Mellars P. and Grün R. (1991): A comparison of the electron spin resonance and thermoluminescence dating methods: the results of ESR dating at Le Moustier (France). *Cambridge Archaeo. J.* **1**, 269-276.

Mengeot M., Bartram R.H. and Gilliam O.R. (1975): Paramagnetic hole-like defect in irradiated calcium hydroxyapatite single crystals. *Phys. Rev.* B **11**, 4110-4124.

Michel V, Falgueres C, Yokoyama Y (1997): Uranium content and U-Th dating of fossil bones and dental tissues from Lazaret cave. *CR Acad. Sci.* II A **325**, 381-387.

Moens P.D.W., Callens F.J., Verbeeck R.M.H. and Naessens D.E. (1993): An EPR spectrum decomposition study of Na^+-containing precipitated carbonated apatite powders. *Appl. Radiat. Isot.* **44**, 279-285.

Morinaga H., Inokuchi H. and Yaskawa K. (1989): Growth history of phosphorite nodules inferred from their remanent magnetization. *Earth Planet. Sci. Lett.* **91**, 374-380.

Murata T., Shiraishi K., Ebina Y. and Miki T.Å@(1996): An ESR study of defects in irradiated hydroxyapatite. *Appl. Radiat. Isot.* **47**, 1527-1531.

Oakley J. W. (1980): Relative dating of fossil hominids of Europe. *Bull. British Mus. Nat. Hist.* **34**, 1-63.

Oduwole A.D. and Sales K.D. (1991): ESR signals in bones: interference from Fe3+ ions and a new method of dating. *Nucl. Tracks* **18**, 213-222.

Oduwole A.D. and Sales K.D. (1994): Transient ESR signalsinduced by g-irradiation in tooth enamel and in bone. *Quater. Geochr.* **13**, 647-650.

Oduwole A.D., Sales K.D. and Dennison K.J. (1993): Some ESR observations on bone, tooth enamel and eggshell. *Appl. Radiat. Isot.* **44**, 261-266.

Oka T., Grun R., Tani A., Yamanaka C., Ikeya M. and Huang PH. (1997): ESR microscopy of fossil teeth. *Radiat. Meas.* **27**, 331-337.

Ostrowski K., Goctawska A.D. and Stachowics W. (1980): Stable radiation-induced paramagnetic entities in tissue minerals. *Free Radicals in Biology* **4**, 321-344.

Panepucci H. and Farach H.A. (1977): ESR spectra of quasirandomly oriented centers: application to radiation damage centers in bone. *Med. Phys.* **4**, 321-344.

Papastefanou C. and Charalambous S. (1978): 226Ra leaching in fossil bones. *Nucl. Instr. Meth.* **151**, 599-601.

Papastefanou C., Manolopoulou M., Savvides E. and Charalambous S. (1986): Dose rate measurements in Petralona Cave for Archanthropus dating. *Anthropos.* **11**, 41-48.

Piper W. W., Kravitz L. C. and Swank R. K. (1965): Axially symmetric color centers in fluorapatite. *Phys. Rev.* **138**, 1602-1814.

Porat N. and Schwarcz H. P. (1994): ESR dating of tooth enamel - A universal growth curve. *Integrative Path to the Past* ed. Corruchini R.S. and Ciochon R.L., Prentice Hall 521-530. *

Porat N. and Schwarcz H.P. (1994). ESR dating of tooth enamel: a universal growth curve. In: Corruccini R.S. and Ciochon R.L. (eds) *Integrative Paths to the Past* , (Prentice Hall, New Jersey), 521-530.

Poulianos A. (1971): Petralona-A middle Pleistocene Cave in Greece. *Archaeo.* 24, 6-11.

Poulianos A. (1982): Petralona cave dating controversy. *Nature* 299, 281.

Rink W.J. and schwarcz H.P. (1994): Dose response of ESR signals in tooth enamel. *Radoat. Meas.* **23**,481-484.

Rossi A. M. and Poupeau G. (1990): Radiation damage in bioapatites: the ESR spectrum of irradiated dental enamel revisited. *Nucl. Tracks* **17**, 537-545.

Roufosse A., Richelle L.J. and Gilliam O.R. (1976): Electron spin resonance of organic free radicals in dental enamel and other calcified tissues. *Arch. Oral Biol.* **24**, 227-232.

Schwarcz H.P. (1985): ESR studies of tooth enamel. *Nucl. Tracks* **10**, 865-867.

Schwarcz H.P. and Blackwell B. (1983): ^{230}Th/^{234}U age of a Mousterian site in France. *Nature* **301**, 236-237.

Schwarcz H.P. and Zymela S. (1985): ESR dating of Pleistocene teeth from Alberta, Canada. *ESR Dating and Dosimetry*, 325-333.

Schwarcz H.P. and Grün R. (1988): ESR dating of level L2/3 at La Micoque (Dordogne), France: excavation Debenath and Rigaud. *Geoarchaeology* **3**, 293-296.

Schwarcz H.P., Grün R., Latham A. G., Mania D. and Brunnacker K. (1988a). New evidence for the age of Bilzingsleben archaeological site. *Archaeometry* **30**, 5-17.

Schwarcz H.P., Grün R., Vandermeersch B., Bar-Yosef O., Valladas H. and Tchernov E. (1988b): ESR dates for the hominid burial site of Qafzeh in Israel. *J. Human Evolution* **17**, 733-737.

Schwarcz H.P., Buhay W.M., Grün R., Valladas H., Tchernov E., Bar-Yosef O. and Vandermeersch B. (1989): ESR dating of Neanderthal site, Kebara cave, Israel. *J. Archaeo. Sci.* **16**, 653-661.

Schwarcz H.P., Bietti A., Buhay W.M., Stiner M., Grün R. and Segre A. (1991): U-series and ESR age data for Neanderthal site of Monte Circeo, Italy. *Curr. Anthro.* **32**, 313-316.

Schwarcz H.P. and Grün R. (1992): ESR dating and origin of modern man. *Proc. Roy. Soc. Lond.* **B337**, 145-148.

Setin O., Ozer A.M. and Wieser A. (1994): ESR dating of tooth enamel from karain excavation (Antalya Turkey). *Quatern. Geochr.* **13**, 661-669.

Shackley M., Komura K., Hayashi T., Ikeya M., Matsuura S. and Ueta N. (1985): Chronometric dating of bone from Namib IV site, South-West Africa/Namibia. *Bull. Nat. Sci. Mus. Tokyo* **11**, 7-12.

Skinner A.R., Blackwell B.A.B., Dennis C.N., Shao J. and Min S.S. (2000): Improvements in dating tooth enamel by ESR. *Appl. Radiat. Isot.* **52**, 1337-1344.

Skinner A.R., Chasteen N.D., Shao J.L., et al. (2001): Q-band studies of the ESR signal in tooth enamel. *Quat. Sci. Rev.* **20**, 1027-1030.

Skinner A.R, Blackwell B.A.B, Lothian V.S.:(2001): Calibrating ESR dating using 2 Myr-old teeth *J. Human Evol.* **40**, A22-A23.

Stringer C.B., Howell F.C. and Melentis J.K. (1979): The significance of fossil hominid skull from Petralona, Greece. *J. Archaeo. Sci.* **6**, 235-253.

Stringer C.B., Grün R., Schwarcz H.P. and Goldberg P. (1989): ESR dates for the hominid burial site of Es Skhul in Israel. *Nature* **338**, 756-758.

Tiemei C, Qi C, Quan Y, et al. (2001): The problems in ESR dating of tooth enamel of Early Pleistocene and the age of Longgupo hominid, Wushan, China. *Quat. Sci. Rev.* **20**, 1041-1045.

Wieser A., Göksu H.Y., Regulla D.F. and Vogenauer A. (1988): Dose rate assessment in tooth enamel. *Quat. Sci. Rev.* **7**, 491-495.

Wu ZH, Jiang W, Blisniuk P, et al. (1999): ESR dating of the evolution of the Shuanghu basin in the northern Tibetan Plateau. *Acta Geol. Sin-Engl.* **73**, 289-293.

Yokoyama Y., Nguyen H.V., Quaegebeur J.P. and Poupeau G. (1982): Some problems encountered in the estimation of annual dose-rate in the electron spin resonance dating of fossil bones. *PACT* **6**, 103-115.

Zymela S., Schwarcz H.P., Ford D.C. and Hentsch B. (1988): ESR dating of Pleistocene fossil teeth from Alberta and Saskatchewan. *Can. J. Earth Sci.* **25**, 235-245.

Zhou L.P., McDermott F., Rhodes E.J.,Marseglia E.A. and Mellars P.A. (1997): ESR and mass-spectrometric uranium-series dating studies of a mammoth tooth from Stanton Harcort, Oxfordshire, Engrand. *Quatern. Geol.* **16**, 445-454.

Chapter 9

SiO$_2$

– Rocks, Faults and Sediments –

Quartz (SiO$_2$) is the most abundant mineral that can be dated with ESR. Defects in quartz can be used determine the ages of volcanic materials, fault formation, and presumably sedimentation, based on different mechanisms.

9.1 Introduction

Silicates constitute the commonest class of minerals. The silicon atom, Si is inside four oxygen atoms forming a tetrahedral SiO$_4$. The tetrahedron is either isolated or joined with one or more oxygen atoms forming a one–, two– or three–dimensional network. A special type of silicate is **silica** (SiO$_2$) or **quartz** where each oxygen atom is bonded with two Si atoms to form a three–dimensional network in a crystalline form. Defects, both in synthetic crystalline quartz and in amorphous silica glass have been studied extensively because of the technological importance of SiO$_2$ in electronics, optics and telecommunications. Review articles are available concerning the physics of this important insulating material (Griscom 1990, Halliburton 1989, Weil 1984).

Quartz is the most common rock–forming mineral next to feldspars and exists in most rocks of sedimentary, igneous and metamorphic origins. Intrafault gouges (clays) also contain abundant quartz grains. ESR dating of quartz was proposed and called *"trapped electron dating"* (McMorris 1971) and as *"paleo–dosimetry"* (Moiseyev and Rakov 1977, Moiseyev 1980). However, the sensitivity of the ESR spectrometers at that time was not sufficient as to get meaningful ages (see Chapter 3).

ESR dating of flint by fast neutron irradiation (Garrison *et al.* 1981) and of quartz grains from geological faults (Ikeya *et al.* 1982, 1983) and from volcanic materials (Ikeya 1983, Imai *et al.* 1985, Yokoyama *et al.* 1985a, b) revived the ESR dating of quartz. The observation that the relative signal intensity of the E$'_1$ center (an electron at an oxygen vacancy) is correlated to the geologic age up to billions of years (Ga) is an attractive proposition in geochronology since ESR dating may cover the time range extending to the formation age of Earth itself (Odom and Rink 1988).

Geological faults are accompanied by rocks with slickensides and grooving, gouge, breccia and mylonite produced by the high strains of faulting motion. ESR dating of geological fault movement has been done using quartz grains abundant in intrafault gouge (Ikeya *et al.* 1982). The hypothesis is that trapped electrons produced by natural radiation in quartz grains have been annihilated by mechanical annealing (high shear stresses and a temperature rise) around the fault plane at the time of fault formation or movement (Miki and Ikeya 1982). Theoretical studies (McKenzie and Brune 1972) indicated the temperature rise around a fault plane up to 1200°C due to frictional heating. Diffusion of heat and thermal annealing of defects were assessed for a geological fault (Fukuchi 1989a). Natural radiation after that time produces additional defects at higher efficiency than before, owing to deformation–induced dislocations in the lattice (Ikeya *et al.* 1983).

Sediment dating based on the hypothesis of sunlight bleaching has been studied in TL dating (see Chapters 1 and 3). Quartz grain abundant in sediments could be dated with ESR based on sunlight bleaching since ESR detects particular signals sensitive to sunlight or UV–light (Tanaka *et al.* 1985, Yokoyama *et al.* 1985a).

Shock–impact phase transition is an additional mechanism of age zeroing. The high pressure crystalline polymorphs of SiO_2 is **stishovite** and **coesite** which can be described in terms of irregular rings of distorted tetrahedra (Jeanloz 1988). Sandstones at *meteorite impact craters* contain stishovite (and coesite) which might tell the age of impact crater formation.

In this chapter, the principles of ESR dating of geological faults, sunbleached sediments and shock metamorphism (impact crater) are described following description of defects in SiO_2. Stress powdering, shock by impact and optical exposure cause detrapping of electrons. The E'_1 center and oxygen vacancies have been used to date old quartz of Ga ages.

9.2 Crystal Structure and Defects in SiO_2

9.2.1 *Crystal Structure of Quartz*

The Si atom with sp^3 hybrid orbitals forms a SiO_4 tetrahedron with four oxygen atoms with p orbitals. Each oxygen atom is bonded with two nearly equivalent Si atoms and each tetrahedron joins together sharing all four oxygen atoms with neighboring ones and thus forms a three–dimensional network. SiO_2 is a half–covalent and half–ionic crystal and polymorphous. It appears as α–quartz, β–quartz, tridymite, cristobalite, coesite and stishovite.

α–**quartz** with a trigonal structure is stable at atmospheric pressure and temperature up to 573°C. It changes into hexagonal β–**quartz** which is stable between 570°C and 870°C with minor atomic movements involving no breakage of Si–O bonds (Deer *et al.* 1966). **Tridymite** and **cristobalite** are produced by heating at 870°C and 1470°C, respectively. **Coesite** and **stishovite** are produced by high pressure in a laboratory and in meteor craters.

The crystal structure of α–quartz is shown in Figure 9.1 (a). Each Si atom has four oxygen neighbors, approximately tetrahedrally positioned. Each tetrahedron is bonded together sharing all four O atoms with nearby ones. One oxygen has *two bonds* whose lengths are **1.62 Å (long bond)** and **1.60 Å (short bond)**. The distances between Si and the two interstitial sites are different due to a slight difference in the bond length as shown in (b).

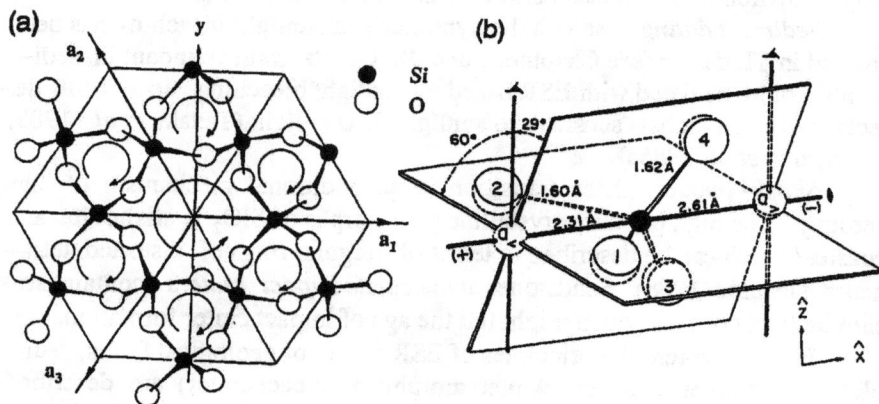

Figure 9.1 (a) Crystal structure of α–quartz (SiO$_2$) with a trigonal lattice $(a_1 = a_2 = a_3 = 4.9134$ Å, $c = 5.4152$ Å). The Si–O–Si angle is $144°$ in α–quartz ($\sim 153°$ in amorphous glass) (after Weil 1984). (b) SiO$_4$ tetrahedron in quartz. Symbols $a_<$ and $a_>$ indicate possible sites for an interstitial atom/ion in the middle of a large c–axis channel at its intersection with two–fold axis a_1 (Isoya et al. 1988).

9.2.2 Defects in SiO$_2$: Intrinsic Defects

The nature of intrinsic and impurity–associated defects in *crystalline* SiO$_2$ (cr–SiO$_2$: α–quartz) and *amorphous* SiO$_2$ (am–SiO$_2$: silica glass) with and without OH$^-$ ions (wet and dry am–SiO$_2$, respectively) are summarized in review papers (Weil 1984, Halliburton 1989, Griscom 1990).

Models and ESR parameters of both intrinsic and impurity–related defects in cr– and am–SiO$_2$ are summarized in Table 9.1. Among them, stable defects relevant to ESR dating are explained below and their ESR parameters and characteristics are summarized in Table 9.2. Notations of defects differ from paper to paper.

Normal lattice	: [SiO$_4$]^o,	SiO$_4$$^{4-}$ or $\equiv$Si–O–Si$\equiv$
Intrinsic defects		
(a) E$'_1$ center	: [SiO$_3$]$^+$,	SiO$_3$$^{3-}$ or $\equiv$Si$\cdot$Si$\equiv$
(b) peroxy center	: [SiO$_5$]$^-$,	SiO$_5$$^{5-}$ or $\equiv$Si–O(–O$\cdot$)–Si$\equiv$
(c) NBOHC	: [SiO$_4$]$^+$,	SiO$_4$$^{3-}$ or $\equiv$Si–O$\cdot$
(d) OHC (O$_3^-$?)	:	O–O–O$\cdot$
(e) oxygen vacancies	: [SiO$_3$]$^{2+}$,	SiO$_3$$^{2-}$ or [V$_-$]$^{2+}$
	[SiO$_3$]^o,	SiO$_3$$^{4-}$ or [V$_-$2e$^-$]^o

Impurity–related defects (see the next section)
(a) Al center : Al–related hole center
     $[AlO_4]^0$, $[Al^{3+}h^+]^0$, AlO_4^{4-} or $=Al–O\cdot$
(b) Ge center : Ge–related electron center
     $[GeO_4/Li]_C^0$, $[Ge^{4+}e^-/Li^+]_C^0$, $GeLiO_4^{4-}$ or $(=Ge–O–)^-$----Li^+
(c) Ti center : Ti–related electron center
     $[TiO_4/Li]_A^0$, $[Ti^{4+}e^-/Li^+]_A^0$, $TiLiO_4^{4-}$ or $(=Ti–O–)^-$----Li^+
(d) Other centers

First we consider a formation mechanism of vacancy– and interstitial–type defects in SiO_2. A vacancy and an interstitial–type defect are together called a "*Frenkel pair*". The following three pairs are considered in SiO_2:

(1) A paired *oxygen vacancy* $[V_-]$ and *interstitial* O^{2-} is not stable at room temperature. These are shown in Figure 9.2 (b) and (d).
(2) A paired E'_1 center (an electron at an oxygen vacancy) and **peroxy center** (interstitial O^-) is stable and is observed in natural quartz and irradiated am–SiO_2. These are shown in (c) and (e).
(3) A paired Si vacancy and Si interstitial is energetically not stable.

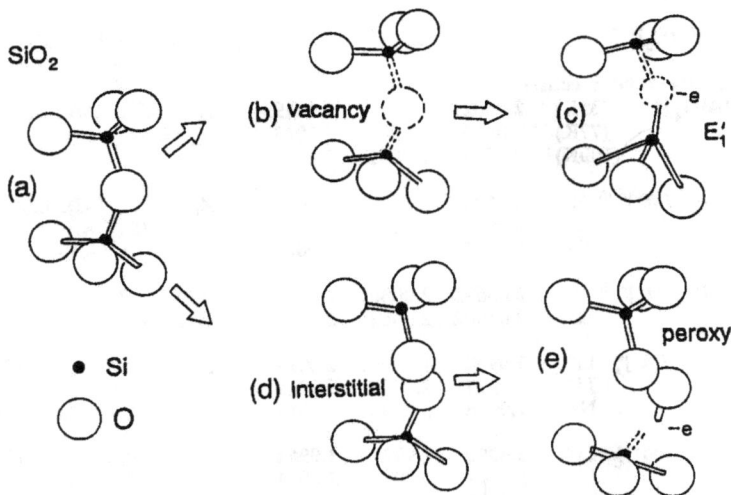

Figure 9.2 A formation scheme of the E'_1 center and peroxy radical: (a) a normal SiO_2 sublattice, (b) oxygen vacancy $[V_-]$, (c) E'_1 center, (d) interstitial O^{2-} and (e) peroxy center (after Griscom 1990).

Table 9.1   ESR parameters of defects observed in irradiated quartz.

Defect (model)	g factor			A tensor in mT			Ref.
	g_1	g_2	g_3	A_1	A_2	A_3	
Oxygen-related center							
E'_1 (≡Si·)	2.00179	2.00053	2.00030	45.3	39.1	39.1 (^{29}Si)	1
E'_2 [a]	$g_{||}$=2.0022	$g_\perp$ = 2.0006		41.2 (^{29}Si),		0.04 (^{1}H)	2
E'_4 [a]	2.00154	2.00065	2.00060	1.886	0.065	0.005	3
E'_1-type Ge(III) [b]	2.0011	1.9950	1.9939	$A_{||}$=1.05 $A_\perp$=0.82 (^{29}Si)			4
Ge(IV) [b]	2.0010	1.9942	1.9935	$A_{||}$=1.03 $A_\perp$=0.31 (^{29}Si)			4
				$A_{||}$=28.7 $A_\perp$=25.9 (^{73}Ge)			
Surface E'	$g_{||}$=2.0017	$g_\perp$=2.0003 (center g = 2.0007)					5
Peroxy (≡Si-O-O·)	2.0014	2.0074	2.067 (av.)	A_1=0.36	A_2=0.42 (^{29}Si)		6,7
NBOHC (≡Si-O·)	2.0010	2.0095	2.078 (av.)	$A_{||}$=10.95 $A_\perp$=1.60 (^{17}O)			6
				A_1=A_2=1.44 (^{29}Si)			7
OHC (O$_3^-$?)	g_{av}=2.012						8
O$_2^-$	2.318	1.992	1.959				9
Hydrogen center (20K)							
[(OH)$_4$]$^+$ [c]	2.0911	2.0103	2.0002	0.068	0.111	0.131 (^{1}H)	10
[O(OH)$_3$]^o [c]	2.1351	2.0047	1.9962	0.067	0.138	0.221 (^{1}H)	10
Impurity-related center							
[AlO$_4$]^o (35K)	2.06021	2.00854	2.00195	A_{av}=0.6 (^{27}Al), 1.1 (^{29}Si)			11
(77K)	2.0590	2.0045	2.0045	A_{iso}=0.59 (^{27}Al)			12
(100K)	2.0568	2.0091	2.0026				13
[AlO$_4$/M$^+$]^o [d] H$^+$	2.0581	2.0093	2.0034	A_{iso}=0.89 (^{27}Al), 0.001 (^{1}H)			12
Li$^+$	2.0618	2.0083	2.0033	A_{iso}=0.75 (^{27}Al)			12
Na$^+$	2.0437	2.0096	2.0021	A_{av}=0.81 (^{27}Al)			12
[AlO$_4$/M$^+$]$^+$ [d] H$^+$	2.05693	2.00806	2.00249	0.94	0.95	0.82 (^{27}Al)	14
Li$^+$	2.01602	2.00826	2.00196	0.81	0.81	0.69 (^{27}Al)	14
[GeO$_4$/M$^+$]^o_A Li$^+$	1.9913	1.99965	2.0014	27.9	29.6	28.2 (^{73}Ge)	15
Li$^+$	1.9907	2.0003	2.0019	0.04	0.10	0.05 (^{7}Li)	16
Na$^+$	1.9918	2.0002	2.0015	0.06	0.10	0.06 (^{23}Na)	16
[GeO$_4$/M$^+$]^o_C Li$^+$	2.0000	1.9973	1.9962	30.9	29.4	29.5 (^{73}Ge)	15
Li$^+$	1.9947	1.9983	2.0014	0.08	0.01	0.002 (^{7}Li)	16
Na$^+$	1.9959	1.9970	2.0005	0.07	0.09	0.11 (^{23}Na)	16

a) E'_2 and E'_4 are H^o in an O^{2-} vacancy with different relaxation (Rudra and Fowler 1985).
b) Ge(III) and Ge(IV) are E'_1 centers perturbed by Ge.
c) Three or four H$^+$ at a Si vacancy.   d) Unstable at room temperature.

Table 9.1 (continued)

Defect (model)		g factor			A tensor in mT			Ref.
		g_1	g_2	g_3	A_1	A_2	A_3	
[TiO$_4$/M$^+$]$^\circ_A$	H$^+$	1.9856	1.9310	1.9151	0.46 0.19	0.94 0.48	0.43 (^{1}H) 2.18 (^{47}Ti)	17
	H$^+$	1.9856	1.9310	1.9151	A_{av}=1.08 (^{47}Ti),		0.50 (^{1}H)	18
	Li$^+$	1.9789	1.9309	1.9119	A_{av}=1.17 (^{47}Ti),		0.13 (^{7}Li)	19
	Na$^+$	1.9675	1.9536	1.8994	A_{av}=1.49 (^{47}Ti),		0.09 (^{23}Na)	18
[TiO$_4$/M$^+$]$^\circ_B$ $^{d)}$	H$^+$	1.9891	1.9138	1.9034	A_{av}=1.98 (^{47}Ti),		0.56 (^{1}H)	18
	Li$^+$	1.9802	1.9298	1.9104	A_{av}=2.62 (^{47}Ti),		<0.03 (^{7}Li)	18
[TiO$_4$]$^+$ ^{d}		1.9910	1.9242	1.9224				20

Reference: 1) Jani *et al.* 1983, Silsbee 1961; 2) Weeks 1963; 3) Halliburton *et al.* 1979; 4) Feigl and Anderson 1970; 5) Arends *et al.* 1963; 6) Stapelbroek *et al.* 1979; 7) Griscom 1990; 8) see text (p. 282); 9) Baker and Robinson 1983; 10) Nuttall and Weil 1980; 11) Nuttall and Weil 1981a; 12) Mackey *et al.* 1970; 13) Schnadt and Schneider 1970; 14) Nuttall and Weil 1981b; 15) Weil 1971; 16) Mackey 1963; 17) Rinneberg and Weil 1972; 18) Okada *et al.* 1971; 19) Isoya *et al.* 1988; 20) Isoya and Weil 1979.
Nuclear spin and abundance: ^{1}H (I = 1/2, 99.985%); ^{7}Li (I = 3/2, 92.5%); ^{17}O (I = 5/2, 0.038%); ^{23}Na (I = 3/2, 100%); ^{27}Al (I = 5/2, 100%); ^{29}Si (I = 1/2, 4.67%); ^{47}Ti (I = 5/2, 7.3%); ^{73}Ge (I = 9/2, 7.73%).

A paired anion (O^{2-}) vacancy and cation (Si^{4+}) vacancy called a "Schottky pair" is negligible around room temperature unless SiO$_2$ is quenched from high temperature.

(a) E$'_1$ center: $g_1 = 2.00179$ $g_2 = 2.00053$ $g_3 = 2.00030$

Model :   The simplest point defect in quartz is an electron at the oxygen vacancy called the "E$'_1$ center". ESR spectra of the E$'_1$ center are shown in Figure 9.3 (a) and Figure 9.5 (d). The weak doublet spectrum is due to hf splitting (A = 40.4 mT) of ^{29}Si (I = 1/2, 4.67%) (Jani *et al.* 1983). The model for the E$'_1$ center in am– and cr–SiO$_2$ is an unpaired electron in a dangling sp^3 hybrid orbital of pyramidal SiO$_3^{3-}$ where an oxygen ion, O$^-$, is extracted from a SiO$_4$ tetrahedron (Silsbee 1961, Jani *et al.* 1983). The E$'_1$ center is expressed in the form of ≡Si· or of [SiO$_3$]$^+$ (SiO$_3^{3-}$). The lattice relaxation or atomic movement occurs to minimize the distortion energy as shown in Figure 9.2 (c). The E$'_1$ center in distorted SiO$_2$ shows a broad linewidth as described in Chapter 2. A formation scheme of the E$'_1$ center is shown in Figure 9.2 (a)–(c).

Table 9.2 ESR parameters and characteristics of defects in α-quartz (cr-SiO$_2$) used for ESR dating.

Defect	g factor	Microwave power (mW)	Temp.[a]	Thermal property	Characteristics	Reference
E'$_1$ center	2.00179 2.00053 2.00030	10^{-3} ~ 0.1	RT	increases above 200°C up to 300°C annealed at ~ 400°C (T$_{a10}$ =330°C)[b] regeneration efficiency is reduced by annealing at 500°C	not created by weak γ-irradiation but created by neutron-irradiation created by γ-irradiation in am-SiO$_2$	1
Peroxy center (dry OHC)	2.0014 2.0074 2.067 (average)	2 – 200	RT	increases above 170°C up to 220°C annealed at 350°C (T$_{a10}$ =270°C)	not created in cr-SiO$_2$ by weak γ-irradiation created in OH⁻-free am-SiO$_2$	2
NBOHC (wet OHC)	2.0010 2.0095 2.078 (average)	10 – 50	RT		not clear but overlaps the peroxy signal in natural quartz created in OH⁻-doped silica glass	2
OHC (O$_3^-$?)	2.011 −2.012	5 – 15	RT	increases above 200°C annealed at 400°C (T$_{a10}$ =280°C)	simply referred to as OHC in this book	3
Al center [AlO$_4$]$_o$	2.060208 2.008325 2.001948	10 ~ 100	77K	decreases above 200°C annealed at 400°C (T$_{a10}$ =230°C)	enhanced by γ-irradiation slightly bleached by light	4
Ge center [GeO$_4$/Li$^+$]^o_c	2.0000 1.9973 1.9962	5 – 100	RT	decreases above 150°C annealed at 220°C (T$_{a10}$ =170°C)	enhanced by γ-irradiation observed in pegmatic quartz optically bleached	5 6 7
Ti center [TiO$_4$/Li$^+$]$_A$	1.9789 1.9309 1.9119	>10 ~100	77K	decreases above 170°C annealed at 250°C (T$_{a10}$ =180°C)	enhanced by γ-irradiation	8

a) ESR measurement at room temperature (RT) or 77K.
b) T$_{a10}$ is annealing temperature at which the signal intensity is reduced by 10%.
Reference: 1) Jani et al. 1983; 2) Stapelbrock et al. 1979; 3) see text (p.282); 4) Nuttall and Weil 1981a; 5) Weil 1971; 6) Yokoyama et al. 1985; 7) Tanaka et al. 1985; 8) Isoya et al. 1988.

Production : The E'_1 center is produced at room temperature by γ–irradiation in am–SiO_2, but not created in cr–SiO_2. If there are oxygen vacancies or dislocations created by mechanical deformation, the E'_1 center is created even in cr–SiO_2 by γ–irradiation. It is detected in neutron–irradiated cr– and am–SiO_2 (Weeks 1956) and in natural quartz (McMorris 1970). Superlinear growth of E'_1 center at a low formation efficiency was detected after heavy γ–irradiation (Wieser *et al.* 1991).

Thermal properties : The signal intensity of the E'_1 center is increased above 200°C at the expense of the hole–type Al center (Jani *et al.* 1983). It is annealed out at 360 ~ 380°C and the regeneration efficiency is strongly reduced by annealing at 500°C. Presumably, oxygen vacancies around dislocations or in α–recoil tracks are annihilated by high temperature annealing. From the annealing and saturation properties, two types of E'_1 centers have been proposed to exist in cherts. One is annealed around 200°C in an artificially irradiated sample, while the other is stable up to about 360°C (Griffiths *et al.* 1983). In am–SiO_2, the E'_1 center is annealed out above 100°C (Stapelbroek *et al.* 1979).

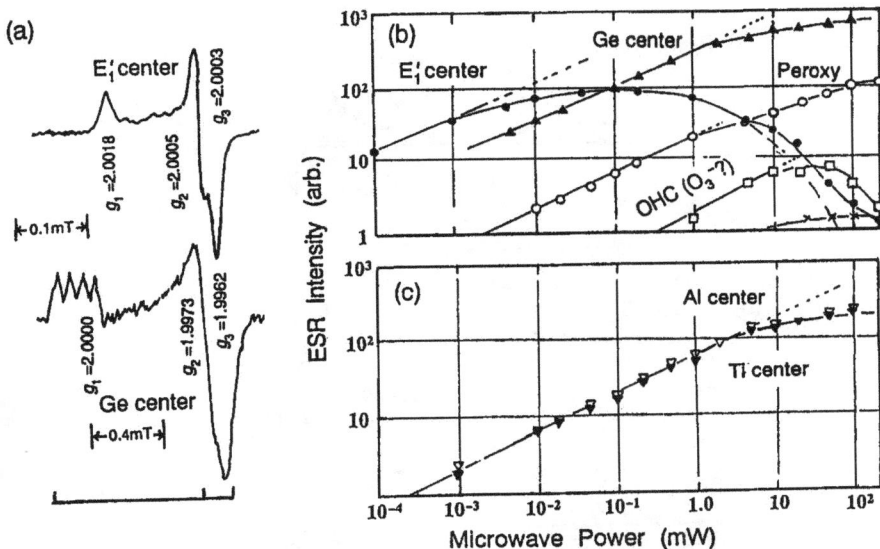

Figure 9.3 (a) ESR spectra of the E'_1 and Ge centers. (b) and (c) Micro–wave power dependence of the signal intensity; (b) E'_1, peroxy plus NBOHC, OHC (O_3^- ?) and Ge centers at room temperature, and (c) Al and Ti centers at 77K.

Microwave power dependence : The signal intensity of the E$'_1$ center is easily saturated at a microwave power of 10^{-2} mW as shown in Figure 9.3 (b). The signal which still remains at $10 \sim 100$ mW is a part of NBOHC described later.

Optical Properties :   Optical absorption occurs at 5.85 eV (212 nm) [5.5 eV (225 nm) for the E$'_2$ center]. The E$'_1$ center is not bleached by ordinary light (Jin *et al.* 1993). Neither luminescence nor photoconductivity is observed by light excitation at 5.85 eV. Sun–light bleaching is not expected.

(b) **Peroxy center (dry OHC):** $g_1 = 2.0014$ $g_2 = 2.0074$ $g_3 = 2.067$

Two types of oxygen–associated trapped–hole centers have been identified in highly purified fused silica (am–SiO$_2$) and in silica with a high OH$^-$ content after γ–irradiation. They are denoted *"dry OHC"* and *"wet OHC"*, respectively (Stapelbroek *et al.* 1979). The ESR spectra are shown in Figure 9.4. The dry OHC is actually a peroxy radical (oxygen interstitial) in the form of ≡Si–O–O· (Friebele *et al.* 1979), while the wet OHC is an oxygen hole in the form of ≡Si–O·, i.e., non–bonding oxygen hole center (NBOHC).

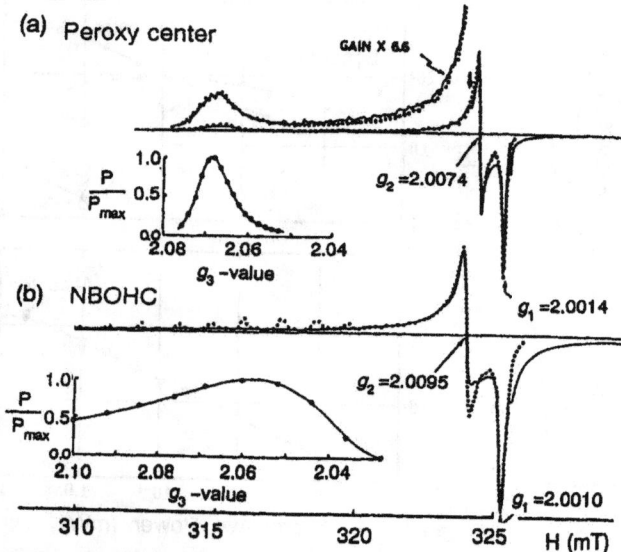

Figure 9.4 ESR spectra at 100K and computer simulation (dots) of (a) the peroxy center in dry silica annealed at 500°C after γ–irradiation and (b) NBOHC in wet silica several months after γ–irradiation. The g_3–distributions used for the simulation are shown in the insets (Stapelbroek *et al.* 1979).

In this book, we will refer to the former dry–type as a "**peroxy center**" and to the latter wet–type as "**NBOHC**" as in the review by Griscom (1990). NBOHC is described in the next item (c).

Model : In am–SiO$_2$, a model for the peroxy center is an antimorph (a complimentary center) of the E$'_1$ center in a Frenkel pair and is actually an interstitial oxygen ion O$^-$ which bonds with O^{2-}, as shown in Figure 9.2 (d). It may be considered a hole trapped by two O^{2-} at an O^{2-} site, i.e., O$_2^{3-}$ in the form of ≡Si–O–O· or [SiO$_5$]$^-$ (SiO$_5^{5-}$). On the other hand, a model for the peroxy center in cr–SiO$_2$ has not yet been clearly elucidated. An ESR spectrum of the peroxy center in am–SiO$_2$ is shown in Figure 9.4 (a).

Production : A peroxy center is produced by γ–irradiation in am–SiO$_2$ with a low OH$^-$ content but not in cr–SiO$_2$ though neutron irradiation generates similar centers in a crystal (Weeks 1956). Alpha–recoils and α–rays may produce the peroxy center in nature (McMorris 1971). The fact that the complimentary E$'_1$ center is also not formed by weak γ–irradiation in cr–SiO$_2$ indicates that the peroxy center is not formed by ionizing radiation. However, recent results showing that the intensity of the E$'_1$ center is increased by heavy γ–irradiation may indicate that the peroxy center is formed around dislocations.

Optical and thermal properties : Optical absorption at 325 nm. The signal intensity increases above 170°C up to 220°C and anneals out at 300°C.

(c) **NBOHC** (wet OHC): $g_1 = 2.0010$ $g_2 = 2.0095$ $g_3 = 2.078$
$$A_\| = 10.95 \quad A_\perp = 1.60 \text{ mT } (^{17}\text{O}; I = 5/2, 0.038\%)$$
$$A_1 = A_2 = 1.44 \text{ mT } (^{29}\text{Si}; I = 1/2, 4.67\%)$$

Model : A model for NBOHC (non–bonding oxygen hole center) in am–SiO$_2$ is ≡Si–O·. A proton is released from Si–OH by ionizing radiation and a hole is trapped at the remaining oxygen in the non–bonding $2p$ orbital. It may be expressed as [SiO$_4$]$^+$ or SiO$_4^{3-}$. The value of g_3 distributes and thus the signal is broadened beyond detection as shown in Figure 9.4 (b).

Production : This center is observed in γ–irradiated am–SiO$_2$ but is not clear in synthetic cr–SiO$_2$. NBOHC is observed in natural quartz as the shoulder signal of the peroxy center indicated in Figure 9.5 (a)–(c). It is considered that the region around lattice dislocations is close to the amorphous state. Since weathering allows the diffusion of H$_2$O into cr–SiO$_2$, NBOHC might be formed in natural quartz.

Optical and thermal properties : Optical absorption occurs at 260 nm and 630 nm. The possibility of optical bleaching is high. This center is annealed out at 300°C, but cannot be regenerated by γ–rays.

(d) OHC (O$_3^-$?) : g = 2.011 ~ 2.012

   Model :   The signal at g = 2.012 has been often termed an "OHC" because the g factor is close to the g_2 = 2.0095 of **NBOHC** in am–SiO$_2$ and has been used in ESR dating of quartz grains from geological faults or volcanic ash (Fukuchi _et al._ 1986, Fukuchi 1988, Shimokawa and Imai 1987). The signal termed OHC in _chert_ by Griffiths _et al._ (1983) is actually **"peroxy"** at g = 2.0075 and not this signal. Further confusion comes from the fact that the second signal of the quartet associated with methyl radicals (·CH$_3$) in altered SiO$_2$ falls on tne g factor at g = 2.012 at the X–band frequency (the third line at g = 1.997 falls on the Ge center: see Chapter 10). A wide range sweep (at least ± 5 mT) is necessary to distinguish this signal from that of · CH$_3$.

   The ESR intensity saturates at a microwave power above 50 mW and is

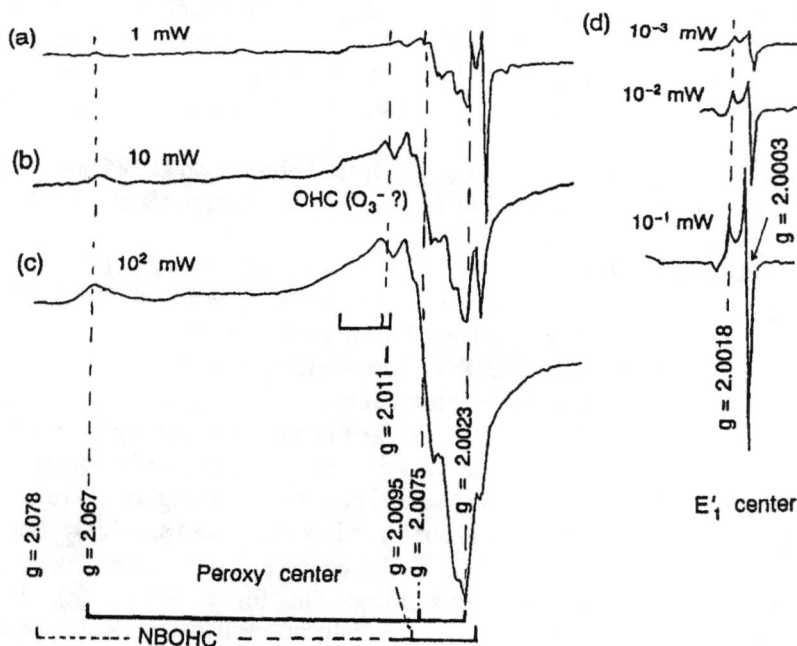

Figure 9.5   ESR spectra at room temperature of natural quartz grains (20 ~ 60 μm) from Scotland wind–blown sediment at different microwave powers; (a) 1 mW, (b) 10 mW and (c) 10^2 mW, and (d) signal of the E$'_1$ center at 10^{-3}, 10^{-2} and 10^{-1} mW The g factors of the peroxy center, NBOHC and tentatively called OHC (O$_3^-$) are indicated.

enhanced by γ–irradiation. A possible model for this signal is rotating O_3^- (ozonide ion) giving $g_{av} = 2.011~2.012$ in anhydrite (CaSO$_4$·2H$_2$O) (Bershov et al. 1975). Since a weak broad signal is lasting from $g = 2.019$, an orthorhombic O_3^- with $g_1 = 2.0183$ $g_2 = 2.0105$ and overlapping signal of $g_3 = 2.0051$ in perborate glass may correspond to this signal. This signal is common in quartz grains from fault and was often is referred to as "OHC".

(e) **Oxygen vacancies :** $[V_-]^{2+}$ and $[V_-\ 2e^-]^o$

Model : Oxygen vacancies without any electron, $[V_-]^{2+}$ or $[SiO_3]^{2+}$ (SiO$_3^{2-}$), have a net charge of +2 and those with two electrons, $[V_-\ 2e^-]^o$ or $[SiO_3]^o$ (SiO$_3^{4-}$), are neutral in the SiO$_2$ lattice (Rudra and Fowler 1987). The former is energetically unstable and both cannot be detected with ESR. Gamma–irradiation (a few kGy) fills vacancies with electrons.

Optical and thermal properties : No optical absorption band is reported. Holes trapped at Al^{3+} or at other impurity ions may be released by annealing at 250°C and will recombine with $[V_-\ 2e^-]^o$, converting to the E$'_1$ center, $[V_-\ e^-]^+$. A hole does not recombine with an electron at the E$'_1$ center since the net charge of the E$'_1$ center is positive in the SiO$_2$ lattice. Thus, thermal annealing at 250°C enhances the concentration of the E$'_1$ center. Although it is not clear whether $[V_-\ 2e^-]^o$ is totally converted into the E$'_1$ center, one can obtain the number of oxygen vacancies $[V_-]$ (Toyoda and Ikeya 1991). No E$'_1$ center is produced by weak γ–irradiation after annealing above 600°C. Oxygen vacancies will be annealed out at 600°C.

9.2.3   _Impurity–Associated Defects_

An ESR spectrum of natural cr–SiO$_2$ powder shows signals associated with impurities. Electrons (or holes) trapped by impurities give ESR signals associated with impurities such as substitutional tetravalent ions (Ge^{4+} and Ti^{4+}) and trivalent ion (Al^{3+}) as well as interstitial monovalent cations, M$^+$ (H$^+$, Li$^+$ and Na$^+$).

ESR spectra of the Al–related hole center and the Ge– and Ti–related electron centers in natural cr–SiO$_2$ are shown in Figure 9.6 and their properties are summarized in Table 9.2 together with microwave power dependence (Figure 9.3) and thermal annealing curves (Figure 9.7).

(a) **Al center :** $[AlO_4]^o$: $g_1 = 2.060208$ $g_2 = 2.008325$ $g_3 = 2.001948$
$\qquad\qquad A_1 = 0.61$ $A_2 = 0.62$ $A_3 = 0.50$ mT (^{27}Al; $I = 5/2$, 100%)
$\qquad\qquad A_1 = 1.53$ $A_2 = 1.79$ $A_3 = 11.1$ mT (^{17}O; $I = 5/2$, 0.04%)
$\qquad\qquad A_1 = 1.14$ $A_2 = 1.15$ $A_3 = 1.08$ mT (^{29}Si; $I = 1/2$, 4.67%)

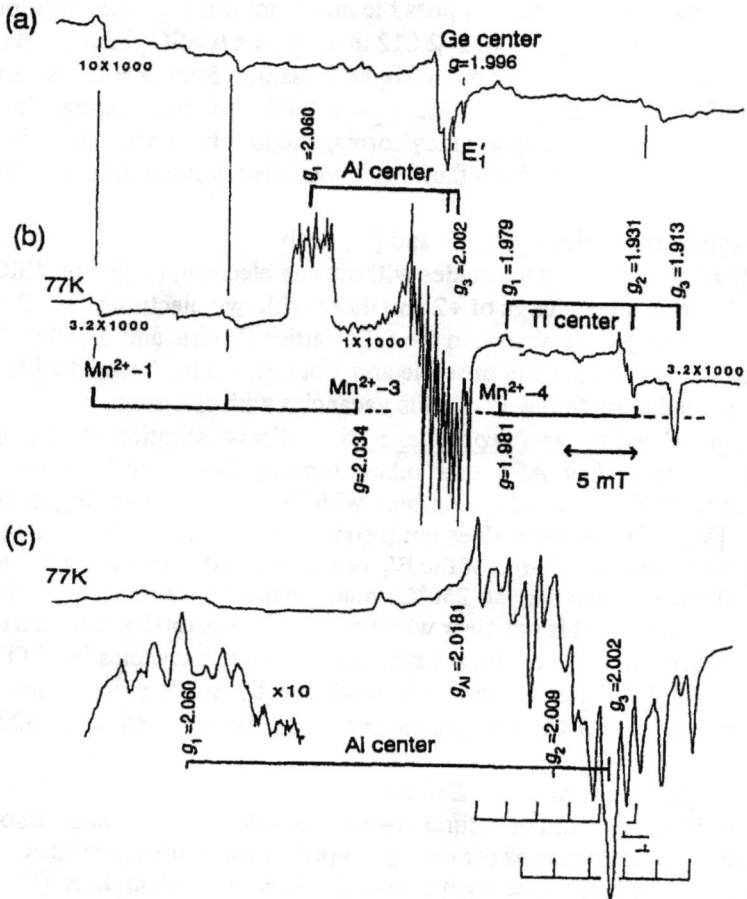

Figure 9.6   ESR spectra of quartz grains from the Fujioka fault, central Japan.   (a) Ge center and Mn^{2+} at room temperature,   (b) Al and Ti centers at 77K, and (c) an expanded spectrum of the Al center with hf and quadrupole splitting due to ^{27}Al ($I = 5/2$, 100%).

Model and production :   A trivalent aluminum ion Al^{3+} is easily substituted at a Si^{4+} site in the ionic crystal scheme of the SiO_2 lattice. The principle of ionic charge neutrality in an ionic crystal requires a charge compensating monovalent cation interstitial. In this case, a monovalent cation M^+ such as Li^+, Na^+ and H^+ at the interstitial site is associated with

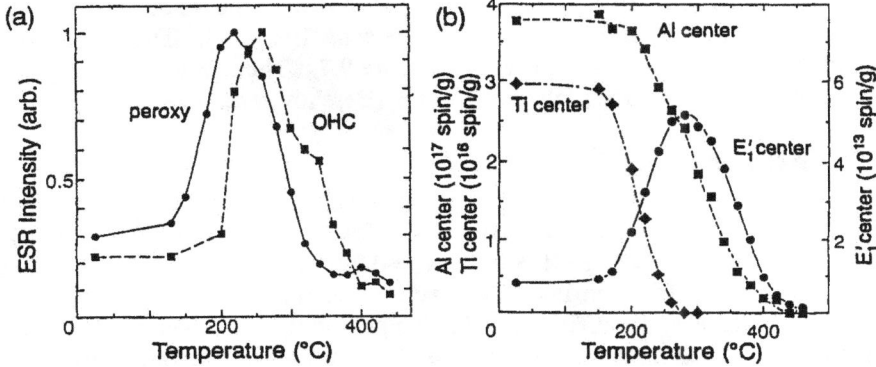

Figure 9.7 Isochronal annealing curves of (a) peroxy and OHC and (b) E'_1, Al, Ti centers. Annealing time t_a is 15 minutes. Quartz grains were extracted from Mannari granite, Okayama, Japan (Toyoda and Ikeya 1991).

Al^{3+} as a charge compensator and gives rise to a $[AlO_4/M^+]^0$ center. When the charge compensated $[AlO_4/M^+]$ traps a hole created by ionizing radiation at room temperature on one of the two long–bond oxygen neighbors, the cation M^+ diffuses away leading to $[AlO_4/h]^0$, where h is a hole trapped in a non–bonding $2p$–orbital of oxygen located adjacent to the substitutional aluminum (O'Brien 1955, Hitt and Martin 1983). The process is summarized as follows.

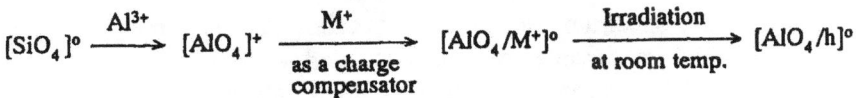

$$[SiO_4]^0 \xrightarrow{\ Al^{3+}\ } [AlO_4]^+ \xrightarrow[\substack{\text{as a charge} \\ \text{compensator}}]{\ M^+\ } [AlO_4/M^+]^0 \xrightarrow[\text{at room temp.}]{\ \text{Irradiation}\ } [AlO_4/h]^0$$

The hole hops rapidly among the long– and short–bond oxygen neighbors (Schnadt and Schneider 1970). Measurement at 77K is required so that the hole is localized on one of the two long–bond oxygens as shown in Figure 9.1 (b).

A typical ESR spectrum of the Al center at 77K is shown in Figure 9.6 (b) and its expanded spectrum in (c). A number of hf lines are observed due to the nuclear spins and quadrupole splitting $Q[I_z^2 - (I/3)(I + 1)]$ of ^{27}Al.

Microwave power dependence : Saturation is barely reached up to 100 mW as shown in Figure 9.3 (b).

Optical and thermal properties :   Optical absorption occurs at 2.9 eV (427 nm).   Some Al centers might be localized around dislocations where Al^{3+} impurities tend to coagulate. Only about 10% of the Al center is optically bleached in natural quartz (Yokoyama *et al.* 1985a, b). Thermal stability is shown by the annealing curve in Figure 9.7. The content of Al is high when quartz is formed at high temperature (Bershov *et al.* 1975).

(b) **Ge center : [GeO$_4$/M$^+$]$_C^o$:**

$$\mathbf{M = Li}\ (C):\quad g_1 = 2.0000 \quad g_2 = 1.9973 \quad g_3 = 1.9962$$
$$A_1 = 30.9 \quad A_2 = 29.4 \quad A_3 = 29.5 \text{ mT} \quad (^{73}\text{Ge};\ I = 9/2,\ 7.73\%)$$
$$A_1 = 0.08 \quad A_2 = 0.01 \quad A_3 = 0.002 \text{ mT} \quad (^{7}\text{Li};\ I = 3/2,\ 92.5\%)$$

Model :   Isoelectronic Ge^{4+} substitutes for a Si^{4+} site in the SiO$_2$ lattice but its electron affinity (ionization potential) is considerably larger, resulting in trapping of an electron created by ionizing irradiation to form [GeO$_4$·e$^-$]$^-$ in the SiO$_2$ lattice. The total charge of the center is −1 and thus Ge^{3+} at a Si^{4+} site in an SiO$_2$ structure has a negative effective charge, attracting an interstitial monovalent cation M$^+$ such as H$^+$, Li$^+$ or Na$^+$, to form stable [GeO$_4$/M$^+$]^o at room temperature (Anderson *et al.* 1974). The process is summarized as follows.

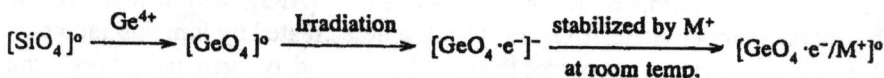

$$[\text{SiO}_4]^o \xrightarrow{\text{Ge}^{4+}} [\text{GeO}_4]^o \xrightarrow{\text{Irradiation}} [\text{GeO}_4 \cdot \text{e}^-]^- \xrightarrow[\text{at room temp.}]{\text{stabilized by M}^+} [\text{GeO}_4 \cdot \text{e}^-/\text{M}^+]^o$$

Two distinct types of Ge centers (M$^+$ = Li$^+$) with anisotropic *g* factors have been reported and designated as A and C (Weil 1971) and it is postulated that the Li$^+$ ions in the two centers occur at opposite sides of the tetrahedron at or near the channel sites a$_>$ and a$_<$ (see Figure 9.1 (b)).

The ESR spectrum after γ–irradiation shows an asymmetric signal at *g* = 1.997 which is considered to be associated with the [Ge^{4+}/Li]$_C^-$ center. ESR signals of the Ge center appear at the high–field side of the E'$_1$ center. The Ge center is frequently found in quartz from pegmatite which is formed by slow cooling of magma materials. Quartz veins from gold mines clearly show the signal of the Ge center in addition to that of the peroxy center (Van Moort and Russell 1987).

Microwave power dependence : The Ge center saturates at high microwave power (> 100 mW) as shown in Figure 9.3 (b).

Optical and thermal properties : Optical absorption occurs at 4.43 eV (280 nm). The Ge center is easily bleached by sunlight (Tanaka *et al.* 1985). It is annealed out at 250 − 280°C, indicating a lifetime of − 100 ka at 290K.

(c) **Ti center :** $[TiO_4/M^+]_A^0$

$\quad$ **M = Li (A) :** $\quad g_1 = 1.9789 \quad g_2 = 1.9309 \quad g_3 = 1.9119$

$\qquad\qquad\qquad A_{av} = 1.17$ mT (^{47}Ti; $I = 5/2$, 7.3%)

$\qquad\qquad\qquad A_{av} = 0.13$ mT (^{7}Li; $I = 5/2$, 92.5%)

$\quad$ *Model and production* : The model is the same as that of the Ge center (Weil 1984, Wright et al. 1963) and so the formation mechanism. M^+ is a charge compensator and H^+ acts as a charge compensator in irradiated α–quartz. There are two types of Ti centers (A and B) as for Ge centers, but B type is not stable at room temperature. An ESR spectrum of the Ti center is shown in Figure 9.6 (b). The measurement should be at 77K.

$\quad$ *Thermal properties* : Thermal stability and microwave power depend-ence are shown in Figures 9.7 and 9.3 (c), respectively.

(d) *Other Centers* :

$\quad$ *Hydrogen centers* : Two types of hydrogenic hole centers having the forms of $[(HO)_4]^+$ and $[O(HO)_3]^+$ have been identified by their complex hf structures due to hydrogen (Nuttall and Weil 1980). The former is formed by trapping a hole in a p–orbital of oxygen associated with one of four hydrogen atoms replacing a Si site as $[H_4O_4]$. The latter is formed when the precursor state traps a hole at one of the oxygen atoms and releases the proton associat-ed with the oxygen atom. These centers have not yet been found in natural samples although the latter center is still observed in a sample kept at room temperature for about one year after irradiation. It will be interesting to use these signals to study weathering of both cr– and am–SiO_2 in nature.

$\quad$ *Surface defects* :

$\qquad$ **Surface E'–like center :** $\quad g_\parallel = 2.0017 \quad g_\perp = 2.0003$

$\qquad$ **CO_3^-–type center** $\quad$: $\quad g = 2.0048 \quad g = 2.0063 \quad g = 2.0248$

$\quad$ A surface E'–like center observed in crushed SiO_2 is not stable at room temperature (Arends et al. 1963), especially when the sample contacts with water. This contact with H_2O leads to formation of gaseous H_2, which may be the reason for H_2 emanation in active fault zones (Kita et al. 1982). Atmospheric CO_2 adsorbed on the surface of SiO_2, presumably at the sur-face E' site, results in a CO_3^- type center (Ebert and Hennig 1974).

$\quad$ There is a proposal to use surface defect centers in ESR dating of faults (Toyoda et al. 1991, 1993, Fukuchi 1993), but the stability is a major limi-tation for the use of surface centers in geology. It is defects at the near sur-face region that might be used for dating a fault.

9.3   ESR Dating of SiO₂ from Fault Gouge
9.3.1   ESR Dating of Quartz

Several attempts to use ESR signals in SiO_2 for dating were made after the proposal of "trapped electron dating" by McMorris (1971). However, *ceramic dating* using quartz grains was not successful since the ESR sensitivity was not high enough to detect trapped electrons produced in an archaeological time scale of a few ka.

Garrison *et al.* (1981) used *neutrons* from a nuclear reactor as a source of artificial additive irradiation for natural α–rays and obtained an ESR age of 300 Ma for quartz samples. Use of a nuclear reactor as a neutron source is not practical to simulate α–rays and α–recoils since the energy spectrum of neutrons is different at different positions for different reactors. The equivalent dose (*ED*) is dependent on nuclear reactors. Alpha–irradiation using a radioactive source of ^{241}Am or by an accelerator will be better for the additive dose method.

9.3.2   Formation of Fault Gouge and Intrafault Minerals

In structural geology, a fault is a fracture or a zone of fractures cutting any rock and showing displacement relative to one another. A fault zone generally consists of the fault gouge or soft uncemented pulverized clay or claylike minerals and of fault breccia formed by the crushing, shattering and shearing of rocks. The age of a fault formation or movement is vitally important for the risk assessment of future fault movements for consideration in structural or engineering geology (Billing 1972). A fault that has moved in the last 35,000 years is considered to be capable of movement in the near future and is designated as a "*capable fault*" by the US Nuclear Regulatory Commission for the siting of a nuclear power plant. The age assessment of fault movement is based on geological observation of the cutting and noncutting of rocks at fault outcrops. Trenches across a fault region are excavated to establish the chronology of relative displacement of the rocks.

Attempts have been made to determine the absolute age of faults by radiometric dating techniques. *Illites* in intrafault materials have been dated by the K–Ar method (Lyons and Snellenburg 1971, Kralik *et al.* 1987). Leakage of radiogenic argon might occur at the time of fault gouge formation, which leads to K–Ar ages. The age of a secondary mineral, epidote, formed along a fault plane has been determined by fission track dating (Bar *et al.* 1974) and of gypsum at the San Andreas fault by ESR (see Chapter 7). However, these materials are less common in faults. So far, attempts to provide absolute ages of fault movements have been generally unsuccessful.

9.3.3   Basic Considerations in Dating of Fault Movement

It is considered that small quartz grains in the fault gouge, brecciated rocks and mylonite must have lost their geologically accumulated defect concentration at the time of fault movement (Ikeya et al. 1982). Experimental results indicate that the stresses followed by fracturing of quartz grains reduce the ESR signal intensities of several defects (Miki and Ikeya 1982, Ariyama 1985) and enhance the efficiency of defect formation due to the introduction of defects and dislocations (Ikeya et al. 1983, Sato et al. 1985).

Figure 9.8 shows the hypothesis of fault dating schematically. The defect concentration is reduced by a mechanism operating during faulting. The defect formation efficiency is enhanced by formation of dislocations and of the E'_1 center as well as by dispersion or diffusion of Al and Ti impurities for the Al and Ti centers. Zeroing of the age partially gives an old age for a young fault. The basic premise of dating geologic faults with ESR is the need for a mechanism that could either cause zeroing of geological ESR signal during fault movement or create new defects which could then become paramagnetic centers with capture of a charge. Thus, the following mechanisms are considered for fault dating. Attempts to date fault have been done based on the mechanisms (a) and (b), while (c) ~ (e) are proposals.

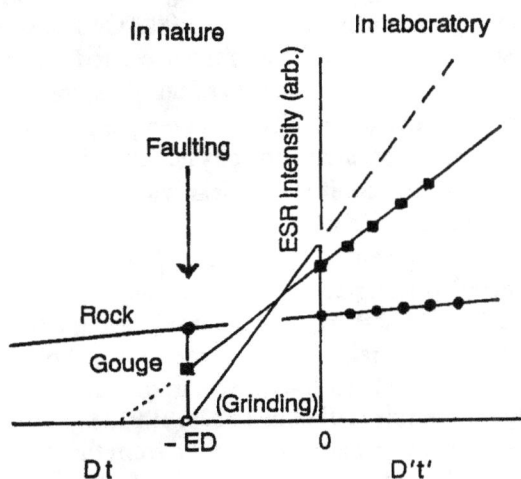

Figure 9.8 A proposed mechanism for dating a geological fault based on mechanical annealing (or frictional heating) and the enhanced efficiency of defect formation.

(a) *Mechanical bleaching and efficiency enhancement* : Decrease of the ESR signal intensity by grinding the samples is often experienced during sample preparation. This may be attributed to the introduction of dislocation by plastic deformation and cracking as well as local heating, and may occur as a mechanism of zeroing for fault dating (Ikeya *et al.* 1982). The phenomenon is known in terms of *"mechanical bleaching"* or *"mechanical annealing"* in the physics of color centers in alkali halides (Bauer and Gordon 1962). We are not sure whether it is due to the local temperature rise or to release of electron (hole) by lattice dislocation movements.

Deformation luminescence (DL) is observed during plastic deformation due to the ionization and recombination of trapped electrons and holes. This indicates that electron and hole centers produced by radiation are released by plastic deformation. Speculation can be extended to the lightning at the time of an earthquake that might come from this deformation luminescence.

It is considered that natural radiation after the mechanical bleaching at the time of faulting can create the E'_1 center at a high efficiency. The stress and lattice deformation of quartz produces either oxygen vacancies as a source of the E'_1 center or dislocations that can stabilize interstitial oxygen atoms which cause enhanced formation of the E'_1 center (Ikeya *et al.* 1982, 1983, Fukuchi 1989b).

(b) *Temperature rise by frictional heating* : Theoretical calculations suggest that the temperature can rise up to 1,000°C at the fault plane due to high shearing stresses and frictional forces caused by the displacement (McKenzie and Brune 1972). The role of temperature rise in mechanical annealing can not be excluded. The local temperature at the defect site, however, can neither be measured nor physically defined during the lattice dislocation movement. Hence, it is not clear whether the mechanical annealing is identical to thermal annealing.

The energy release caused by frictional heating and thermal diffusion have been estimated to examine if the defects such as E'_1, OHC, Al, Ti and Ge centers in SiO_2 are thermally annealed during a fault movement. If the entire energy is used to raise the temperature in a short duration movement, it has been concluded that the defects in quartz grains are annealed out by the bulk temperature rise (Fukuchi 1989a, 1991, 1992). The temperature rise as a function of time and also of distance from the fault plane suggests different annealing characteristics for several defects. A physico–chemical reaction is also proposed for fault gouge formation by frictional heat in the presence of groundwater rather than crushing and powdering by mechanical faulting forces. Thus, ESR dating of minerals around the fault plane should provide the time that has elapsed after the fault movement using the defects that have been accumulated since that time. My minor comments to this paper are;

(1) Appreciable energy may also be stored in forms of lattice defects and dislocations. Temperature rise might be overestimated unless this is considered.

(2) If the temperature was really raised close to 1,000°C or to the melting point, the oxygen vacancies in thermal equilibrium comprising Schottky defects (V_+ and V_-) at high temperature may be quenched. This can be attributed to the observed high efficiency of defect formation.

(3) If the temperature is as high as 600°C, the formation efficiency of the E'_1 center is reduced since oxygen vacancies are annealed out at 600°C. This is not observed in laboratory and in nature.

(4) The 2nd and 3rd lines of $\cdot CH_3$ might have been measured as OHC and the Ge center, respectively. Concordance of the *ED*'s suggests the same centers. An ESR spectrum with wide–range sweeping and a microwave power dependence must be measured. In many cases, it is quartet hf with intensity ratios of $1:3:3:1$ which is confirmed by the K–band spectrum. It is noted that geothermal quartz contains abundant methyl radicals.

(c) *Crystallization at a fault plane* : It has been suggested that shear-induced melting may occur during steady sliding on a fault plane. Should this be so, *recrystallization of minerals* at the fault plane at the time of fault movement can be used for ESR dating with the assumption that the ESR signal of the recrystallized mineral is *a priori* zero. *Mineral precipitation* at the fault plane can be used to assess the age of fault movement though it is post–depositional for gypsum at the San Andreas fault (see Chapter 7).

(d) *External α–damage at the surface* : *a new proposal* : Some quartz grains are quite large, a few mm in diameter, in granite and in pegmatite. The uranium content in quartz is extremely small relative to that in other minerals. Hence, the radiation damage by α–particles is predominantly limited to the surface of the grains. Faulting and splitting of the large grains into finer grains in gouge create new surfaces that will be damaged by α–particles. Thus, the fine grains will have a full α–range. If so, the annual dose rate by external α–rays ($D_{\alpha ex}$) is increased from that time since the attenuation factor (see Chapter 4) changes after powdering at the time of faulting. Note that α–rays produce oxygen vacancies.

   If the surface layer of a few μm, for example, $d = 10 \sim 20\ \mu$m of quartz grains with a radius r, is damaged by external α–rays from the enclosing rock, the effective concentration of defects is $n = 4\pi r^2 d/(4\pi r^3/3) = 3d/r$. Hence, the ESR intensity of defects in natural samples should be large for small grains. Experimental results indicate a small ED for small grains (see Figure 9.14). The dependence of the ED on the reciprocal radius $(1/r)$ should be studied rather than on the radius itself.

This mechanism to study the near–surface defects by external α–rays might be used to date lithic tools after cleaving. Future research should involve archaeological applications by studying the surface α–damage. These defects would be different from the unstable surface E' or CO$_3^-$ center.

(e) *E'$_1$ center around induced dislocations* : Faulting can also introduce dislocations in the crystalline lattice of quartz grains through grinding stress. The E'$_1$ center produced around dislocations is distorted and has a large linewidth (Toyoda *et al.* 1993) and a short spin–spin relaxation time (T_2) as shown in Figure 2.29 (Ikeya *et al.* 1992). If this component of the signal can be separated from the relaxation time, the age of faulting can be estimated with pulsed ESR.

9.3.4   *Laboratory Simulation of Faulting*

(a) *Crushing experiment* : Quartz grains (250 ~ 840 μm in diameter) separated from Cretaceous gneiss were fractured using an Instron machine (Miki and Ikeya 1982). Figure 9.9 (a) shows a decrease in the signal intensity for the E'$_1$ center by grinding the quartz grains, together with the weight percent for the sieved fractions of different grain sizes. The intensity of the E'$_1$ center in a bulk fraction is decreased linearly up to 40% of the initial intensity by a stress of about 100 MPa. Further pressure was not effective both for further grinding and for the destruction of the E'$_1$ center. The fractured grain fraction of 250 ~ 840 μm in diameter shows little decrease in the signal intensity, while that below 74 μm shows an intensity of about 10% of the initial value. These observation indicate that the effect of the bulk temperature rise is small in mechanical annealing.

The crushing experiment was also carried out using both quartz from Cretaceous granite and pure synthetic quartz. Different crushing rates or differential stresses using a high compressive triaxial testing machine were employed (Tanaka and Shidahara 1985). The intensities of the E'$_1$ center and OHC decreased by up to 55% and 60% following stress at 500 MPa, respectively. Since the intensity of the E'$_1$ center returns to the initial value after etching with hydrofluoric acid (HF), the E'$_1$ center reduced by crushing may be close to the surface etched along the dislocation lines.

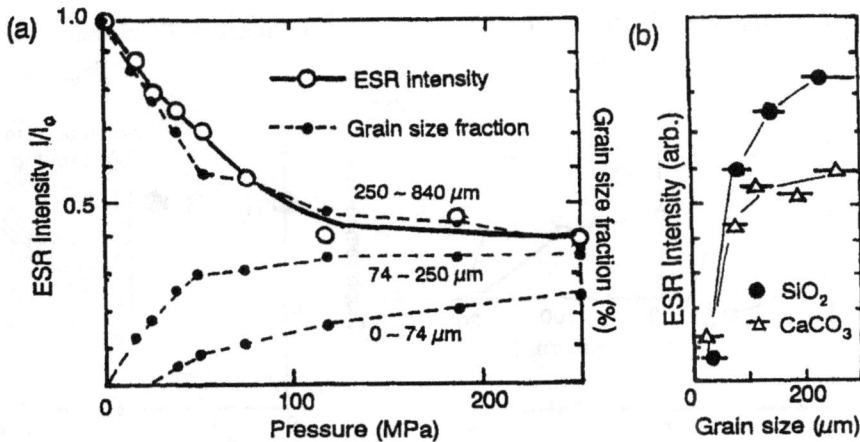

Figure 9.9 (a) Decrease in the signal intensity of the E$'_1$ center by a crushing experiment and the weight percent of fractured fractions of different size. (b) Grain size dependence of the signal intensity of the E$'_1$ center in ground SiO$_2$ and CO$_2^-$ ($g = 2.0007$) in CaCO$_3$.

The proposal to use the surface crushing signals for dating a geological fault may be difficult as they are only stable for a few days as has already been noted by Tanaka and Shidahara (1985). The signal disappears when the sample contacts with water. The E$'_1$ center with a short T_2 is observed in crushed SiO$_2$ but unfortunately is not observed in quartz grains at the Rokko fault (Ikeya et al. 1992).

(b)   *Shearing experiment* :   Quartz grains (74 – 250 μm in diameter)   mixed with kaolinite power were ground by shear stresses of a simulated near-surface fault under high pressure and sliding motion to study the condition of resetting the ESR signal clock during faulting (Ariyama 1985). The pressure normal to the fault plane is more than 10 MPa near the ground surface.   The intensities of two signals at $g = 2.001$ (E$'_1$ center) and 1.996 (Ge center) decrease drastically as shown in Figure 9.10. In one experiment shown in (a), the pressure was fixed at 2 MPa and the displacement was changed from 0 to 100 cm with a speed of 10 cm/s. In the other experiment shown in (b), the displacement was confined to 100 cm and the pressure was changed from 0 to 2.5 MPa. The *displacement–pressure diagram* of complete and incomplete resetting is shown in (c). The result suggests that zeroing of the ESR age can

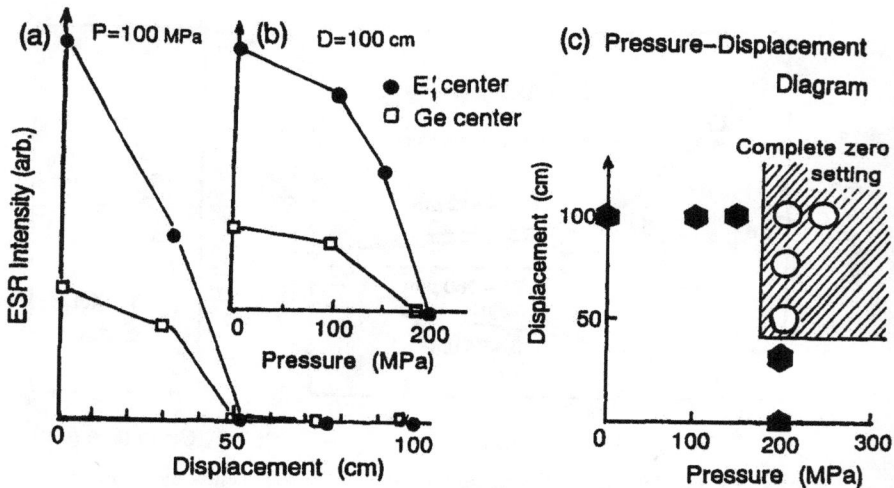

Figure 9.10 (a) and (b) Decrease in signal intensities by experimental faulting. The conditions for the two experiments are described in the text.   (c) A
displacement–pressure (D–P) diagram. The hatched area shows complete
setting (Ariyama 1985).

occur for near–surface faulting. Similar experiments indicate that signals of
the E$'_1$ and Al centers are zeroed by the stress of 5 – 10 MPa, while those of
the OHC, Ge and Ti centers are not effectively zeroed. Enhancement of the
E$'_1$ center formation efficiency is also confirmed (Lee and Schwarcz 1993).

9.4   Application to Fault Dating

9.4.1 *Initial Work using the E$'_1$ Center*

(a)   *Atotsugawa fault (ED's around the fault plane)* : The first attempt at
dating geological faults with ESR was made for the Atotsugawa fault, a well
known active fault located in Toyama prefecture, Central Japan (Ikeya *et al.*
1982). The most recent movement was about 200 years ago according to the
old historical record of an earthquake in the vicinity. The fault zone is about
30 m wide at the outcrop, consisting of five brecciated (fractured) zones.
The intrafault gouge (clay) zone of a few cm is sandwiched by rocks with
slickensides and a brecciated zone. Minerals from fault gouge, especially
quartz grains and feldspars have been studied using optical and electron
microscopy (Kanaori *et al.* 1985).

An ESR spectrum of quartz grains (74 ~ 250 μm in diameter) showed the E'_1 center but the Ge center was somehow not observed even after artificial irradiation. The ED of the sample was obtained using the E'_1 center by the additive dose method. The ED's and ESR ages of quartz grains and the sampling locations in the outcrop are shown in Figure 9.11. The annual dose rate was estimated to be 3.08 mGy/a from the content of ^{238}U (1.5 ± 0.1 ppm), ^{232}Th (7.9 ± 0.4 ppm) and K_2O (1.8 ± 0.1%) in the intrafault material.

The quartz grains at the fault plane showed a small ED of about 200 Gy corresponding to the age of 65 ka as indicated by an arrow in the figure. On the other hand, the ED's at the other points of the fault zone were 2 to 5 kGy (the ages of 0.65 Ma to 1.63 Ma). The result suggests that the last fault movement at the site occurred 65 ka BP at the edge of the fault zone.

There has been controversy over the age of Atotsugawa fault among geologists. The ESR age (65 ka) is considerably older than some geologist's estimates (200 a) or younger than some others (more than 1.5 Ma). The ESR age either represents (a) the mixed ages of gouges formed at different times, or (b) an imperfect zeroing of the ESR signal. The surface sampling might not be appropriate in the latter case.

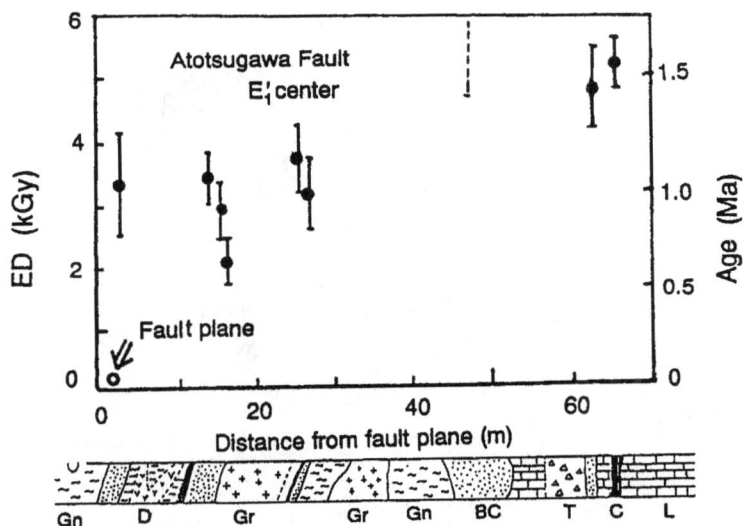

Figure 9.11 ED at several sites around the Atotsugawa fault zone. The ED is the smallest at the fault zone indicated by an arrow. The ESR age is shown on the right ordinate. BC: breccia and clay, C: black clay, Gr: granites, Gn: gneiss, L: limestone and T: talus.

(b) *Rokko fault (formation efficiency around the fault plane)* :   Quartz
grains were obtained from the outcrop of the Rokko fault close to Osaka,
Japan as shown in Figure 9.12.   The enhancement of the signal intensity of
the E'_1 center by γ-irradiation at several sites with different distances from
the fault plane is shown in Figure 9.13 (Ikeya *et al.* 1983). Both the defect
production efficiency and the saturation value of the defect concentration are
generally higher with increasing proximity to the fault plane. The enhancement is negligibly small for quartz grains away from the main fault (No.5 and
No.6).   This was confirmed by the subsequent work on this fault (Toyokura
et al. 1985) and by the study on other faults (Fukuchi *et al.* 1985, 1989a, b?).

     The results indicate that the closer the quartz grains are to a fault,   the
more should be the strain involved. This is consistent with the radiation
sensitivity of color center formation by the *mechanical deformation* of alkali
halide crystals, i.e., the more the crystal is deformed, the higher becomes the
defect formation efficiency (Bauer and Gordon 1962). In fact, X-ray diffraction studies of the lattice parameters for the quartz grains indicate that
large *lattice distortions* are involved for the sample close to the fault plane.

     In extreme proximity to the fault, recovery of the lattice distortion is

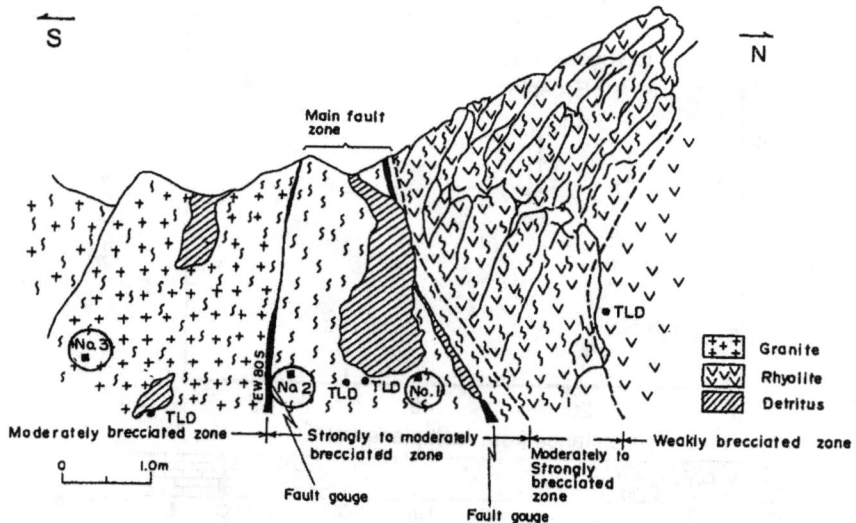

Figure 9.12   A geological fault map of the outcrop at the Rokko fault. The
sampling sites (No.1, No.2 and No.3) are indicated by squares while those of
TLD embedding by black dots. The sites No.4, No.5 and No.6 are out of this
map, located 10, 20 and 30 m away from No.2, respectively.

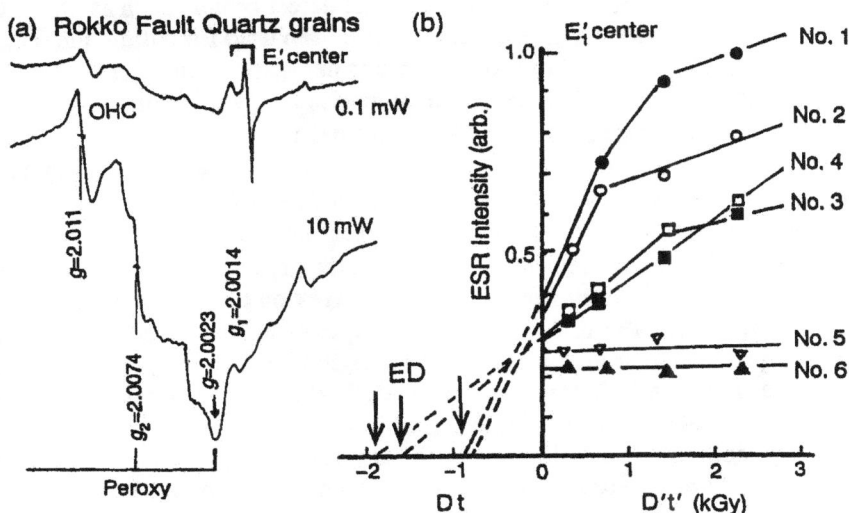

Figure 9.13   Enhancement of the E'_1 center intensity by γ–irradiation. As the distance of the sampling site from the fault plane (No.1) increases, the ED is increased. Quartz from granite away from the fault did not show any enhancement of the signal intensity by γ–irradiation (No.5 and No.6).

found in some cases, presumably due to partial melting and recrystallization by a temperature rise caused by frictional heating. No decrease in the formation efficiency of the E'_1 center was observed for the sample at the fault plane. If recrystallization occurs, the formation efficiency may be as small as that of No.5 and No.6. The *ED*'s obtained from the initial enhancement are 0.7 (No.1), 0.8 (No.2), 1.6 (No.3) and 1.9 kGy (No.4), corresponding to the ages of 150, 180, 360 and 430 ka, respectively using an annual dose rate of 7.2 mGy/a. These concord with the ages estimated by geologists.

9.4.2   *Dating with Various Defects*

(a) E'_1 *and Ge Centers* : Quartz grains from fault gouges show an ESR signal of the Ge center (g = 1.997). ESR dating using signals of the E'_1 and Ge centers was made for the Atotsugawa fault (Ito and Sawada 1985) and the Tsurukawa fault in Japan (Kosaka and Sawada 1985) as well as for the San–Jian fault in China (Lin *et al.* 1985). Pleistocene fault gouges using the Ge center gave considerably younger ages than the E'_1 center. The obtained age of 200 ~ 300 ka may be the upper limit due to the lifetime of the Ge center.

(b) *Temperature rise thermometer* : Several defect centers such as the Ge, Al, OHC and Ti as well as E$'_1$ centers have been used for dating the large fault of Fossa Magna and several other faults in Japan (Fukuchi *et al.* 1985, 1986, Fukuchi 1988). If local temperature rise is involved in mechanical annealing, several defects with different thermal stabilities may show differ- ent degrees of signal decrease. The result indicates that the E$'_1$, Al, Ge and OHC centers are easily resettable (Fukuchi 1989a). A good concordance of ESR ages using Al, Ti and Ge centers has been obtained.

The *ED*'s for these centers increase in the order of E$'_1$, Al, Ge and OHC, somewhat different from the previous work on the Ge and E$'_1$ centers (Lin *et al.* 1985). The lifetimes of defects were estimated based on thermal annealing experiments and the decay of the signal intensities by the tempera- ture rise around the fault plane was calculated. However, regression analysis of annealing data using an empirical function is not persuasive.   The decay curves by isothermal annealing should be analyzed using the general form of the first– and second–order decay kinetics given in Appendix A1.

(c) *Dislocation movement*: *grain size plateau* : A plateau test for the grain size dependence is proposed using fault gouges from the San Andreas fault in California (Buhay *et al.* 1988). The *ED*'s for the Al, OHC and Ti centers were obtained as a function of grain size, 75 to ~ 450 μm as shown in Figure 9.14 (a). The *ED* decreases for the Al center but rather increases for the Ti center as the grain size decreases. Only the *ED* of OHC shows a plateau and converges with that of the Al center for the small grain size. The Al center in large grains might be incompletely zeroed, while OHC might be completely zeroed. The easy order of thermal annealing is OHC > Al > Ti centers contrary to that of Fukuchi (1989a). It must be noted again that sometimes the peroxy center (*g* = 2.0075) is ascribed to OHC (*g* = 2.011).

The dependence of the degree of *age resetting* on the grain size sup- ports the model of mechanical bleaching by the movement of dislocations in the crystalline lattice. The higher *stress sensitivity* of OHC is explained by *hole detrapping*. Considering the tetragonal distortion, OHC will interact with stress fields induced by dislocation movement. On the other hand, the Al and Ti centers would distort the lattice isotropically and will not interact with dislocation so much as OHC. The *ED* of the Al center decreases only for the small grain size. This tendency cannot be explained by a large bulk temperature rise as a clock zeroing mechanism.

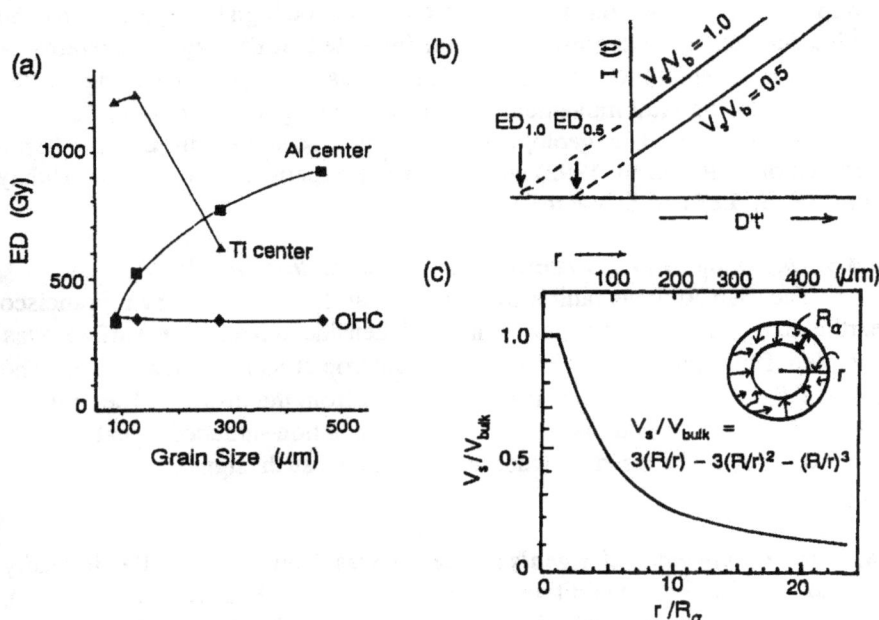

Figure 9.14 (a) Grain size plateau for ESR dating of quartz grains for geologic fault movement (Buhay *et al.* 1988). (b) Schematic growth by an additive dose for samples with ratios of a near–surface volume (V_s) and a bulk volume (V_{bulk}), $V_s/V_{bulk} = 1.0$ and 0.5. (c) V_s/V_{bulk} (i.e., *ED* ratio) as a function of grain radius, r/R_α. R_α is an external α–ray range.

(d) *Surface E' and distorted E'$_1$ centers*: Dislocations and lattice distortions are introduced by stress and shearing caused by geological faulting. Although surface defects such as a surface E' center are not stable, the E'$_1$ center at the distorted lattice site close to the surface or around cracks and dislocations might be used by taking a ratio of the near–surface volume (V_s) to the bulk volume (V_{bulk}) as shown in Figure 9.14 (b); near–surface defects by the external α–rays should show the same tendency. Experimental plotting of the *ED* against the reciprocal grain radius is better to analyze near–surface defects. The separation of such centers with spin–spin relaxation time (T_2) using pulsed (spin–echo) ESR will be the subject of studies (Ikeya *et al.* 1992, Toyoda *et al.* 1993).

An obvious way to examine the validity of dating faults is to show that ESR ages of historically known faults correlate with their historic ages. Most samples are collected at outcrops close to the ground where stresses and strains as well as the raised temperature may not be high enough to zero the lattice defects. The experimental results indicate that ESR ages are overestimated for recently moved faults. This suggests either partial zeroing of the ESR ages during fault movement or mixing of new gouge with old gouge.

Attempts to date geological faults have also started in Europe (Grün and Fenton 1991, Grün 1992) with laboratory experiments in Canada (Buhay *et al.* 1988, Lee and Schwarcz 1993).

9.4.3 *Mylonite with Slickensides* : *Recrystallization or Melting*

The San Andreas fault moved in 1906 at the time of the San Francisco earthquake. A block of rock with a slickenside mirror–like surface was sampled from the brecciated zone at an outcrop at San Andreas lake. The surface 0.5 mm looked significantly different from the inside. The mirror–like surface portion and the inside showed a radiation–induced signal around $g = 2.0$, while fractured powder did not have such signal, as shown in Figure 9.15.

(a) *Radiation–induced signals* : The age was deduced from the intensity around $g = 2.0$ by the additive dose method and the ^{238}U content (20 ppm), to be 6 ka for (a) and (b). Since the site is overwhelmed by the massive movements, some rocks may be ground into fine powder to give an ESR age of zero or younger than a few hundred years. Some may have escaped grinding by the recent movement. The temperature may have not been raised at the site and mixing of new gouge with old gouge may occur. The spectrum in (c) gave nearly zero age consistent with the earthquake in 1906.

(b) *Linewidth of Mn^{2+} signals* : The Mn^{2+} signals are different in these samples. Mn^{2+} in a crystal shows a sharp hf septet but if the lattice is distorted, the line becomes broad. Thus, the broad sextet observed in Figure 9.15 (b) is due to the *lattice distortion* of polycrystalline mylonite. On the other hand, a sharp sextet signal of Mn^{2+} which overlaps the broad one in (a) indicates the *recrystallization* of the surface at the time of *slickenside* formation. A partial *melting or phase changes* by the heat and stress at the fault plane may be a mechanism of zeroing the age.

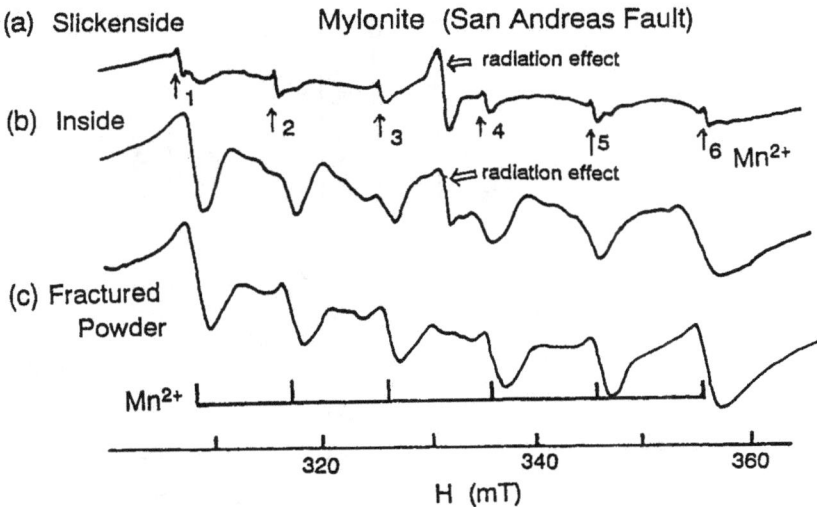

Figure 9.15 ESR spectra of mylonite at an outcrop of the San Andreas fault at San Andreas Lake that had moved in 1906. (a) Mirror–like slickenside surface 0.5 mm. The sharp sextet overlapping with the broad one is due to Mn^{2+} recrystallized by high stresses. (b) Inside the same piece of mylonite with a broad sextet due to Mn^{2+} in the distorted lattice. Radiation–induced signal at $g = 2.0$ is present in both sample. (c) Fractured powder obtained at the outcrop does not show the radiation–induced signal at $g = 2.0$.

9.5 Sediment Age

9.5.1 *Optical Bleaching ?*

If some defect centers are optically excited by light, electrons in the excited state are released and recombine with holes or are retrapped by some other impurity. The TL dating of sediments, loess and deep–sea core sediments are under investigation based on the assumption that *sunlight* has bleached the geological TL centers. Natural radiation, thereafter, has created TL responsible centers. The hypothesis is schematically shown in Figure 9.16. One of the problems for TL studies is that defects sensitive to light are unknown. The degree of age–zeroing is another problem. It is not yet clear whether the sunlight bleached defects before sedimentation. Extensive studies with TL give evidence that river–transported sedimentation is out of the scope of sediment dating. In principle, ESR can offer an advantageous method of sediment dating as *light–sensitive centers* can be identified and used to determine the *ED*.

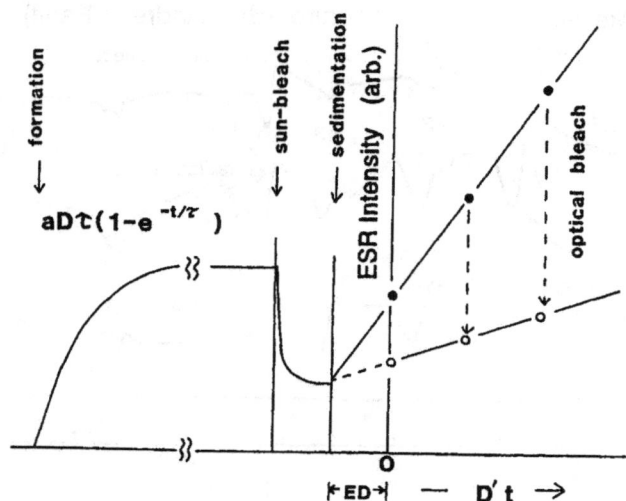

Figure 9.16 A hypothesis of the process of sunlight bleaching in nature and the procedure to obtain the *ED* following TL and OSL dating of sediments. ESR can identify light–sensitive centers for sediment dating.

The author's attempts to date beach sands in Brazil and Japan in 1981 were not successful because of the mixture of quartz grains of different origins. This is common to almost all geological samples of sediments. No two quartz grains are identical, some grains giving an intense signal while others no signal at all. They have different histories due to repeated reworking. The signal intensity cannot be reproduced once the remeasurement is made by taking sands out of the sample tube. Redistribution of the grains in a sample holder for ESR measurements where the sensitivity depends on the position may result in a different signal intensity.

The origin of the sands and the degree of mixing may be traced back by checking impurity–related centers. Thus, special precautions are required for sampling sands from geological field locations. Unless dating of single grains is made with ESR, it is difficult to date sediments at reworked areas where the degrees of sunlight bleaching are different.

9.5.2 *Al Center*

Bleaching of the signal intensity of the Al center has been studied for quartz grains from paleolithic sites in France (Yokoyama *et al.* 1985a).

Only about 10% of the total intensity of the Al center was bleached in the laboratory. The *ED*'s were obtained as a function of the magnetic field using each hf line of the Al center following a similar "*ED plateau*" procedure at different magnetic fields (Miki and Ikeya 1985). ESR ages of 430 ± 85 ka and 380 ± 80 ka obtained for quartz grains from the sites of Arays and Terra Amata, respectively, are in agreement with the faunal studies of the sites (Li *et al.* 1993).

9.5.3 Ge Center

Near–shore terrace sands and modern beach sands were studied with ESR using the light–sensitive Ge center at $g = 1.997$ (Tanaka *et al.* 1985). The signal intensity of the Ge center is significant in terrace sands. The intensity at $g = 1.997$ of γ–irradiated sands is reduced in several hours almost completely by sunlight bleaching. Sands in beach, deltaic deposits and some channel floors of rivers are considered to be exposed to sunlight. The surface sands on land terraces are also exposed to sunlight until other deposits such as volcanic ash overlie and inhibit further sunlight exposure. Using a tentative dose rate, ESR ages of $50 \sim 400$ ka for Pleistocene sands have been obtained, which are consistent with the stratigraphic estimation.

It should be noted that the signal shape of the Ge center in quartz grains becomes asymmetric after γ–irradiation. The amplitude in irradiated samples is thus overestimated with consequent underestimation of the intensity of the Ge center in nature and of the obtained *ED*. The sunlight bleaching experiment with γ–irradiated quartz grains does not indicate the case in nature. Experiments must be repeated with Pleistocene sands. A bleaching experiment with UV–light and with sunlight has also been done for the Ge, Al and Ti centers and OHC (Buhay *et al.* 1988) but here again with quartz grains exposed to an artificial dose of 1000 Gy.

9.5.4 Change in the Dose Rate: A New Sediment Dating Method

Reworking of the sediments is common in nature. Reversed ESR stratigraphy has been observed in our laboratory for wind–blown glacier sand sediments in Scotland and desert sediments in China. However, involved geologists who collected the samples consider that reworking did not occur. If this is correct, one possible explanation is the change in the radiation environment. We have already developed a general theory of defect formation based on first– and second–order natural annealing in Chapter 3. One can solve the general equation to obtain the age when the physical environment, or the parameters such as the dose rate D, lifetime τ and second–order decay constant λ were changed.

Figure 9.17 shows a theoretical calculation of defect formation in quartz grains in a granite rock and in sediments. The *weathering* of a granite released the sands at different times from $T_0, T_1, T_2, \cdots$ to the present time. The radiation environment may be changed from a high dose rate in a rock $D_R = 5$ mGy/a to a low dose rate in the sediment $D_S = 0.5$ mGy/a. Then the defect concentration or the *ED* by the conventional additive dose method would be lower for older samples, completely opposite to the stratigraphy if a sunlight bleaching mechanism is not operative. Nevertheless, one can determine the age using the new additive dose method assuming the general equation of defect formation.

Some geologists are pessimistic about sediment dating using the drastic change in the dose rate and the resultant reversed *ED* to the stratigraphy due to reworking or mixing of sands in nature. Continuous deposition from mother rocks to sediments rarely happens. However, this proposal gives a new concept to date some geologic events occurred in a material for which the annual dose rate was changed in the past.

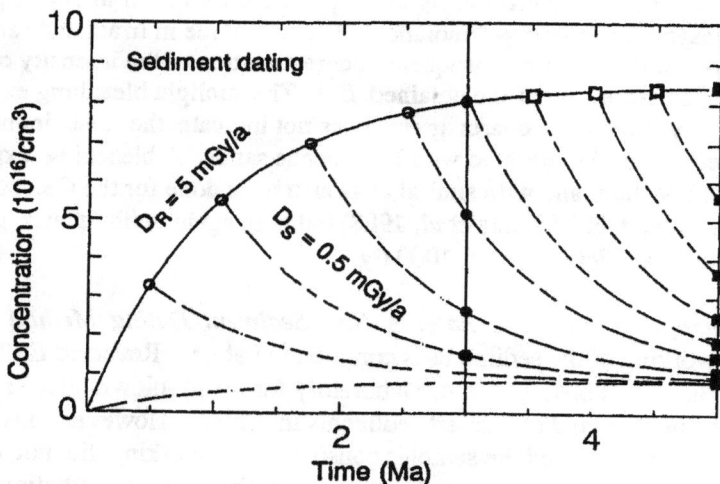

Figure 9.17 A new method of dating sediments using quartz sands from one area assuming the change in the radiation environment from a high dose rate in granite rock, $D_R = 5$ mGy/a to a low dose rate in the sandy sediments, $D_S = 0.5$ mGy/a at different times in the past. The *ED* is the reverse of the stratigraphic order. The ambient temperature must be constant.

9.6 E'_1 Center for Gigaannee (Ga) Age

9.6.1   *Age versus E_1' Intensity*

It was mentioned previously that the E'_1 center is not created by γ–ray irradiation (Garrison *et al.* 1981) except in *distorted quartz* grains in fault gauges (Ikeya *et al.* 1982). *Oxygen vacancies* are not produced by simple ionizing radiation at the normal lattice site of crystalline SiO_2 but only in the vicinity of dislocations or glass either due to the presence of oxygen vacancies near dislocations or to *interstitial oxygen trapping* (peroxy radical formation) by dislocations. The accumulated amount of the E'_1 center was considered due to α–rays or α–recoil atoms in SiO_2. If this is the case, the by intensity is increased as the geological age of quartz increases.

Use of the E'_1 center for dating old quartz of Ga age has been proposed based on the observation that the intensity is grossly related to the geologic age of quartz (Odom and Rink 1989, Grün 1989, Rink and Odom 1991). Since the content of uranium and other radioactive elements is very small in quartz grains, saturation of the defect concentration is rarely established even for old quartz. The intensity of the E'_1 center is plotted as a function of age in Figure 9.18 (a) (Toyoda *et al.* 1992). The observed oxygen vacancies including the E'_1 center are $0.2 \sim 20\%$ of the total oxygen vacancies estimated from the α–ray dose. Some old samples did not show the detectable E'_1 center and so are not included in the figure.

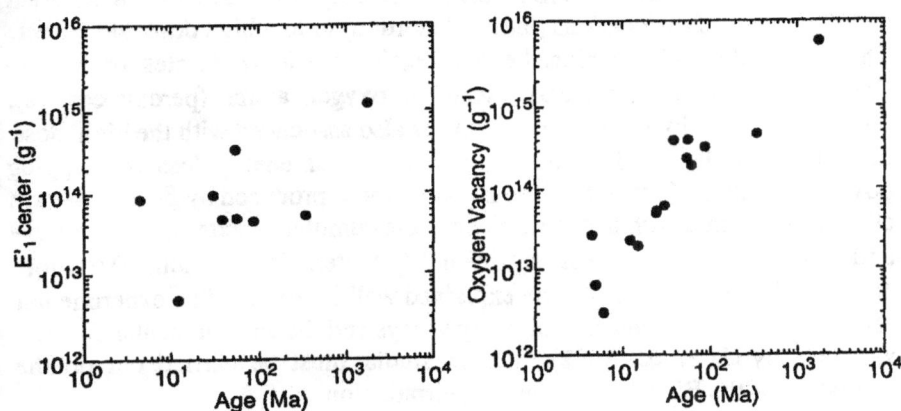

Figure 9.18   The relation of the intensity of (a) the E'_1 center and (b) oxygen vacancies with geologic ages of quartz grains in a logarithmic scale. The concentration of oxygen vacancies is obtained through formation of the E'_1 center by annealing at 300°C following weak γ–irradiation.

9.6.2 Oxygen Vacancies and Age

(a) *Evaluation of the amount of the E'$_1$ center by annealing* : If the sample is annealed at 300°C, the intensity of the E'$_1$ center increases as holes released from the Al center convert oxygen vacancies with two electrons into the E'$_1$ center (Medvedev 1979). Thus, *weak γ–irradiation* to put electrons into oxygen vacancies and the E'$_1$ center and *thermal annealing at 300°C* will give an enhanced intensity of the E'$_1$ center which may be more closely related with the content of oxygen vacancies. Toyoda *et al.* (1992) used this technique to correlate the intensity of the E'$_1$ center with the geologic ages as shown in Figure 9.18 (b).

It should be noted that a saturation level occurs in the time range corresponding to 100 Ma. This might be either due to saturation of the trapping of interstitial oxygens stabilization around dislocation or to some geological events such as volcanic activity, intrusion, faulting and meteorite impact (Smolyanskiy and Masaytis 1981). Recent studies indicate that the plateau observed around 100 Ma is due to an apparent result coming from two different mechanisms of vacancy formation by β– and γ–rays at extremely low efficiencies of $G = 10^{-4}$ and 10^{-5}.

(b) *Alpha–recoil or β– and γ–rays* ? : Quartz was also studied as an ESR dosimeter for giga–rad (= 10 MGy) high–dose detection using the intensity of the E'$_1$ center (Wieser and Regulla 1989). *Heavy irradiation* creates the E'$_1$ center at a low production efficiency presumably either due to the inherent small number of dislocations (lattice distortion) or to a high dose rate effect. The latter is due to interactions between excitons or ionized states for formation of oxygen vacancies and interstitial oxygen atoms (peroxy centers). Defect formation by α–recoil or α–rays is also associated with the high dose rate effect at the local site due to the large rate of energy loss or stopping power (dE/dx). However, if the E'$_1$ center was produced by β– and γ–rays at a low dose rate with a low efficiency, the cumulative external dose of β– and γ–rays needed to create sufficient E'$_1$ centers is enormous (MGy for 10 Ga). The geologic age can be explained well if one uses the experimental efficiency for E'$_1$ center formation by γ–rays and the environmental β– and γ–rays (Toyoda *et al.* 1992). Further studies must be carried out on the formation of the E'$_1$ center by heavy γ–irradiation.

ESR dating has been a method for the last few Ma, mostly the Quaternary era. If one uses stable defects, the time range can be expanded further. The stability of the E'$_1$ center is affected by that of hole centers such as the Al center. However, if we could convert oxygen vacancies into the E'$_1$ center, dating of Ga ages would be possible using the signal intensity indicative of oxygen vacancies stable up to 600°C (Toyoda *et al.* 1992).

9.7 Shock–induced Metamorphism – Impact Crater

9.7.1 *Meteorite Impact Crater*

Stishovite and *coesite* are high–pressure crystalline polymorphs of SiO_2 described in terms of irregular rings of distorted tetrahedral SiO_4^{4-} (Jeanloz 1988). Defects in natural stishovite (Devine and Hübner 1989) as well as in synthetic stishovite and coesite (Ogoh *et al.* in press) irradiated by γ–rays have been studied with ESR. Although the identification and assignment of signals are still tentative, observed signals and suggested models are summarized in Table 9.3. It is suggested that two different oxygen sites in

Table 9.3 Observed signals and their suggested models in coesite and stishovite.

Material	g factor (A–tensor in mT)	Suggested model	Ref.
Synthetic coesite	$g_\parallel$ =1.9993 $g_\perp$ =2.0027	axial SiO_3^{3-} [a]	1
(c.f.) Zircon	g_{zz} =2.0004 g_{xx} =2.0047 g_{yy} =1.9993 g_{zz} =2.0004 g_{xx} =2.0048 g_{yy} =1.9993	orthorhombic SiO_3^{3-} [a] orthorhombic SiO_3^{3-}	1 2
Synthetic stishovite	$g_\parallel$ =2.0023 $g_\perp$ =2.0052	E'_1 center [a] ?	1
	g =2.0095	NBOHC	1
	g =2.0081	peroxy center	1
	g_{iso} =2.0011	isotropic SiO_4^{5-} ?	1
Coesite & stishovite	g =2.002 ($A_\parallel$ =59 $A_\perp$ =46) ($A_\parallel$ =53 $A_\perp$ =50)	Interstitial $H°$ [b]	1
Natural stishovite	g = 2.0017 g = 2.0024	?	3
	g_1 =2.0026 g_2 =2.0028 g_3 =2.0031	?	3
	g = 2.002 $A_\parallel$ =59.0 $A_\perp$ =45.7 (sharp) $A_\parallel$ =52.5 $A_\perp$ =50.3 (broad)	E'_1 center [b] E' center in am–SiO_2 [b]	3 3

a) Coesite has two different oxygen sites which give two (axial and orthorhombic) E'_1–type centers, while stishovite has only one type.
b) Note that these are the same center but different models are suggested.
Reference: 1) Ogoh *et al.* in press; 2) Solntsev *et al.* 1974; 3) Devine and Hübner 1989.

coesite lead to an axial and an orthorhombic E'_1-type center, while stishovite has only one type of (octahedral) oxygen site which causes an orthorhombic E_{s_1} center. An isotropic broad line with $g = 2.0028$ is attributed to coal-like macroregions in a quartz crystal (Rao et al. 1989). Models of defects in synthetic coesite and stishovite were studied (Ogoh et al.,1996a,b; Tani et al., 2000).

9.7.2 Impact Experiment on SiO_2 and Crater Sandstone

High pressure induced by shock or impact causes a phase change and erases radiation-induced defects such (Smolyanskiy and Masaytis 1981). Reduction of the E'_1 center concentration by an impact has been studied using a plastic bullet with a speed of 5 km/s from a rail-gun at the Institute of Space and Astronautical Science, Japan. A block of SiO_2 glass was irradiated by γ-rays prior to the impact. The concentration of the E'_1 centers was reduced by about 50% for scattered SiO_2 and completely erased for SiO_2 pieces recrystallized by impact melting (Yamanaka et al., 1996). Thus, complete resetting of the ESR clock time occurs for the E'_1 center in recrystallized SiO_2. If melted SiO_2 was selected for ESR dating, a precise age could be obtained from E'_1 centers.

Coconino sandstone from the meteor crater in Arizona was treated with HF to reduce the fraction of quartz and glassy SiO_2 and the residue was measured with ESR. The ED of about 11 Gy gives a tentative ESR age of about 34 ka using a dose rate of 0.32 mGy/a, slightly younger than the TL age of 49 ka (Sutton 1985). Yamanaka discovered signals of CO_3- centers in SiO_2 in Australian impact crater presumably because of carbonaceous metorite impact or forest carbon. Griscomb investigated crators and carbonates at K-T boundary.

9.7.3 Technical Note

(a) *Sampling and preparation:* Since the Ge center and partly the Al center are sensitive to light, exposure to light should be avoided. Collect samples as a function of distance from the fault plane to determine variations. This was done for gouges from the Nojima Fault that moved in the Kobe earthquake in 1995. Fault gouge close to the fault plane at the edge of the fault zone should be used.

Quartz grains are separated using heavy liquid and a magnetic separator after washing and sieving. Removal of carbonates by etching with HCl is necessary. Selection of grain size is preferable.

(b) *ESR measurement aaand radiation assessment*:   Select an appropriate microwave power for defects of interests (see Figure 9.3). Dating centers are E'_1, Al, Ge, Ti and peroxy centers as well as OHC (O_3^- at $g = 2.012$). The NBOHC (wet OHC) is rarely used for dating as the signal overlaps that of the peroxy center. Measure at 77K for the Al and Ti centers in a range of 10 mT without the Mn^{2+} standard so that the second and third signals of the quartet due to $\cdot CH_3$ should not be taken for that of the OHC and Ge centers, respectively.

Whole fault gouge and breccia should be analyzed to determine the content of ^{238}U, ^{232}Th and K_2O. Embed TLD to obtain the D_{ex}. Grain size attenuation factors should be considered for α- and γ-rays as in Chapter 4.

9.8 Paramagnetic Centers at the Boundary of Si and SiO₂

Paramagnetic centers at the interface of Si and SiO_2 on the silicon wafer is a topic in semiconductor technology as   tabulated in Table 9.4,

Table 9.4 Paramagnetic centers in $Si_{1-x}O_x$ (H) at the inetrface of $Si-SiO_2$ of Si-wafer.

Materials	Center	g-factor	Structure	Reference
a-Si:H (amorphous silicon)	Isotropic dangling bond (D center)	$g = g_{xx} = g_{yy} = g_{zz}$ $= 2.0055$	•Si≡Si₃	1
(111) c-Si Si/SiO₂ interface	P_b center	$g_{//} = g_{zz}$ $= 2.0012,$ $g_{\perp} = g_{xx} = g_{yy}$ $= 2.0081$	Pb •Si≡Si₃	2
(100) c-Si Si/SiO₂ interface	P_{b0} center	$g_{xx} = 2.0015,$ $g_{yy} = 2.0080,$ $g_{zz} = 2.0087$	Pb0 •Si≡Si₃	3
	P_{b1} center	$g_{xx} = 2.0012,$ $g_{yy} = 2.0076,$ $g_{zz} = 2.0052$	Pb1 •Si≡Si₂O	3
a-SiO₂ (amorphous quartz)	E_1' center	$g_{xx} = 2.00179,$ $g_{yy} = 2.00053,$ $g_{zz} = 2.00030$	•Si≡O₃	4
a-SiO₂ with high OH⁻ content	NBOHC (non-bridging oxygen hole center; wet-OHC)	$g_{xx} = 2.0010,$ $g_{yy} = 2.0095,$ $g_{zz} = 2.0078$	•O–Si≡O₃	5
a-SiO₂ ; highly purified fused silica	PR (peroxy radical; dry-OHC)	$g_{xx} = 2.0014,$ $g_{yy} = 2.0074,$ $g_{zz} = 2.067$	•O–O–Si≡O₃	6

1) Pantelides, 1986; Ishii et al., 1990; 2) Caplan et al., 1979; 3) Poindexter et al., 1981; 4) Jani et al., 1983; Silsbee, 1961; 5) Stapelbroek et al., 1979; 6) Friebele et al., 1979

(After H. Furuta, 2000: *PhD thesis*, Osaka Univ.)

9.9 Summary

Both intrinsic and extrinsic (impurity-related) defects in SiO_2 and ESR dating of quartz grains are described. Signal intensities of some defects are reduced strongly by mechanical grinding of the materials. This effect is either due to local temperature rise by the release of mechanical energy or to detrapping of electrons from defect centers by the movement of dislocations. The defect formation efficiency is enhanced since dislocations and defects are introduced by deformation.

A method of sediment dating has been proposed utilizing a change in the radiation environment from a high dose rate in the source rock to a low one in the sandy sediment. Quartz is a common mineral that cannot be an object of dating. ESR can tell the age of formation and events that occurred during geologic time.

Reference

Anderson J. H., Feigl F. J. and Schlesinger M. (1974): The effect of heating on color centers in germanium-doped quartz. *J. Phys. Chem. Solids* **35**, 1425-1428.

Arends J., Dekker A. J. and Perdok W. G. (1963): Color centers in quartz produced by crushing. Phys. Status Solidi **3**, 2275-2279.

Ariyama T. (1985): Conditions of resetting the ESR clock during faulting. ESR Dating and Dosimetry (Ionics, Tokyo) 251-258.

Baker J. M. and Robinson P. T. (1983): ESR of a new defect in natural quartz: possibly O_2^- . *Solid State Commun.* **48**, 551-554.

Bar M., Kolodny Y. and Bentor Y. K. (1974): Dating faults by fission track dating of epidotes - an attempt. *Earth Planet. Sci. Lett.* **22**, 157-162.

Bauer C. L. and Gordon R. B. (1962): Structure sensitivity of F center generation by X-rays at low temperature. *Phys. Rev.* **126**, 73-78.

Bershov L. V., Martirosyan V. O., Marfunin A. S. and Spernaskii A. V. (1975): EPR and structure models for radical ions in anhydrite crystals. *Forshr. Mineral.* **52**, 591-604.

Billing M. P. (1972): *Structural Geology* (Prentice-Hall, N. J.) 606.

Brumby S. and Yoshida H. (1994): An investigation of the effect of sunlight on the ESR spectra of quartz centers:implications for dating. *Quater. Geochr.* **13**, 615-618.

Buhay W. M., Schwarcz H. P. and Gr n R. (1988): ESR dating of fault gouge: the effect of grain size. *Quat. Sci. Rev.* **7**, 515-522.

Chen Y., Feng J., Gao J., Taylor L. and Grun R (1997): Observations on the Micro texture and ESR spectra of quartz from fault gouge. *Quatern. Geol.* **16**, 487-494

Chen Y., Feng J., Gao J. and Grun R. Äi1997Aj: Investigation of the potential use of ESR signals in quartz for palaeothermometry. *Quatern. Geol.* **16**, 495-

Deer F. R. S., Howie R. A. and Zussman J. (1966): *An Introduction to the Rock- Forming Minerals* (Longman, Essex).

Devine R. A. B. and H bner K. (1989): Radiation-induced defects in dense phases of crystalline and amorphous SiO2 . *Phys. Rev.* **B40**, 7281-7283.

Ebert I. and Hennig H. P. (1974): Elektronspinrezonanz von mechanische aktiviertem Quartz. *Z. Phys. Chem. Leipzig* **255**, 812-814.

Feigl F. J. and Anderson J. H. (1970): Defects in crystalline quartz: electron paramagnetic resonance of E' vacancy centers associated with germanium impurities. *J. Phys. Chem. Solids* **31**, 575-596.

Friebele E. J., Griscom D. L. and Stapelbroek M. (1979): Fundamental defect centers in glass: the peroxy radical in irradiated high-purity fused silica. *Phys. Rev. Lett.* **42**, 1346-1349.

Fukuchi T., Imai N. and Shimokawa K. (1985): Dating of the fault movement by various ESR signal in quartz: cases of the faults in the south Fossa Magna, Japan. *ESR Dating and Dosimetry*, 211-217.

Fukuchi T., Imai N. and Shimokawa K. (1986): Dating of the fault movement using various ESR signal in quartz: the case of the western South Fossa Magna, Japan. *Earth Planet. Sci. Lett.* **78**, 121-128.

Fukuchi T. (1988): Application of ESR dating using multiple centers to fault movement: the case of the Itoigawa-Shizuoka tectonic line, a major fault in Japan. *Quat. Sci. Rev.* **7**, 509-514.

Fukuchi T. (1989a): Theoretical study on frictional heat by faulting using ESR. *Appl. Radiat. Isot.* **40**, 1181-1193.

Fukuchi T. (1989b): Increase of radiation sensitivity of ESR centers by faulting and criteria of fault dates. *Earth Planet. Sci. Lett.* **94**, 109-122.

Fukuchi T. (1991): The Itoigawa-Shizuoka tectonic line at the western edge of the south fossa magna, Japan. *Modern Geol.* **15**, 347-366.

Fukuchi T. (1992): ESR studies for absolute dating of fault movements. *J. Geol. Soc. London* **149**, 265-272.

Fukuchi T. (1993): Vacancy-associated type ESR centers observed in natural silica and their application to geology. *Appl. Radiat. Isot.* **44**, 179-184.

Fukuchi T. (1996): A mechanism of the formation of E' and peroxy centers in natural deformed quartz. *Appl. Radiat. Isot.* **47**, 1509-1521.

Fukuchi T. (2001): Assessment of fault activity by ESR dating of fault gouge; an example of the 500 m core samples drilled into the Nojima earthquake fault in Japan. *Quat. Sci. Rev* **20**,1005-1008.

Garrison E.G., Rowlett R.M., Cowan D.L. and Holroyd L.V. (1981): ESR dating of ancient flints. *Nature* **290**, 44-45.

Griffiths D. R., Seeley N J., Chandra H. and Symons M.C.R. (1983): ESR dating of heated chert. *ACT* **9**, 399-409.

Griscom D.L. (1978): Defects in amorphous insulators. *J. Non-Cryst. Solids* **31**, 241-266.

Griscom D.L. and Friebele E.J. (1981): Fundamental defect centers in glass: ^{29}Si hyperfine structure of the nonbridging oxygen hole center and the peroxy radical in a-SiO_2. *Phys. Rev.* **B 24**, 4986-4898.

Griscom D. L. (1990): Electron spin resonance investigations of defects and defect processes in amorphous silicon dioxide. *Rev. Solid State Sci.* **4**, 565-599.

Grün R. (1989): ESR dating for the early Earth. *Nature* **338**, 543-544.

Grün R. and Fenton C. (1991): An ESR study of fault gouge from Holocene fault system in Scotland.

Grün R. (1992): Remarks on ESR dating of fault movements. *J. Geol. Soc. London* **149**, 261-264.

Halliburton L.E., Perlson B.D., Weeks R.A., Weil J.A. and Wintersgill M.C. (1979): EPR study of the E'$_4$ center in quartz. *Solid State Commun.* **30**, 575- 579.

Halliburton L. E. (1989): ESR and optical characterization of point defects in quartz. *Appl. Radiat. Isot.* **40**, 859-863.

Hataya R., Tanaka K. and Miki T(1997): Studies on a new ESR signal (R signal) of fault gouges for fault dating. *Quatern. Geol.* **16**, 477-482

Hirai M. and Ikeya M. (1998): Narrowing of ESR spectra of the E' center in crushed quarts by thermal annealing. *Phys. stat. solidi* (b) **209**, 449-462.

Hutt K.B. and Martin J.J. (1983): Radiation-induced mobility of lithium and sodium in α-quartz. *J. Appl. Phys.* 54, 5030-5031.

Ikeya M., Miki T. and Tanaka K. (1982): Dating of a fault by ESR on intrafault materials. *Science* **215**, 1392-1393.

Ikeya M (1983): ESR studies of geothermal boring cores at Hachobaru power station. *Jpn. J. Apply. Phys. Letters* **22**, L763-769.

Ikeya M., Miki T., Tanaka K., Ohmura K. and Sakuramoto Y. (1983): ESR dating of faults at Rokko and Atotsugawa. *PACT* **9**, 411-419.

Ikeya M., Kohno H., Toyoda S. and Mizuta Y. (1992): Spin-spin relaxation time of E' centers in neutron-irradiated quartz (SiO2) and in fault gouge. *Jpn. J. Appl. Phys.* **31**, L1539-L1541.

Imai N., Shimokawa K and Hirota M. (1985): ESR dating of volcanic ash. *Nature* **314**, 81.

Isoya J. and Weil J. A. (1979): Uncompensated titanium(3+) center in α-quartz. *Phys. Status Solidi* **A 52**, K193-K196.

Isoya J., Tennant W.C. and Weil J.A. (1988): EPR of the TiO_4 /Li center in crystalline quartz. *J. Mag. Res.* **70**, 90-98.

Ito T. and Sawada S. (1985): Reliable criteria for selection of sampling points for ESR fault dating. *ESR Dating and Dosimetry* (Ionics, Tokyo) 229-237.

Jani M. G., Bossoli R. B. and Halliburton L. E. (1983): Further characterization of the E'$_1$ center in crystalline SiO_2. Phys. Rev. B 27, 2285-2293.

Jeanloz R. (1988): 'Easy transformation' in glasses. *Nature* **332**, 207.

Jin S. Z., Deng Z. and Huang P. H. (1993): A comparative study on optical effects of E' center in quartz grains from loess and fault. *Appl. Radiat. Isot.* **44**, 175-178.

Kanaori Y., Tanaka K. and Miyakoshi K. (1985): Further studies on the use of quartz grains from fault gouges to establish the age of faulting. *Eng. Geol.* **21**, 175-194.

Kita I., Matsuo S. and Wakita H. (1982): H_2 generation by reaction between H_2O and crushed rock: An experimental study of H_2 degassing from the active fault zone. *J. Geophys. Res.* **87**, 10789-10795.

Kosaka K. and Sawada S. (1985): Fault gouge analysis and ESR dating of the Tsurukawa fault, west of Tokyo. *ESR Dating and Dosimetry,* 259-268.

Kralik M., Kimura K. and Riedmuller G. (1987): Dating of fault gouges. *Nature,* **327**, 315-317.

Lee H. K. and Schwarcz H.P. (1993): An experimental study of shear-induced zeroing of ESR signals in quartz. *Appl. Radiat. Isot.* **44**, 191-195.

Lee H. K. and Schwarcz H.P. (1994): ESR plateau dating of fault gouge. *Quater. Geochr.* **13**, 629-634.

Lee H.K., Schwarcz H.P. (2001): ESR dating of the subsidiary faults in the Yangsan fault system, Korea. *Quat. Sci. Rev.* **20**, 999-1003.

Li D., Zhao X., Ding Z., Yang B. and.Gao H. (1993): A. study of the clock zero of sedimentary loess for ESR dating. *Appl. Radiat. Isot.* **44**, 203-206.

Lin Z., Yang M., Fan Q. and Ikeya M. (1985): ESR Dating of the age of the fault movement at San-Jiang fault zone in west China. *ESR Dating and Dosimetry* 205-210.

Lyons J. B. and Snellenburg J. (1971): Dating faults. *Geol. Soc. Am. Bull.* **82**, 1749-1752.

Mackey J.H. Jr. (1963): EPR study of impurity-related color centers in germanium- doped quartz. *J. Chem. Phys.* **39**, 74-83.

Mackey J.H. Jr., Boss J.W. and Wood D.E. (1970): EPR study of substitutional- aluminum-related hole centers in synthetic _-quartz. *J. Mag. Res.* **3**, 44-54.

McKenzie D. and Brune N. (1972): Melting on fault planes during large earthquakes. *Geophys. F. Res. Astr. Spc.* **29**, 65-68.

McMorris D.W. (1970): ESR detection of fossil alpha damage in quartz. *Nature* **226**,146.

McMorris D.W. (1971): Impurity color centers in quartz and trapped electron dating: electron spin resonance, thermoluminescence studies. *J. Geoghys. Res.* **76**, 7875-7887.

Medvedev E.M. (1979): Thermal characteristics of E' centers in natural quartz and the role of ionizing radiation in the production of defects. Geochem. Intern. 16, 71-73.

Miallier D., Sanzelle S., Falgueres C., Fain J., Pilleyre Th. and Viccent P. M. Åi1997Åj: TL and ESR of quartz from the Astrobleme of Aorounga (Sahara of Chad) *Quatern.Geol.* **16**, 265-274

Miki T. and Ikeya M. (1982): Physical basis of a fault dating with ESR. *Naturwissen.* **69**, 390-391.

Miki T. and Ikeya M. (1985): A plateau method for total dose evaluation in ESR dating with a digital data processing. *Nucl. Tracks* **10**, 913-919.

Moiseyev B. M. and Rakov L. T. (1977): Paleodosimetry properties of E' centers in quartz. *Dokladi Akad. Nauk SSSR* **223**, 679-683.

Moiseyev B. M. (1980): Paleodosimetric method for determining the age of deposits of radioactive elements. *Dokladi Akad. Nauk SSSR.* **254**, 1227-1229.

Nuttall R. H. D. and Weil J. A. (1980): Two hydrogenic trapped-hole species in a quartz. *Solid State Commun.* 33, 99-102.

Nuttall R.H.D. and Weil J.A. (1981a): The magnetic properties of the oxygen-hole aluminum centers in crystalline SiO_2: I. [AlO_4]. *Can. J. Phys.* **59**, 1696- 1708.

Nuttall R. H. D. and Weil J.A. (1981b): The magnetic properties of the oxygen-hole aluminum centers in crystalline SiO_2: II. [AlO_4/H^+]$^+$ and [AlO_4/Li^+]$^+$. Can. J. Phys. 59, 1709-1718.

O'Brien M.C.M. (1955): The structure of the color centers in smoky quartz. Proc. Roy. Soc. A.**231**, 404-414.

Odom A.L. and Rink W.J. (1988): Natural accumulation of Schottky-Frenkel defects : implications for a quartz geochronometer. *Geology* 17, 55-58.

Ogoh K, Yamanaka C., Ikeya M. and Ito E. (1994): ESR studies of radiation-induced defects in high pressure phase SiO_2 for dating of meteorite impact craters. *Nucl. Instr. Meth.* **B91**, 331-333.

Ogoh K., Yamanaka C., Ikeya M. and Ito E. (1995): Hyperfine interaction of electron at oxygen vacancy with nearest and next-nearest 29Si in high-pressure phase SiO2: Stishovite. *J. Phys.Soc. Jpn* **64**, 4109-4112.

Ogoh K., Takaki S., Yamanaka C., Ikeya M. and Ito E. (1996a): Thermoluminescence and electron spin resonance of atomic hydrogens in coesite and stishovite, high pressure phase of SiO_2. *J. Phys. Soc. Jpn.* **65**, 884-847.

Ogoh K., Yamanaka C., Ikeya M. and Ito E. (1996b): Two center model for radiation-induced aluminum hole center in stishovite. *J. Phys. Chem. Solids* **57**, 85-88.

Okada M., Rinnenberg J., Weil J. A. and Wright P. M. (1971): EPR of Ti^{3+} centers in a quartz. *Chem. Phys. Letters* **11**, 275-276.

Porat N. and Schwarcz H.P. (1991): Use of signal subtravtion methods in ESR dating of burnt flint. *Nucl. Tracks Radiat. Meas.* **18**, 203-212.

Rao P.S., Weil J. and Williams J.A.S. (1989): EPR investigation of carbonaceous natural quartz single crystal. *Canad. Mineralogist* **27**, 219-224.

Rink W.J. and Odom A.L. (1991): Natural α-recoil particle radiation and ionizing radiation sensitivities in quartz detected with EPR: implication for geochronometry. *Nucl. Tracks* **18**, 163-173.

Rink WJ. (1997) Electron spin resonance (ESR) dating ansd applications in Quaternary science and aarchaeometry. *Radiat. Meas.* **27**, 975-1022.

Rinnenberg H. and Weil J.A. (1972): EPR studies of Ti^{3+}-H^+ centers in X-irradiated _-quartz. *J. Chem. Phys.* **56**, 2019-2028.

Rudra J.K. and Fowler W.B. (1987): Oxygen vacancy and the E'_1 center in crystalline $SiO2$. *Phys. Rev.* **B 35**, 8223-8230.

Rudra J.K., Fowler W.B. and Feigl F.J. (1985): Model for the E'_2 center in a-quartz. *Phys. Rev. Lett.* **55**, 2614-2617.

Sato T., Suito K. and Ichikawa Y. (1985): Characteristics of ESR and TL signals on quartz from fault regions. *ESR Dating and Dosimetry*, 267-275.

Schnadt R. and Schneider J. (1970): The electronic structure of the trapped-hole center in smoky quartz. *Phys. Condens. Mater.* **11**, 19-42.

Shimokawa K.and Imai N. (1987): Simultaneous determination of alteration and eruption ages of volcanic rocks by ESR. *Geochim.Cosmochim. Acta* **51**, 115-119.

Silsbee R.H. (1961): Electron spin resonance in neutron-irradiated quartz. *J. Appl. Phys.* **32**, 1459-1462.

Skinner A. R. and Rudolph M. N. (1996): The use of the E' signal in flint for ESR dating. *Appl. Radiat. Isot.* **47**, 1399-1404.

Smolyanskiy P. L. and Masaytis V. L. (1981): Reconstruction of paleotemperature anomalies of old rocks from radiation-induced defects in quartz. *Doklady Akad. Nauk SSSR* **248**, 184-186.

Solntsev V. P., Shcherbakova M. Y. and Dvornikov E. V. (1974): Radicals SiO2-, SiO_3^{3-} and SiO_4^{5-} in $ZrSiO_4$ structure according to EPR data. *Zh. Strukt. Chim.* **15**, 217-222.

Stapelbroek M., Griscom DL., Friebele E.J. and Sigel G.H. Jr. (1979): Oxygen-associated trapped-hole centers in high-purity fused silicas. *J. Non-crystal. Solids.* **32**, 313-326.

Sutton S. R. (1985): Thermoluminescence measurements on shock-metamorphosed sandstone and dolomite from meteor crater, Arizona: 2. Thermoluminescence age of meteor crater. *J. Geophys. Res.* **90**, 3690-3700.

Tanaka K. and Shidahara T. (1985): Fracturing, crushing and grinding effects on ESR signal of quartz. *ESR Dating and Dosimetry*, 239-247.

Tanaka T., Sawada S. and Ito T. (1985): ESR Dating of late Pleistocene nearshore and terrace sands, in southern Kanto Japan. *ESR Dating and Dosimetry*, 275-280.

Tani A., Kohno H., Yamanaka C. and Ikeya M. (1996): ESR dating of geological fault with a new isochrone method granite fractured on the earthquake in 1995. *Appl. Radiat. Isot.* **47**, 1423-1426.

Tani A. yamanaka C. and Ikeya M. (2000): ESR study of a new electron center in synthetic stishovite, a high oressure polymorph of silica. *Appl. Magn. Reson.* **18**, 559-564.

Toyoda S. and Ikeya M. (1991): Thermal stabilities of paramagnetic defects and impurity centers in quartz: basis for ESR dating of thermal history. *Geochem. J.* **25**, 437-445.

Toyoda S. and Ikeya M. (1994): Formation of oxygen vacacies in quartz and its application to dating. *Quater. Geochr.* **13**, 607-609.

Toyoda S. and Ikeya M. (1994): ESR dating of quartz with stable component of impurity centers. *Quater. Geochr.* 13, 625-628.

Toyoda S., Ikeya M., Morikawa J. and Nagatomo T. (1992): Enhancement of oxygen vacancies in quartz by natural external α and γ-ray dose: a possible ESR geochronometer of Ma-Ga range. *Geochem. J.* 26, 111-115.

Toyoda S., Kohno H. and Ikeya M. (1993): Distorted E$_1$ centers in crystalline quartz: an application to ESR dating of fault movements. *Appl. Radiat. Isot.* 44, 215- 220.

Toyoda S.and Ikeya M. (1994) : Formation of oxygen vacancies in quartz and its application to dating. *Quatern. Geochr.* 13, 607-609

Toyoda S. and Ikeya M.(1994) : ESR dating of quartz with stable component of impurity centers. *Quatern. Geochr.* 13, 625-628

Toyoda S., Rink J.W., Schwarcz H.P. and Ikeya M. (1996): Formation of E1' precursors in quartz: applications to dosimetry and dating. *Appl. Radiat. Isot.* 47, 1393-1398.

Toyoda S. and Schwarcz H.P. (1997) : The hazard of the counterfeit E' signal in quartz to the ESR dating of fault movements. *Quatern. Geol.* 16, 483-486.

Toyoda S. and Schwarcz H.P. (1997) : Counterfeit E' signal in quartz. *Radiat. Meas.* 27, 59-66.

Toyoda S. and Schwarcz H. P. (1996): The spatial distribution of ESR signals in fault gouge revealed by abrading technique. *Appl. Radiat. Isot.* 47, 1409-1413.

Toyoda S. and Hattori W. (2000): Formation and decay of the E$_1$' centerand of its precusor. *Appl. Radiat. Isot.* 52, 1351-1356.

Toyoda S., Voinchet P., Falgueres C., Dolo J. M. and Laurent M. (2000): Bleaching of ESR signals by the sunlight: a laboratory experiment for establishing the ESR dating of sediments. *Appl. Radiat. Isot.* 52 No.5, 1357-1362.

Toyokura I., Sakuramoto Y., Ohmura K., Iwasaki E. and Ishiguchi M. (1985): Determination of the age of fault movement at the Rokko . *ESR Dating and Dosimetry*, 219-228.

Ulusoy Ü. and Apaydin F. (1996): ESR studies and ESR dating of quartz collected from Kapadokya, Turkey. *Appl. Radiat. Isot.* 47, 1405-1407.

Ulusoy U.(2000): ESR dating of a quartz single crystal from the Menderes Massif in Turkey. *Appl. Radiat. Isot.* 52, 1363-1370.

Van Moort J. C. and Russell D.W. (1987): Electron spin resonance of auriferous and barren quatz at Beaconfield, northern Tasmania. *J. Geoch. Explor.* 27, 227-237.

Walther R. and Zilles D. (1994): ESR studies on bleached sedimentary quartz. *Quanter, Geochr.* 13, 611-614.

Weeks R.A. (1956): Paramagnetic resonance of lattice defects in irradiated quartz. *J. Appl. Phys.* 27, 1376-1381.

Weeks R.A. (1963): Paramagnetic spectra of E'$_2$ center in crystalline quartz. *Phys. Rev.* 130, 570-57.

Weil J.A. (1971): Germanium-hydrogen-lithium center in a quartz. *J. Chem. Phys.* 55, 4685-4698.

Weil J.A. (1984): A review of electron spin spectroscopy and its applications to the study of paramagnetic defects in crystalline quartz. Phys. Chem. Minerals 10, 149-165.

Wieser H.D. and Regulla D.F. (1989): ESR dosimetry in the "Giga-rad" range. *Appl. Rad. Isot.* 40, 911-913.

Wieser H. D., Goekdu Y., Regulla D.F. and Waibel A. (1991): Unexpected superlinear dose response of the E' centers in fused silica. *Nucl. Tracks* 18 175-178.

Wright P., Weil J. A., Buch T. and Anderson J. H. (1963): Titanium color centers in rose quartz. *Nature* 197, 246-248.

Yamanka C., Kohno H. and Ikeya M. (1996): Pulsed ESR measurements of oxygeb deficient type centers in various quartz. *Appl. Radiat. Isot.* 47, 1573-1577.

Yokoyama Y., Falgueres C. and Quaegebeur J.P. (1985a): ESR dating of quartz from Quaternary sediments: first attempt. *Nucl. Tracks* 10, 921-928.

Yokoyama Y., Falgueres C. and Quaegebeur J.P. (1985b): ESR dating of sediment baked by lava flows: comparison of paleodoses for Al and Ti centers. *ESR Dating and Dosimetry*, 197-204.

Yu-Guang Ye, Shao-Bo Diao, Jie He and Jun-Cheng Gao (1996): Abnormal resoponse to dose of E'center of in coastal eolian sand. *Appl. Radiat. Isot.* 47, 1457-1458.

Chapter 10

Silica and Silicates

– Geotherm and Volcanism –

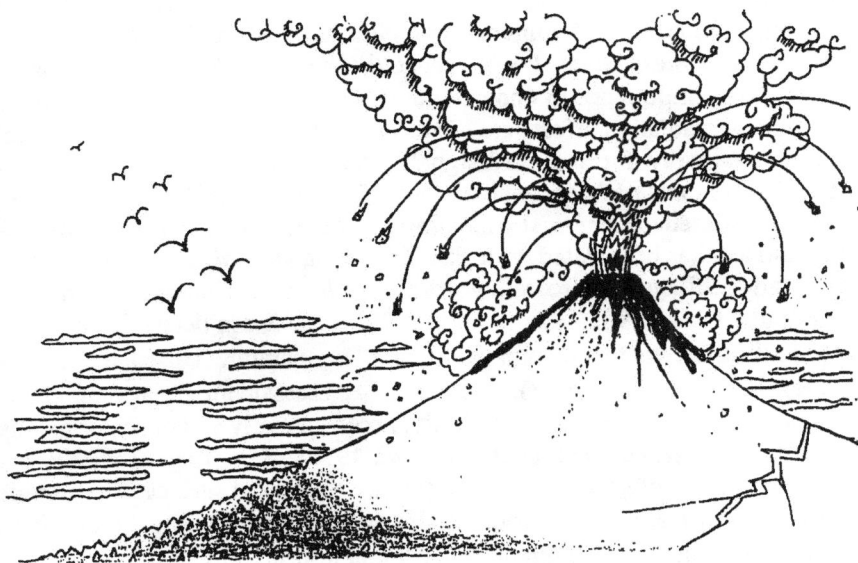

The history of geothermal and volcanic activity and of rock alteration can be assessed using ESR signals of defects and paramagnetic ions in silica and secondary minerals ubiquitous in a geothermal area. ESR dating of volcanic ash, tephra and metamorphic rocks is described with ESR studies of Fe^{3+} in silicate minerals. ESR spectra of clays and magnetic impurities are discussed.

10.1   Introduction

The possible use of geothermal heat is examined throughout countries having volcanic features as an energy source alongside nuclear energy. The temperature and the cooling rate of a magma reservoir are of concern in the basic study of geothermal fields. Physical, chemical and mineralogical studies on geothermal alteration of borehole cores have been done extensively, together with the measurement of geothermal underground temperatures. Geothermal studies by thermoluminescence (TL) (Johnson 1966, Takashima 1985, Berger 1991), fission track (FT) (Calk and Naeser 1973) and K–Ar methods (Berger and York 1981) have been done to determine the age of rock alteration or cooling. Crystallization of a mineral or its cooling marks the start of the geochronological clock time as described in previous chapters.

Silica (SiO_2) and silicate minerals such as zircon $(ZrSiO_4)$, feldspar and clay minerals are abundant in volcanic and geothermal areas due to rock alteration, sintering and mineral precipitation from hydrothermal fluids. *When and how these minerals were crystallized and cooled* are major issues because of the importance in understanding volcanic and geothermal history.

Natural radiation effects are annealed out by a temperature rise, which erases geologically formed ESR signals, thereby resetting the ESR age to zero. Volcanic materials can be dated with ESR using the time zero setting mechanism based on thermal annealing (Ikeya 1983). Quartz in volcanic rocks and ash has been dated with ESR (Shimokawa *et al.* 1984, Imai *et al.* 1985) and ESR ages of zircon in volcanic ash have been compared with FT ages (Taguchi *et al.* 1985, Kasuya *et al.* 1992). ESR can determine the time since cooling down to a certain geothermal temperature called a "closure temperature" (Dodson 1973). Hence, ESR assessment of a *geothermal heat source* and *cooling history* is possible both spatially and chronologically by dating geothermal minerals (Ogoh *et al.* 1993). Sampling and ESR assessment of traces of geothermal alteration at the ground surface can provide important information on geothermal areas without boring. Additional borehole core samples give three–dimensional information.

Intense signals associated with paramagnetic Fe^{3+} or *ferromagnetic resonance* mostly interfere with the ESR signals associated with radiation–induced defects in silicates (Poole *et al.* 1978). The state of paramagnetic impurity ions such as Fe^{3+} and Mn^{2+}, their valency and distribution in the host lattice have been studied (Stucki and Banwart 1980, Perry 1990). Studies of these ions and radiation–induced defects with ESR will clarify the thermal history or environmental changes.

In this chapter, ESR studies on silica and silicate minerals, mostly volcanic and metamorphic minerals as well as on clay minerals are described together with paramagnetic and ferromagnetic resonance.

10.2   Growth of Defect Concentrations in Geothermal Areas

10.2.1   *Saturation of Defect Concentration*

The stability of a lattice defect at the present *geothermal temperature*, T_g, has been calculated theoretically in Chapter 3. Theoretical analysis of the growth curve of defects has already been described (Sections 3.5.2 and 3.5.3). Equations (3.11) and (3.33) for first–order decay and growth of the defect, respectively, can be used and thus the saturation value is proportional to the lifetime of the defect and the dose rate. The behavior of defect formation and the lifetime τ are shown in Figure 10.1 for several geothermal temperatures. If the ESR age is less than the lifetime τ_g at a geothermal temperature T_g, it is related either to the time after cooling down from the closure temperature (T_c) to the present T_g or to that after sintering (or mineralization) in geothermal areas. If several different defects with different life–times are used, further information on "*when the mineral was at what*

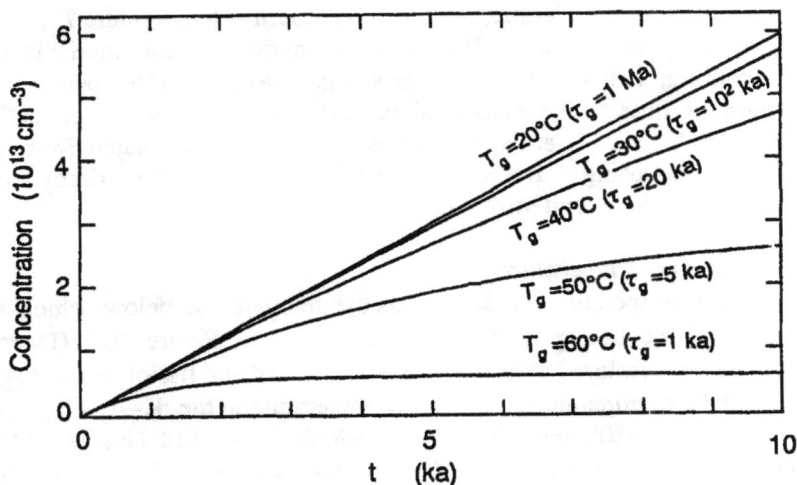

Figure 10.1 Formation of defects in a mineral at various geothermal temperatures. An annual dose rate, $D = 1$ mGy/a and the lifetime of defects $\tau = (1/\nu_o) \exp(-E/k_B T_g)$ for $CaCO_3$ in Figure 3.7 are used for calculation.

temperature" becomes available. Studies using defects in geothermally altered or sintered minerals can provide detailed information on the cooling history.

Second–order decay has to be considered for defects in silica (Toyoda and Ikeya 1991b). In this case, Eq. (3.22) and Eq. (3.40) can be used and one may consider that the lifetime of defects depends not only on the temperature but also on the defect concentration ($\tau = 1/\lambda n$).

10.2.2 *Slow Cooling to the Present Temperature T_g*

A calculation of defect formation assuming a gradual decrease of the temperature to the present temperature T_g, is done using Eq. (3.28) or Eq. (3.33). An equilibrium state arises for slow cooling. For first–order decay, the defect concentration is written as

$$n_{sat} = aD\tau = (1/\nu_0)aDe^{E/k_B T} , \tag{10.1}$$

and therefore,

$$ED = (1/\nu_0)De^{E/k_B T} , \tag{10.2}$$

where the parameters are explained in Sections 3.4 and 3.5 of Chapter 3. If the ESR age is equal to the lifetime at the present temperature, no information on the age or on past temperature variation can be deduced. If the ESR age is younger than the lifetime, τ_g, at the geothermal temperature T_g, for a defect, an abrupt temperature fall is a possible model. Another model is that of slow cooling, either a linear or an exponential decrease of temperature as a function of time. The numerical calculation is possible if annealing parameters τ and/or λ at various temperatures can be estimated from the Arrhenius plot using $1/\tau = \nu_0 e^{-E/k_B T}$ for first–order decay or $\lambda = \lambda_0 e^{-E/k_B T}$ for second–order decay.

10.2.3 *Closure Temperature*

Closure temperature is defined as the temperature below which the clock time of the dating method starts as shown in Figure 10.2 (Dodson 1973). The temperature below which Ar is confined and fission tracks begin to accumulate in minerals is the closure temperature in the K–Ar (Harrison and McDougall 1980) and FT dating methods (Calk and Naeser 1973), respectively. The *closure temperature in ESR dating* is the temperature below which minerals begin to accumulate the ESR signal. The concept of the closure temperature is too naive as the lifetime is temperature dependent. It depends heavily on the cooling rate. Different defects in the same material

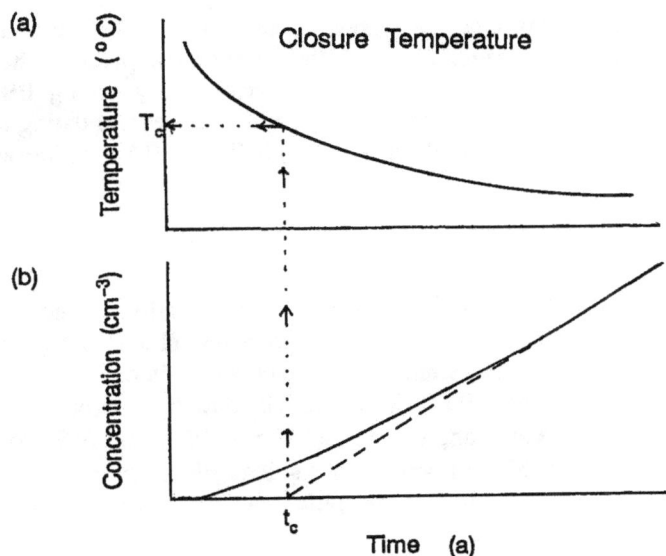

Figure 10.2   Definition of closure temperature. (a) Cooling of geothermal temperature and (b) growth of the defect concentration as a function of time. The extrapolation of the linear part of the growth curve to zero ordinate gives an apparent age (t_c) and a closure temperature (T_c) (Dodson 1973).

Table 10.1   Closure temperatures, T_c (°C), of minerals with several dating methods.

Mineral [a]	K–Ar	Rb–Sr	U–Pb	FT
Muscovite	350	500		
Biotite	300	300	(all ± 50°C)	
K–feldspar	150	350		
Zircon			700 ± 50	175 ± 25
Apatite				120 ± 20
Quartz with ESR [b]				
Cooling rate (°C/Ma)	10	100	1000	
b_c (/°C·Ma) [c]	10^{-4}	10^{-3}	10^{-2}	
E_1' center	91	123	162	
Al center	78	110	149	
Ti center	31	55	82	

[a] : These data are tabulated by Shibata (1991).
[b] : Toyoda and Ikeya (1991b).
[c] : The reciprocal of absolute temperature is linear to the time according to the theory by Dodson (1973), i.e., $1/T - 1/T_c = b_c(t - t_c)$.

have also different closure temperatures. Thus, a study of closure temperature considering the cooling history (cooling rate) should precede the dating
effort. Failure to do this may lead to an erroneously young ESR age.
Closure temperatures of some minerals determined by several dating methods
are summarized according to Shibata (1991) in Table 10.1 together with the
ESR result.

10.3 Silica (SiO_2)

The crystal structure of silica (SiO_2) and its radiation–induced defects
have already been described in Chapter 9. However, silica from *geothermal
samples* often shows signals similar to those observed in calcite and arago–
nite, i.e., CO_2^-, SO_2^- and SO_3^-. This might be due to carbonate inclusions
in the sample. In this section, several pioneering trials on ESR dating of
crystalline silica (*quartz*) and amorphous silica (*glass*) in volcanic rocks,
geothermal borehole cores and metamorphic rocks are described.

10.3.1 *Volcanic Rocks*

(a) *Sediment baked by lava flow* :   Quartz grains extracted from sediments
baked by lava flow have been studied with ESR (Yokoyama *et al.* 1985,
Shimokawa *et al.* 1988, Falgueres *et al.* 1991). The signals of Al and Ti
centers have been used to determine the age. The *ED*'s obtained for these
two centers are in good agreement. On the other hand, no Ge center signal
was detected in natural samples, though it appeared after γ–irradiation. One
possibility is that the Ge center with a low annealing temperature has been
annealed out, but the Al and Ti centers with high annealing temperatures
have not completely been annealed by the lava flow; the clock for Al and Ti
centers started much earlier than that of the Ge center. If all defects had
been annealed out by the lava flow, the temperature of the area might have
been sufficiently high for a long time enough to prevent the build–up of the
relatively unstable Ge center. It is common that ESR ages of minerals in
geothermal areas are younger than those expected from other dating methods.

(b) *Tephra* :   ESR signals at $g = 2.011$ and $g = 1.997$ identified as the
oxygen hole center (OHC or O_3^-) and the Ge center, respectively, were
observed at room temperature and additional signals of the Al center ($g = 2.061$) at 77K in quartz grains hand–picked from welded tuff (Shimokawa
and Imai 1987). The enhancement of the signal intensity of these centers by
γ–irradiation gives the *ED* and the ESR age. Table 10.2 summarizes the

obtained ESR ages compared with the TL and FT ages. A fairly good agreement was obtained between the ESR ages using the Al center and the TL ages which have been suggested to indicate the *age of alteration* (Taka-shima 1985). The ESR ages obtained using the Ge center and OHC are generally older than the alteration age using the Al center. ESR ages close to the FT ages are considered to indicate the age of *volcanic eruption*.

Quaternary tephra and volcanic ash have been dated using the ESR signals of the Al, Ti and E'_1 centers (Imai *et al.* 1985, 1992, Imai and Shimokawa 1988, 1989, Toyoda and Ikeya 1991a). The signal intensity of the Ti center is initially enhanced by isochronal annealing whereas that of the Al center is decreased. The isochronal annealing experiment indicates that the lifetime of the Al center is of the order of 10^2 ka at room temperature, while that of the Ti center is of the order of Ma based on first–order kinetics.

(c)   *Volcanic breccia* :   Quartz grains were separated from ignimbrite at the Rototi Breccia in the Taupo geothermal zone, New Zealand. The *ED* was obtained for the Al and Ti centers by an additive dose method (Buhay *et al.* 1992) and the ESR ages were calculated from the *ED* and the contents of

Table 10.2   ESR, TL and FT ages of quartz in some volcanic materials.

Sample		U (ppm)	Th (ppm)	K$_2$O (%)	D (mGy/a)	ED (Gy)	ESR age (Ma)	TL–age [a] (Ma)	FT–age [b] (Ma)
Welded tuff [c]									
TM–6	Ge					3580	1.40		
	OHC	1.3	4.0	2.13	2.56	3990	1.56	0.38	2.0
	Al					1110	0.43		
TK–32	Ge					3280	1.46		
	OHC	1.3	4.5	1.73	2.45	2990	1.33	0.26	
	Al					960	0.44		
TM–1	Al	1.2	4.0	1.48	1.96	1830	0.94	1.04	1.1
TM–4	Al	1.4	4.7	1.48	2.07	2820	1.36	1.86	1.7
TK–24	Al	1.9	6.3	2.39	3.14	240	0.077	0.11	
TK–48A	Al	0.5	1.8	1.77	1.85	110	0.059	0.12	
Volcanic ash [d]									
Yamada Ly.	OHC					5100	0.61		
	E'_1	9.6	6.1	2.54	8.3	7800	0.94		
	Ge					3400	0.47		

Reference: a) Takashima 1985; b) Tamanyu and Suto 1978; c) Shimokawa and Imai 1987; d) Toyoda and Ikeya 1989.

[238]U $(1.7 \sim 2.5$ ppm), [232]Th $(7.9 \sim 13$ ppm) and K $(2.9 \sim 3.9\%)$ by taking α- and β-attenuation into account using the α-ray efficiency, $k = 0.07$ in quartz. The external β- and γ-dose rates from the sediments were $D_\beta = 1.4 \sim 2.7$ mGy/a and $D_\gamma = 1.5 \sim 3.6$ mGy/a. The combined dose rate for seven samples ranges from 3.75 ± 0.61 to 5.61 ± 0.26 mGy/a. Average ESR ages for the Al and Ti centers are 48.6 ± 8.8 and 41.8 ± 6.3 ka, respectively, which are in consistent with ^{14}C-ages of > 41.6 ka, but are younger than an age of 71 ± 6 ka based on [238]U-series disequilibrium. Presumably, the geothermal temperature might have been high and the ESR age indicates the time after cooling down to T_c.

10.3.2 Geothermal Silica

(a) *Borehole Core* : Borehole core specimens from the Hachobaru Geo-thermal Power Station close to the active Aso volcano, Japan and from Broadlands, New Zealand have been studied (Ikeya 1983, Ikeya et al. 1986). Samples were obtained from a core in which the temperature versus depth profile is known. Figure 10.3 shows ESR spectra of a geothermally altered white-gray core sample obtained from a depth of 230 m (75°C) at the Hachobaru Station before and after γ-irradiation. Sharp signals are observed

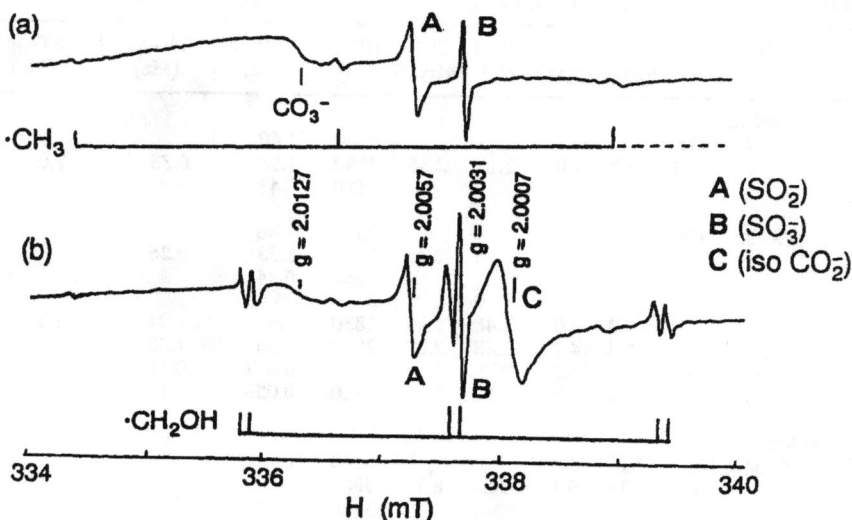

Figure 10.3 ESR spectra of a geothermal sample from the Hachobaru Geothermal Power Station (a) before and (b) after γ-irradiation.

at $g = 2.0057$ and $g = 2.0031$ with respective linewidths of 0.07 and 0.04 mT before γ–irradiation.   These signals are tentatively attributed to SO$_2^-$ and SO$_3^-$, respectively, in carbonate inclusions or in the geothermal silica matrix.   The overall spectrum has a broad band with a width of 70 mT around $g = 2.0$ indicating the presence of iron impurities.

After γ–irradiation, a signal around $g = 2.0007$ with a linewidth of 0.2 mT appeared in addition to a sharp triplet–doublet signal.   From the g factor, the former was first identified as the E$'_1$ center in silica (Ikeya 1983), but may be associated with the signal due to freely rotating CO$_2^-$ ($g = 2.0007$) in SiO$_2$ from its characteristics of microwave power saturation. The linewidth of 0.2 mT is larger and its thermal stability (200°C) is much lower than those of the E$'_1$ center (400°C). The triplet with hf splitting of 1.7 mT and further doublet splitting of 0.1 mT indicates the coupling with two equi-valent α protons and with an additional proton. These splitting parameters led to the identification of *methanol radicals* ($\cdot$CH$_2$OH) (Ikeya 1983, Livingston and Zeldes 1966).

Figure 10.4 shows (a) the *ED* versus depth profile of the borehole core samples from the Hachobaru Power Station and (b) the temperature–depth profile. The ESR signal at $g = 2.0057$ (SO$_2^-$?) and an assumed dose rate of 5 mGy/a were use to obtain the age (Ikeya 1983). The approximate estimation of lifetimes τ using the temperature–depth profile and Eq. (3.18) for two different types of defects with annealing temperatures of $T_a = 360°C$ and 400°C is also shown by dashed curves in (a). The *ED* decreases from about 400 Gy (80 ka) down to 300 m to nearly zero at a depth of about 500 m (200°C). On the other hand, the *ED* for the signal at $g = 2.0007$ (freely rotating CO$_2^-$) is zero throughout the core, suggesting that the temperature remains too high to allow accumulation of the defect.

No ESR signal of defects is detected at a depth of about 570 m indicat-ed by site 8 in the figure, where the temperature is low and thus the lifetimes of defects are expected to be long enough for the defects to survive. This indicates that the temperature has dropped quite recently by some event such as cold water inflow into fissures. The temperature drop at site 8 is actually an artifact of drilling, while the small dips in temperature at sites 2 and 5 seem to have some correlation with the increase in the *ED*. These would be due to the equilibrium between formation and fading of the defect at a par-ticular temperature; variation in the temperature would have lasted for a long time, longer than the lifetime at site 2 (20 ka) and site 5 (15 ka).

The volcanic rocks in this region were formed a few tens of thousands of years ago.   The age of 80 ka (site 1) would be the age of volcanic rock

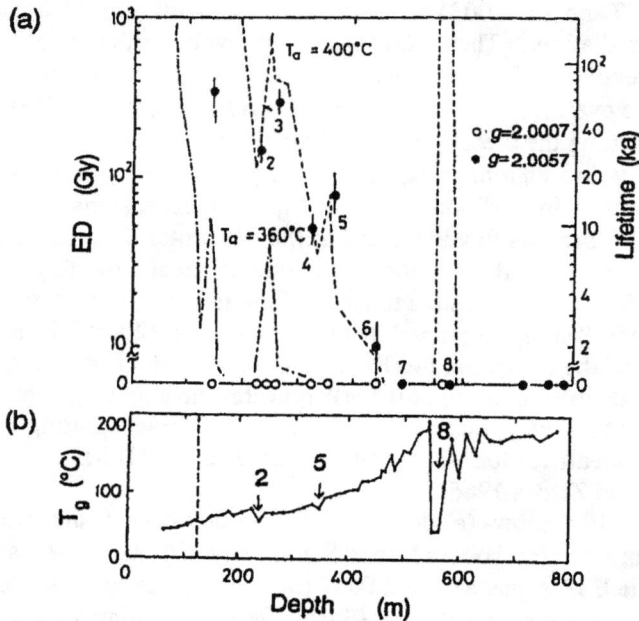

Figure 10.4 (a) The *ED* for the signal at *g* = 2.0057 versus depth for core samples at the Hachobaru geothermal power station and (b) the temperature–depth profile. The theoretical lifetimes of two different defects with T_a =360°C and 400°C are shown by dashed curves.

formation because it is younger than the lifetime of defects. Other ages indicate the lifetimes at the geothermal temperature except for site 8.

(b) *Porous silica* : Our preliminary work on porous silica from *dacite lava* from Beppu, Kyushu, Japan shows a clear ESR spectrum as shown in Figure 10.5. The observed signals are those typical of aragonitic corals and shells i.e., signals A (SO_2^-), B (SO_3^-), C (freely rotating CO_2^-) and D (g_{yy} of orthorhombic CO_2^-). Dating is possible using signal C at *g* = 2.0007. Whether CO_2^-, SO_3^- and SO_2^- are in the carbonate or fluid inclusion of porous silica or embedded in the silica matrix is an issue to be investigated. If the carbonate inclusion is responsible for these signals, Mn^{2+} signals should be observed because most carbonates and gypsum in geothermal fields contain a large amount of Mn^{2+}. Additional triplet–doublet signals due to *methanol radicals* are also observed.

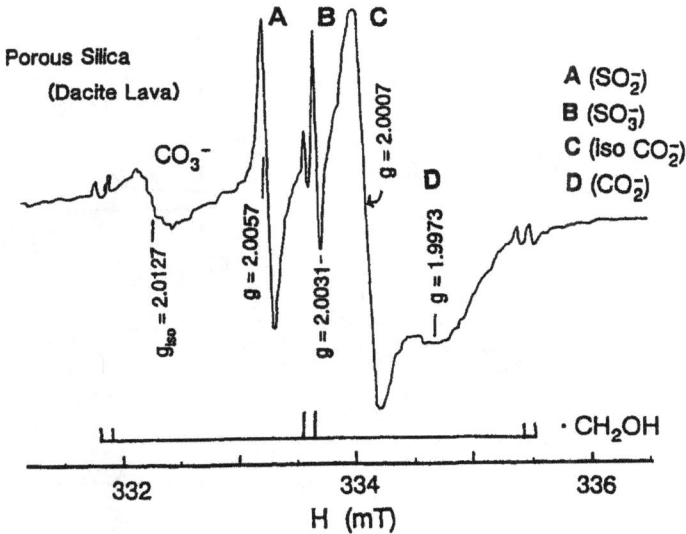

Figure 10.5 ESR spectrum of porous silica from dacite lava. The spectrum is almost the same as that of aragonitic corals. The doublet–triplet is ascribed to the methanol radical [·CH$_2$OH].

10.3.3 Hornfels and Igneous Metamorphism : Intrusive Rock

(a) *Heat diffusion* : Metamorphic rocks are those which have been changed by heat, pressure and permeation of other substances. Simple metamorphic rocks are formed directly from alterations of sedimentary rocks, sandstones and mudstones. *Hornfels* are clays or shales metamorphosed through the action of heat from nearby igneous rocks. Figure 10.6 shows the result of model heat analysis on *metamorphism by dike intrusion* of igneous rock (magma) with a width of 60 m. A one–dimensional heat diffusion equation was solved assuming an initial temperature of 1200°C at the intrusive igneous rock. Temperature profiles at several ages obtained as a function of distance from the boundary between the igneous rock and hornfels are shown in (a) and temperature variation at several sites as a function of time in (b). The accumulated dose (*TD*) is shown as a function of time for two types of defects with different stability in (c) and (d) (Ikeya and Toyoda 1991).

One can evaluate the age when the temperature of the hornfels reached the closure temperature. Thus, additional information on the thermal history is obtained as already deduced from mineralogical, FT (Naeser *et al.* 1970,

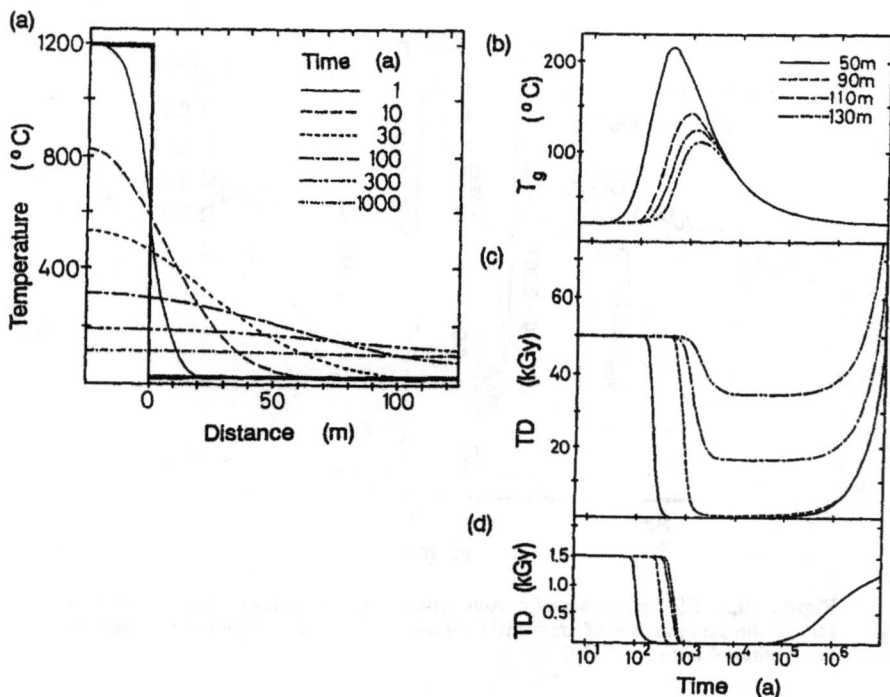

Figure 10.6 A simplified solution of temperature rise using a model of one–dimensional *heat diffusion* in mudstone by intrusion of igneous lava of 1200°C with a width of 60 m using a heat diffusion constant of 1.5×10^{-6} cm^2/s for mudstone. (a) and (b) Temperature variation as a function of distance from the intrusive rock and of time, respectively. (c) and (d) The *TD* as a function of time for two different types of defects with different thermal stability.

Calk and Naeser 1973) and TL studies on contact metamorphic rocks (Johnson 1966, Takashima 1985).

(b) *ESR spectra* : The size of quartz grains in hornfels at Mino National Park, Japan is significantly large, suggesting the growth of SiO_2 microcrystals by metamorphism. ESR spectra at room temperature and 77K of hornfels are shown in Figure 10.7. The signals of an E'_1 center ($g = 2.0010$) and a quartet due to *methyl radicals* ($\cdot CH_3$) is observed in addition to an unidentified signal at $g = 2.0042$. The quartet observed at room temperature is no longer observed at 77K. Instead, a prominent signal at $g = 2.0042$ and hf lines of ^{27}Al ($I = 5/2$, 100%) associated with the Al center are observed.

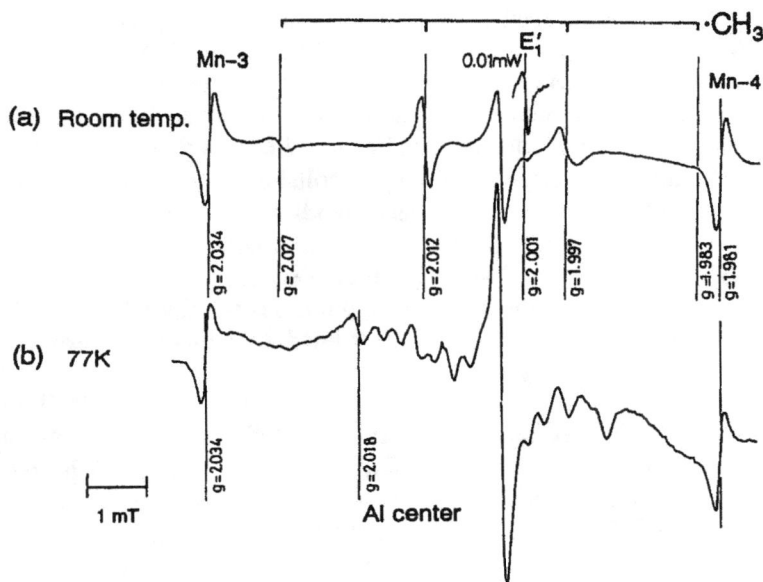

Figure 10.7 ESR spectra of metamorphic rock near the igneous intrusive rock measured at (a) room temperature and (b) 77K. Note that the quartet is not observed at 77K. Instead, the central signal (g = 2.0027) is dominant and hf lines of ^{27}Al are observed.

The ESR intensities of the E$'_1$ center and methyl radicals increase as a function of distance from the boundary between the igneous intrusion and the hornfels. The ESR age obtained using the E$'_1$ center is about 10 Ma for igneous rocks and hornfels and 30 ~ 40 Ma for mudstones which show an intensity close to saturation (Ikeya and Toyoda 1991).

The geothermal history of Kurobe in Central Japan has been studied using ESR of quartz bearing diorite. Two hydrothermal alteration were discussed from signals of OHC ($g = 2.011$) and the Ge center ($g = 1.997$) in quartz grains from an altered zone (Miyakawa and Tanaka 1985).

10.3.4 *Organic Radicals in Chert*

Chert is an extremely hard microcrystalline sedimentary rock, composed of almost pure SiO$_2$, and it has been used for *flint* tools in prehistoric times. ESR spectra of chert and flint show signals consisting of the E$'_1$ center, OHC (supposed to be a peroxy center; see Chapter 9) and methyl radicals (Griffiths *et al.* 1982, 1983, Porat and Schwarcz 1991). Organic radicals (g = 2.0042) are observed in *silcrete* (Radtke and Brückner 1991).

Methyl radicals detected in heated flints have been used to investigate the *heat treatment* of lithic tools to improve the hardness. The heat treatment of prehistoric lithic tools has also been studied using the thermal behavior of the E'_1 center and oxygen vacancy in collaboration with archaeologists to clarify the trade of American Indians (Toyoda *et al.* 1993).

When flints are heated to 400°C, a septet–quartet spectrum appears at g = 2.00268 and A = 0.63 mT with six–equivalent strongly coupled protons and A = 0.18 mT with three weakly coupled protons as shown in Figure 10.8. From the spin Hamiltonian parameters studied in radical chemistry, this spectrum was identified as *perinaphthenyl radicals* with three benzene rings fused together and the unpaired electron delocalized symmetrically through the π–electron system (Chandra *et al.* 1988). Similar organic radicals have been detected in sponge (silicate fossils) and stalactite ($CaCO_3$) heated to 400°C and assumed to be isobutyl radicals [$(CH_3)_3 C\cdot$] or a trimer of methyl radicals [$(\cdot CH_3)_3$] (see Section 5.4.2 (b) of Chapter 5). If the latter models are correct, isobutyl radicals or a trimer of methyl radicals might also be stabilized in flints. In fact, formation of this radical accompanies decay of the $\cdot CH_3$ signal. If benzene rings are really involved in the silica matrix, it would be interesting to investigate the origin of organic molecules in the context of inorganic formation of petroleum.

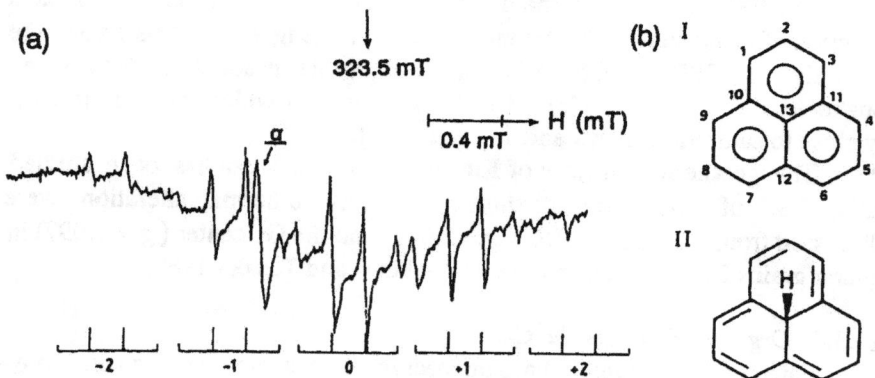

Figure 10.8   (a) ESR spectrum of a flint after heating to 400°C and (b) the structure of the perinaphthenyl radical (I) and its possible precursor (II). The line marked α (g = 2.0054) is possibly due to carbon (Chandra *et al.* 1988).

10.4   Zircon (ZrSiO$_4$)

10.4.1   *Crystal Structure and Models of Defects*

Zircon (ZrSiO$_4$) is a silicate mineral consisting of a Zr^{4+} cation and a tetrahedral SiO$_4{}^{4-}$ anion. The crystal structure is shown in Figure 10.9. It is a common accessory mineral in igneous rocks, schists, gneisses, beach deposits, etc., (Bates and Jackson 1980). Various colors derived from impurities and radioactive bombardment, have been used for age estimation in geology. It dissociates to ZrO$_2$ and SiO$_2$ when heated above 1,540°C and recombines when cooled slowly, but forms a mixture of ZrO$_2$ and vitreous SiO$_2$ when rapidly quenched.

Because of a relatively high content of uranium, FT dating of zircon grains is popular to determine the age of tephra. Although it was suggested that zircon is a suitable mineral for ESR dating (Zeller *et al.* 1967), zircon is a low abundance mineral in igneous rocks. Extraction of a large amount of zircon (from 5 mg for old samples to 300 mg for Quaternary or Holocene samples) from volcanic ash is usually difficult. However, ESR signals are normally detectable for old samples if a single grain has a volume larger than 2×10^{-6} cm^3 ($\sim 10\,\mu$g) (Kasuya *et al.* 1992).

ESR spectra of powder and a single crystal of natural zircon are shown in Figure 10.10, which give anisotropic signals at $g_{zz} = 2.0035$, $g_{xx} = 2.0178$ and $g_{yy} = 2.0089$. These have been identified as a hole center associated with Y^{3+} substituted at Zr^{4+}. A set of signals due to fine and hf structures around

(a) Zircon (ZrSiO$_4$)

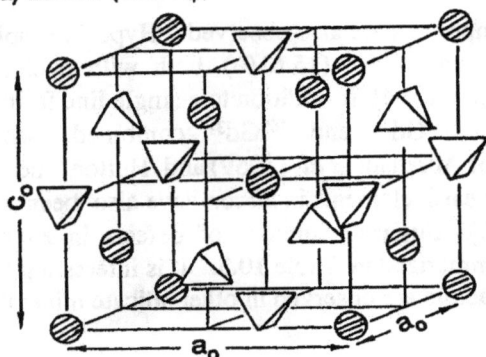

(b) SiO$_4$

Zr
Si
O

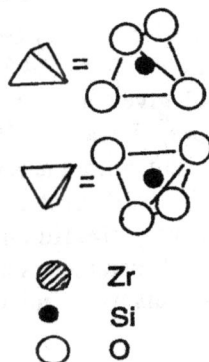

Figure 10.9   (a) Crystal structure of zircon (ZrSiO$_4$) and (b) a tetrahedral SiO$_4{}^{4-}$ anion.

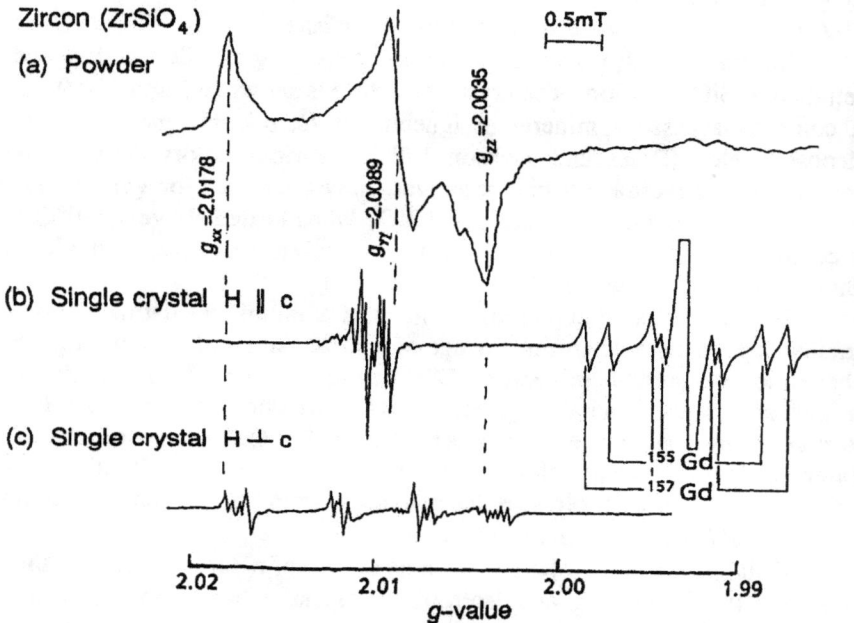

Figure 10.10 ESR spectra of zircon. (a) Powder. A hole type SiO_4^{3-} center is observed. (b) and (c) A single crystal with the direction of the magnetic field (b) parallel and (c) perpendicular to the c-axis. The quartets associated with isotopes of $^{155}Gd^{3+}$ (14.80%) and $^{157}Gd^{3+}$ (15.65%) both with $I = 3/2$ are observed.

$g = 1.992$ caused by Gd^{3+} impurities are also observed. Hyperfine splittings due to $^{155}Gd^{3+}$ (14.80%) and $^{157}Gd^{3+}$ (15.65%) both with nuclear spin $I = 3/2$ give two quartets $(2I + 1 = 4)$ in addition to a single line from other isotopes, $^{155}Gd^{3+}$, $^{156}Gd^{3+}$, $^{158}Gd^{3+}$ and $^{160}Gd^{3+}$ combined, which is consistent with the results of Abraham *et al.* (1969) and Hutton and Troup (1964). ESR spectra of rare earth elements in zircon have also been studied extensively (Marfunin 1979). Suggested models of defects in zircon and their ESR parameters are summarized in Table 10.3. It is interesting to note that signals with similar g-factors are observed in other silicate minerals.

Table 10.3 ESR parameters of defects and paramagnetic impurities in zircon.

Model	g_{zz}	g_{xx}	g_{yy}	g_{av} [a]	A_{zz}	A_{xx}	A_{yy}	Ref.
Electron center								
SiO$_4^{5-}$	$g_\parallel$ = 2.0020	$g_\perp$= 2.0007		2.0011	$A_\parallel$ =45.7 $A_\perp$< 42.0 (^{29}Si)			1
SiO$_3^{3-}$ [b]	2.0004	2.0048	1.9993	2.0015	21.8	16.8	16.8 (^{29}Si)	1
SiO$_2^-$	1.9920	2.005	1.9960	1.9977	20.2	16.1	16.8 (^{29}Si)	1
Hole center								
SiO$_4^{3-}$	2.062	2.003	2.004	2.023	1.9 (^{29}Si)			2
Y^{3+}–SiO$_4^{3-}$	2.0033	2.0168	2.0076	2.0092				3
	2.0036	2.0161	2.0076	2.0091	0.08	0.04	0.04 (^{89}Y)	4
	2.0035	2.0178	2.0089	2.0101				5
(SiO$_3$ SiO$_4$)$^{5-}$ –2Y^{3+}	2.0023	2.0169	2.0080	2.0091	0.9 0.14	0.85 0.21	0.85 (^{29}Si) 0.21 (^{89}Y)	2
Zr^{3+} (I)	1.882	1.934	1.923	1.913	9.3	5.0	5.0 (^{91}Zr)	2
(II)	1.977	1.874	1.913	1.922	2.5	8.5	5.5 (^{91}Zr)	2
(III)	1.879				A=8.8 (^{91}Zr)		0.34 (^{89}Y)	2
Y^{3+} – O$^-$ – Y^{3+}	2.0315	2.0030	2.0278	2.0208	0.13	0.14	0.13 (^{89}Y)	6
	2.0219	2.0045	2.0190	2.0151	0.11	0.12	0.14 (^{89}Y)	6
Impurity center								
AlO$_4^{4-}$	2.051	2.002	2.012	2.022	$A_\parallel$ =0.67 $A_\perp$=0.81 (^{27}Al)			2
Gd^{3+}	$g_\parallel$ = 1.9918	$g_\perp$= 1.9915		1.9916	A=0.35 (^{155}Gd) 0.45 (^{157}Gd)			7
	$g_\parallel$ = 1.9917	$g_\perp$= 1.9912		1.9914	A=0.46 (^{155}Gd) 0.61 (^{157}Gd)			8
Ti^{3+}	$g_\parallel$ = 1.998	$g_\perp$= 1.996		1.997				9
	$g_\parallel$ = 1.927	$g_\perp$= 1.940		1.936	A=3.1 (^{47}Ti) 1.5 (^{49}Ti)			10
V^{4+}	$g_\parallel$ = 1.8930	$g_\perp$= 1.9699		1.9446	$A_\parallel$ =9.28 $A_\perp$=3.15 (^{51}V)			11

a) The g_{av} is calculated using the anisotropic g factors [see Eq. (2.10) and Eq. (2.11)].
b) may be associated with an E$'_1$ center.
Reference: 1) Solntsev *et al.* 1974; 2) Solntsev and Shcherbakova 1973; 3) Vinokurov *et al.* 1971; 4) Baker and Hutton 1973; 5) Kasuya *et al.* 1990; 6) Bershov 1971; 7) Hutton and Troup 1964; 8) Abraham *et al.* 1969; 9) Bershov 1970; 10) Solntsev and Shcherbakova 1972; 11) Ball and Wanklyn 1976.

Nuclear spin and abundance:
^{27}Al (I = 5/2, 100%), ^{29}Si (I = 1/2, 4.67%), ^{47}Ti (I = 5/2, 7.3%),
^{49}Ti (I = 7/2, 5.5%), ^{51}V (I = 7/2, 99.75%), ^{89}Y (I = 1/2, 100%),
^{91}Zr (I = 5/2, 11.22%), ^{155}Gd (I = 3/2, 14.80%), ^{157}Gd (I = 3/2, 15.65%).

10.4.2 *ESR Dating of Zircon*

ESR dating of zircon in volcanic rocks has been done using the signal at $g = 2.0083$ and the obtained ages were compared with FT ages as shown in Figure 10.11 (Taguchi *et al.* 1985). Although the uranium content varies from 20 to 200 ppm, the absolute signal intensity at $g = 2.0083$ seems to be related to the age. The result is somewhat puzzling unless the dose rate for defect formation is constant: the distribution of uranium is inhomogeneous and uranium content in solution in zircon is constant for all samples. It is interesting to note that the signal intensity at $g = 2.0083$ is neither enhanced by γ–ray nor neutron irradiation, but is created in synthetic zircon doped with radioactive elements. Internal α–rays or α–recoil atoms seem to create the signal (Hayashi *et al.* 1990).

A correlation of the signal intensity of the hole center around $g = 2.011$ with the FT density has been observed up to the track density of 1.5×10^6 tracks/cm^2. A new finding is the abrupt drop in the signal intensity to zero as the track density exceeds $4 \sim 5 \times 10^6$ tracks/cm^2 (Kasuya *et al.* 1992). It is considered that hole centers induced by α–recoils have accumulated but FTs with a large volume ($10\,\mu m^3$) destroy them. The large interaction volume of a FT will not allow the hole center to stabilize. The distribution of the hole center and of the Gd^{3+} impurity ion in zircon has been imaged with a microwave scanning ESR microscope (see Figure 14.17).

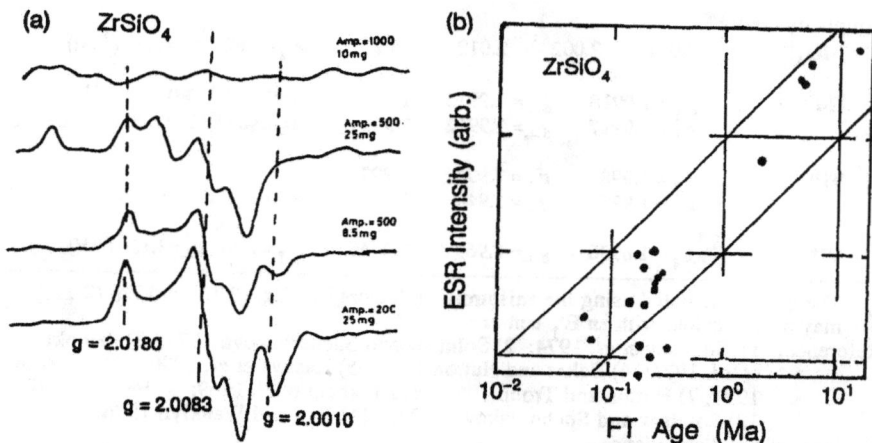

Figure 10.11 (a) ESR spectra of natural zircon crystals with different FT ages. (b) The relation of the absolute intensity of the signal at $g = 2.0083$ and the FT ages (after Taguchi *et al.* 1985).

10.5. Feldspars

10.5.1 *Structures and Models of Defects*

Feldspars are widespread and abundant rock–forming silicate minerals forming a solid solution with a general formula of $MAl(Al, Si)Si_2O_8$, where $M = Na$, K, Ca and Ba. They belong to a structural type of silicate called the "tectosilicate" characterized by the sharing of all four oxygens of SiO_4 tetrahedra with neighboring tetrahedra similar to the structure of SiO_2. Feldspars consist of the following three end members:

(1) Albite (Ab; $NaAlSi_3O_8$)
(2) Anorthite (An; $CaAl_2Si_2O_8$)
(3) Potash feldspar (Or; $KAlSi_3O_8$)

A group of feldspars consisting mainly of Ab and Or is called "alkali feld-spar" and that of Ab and An is called "plagioclase" (Bates and Jackson 1980).

(a) Alkali feldspar: a solid solution of Ab and Or in the form of $Ab_{100-x}Or_x$ (x in %) : orthoclase, microcline, sanidine, adularia and anorthoclase

(b) Plagioclase: a solid solution of Ab and An in the form of $Ab_{100-x}An_x$ oligoclase ($x=10 \sim 30$), andesine ($x=30 \sim 50$), labradorite ($x=50 \sim 70$) and bytownite ($x=70 \sim 90$).

Most feldspars contain paramagnetic Fe^{3+} ions and ferromagnetic microin-clusions which cause broad signals. Sometimes, all such materials are removed by a magnetic separator.

ESR spectra of 15 different feldspars of different compositions were measured after X–ray irradiation and defects were characterized (Speit and Lehmann 1982). Electron and hole centers at the oxygen lattice site as well as vacancies associated with oxygens are basic defects in feldspars as in other silicate and silica (SiO_2). The Al and Ti centers, Si–O⁻...X, Pb–O⁻...X, and NH_3^+ have been identified. It is noted that the Al center in feldspar has a hole shared by two Al atoms, i.e., Al–O⁻–Al while a hole is shared by one Al and four oxygen atoms in the form of tetrahedral $[AlO_4]^0$ in quartz. Models of defects and their ESR parameters are summarized in Table 10.4.

Table 10.4   ESR parameters of defects in feldspars.

Defect/Material	Composition	g_{zz}	g_{xx}	g_{yy}	g_{av}	Ref.
Al center	$[Al-O^--Al]$ $A = 0.9$ mT					
Alkali feldspar						
Microcline	$KAlSi_3O_8$	2.0275	2.0036	2.0067	2.0126	1
		2.0555	2.0043	2.0070	2.0224	2
Adularia	$KAlSi_3O_8$	2.0301	2.0038	2.0094	2.0145	1
Sanidine	$KAlSi_3O_8$	2.0285	2.0044	2.0086	2.0139	1
Orthoclase	$KAlSi_3O_8$	2.0799	2.0036	2.0280	2.0374	2
Anorthoclase	$(Na,K)AlSi_3O_8$	2.0342	2.0043	2.0099	2.0162	1
Plagioclase						
Albite (Ab)	$NaAlSi_3O_8$	2.0347	2.0044	2.0109	2.0167	1
		$g_\parallel = 2.0044$		$g_\perp = 2.0750$	2.0517	2
		2.067	2.003	2.010	2.0269	3
Moonstone (oligoclase)	$Ab_{90-70}An_{10-30}$	2.0291	2.0039	2.0079	2.0137	1
Sunstone	oligoclase + hematite $Ab_{90-70}An_{10-30} + Fe_2O_3$	2.0259	2.0034	2.0081	2.0125	1
Ti center	$[TiO_4]^-$ or TiO_4^{5-}					
Alkali feldspar						
Microcline	$KAlSi_3O_8$	1.9981	1.900	1.9445	1.9479	2
Sanidine	$KAlSi_3O_8$	$g_\parallel = 1.9984$		$g_\perp = 1.9005$	1.9337	4
Orthoclase	$KAlSi_3O_8$	$g_\parallel = 1.9981$		$g_\perp = 1.9016$	1.9343	4
Plagioclase						
Labradorite	$Ab_{50-30}An_{50-70}$	1.9932	1.9134	1.9340	1.9472	1
Andesine	$Ab_{70-50}An_{30-50}$	$g_\parallel = 2.0005$		$g_\perp = 1.904$	1.9367	4
(cf) Quartz $[TiO_4/H^+]$	SiO_2	1.9856	1.9151	1.9310	1.9443	5
Si-O$^-$...X						
Alkali feldspar						
Amazonite (microcline)		2.0123	2.0040	2.0098	2.0087	1
	$KAlSi_3O_8$	2.0599	2.0036	-----	2.0225	2
Plagioclase						
Labradorite	$Ab_{50-30}An_{50-70}$	2.0193	1.9917	2.0101	2.0071	1
Pb-O$^-$...X	$A = 9.4$ mT					
Alkali feldspar						
Amazonite (microcline)		2.0561	1.9883	2.0257	2.0236	1
	$KAlSi_3O_8$	2.0579	1.9890	2.0247	2.0241	2
NH$_3^+$	$A = 0.59,\ 4.2,\ 1.1\ (^{14}N);\ A = 2.3,\ 2.8,\ 2.3$ mT (^1H)					
K–feldspar	$KAlSi_3O_8$				$g_{iso} = 2.003$	6

References: 1) Speit and Lehmann 1982; 2) Marfunin and Bershov 1970; 3) Ioffe and Yanchevskaya 1968; 4) Bershov 1970; 5) Rinneberg and Weil 1972; 6) Bagmut *et al.* 1975.

Al center :   Al–O⁻–Al :   An ESR spectrum of the Al center in feldspars is characterized by the anisotropic g factor and the hf interaction with two ^{27}Al nuclei $(I = 5/2 \times 2 = 5)$ giving $2I+1 = 11$ lines with intensity ratios of $1:2:3:4:5:6:5:4:3:2:1$ (Marfunin 1979, Speit and Lehmann 1982). The spectrum can be measured only at low temperature as the hole is hopping between nearly equivalent oxygen sites at room temperature.   Figure 10.12 shows (a) an ESR spectrum of the Al center in a powder sample of oligoclase $(Ab_{90-70}An_{10-30})$. The powder spectrum broadened by the anisotropic g factors and hf splitting is simulated in (b). The hf splitting for g_{yy} is clearly resolved. A broad signal overlaps the spectrum in (a).

   An isochronal annealing curve of the Al center is shown in Figure 11.12 (c).   One can estimate the lifetime of the center to be as old as 1 Ma (Sasaoka *et al.* in preparation).

Ti center :   [TiO₄]⁻ or TiO₄⁵⁻ :   A Ti electron center is observed in some feldspars such as labradorite and microcline. In labradorite, the z–direction for the anisotropic g factor g_{zz} coincides with the Ti–Al direction.

Si–O⁻...X center : This center was observed in X–ray–irradiated labradorite at 77K and also identified in bytownite, anorthite, microcline and sunstone (Speit and Lehmann 1982). This center was assigned to a Si–O⁻–Si

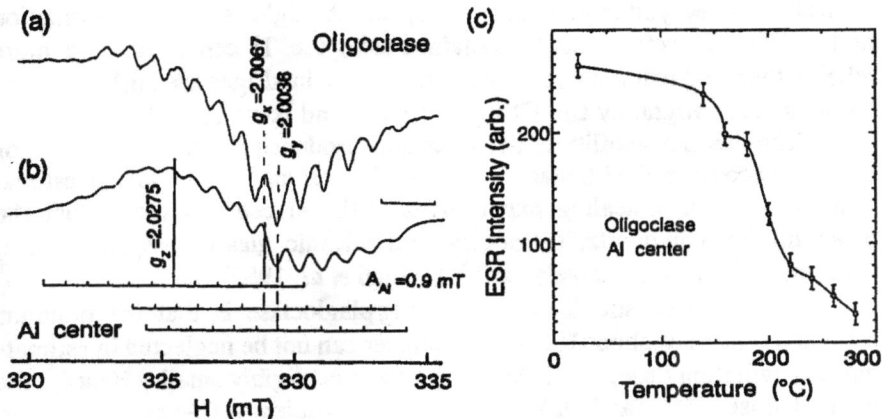

Figure 10.12 (a) ESR spectrum of plagioclase (oligoclase) separated from the volcanic ash shows the Al centers at 77K. (b) Simulation spectrum of the Al center. (c) Isochronal annealing curve of the Al center (Sasaoka unpublished).

center by Marfunin and Bershov (1970), but since holes adjacent to only one Al are already unstable at room temperature, this center must be stabilized by a divalent metal ion X at a site adjacent to the oxygen.

Pb–O⁻...X center : An ESR spectrum at 77K of amazonite shows a characteristic pattern of a strong central signal and two satellites with relative intensities of $1 : 6.8 : 1$ which are consistent with those expected for the isotope ^{207}Pb ($I = 1/2$, 22.6%). The isotropic hf constant is $A = 9.4$ mT.

NH_3^+ : This center isomorphously replaces the alkali metal ion in the feldspar. The hf interaction of the unpaired electron is found to be anisotropic both with the nitrogen nucleus and with protons. This center is stable up to 200°C (Bagmut et al. 1975).

10.5.2   Plagioclase : Volcanic Ash

ESR dating of plagioclase separated from volcanic ash has been done using the Al and Ti centers (Imai and Shimokawa 1985, Toyoda and Ikeya 1991a). The Al and Ti centers in plagioclase separated from the volcanic ash has been studied. Artificial irradiation produced the less stable Al center, while the stability of the Ti center was not changed after irradiation. When irradiated samples were heated at 200°C for 15 min, the Al center in more irradiated samples was decreased more than that in less irradiated ones. The less stable Al center would have also been produced by natural radiation but have already decayed in geological time, which might cause underestimation of the ED. Therefore, the ED obtained using the Ti center may be more reliable than that using the Al center and the obtained ages are fairly consistent with the stratigraphy and FT ages (Toyoda and Ikeya 1991a).

The thermal stability of defect centers produced by artificial irradiation should be compared with that of non–irradiated samples prior to age estimation. Isothermal annealing experiments of the Al center indicate that the mean lifetimes in quartz, plagioclase and volcanic glasses are $18 \sim 38$ Ma, 4 Ma and 10^3 Ma, respectively, at 10°C (Imai et al. 1985).

One characteristic feature in dating plagioclase is that the uranium content in grains of about 100 μm in diameter can not be neglected in estimating the radiation dose, while that in quartz is negligibly small. Hence, the internal doses (D_α and D_β) and the α–ray efficiency (k–value) must be calculated using Eq. (4.28) for ^{238}U–series disequilibrium considering some particular rate of radon loss.

10.6   Clay Minerals

10.6.1   *Structures*

Clay minerals are *sheet silicates* (*phyllosilicates*) formed by *hydro–thermal alteration* or *weathering* of silicate minerals. Crystalline, metacolloidal and amorphous hydrous alumino (magnesium and iron) silicates are called "**clay minerals**". In the layered structure of clay minerals, two different layers combine at a ratio of 1 : 1 or 2 : 1 as shown in Figure 10.13, i.e.,

(1) one or two layers of $(Si, Al)O_4$ tetrahedra and
(2) a layer of cations (Al^{3+}, Fe^{2+}, Fe^{3+}, Mg^{2+}, etc.) octahedrally coordinated with oxygen or hydroxyl.

Several organic molecules are incorporated between the layers of clay minerals. There are four main groups of clay minerals with two layers (1 : 1 type) and three layers (2 : 1 type) (Bates and Jackson 1980).

(a) *Kaolin group* (1 : 1 type) :
Kaolinite [$Al_2Si_2O_5(OH)_4$]: It is a main secondary sheet silicate of the kaolin group. The base material contains a high concentration of aluminum which is less exchangeable with Fe^{3+} or Mg^{2+}. It remains nearly white on firing and is used in the manufacture of ceramics, paper and refractories.

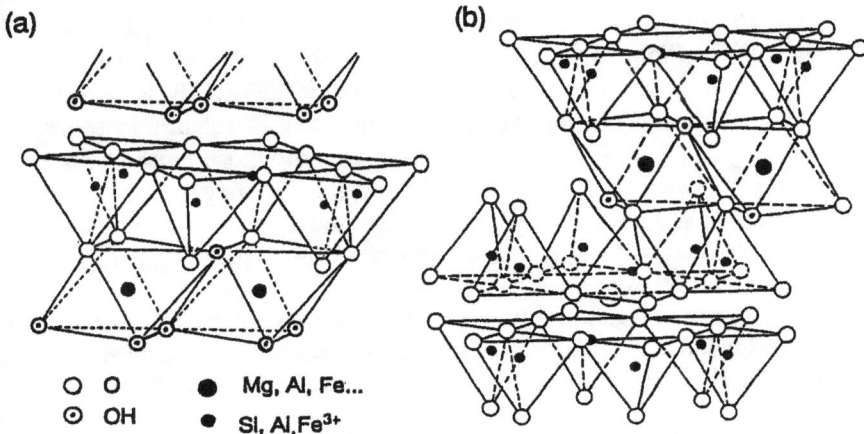

Figure 10.13 Layered structures of clay minerals. Two different layers of $(Si, Al)O_4$ tetrahedra and cations octahedrally coordinated with O or OH combine at (a) 1 : 1 ratio and (b) 2 : 1 ratio (Coey *et al.* 1982, Ballet and Coey 1982).

(b) *Montmorillonite group* (2 : 1 type) :

Montmorillonite $[R_{0.33}Al_2Si_4O_{10}(OH)_2 \cdot nH_2O]$: It has one layer of Al^{3+} and OH^- between two layers of Si^{4+} and O^{2-}. The exchangeable metal cations R ($R = Na^+$, K^+, Mg^{2+}, Ca^{2+}) are involved in the interlayer position to compensate the charge deficiency of the layer. Contrary to the case of kaolinite, it is subject to basic changes by foreign ions.

(c) *Clay mica group* (2 : 1 type) :

Illite or hydromica $[K_yAl_4(Si_{8-y}, Al_y)O_{20}(OH)_4]$: It is intermediate in composition and structure between muscovite and montmorillonite. It contains less K and more water than true micas, and more K than kaolinite and montmorillonite.

(d) *Chlorite group* (1 : 1 type): $(Mg, Fe^{2+}, Fe^{3+}, Al)_6(Si, Al)_4O_{10}(OH)_8$

Chlorite is a greenish plate–like monoclinic silicate mineral resembling mica. It is characterized by prominent ferrous ion and absence of Ca^{2+} and alkali ions. ESR dating of this mineral is difficult as it involves Fe^{3+}. A sheet of chlorite attached at a geologic fault plane shows a dominant ESR spectrum associated with Fe^{3+} and no defect signal.

10.6.2 *ESR Spectra and Models of Defects*

ESR studies on clay minerals, mostly paramagnetic impurities and some radiation–induced signals, have been done extensively (Meads and Malden 1975, Pinnavaia 1980, McBride 1990) following early works (Friedlander *et al.* 1963, Wauchope and Haque 1971). ESR spectra of several standard clay minerals from the American Petroleum Institute are shown in Figure 10.14. Paramagnetic ion signals and those caused by natural radiation are observed. The latter in the silicate part may be analogous to defects in silica and may be used for dating, though the meaning of ages for diagenetic processes is somewhat ambiguous. ESR of paramagnetic impurities are discussed in the next section.

ESR spectra of kaolinite from the Nopal uranium deposit in Mexico are shown in Figure 10.15 (Ildefonse *et.al.* 1990). Defects so far identified are as follows and their ESR parameters are summarized in Table 10.5.

A center : $g_{\parallel} = 2.049$ $g_{\perp} = 2.007$: An ESR spectrum of *hydrothermal kaolinite* from the Nopal uranium deposit in Mexico shows an asymmetric doublet characteristic of powder spectrum of anisotropic g factors with axial symmetry as shown in Figure 10.15 (a). The signal termed the "A center"

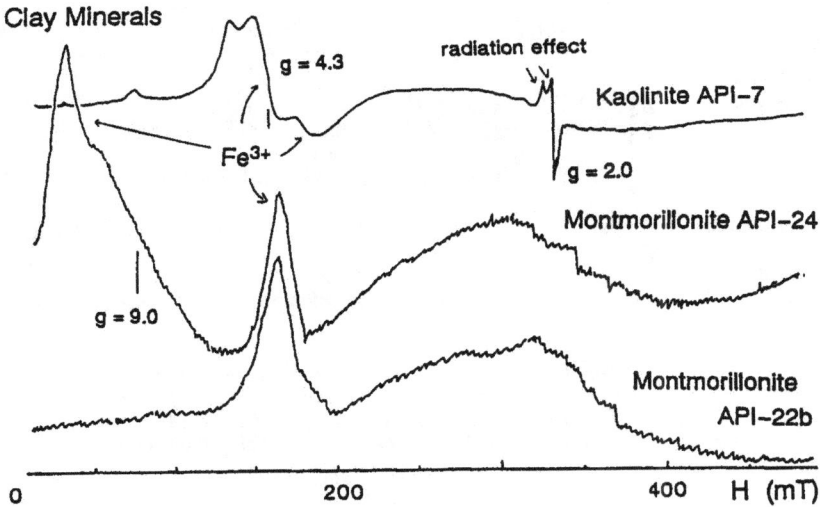

Figure 10.14 ESR spectra of standard clay minerals from the American Petroleum Institute.

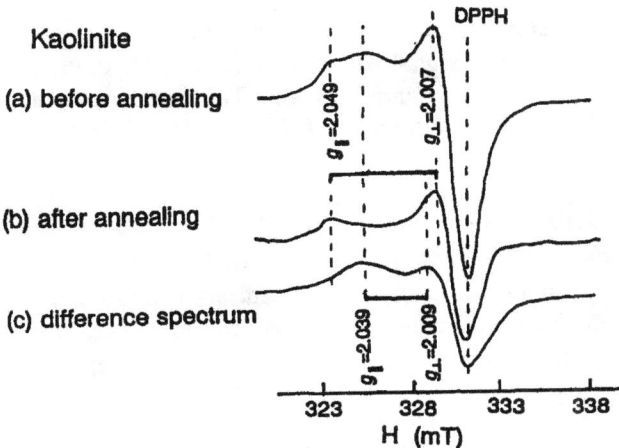

Figure 10.15 ESR spectra of kaolinite from the Nopal uranium deposit in Mexico (a) before and (b) after annealing at 350°C. (c) Spectrum of the A′ center after subtraction of spectrum (b) from spectrum (a) (after Ildefonse *et al.* 1990).

Table 10.5   Radiation–induced centers in clay minerals.

Model	g factor		Comment	Ref.
Kaolinite				
A center	$g_\parallel$=2.049	$g_\perp$=2.003		1
	$g_\parallel$=2.049	$g_\perp$=2.007	remains after annealing at 350°C (τ = 1 Ga)	2, 3
A' center	$g_\parallel$=2.039	$g_\perp$=2.009	annealed out at 350°C (τ = ~1 ka)	2
B center	$g \approx 2.10$		observed after artificial irradiation (τ = ~10 a)	2
Montmorillonite				
E'_1 center ?	$g \approx 2.0$		detected at low microwave power	4
Laterite				
unidentified	g = 2.004		ESR dating was attempted	5

Reference: 1) Jones *et al.* 1974; 2) Ildefonse *et al.* 1990, 1991; 3) Muller and Calas 1989; 4) Wauchope and Haque 1971; 5) Nambi and Sankaran 1985.

(Jones *et al.* 1974) is ascribed to a hole trapped at bridging oxygens stabilized by divalent cations such as Mg^{2+} (Angel *et al.* 1974, Muller and Calas 1989). The annealing temperature is 400°C and the center is stable over a geological period of 1 Ga at room temperature (Ildefonse *et al.* 1991).

A' center : $g_\parallel = 2.039$, $g_\perp = 2.009$: Thermal annealing of kaolinite at 350°C reveals an unstable signal with axial g factors by subtraction of the annealed spectrum from the unannealed one. This was termed the "A' center" (see Figure 10.15 (c)) (Ildefonse *et al.* 1990). The lifetime of this center is 0.1 ~ 1 ka.

B center : $g \approx 2.10$: A hole center trapped at Al–O–Al bonds is annealed out at 250°C and has a lifetime of 1 ~ 10 a. This center is found only in irradiated samples since the saturation concentration is too low due to the short lifetime. Such a center with a short lifetime can be used for estimating the artificial dose in a radiation accident as described in Chapter 13, but not for geological dating.

E'_1 center ? : $g \approx 2.0$: Radicals are formed by ionizing radiation in a clay lattice because organic radicals such as humic acid give signals. An ESR spectrum of montmorillonite gives a broad signal around $g = 2.0$ and a second peak appears at low microwave power (0.5 mW). It saturates easily and disappears completely at 50 mW (Wauchope and Haque 1971). These characteristics are similar to those of the E'_1 center.

10.6.3   *Kaolinite as a Paleodosimeter: Past Contact with Radionuclides*

One of the research projects for a high–level radioactive repository is burial of radionuclides in rocks such as granite, tuff, bedded salt and shale. Migration of radionuclides through space and time at the repository site is a major concern in case of potential leakage since radioactivities with a long half–life must be confined for a geologic period. Natural analogues for the potential leakage of radionuclides are *release* and *migration* of natural radio-nuclides at uranium deposits and *hydrothermal alteration* of acid rocks containing uranium. Radiation–induced defects in kaolinite have been studied to monitor the past contact between rocks and radioactive solutions (Ildefonse *et al..* 1990, 1991, Muller and Calas 1989, Clozel *et al.* 1990).

Natural *ferric gels* at the weathering front absorb uranium and its daughter elements 10^6 times more than ordinary minetals do. The content of Fe_2O_3 has been shown to be indicative of recent mineral assemblages and ore deposits (paragenesis) during weathering processes. Variation in the concentration of the A center in kaolinite has been studied together with the content of Fe_2O_3 for different depths of lateritic weathering profiles from centra–east Cameroon. A general correlation between the content of A center and Fe was observed in large nodules with an inherited rock structure (Muller and Calas 1989).

Figure 10.16 shows the relation of the concentration of the A and A' centers with the uranium content of hydrothermal kaolinite from the Nopal I uranium deposit in Mexico. The concentration of the A center that is stable for ~ 1 Ga is related to the U content only in feldspar pseudomorphs outside the breccia pipe but no correlation was observed in kaolinite. On the other hand, the concentration of the A' center which is stable only for 0.1 ~ 1 ka depends on the present content of U in hydrothermal kaolinite and in kaoli-nite embedded in uranium bearing minerals (Ildefonse *et al.* 1990, 1991).

These studies are stimulating since a memory of uranium transport is stored. Although *weathering* processes are not events that start the clock time among clay minerals, ESR dating of kaolinite is quite encouraging. ESR dating has been carried out for laterite of basaltic origin although the model for the signal is not clear (Nambi and Sankaran 1985).

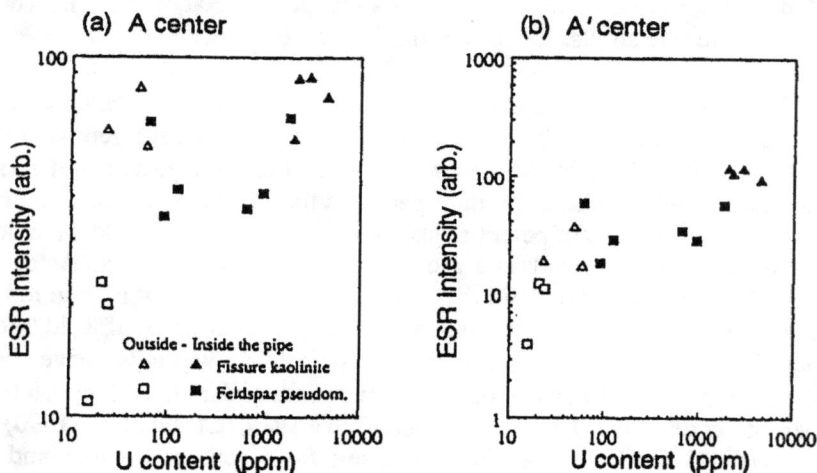

Figure 10.16 The relation of the signal intensity of (a) the A center and the A' center with the uranium content in kaolinite from the Nopal I uranium deposit in Mexico (Ildefonse *et al.* 1991).

10.7 Magnetic Minerals and Impurities

10.7.1 *ESR Spectra of Iron Minerals*

Magnetic impurities, mostly Fe^{3+}, are abundant in geothermal and volcanic materials. Paramagnetic Fe^{3+} ions coagulate and form clusters of ferromagnetic nature. Inclusions, surface coatings and precipitated phases give intense and broad ESR signals. These signals indicate the distribution of paramagnetic impurities and the presence of magnetic minerals, which are related to the geothermal history of volcanic materials. The ESR intensity of *ferromagnetics* (and *ferrimagnetics*) are three orders of magnitude more intense than paramagnetics. ESR spectra of three magnetic minerals, *magnetite* (Fe_3O_4), *hematite* (Fe_2O_3) and *limonite* ($FeO(OH) \cdot nH_2O$) are shown in Figure 10.17. The magnetic and ferromagnetic resonance proper-ties of several iron oxide minerals are summarized in textbooks (Stucki and Banwart 1980, McBride 1990).

(a) *Magnetite* : Fe_3O_4 : It is a strongly ferrimagnetic mineral of the spinel group and is easily removed by a magnetic separator. It constitutes an important ore of iron. An ESR spectrum of magnetite shows an intense signal at a low field less than 100 mT.

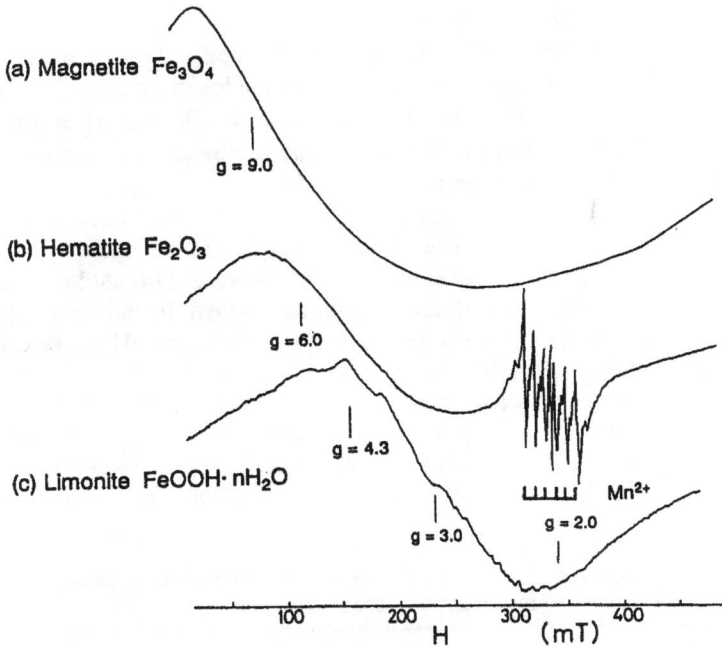

Figure 10. 17   ESR spectra of three iron–associated minerals; (a) magnetite (Fe_3O_4), (b) hematite (Fe_2O_3) and (c) limonite ($FeOOH·nH_2O$).

(b)   *Hematite* : Fe_2O_3 : It is a more oxidized form than magnetite and is weakly ferromagnetic above $-100°C$. It shows a broad low–field signal around $g = 2.3$ and another signal at $g = 4.3$ due to a large orthorhombic crystalline field, $E(S_x^2 - S_y^2)$ (Slichter 1960). It is interesting to note that the signal component of hematite increases at the expense of the magnetite signal as oxidation proceeds upon heating (see paper browning in Chapter 12).

(c)   *Limonite* : $FeO(OH)·nH_2O$: It is antiferromagnetic and has a Néel temperature of $T_N = 120°C$. Limonite was first thought to be $2Fe_2O_3·3H_2O$ but is now considered to have variable compositions including several iron hydroxides, e.g., goethite [α–$FeO(OH)$] and iron oxides with water. It is formed by weathering. A broad ESR signal occurs around $g = 3.0$.

10.7.2   State of Fe^{3+} Impurities

A trivalent iron, Fe^{3+}, has an electron configuration of $3d^5$ and is paramagnetic. A high spin state ($S = 5/2$) and a low spin state ($S = 1/2$) and in some cases intermediate spin states are present. The effective g factors of Fe^{3+} ($S = 5/2$) in a crystalline field of a mineral due to fine structure splitting (zero–field splitting) are summarized in Table 10.6.

In ferromagnetic resonance, the external magnetic field is no longer experienced by electrons. The effective magnetic field including the field induced by other ferromagnetic materials in the vicinity must be considered. Ferromagnetic resonance of an ellipsoidal shaped ferromagnet gives an angular dependent signal due to the magnetic field induced by a ferromagnetic material. Slight angular dependence results in a drastic change of the broad spectrum. One can check the ferromagnetic resonance spectrum by rotating a sample holder in a microwave cavity. Inclusion of ferromagnetic minerals in diamagnetic minerals also gives this orientation–dependent broad signal in spite of the fact that we are dealing with fine powder samples.

Table 10.6 Apparent g factors of Fe^{3+} due to a large zero field splitting, and characteristic signals of ferromagnetic resonance.

Crystalline Field	g_{eff}	Spin Hamiltonian	Models
Axial Field	6.0	DS_z^2	FeO_3^{3-}
Orthorhombic Field	4.3	$E(S_x^2 - S_y^2)$	FeO_2^{2-}
Cubic Field	2.0	$(a/6)(S_x^4 + S_y^4 + S_z^4)$	FeO_4^{5-}
Ferromagnetic Resonance		$h\nu = g\beta(H + 2K_1/M_s) = g\beta H_{eff}$	

* g_{eff} is an apparent (effective) g factor for strong crystalline field parameters. Other parameters are explained in Section 2.4.8 of Chapter 2.

10.7.3   Paramagnetic Impurities in Clay Minerals

Extensive studies have been carried out on ESR spectra of transition metal ions adsorbed on or embedded in clay minerals (Stucki and Banwart 1980). Fully wetted clay minerals show signals of hydrated metal ions such as $[Mn(H_2O)]^{2+}$ tumbling rapidly in a solution–like environment. The ESR spectrum of Mn^{2+} in clay minerals is essentially the same as in powdered $CaCO_3$.

(a) Fe^{3+}: $g = 4.27 \pm E/D$

ESR spectra of kaolinite show two types of substitutional Fe^{3+} in the octahedral sheet (Angel et al. 1974). The signal at $g = 4.27$ indicates substitutional Fe^{3+} with a strong orthorhombic distortion represented by the quadratic potential of $E'(x^2 - y^2)$ presumably associated with a layer stacking disorder. The signals split into two on both sides of the signal at $g = 4.27$ are due to Fe^{3+} in a partial orthorhombic field with an axial field component. The effective g factor is described as $g = 4.27 \pm E/D$. A number of papers on Fe^{3+} in clay minerals, for example, in halloysites (Nagasawa and Noro 1987) have been published. The details should be seen in the literature (Stucki and Banwart 1980, McBride 1990, Perry 1990).

(b) Cu^{2+} $(3d^9)$: $g_\parallel = 2.22 \sim 2.39$ $g_\perp = 2.03 \sim 2.08$
$A_\parallel = 13.5 \sim 20$ $A_\perp \approx 0$ mT

Exchangeable Cu^{2+} in interlayer regions of montmorillonite clay under aqueous coordination has been studied with ESR. The linewidth and spectral shape change depending on the degree of hydration of Cu^{2+}. Recently, a new technique of spin echo modulation has been applied (Brown and Kevan 1988). An ESR spectrum of Cu^{2+} shows axial g factors and anisotropic hf coupling constants in most materials (Corma et al. 1985). An oriented film of Cu^{2+}–montmorillonite forms a complex with adsorbed organic materials such as glycine, alanine and pyridine (Nagai et al. 1974).

(c)   Vanadyl ion (VO_2^+): $A_\parallel = 15 \sim 20$ $A_\perp = 4 \sim 7$ mT

ESR spectra of Mn^{2+} and VO_2^+ in clays and clay minerals have been studied (Hall et al. 1974, Meads and Malden 1975, McBride 1990). VO_2^+ has an unpaired d electron $(S = 1/2)$ and shows axial symmetry. The anisotropic hf coupling constants with nuclear spin $(I = 7/2)$ show a characteristic spectrum of 16 lines due to two octets for $g_\parallel$ and $g_\perp$. A central line is more intense the than the other lines due to overlapping. It is noted that it resembles a signal of radiation–induced defects among the hf lines of Mn^{2+}.

10.7.4   Volcanic Materials and Lunar Soils

(a)   Scoria, lava and pillow basalt : Volcanic eruption ejects scoria with vesicles (cavities) due to escape of volcanic gas. It was oxidized and cooled rapidly. Volcanic lava solidified from molten lava cooled slowly. The degree of oxidation will depend on the position after solidification. Pillow-shaped lava (pillow basalt) is formed in an aqueous environment. The surface in contact with water is cooled rapidly and becomes amorphous volcanic glass, while the inside is basalt.

ESR spectra of volcanic lava–Z1 (from the upper surface of the lava flow, Mt Mihara, Izu–Ohshima, Japan), lava–Z4 (from the center of the flow), scoria and pillow basalt are shown in Figure 10.18 (a) and the difference spectra of lava–Z1 and lava–Z4 after heating at 400, 500 and 600°C are shown in (b) (Toyoda and Ikeya 1989). The characteristic features are:

(1) *Scoria* (cooled rapidly) shows a broad signal around $g = 2.0$.
(2) *Lavas* (cooled slowly) show signal at $g = 9.0$ (magnetite–like).
    (a) The surface (Z1) shows a weak signal at $g = 4.3$,
    (b) while this signal is almost absent in the center (Z4).
    (c) Heating changes the spectrum from magnetite–like to hematite–like.
(3) *Pillow basalt* (cooled rapidly) shows less intense Fe^{3+} signal due to insufficient oxidation but a signal at $g = 4.3$ is relatively intense. Heating at 500°C in air changes into a scoria–like spectrum

Volcanic lava which is *cooled slowly* is sufficiently oxidized and gives a spectrum similar to *hematite* (see Figure 10.17). The lava samples taken from the surface to the inside show different Fe^{3+} signals, which indicates different degrees of oxidation as shown in Figure 10.18 (a) for lava Z1

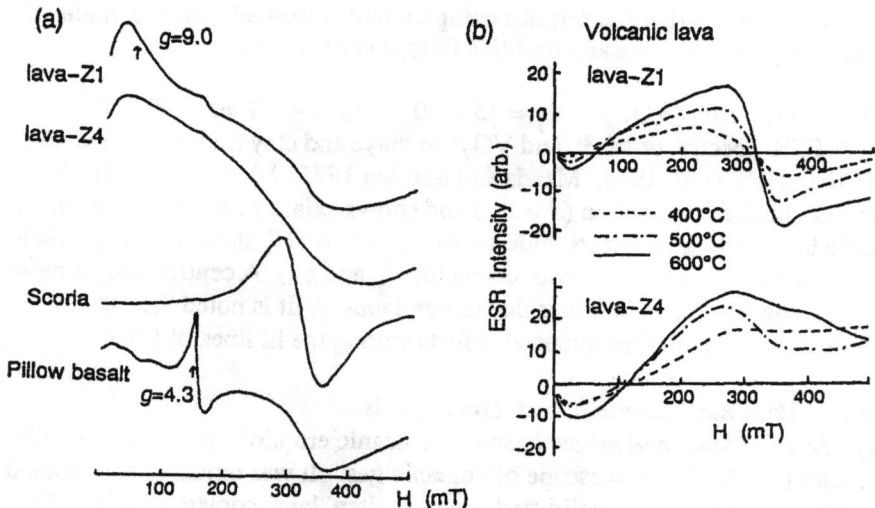

Figure 10.18   ESR spectra of two different portions of lava (Z1 and Z4), scoria and pillow basalt. ESR spectra after heating at 600°C for 20 min are shown by dashed curves. (b) The change in the spectrum (difference) of lava Z1 and Z4 by heating at 400, 500 and 600°C.

(surface) and Z4 (center of the flow).

Hawaiian lava "*aa*" and "*pahoehoe*" with different viscosity in the molten state show different spectra. The former, considered to be cooled rapidly but oxidized sufficiently due to rapid flow, shows a broad signal like hematite, while the latter pahoehoe shows a more magnetite–like spectrum.

Thermal annealing in air indicates that the Fe^{3+} state is frozen at a high temperature around 700°C for scoria, while the temperature is around 400°C for lava. The difference spectra of lava samples from heating at 400°C, 500°C and 600°C indicate a decrease of the intensity close to zero field due to reduction of the magnetite–like signal and increase of the hematite–like signal.

(b) *Lunar soils* : Extensive studies have been done on lunar soils and core samples taken during Apollo missions (Smith 1974). Ferromagnetic resonance of a single domain Fe^{o} has been studied in lunar rocks and soils irradiated by protons in solar wind and impacted by meteorites (Taylor and Cirlin 1985). The erosional and weathering agents on the moon are meteorite and micrometeorite impacts and melting by the impacts. Lunar soils formed by the impact onto lunar rocks under the reducing atmosphere of solar wind protons implanted in the soils contain Fe^{o} spheres with sizes of $4 \sim 30$ nm; this gives a ferromagnetic signal at $g = 2.08$ with a Lorentzian lineshape having a width of about 40 mT (Tsay *et al.* 1972). The relative intensity of this signal (I_s) to the content of FeO in lunar soils is a function of exposure time and indicates the maturity of the soil. It is interesting to note that these samples, microclusters of Fe^{o}, might show "*superparamagnetism*", the property that a small ferromagnetic particle with a volume less than the critical volume (2 nm in radius for Fe$_3$O$_4$) behaves as a paramagnetic species (Bean 1955). In this case, ESR signals will be observed at near–zero field.

ESR spectra of Fe^{3+}, Ti^{3+} and Mn^{2+} in lunar plagioclase have been measured at X– and Q–band frequencies (Weeks 1973). Ti^{3+} gives an anisotropic signal at $g_{zz} = 1.978$, $g_{xx} = 1.936$ and $g_{yy} = 1.910$. Fe^{3+} signals around $g = 4.27$ (3.2, 4.2 and 4.9) are observed.

10.8   Fe^{3+} and Mn^{2+} in Ceramics

10.8.1 *Surface Oxidation*

Mayan and Egyptian pyramid stones clearly show that restored stones are white, while the old ones are brown presumably due to surface weathering or oxidation of iron ions in rocks, i.e., formation of hematite (Fe$_2$O$_3$; $g=$

4.27) from magnetite (Fe_3O_4; $g > 9.0$). It may be possible to estimate the age by studying the oxidation of the rock surface with ESR just as in hydration–rim dating of flint tools in archaeology. An ESR microscope (Chapter 14) will be useful to study the distribution of paramagnetic ions from the surface. The age determination based on weathering is within the framework of *chemical ESR dating* (see Chapter 12). The degree of oxidation depends on the local conditions. Factors such as the water content, acidity and sunshine will also affect weathering. In spite of these drawbacks, it will be of interest to start a systematic study on chemical ESR dating using *impurity oxidation*.

10.8.2 *Diffusion of Mn²⁺ from Environment*

Diffusion of paramagnetic impurity ions such as paramagnetic Mn^{2+} may be used as a method of *chemical ESR dating*. It was discussed that Mn^{2+} concentration in a carbonate fossil *foraminifera* from deep–sea sediments increases as the depth increases up to an age of 3 Ma (Takeuchi and Saeki 1985) presumably due to diffusion at a later time.

The *diffusion coefficient* of Mn^{2+} in feldspars was determined at several temperatures and then extrapolated to the ambient temperature to calculate the diffusion–depth profile. Mn^{2+} in high concentration regions gives a broad signal due to dipolar broadening and exchange interaction and what we measure is the intensity of hf splitting lines of the central $M = 1/2 \leftrightarrow M = -1/2$ transition of isolated Mn^{2+}.

The average concentration in the bulk was calculated as a function of geologic age in Figure 10.19 for local Mn^{2+} concentrations of less than 10^{18} cm^{-3} at 1 mm depth from the surface. The concentration increases in proportion to $t^{0.55}$ which is close to the square–root of the age ($= t^{1/2}$). The initial surface concentration in the environment, A_{Mn} does not appreciably affect the result. Thus, chemical ESR dating using the diffusion of paramagnetic ions from the surface may be possible (Ikeya *et al.* 1985).

10.8.3 *Ancient Technology of Ceramics*

Ancient techniques of firing pottery have been studied using signals associated with Fe^{3+} (Warashina *et al.* 1981). Valency change while firing ceramics is also a chemical reaction although the conditions of firing (the degree of oxidizing or reducing atmosphere) are involved; this reflects the subtle artistic color change of pottery. The signal around $g = 4.27$ is related with orthorhombic field fine structure splitting and the signal around $g = 6.0$ to axial field splitting (see Table 10.6). The signal of magnetite ($g > 9.0$)

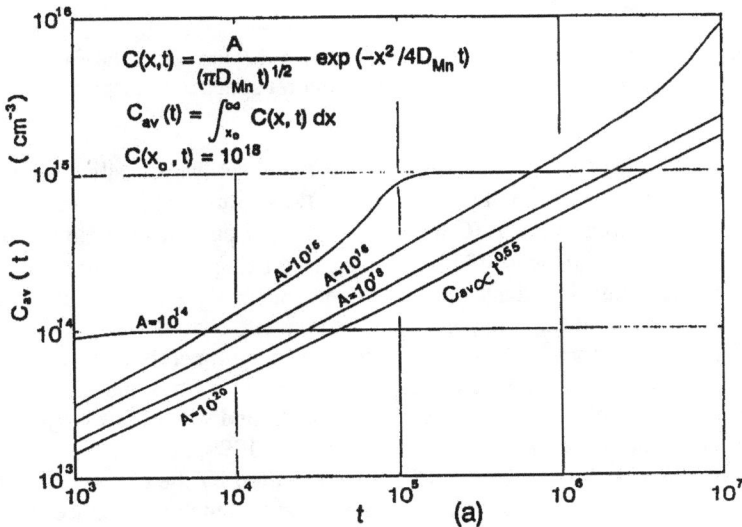

$$C(x,t) = \frac{A}{(\pi D_{Mn} t)^{1/2}} \exp(-x^2/4D_{Mn} t)$$

$$C_{av}(t) = \int_{x_o}^{\infty} C(x, t) dx$$

$$C(x_o, t) = 10^{18}$$

Figure 10.19   The surface average concentration of isolated Mn^{2+} ions diffused from the surface as a function of time, t, using the diffusion coefficient of Mn^{2+} (D_{Mn}) at 20°C in alkali feldspar. D_{Mn} is obtained by extrapolation from the high temperature data in an Arrhenius plot.

changes to that of hematite by firing clay minerals.

Brown coloration of rocks and ceramics by weathering may also be due to oxidation. Formation of hematite (Fe_2O_3) and limonite ($FeOOH \cdot nH_2O$) from magnetite (Fe_3O_4) is an oxidation reaction dealt with by the rate and diffusion equations (see Figure 10.19) from which the time (age) and/or temperature of firing may be assessed.

10.9   Technical Note

Organic radicals such as methyl ($\cdot CH_3$) and methanol ($\cdot CH_2OH$) radicals have been observed in geothermal silica, hornfels and chert. Methyl radicals which show a characteristic quartet spectrum with intensity ratios of $1:3:3:1$ are due to hf lines caused by coupling with three protons, as clearly shown in Figure 10.7. Signals at $g = 1.9975$ and $g = 2.010$ sometimes identified as the Ge center and OHC, respectively, might correspond to the second and third lines of the $\cdot CH_3$ quartet as suggested in Chapter 9. If these are the hf signals of the identical center of $\cdot CH_3$, it is natural to obtain an identical ED for the two signals. Microwave power dependence and thermal properties should be studied to distinguish the signals of the Ge and OHC signals from those of CH_3 at 77K where the quartet spectrum disappears.

10.10 Summary

ESR dating of volcanic ash, igneous rock and geothermal silica is given as a supplementary description of the use of radiation-induced signals in silica. Assessments of geothermal heat (Grun et al., 1999) and paleolithic heat treatment (Toyoda et al., 1993; Tani et al, 1997) are discussed using defect centers in silica. Similar defect centers in silicates related to volcanic processes and in clay minerals are described in addition to paramagnetic impurity signals.

The parameters of defects in feldspars and clay minerals and of paramagnetic ions especially Fe^{3+} in volcanic materials as well as clays and clay minerals were tabulated. Geological assessment of alteration and weathering may be done using these signals (Bartoll et al., 1997; Bensimon, 1999, 2000).

Studies in this field with ESR are still limited by the complexity of the broad signals due to magnetic impurities giving ferromagnetic resonance. Chemical separation of some minerals and desorption of surface impurities must be developed. A high-frequency ESR, the separation or subtraction of impurity signals with a computer, measurement of the relaxation time with spin-echo, etc., will allow us to determine the alteration age of minerals. A vast field within this province of geology and future planetary materials science remains to be explored.

References

Abraham M.M., Clark G.W., Finch C.B., Reynolds R.W. and Zeldes H. (1969): Ground state splitting of trivalent Gd and Cm in $ZrSiO_4$, $HfSiO_4$ and $ThSiO_4$, determined by ESR. J. Chem. Phys. 50, 2057-2062.
Angel B.R., Jones J.P.E. and Hall P.L. (1974): Electron spin resonance studies of doped synthetic kaolinite, I. Clay Miner. 10, 247-255.
Bagmut N.N., Brik A.B., Matyash I.V. and Fedotov Y.V. (1975): Method EPR studies of NH^{3+} radical in feldspar. Akad. Nauk. Ukr. RSR A6, 536-538.
Baker P.R. and Hutton D.R. (1973): A color center in natural zircon. Phys. Stat. Sol. 60, K109-K111.
Ball D. and Wanklyn B.M. (1976): Colored synthetic zircon crystals: flux growth and EPR of V^{4+} impurity centers. Phy. Stat. Sol. 66, 307-316.
Ball D. (1971): Paramagnetic resonance of Er^{3+} and Dy^{3+} in natural single crystals of zircon. Phys. Stat. Sol. (b) 43, 635-641.
Ballet O. and Coey J.M.D. (1982): Magnetic properties of sheet silicates: 2: 1 layer minerals. Phy. Chem. Minerals 8, 218-229.
Bartoll J, Tani A., Ikeya M. and Inada T. (1996): ESR investigation of burnt soil. Appl. Mahn. Reson. 11, 577-586.
Bartoll J, Ikeya M (1997): ESR dating of a pottery: a trial. Appl. Radoat. Iso, 48, 981-984.
Bartoll J, Tani A (1998): Thermal history of archaeological objects, studied by electron spin resonance. Naturwisenschaften 85, 474-481.
Bates and Jackson (1980): Glossary of Geology (AGI, Virginia).

Bean C.P. (1955): Hysteresis loops of mixtures of ferromagnetic micropowders. J. *Appl. Phys.* **26**, 1381-1383.

Bensimon Y, Deroide B, Dijoux F, et al. (2000): Nature and thermal stability of paramagnetic defects in natural clay: a study by electron spin resonance *J Phys. Chem. Solids* **61**, 1623-1632.

Bensimon Y, Deroide B, Martineau M, et al. (1999): Description of fundamental paramagnetic defects observed in a raw clay: an index for the ESR dating of ancient ceramics. *CR Acad. Sci.* II C 2 (2),119-125.

Berger G.W. and York D. (1981): Geothermometry from $^{40}Ar/^{39}Ar$ dating experiments. *Geochim. Cosmochim. Acta* **45**, 795-811.

Berger G.W. (1991): The use of glass for dating volcanic ash by thermoluminescence. *J. Geophys. Res.* **96**, 19,705-19,720.

Bershov L.V. (1970): Isomorphism of titanium in natural minerals. *Izv. Akad. Nauk* **12**, 47-54.

Bershov L.V. (1971): On the isomorphism of Tb^{4+}, Tu^{2+} and Y^{3+} in zircons. Geochimia **1**, 48-53.

Brown D. R. and Kevan L. (1988): Aqueous coordination and location of exchangeable Cu^{2+} cations in montmorillonite clay studied by electron spin resonance and electron spin-echo modulation. *J. Am. Chem. Soc.* **110**, 2743-2748.

Buhay W.M., Clifford P.M. and Schwarcz H.P. (1991): ESR dating of the Rototi Breccia in Taupo volcanic area in New Zealand. *Quat. Sci. Rev.* **11**, 267-271.

Calk L.C. and Naeser C.W. (1973): The thermal effect of a basalt intrusion on fission tracks in quartz monazonite. *J. Geology* **81**, 189-198.

Chandra H., Symmons M.C.R. and Griffiths D.R. (1988): Stable perinaphthenyl radicals in flints. *Nature* **232**, 526-527.

Clozel B., Calas G., Muller J.P., Dran J. C. and Herve A. (1990): Kaolinite as dosimeters: new possibility of tracing radionuclides migration. *Chem. Geol.* **84**, 259- 260.

Coey J.M.D., Ballet O., Moukarika A. and Soubeyroux J.L. (1982): Magnetic properties of sheet silicates: 1:1 layer minerals. *Phy. Chem. Minerals* **7**, 141-148.

Corma A., Per z-Pariente J. and Soria J. (1985): Physico-chemical characterization of Cu^{2+}-exchanged sepiolite. *Clay Miner.* **20**, 467-475.

Deer W.A., Howie R.A. and Zussman J. (1966): *An Introduction to the Rock- forming Minerals* (Longman, Essex)

Dodson M.H. (1973): Closure temperature in cooling geochronological and petrological systems. *Contr. Mineral. and Petrol.* **40**, 259-274.

Dodson M.H. and McClelland-Brown (1985): Isotopic and paleomagnetic evidence for rates of cooling, uplift and erosion. *Contr. Mineral. and Petrol.* **40**, 259-274.

Falgueres C., Miallier D., Sanzelle S., Fain J., Laurent M., Montret M., Pilleyre T. and Bahain J.J. (1994): Potential use of the E'center as indicator of initial resetting in TL/ ESR dating of volcanic materials. Quanter. Geochr. 13, 619-623.

Falgueres C., Yokoyama Y. and Miallier D. (1991): Stability of some centers in quartz. *Nucl. Tracks* **18**, 155-161.

Friedlander H.Z. and Frink C.R. (1963): Electron spin resonance spectra in various clay minerals. *Nature* **199**, 61-62.

Griffiths D.R., Robins G.V., Seeley N. J., Chandra H., McNeil A.C. and Symons M.C.R. (1982): Trapped methyl radicals in chert. *Nature* **300**, 435-436.

Griffiths D.R., Seeley N.J., Chandra H. and Symons M.C.R. (1983): ESR dating of heated chert. *PACT* **9**, 399-403.

Grün R., Tani A., Gurbanov A., Koshchung D., Williams I. and Braun J. (1999): A new method for the estimation of cooling and denudation rates using paramagnetic centers in quartz: A case study on the Eldzhurtinskiy granite, Caucasus. *J. Geophys. Res.* **104**, 17531-17549.

Hall P.L., Angel B.R. and Braven J. (1974): ESR and related studies of lignite and ball clay from South Devon, England. *Chem. Geol.* **13**, 97-113.

Harrison T.M. and McDougall I. (1980): Investigations of an intrusive contact, northwest Nelson, New Zealand - I. Thermal, chronological and isotopic constraints. *Geochim. Cosmochim. Acta* **44**, 1985-2003.

Hayashi M., Shinno I., Taguchi S. and Sugihara S. (1990): ESR signals of zircon irradiated by thermal neutrons and g-rays. *J. Miner. Petro. Econo. Geol.* **85**, 27-33.

Horvth L., Noethig L.V., Bilinski H. and Foerster H. (1990): Transformation of structural defects in heated model clay precursors by ESR spectroscopy. *J. Phys. Chem. Solids* **51**, 1061-1065.

Hutton D. R. and Troup G. J. (1964): Paramagnetic resonance of Gd^{3+} in zircon. *Br. J. Appl. Phys.* **15**, 405-406.

Ikeya M. (1983): ESR studies of geothermal boring cores at Hachobaru Power Station. *Jpn. J. Appl. Phys.* **22**, L763-L765.

Ikeya M., Miura Y., Tanosaka T. and Miura H. (1985): ESR spectroscopy and a possible method of Mn^{2+} dating for minerals and fossils. *ESR Dating and Dosimetry* 477-483.

Ikeya M., Devine S.D., Whitehead N.E. and Hedenquist J.W. (1986): Detection of methane in geothermal quartz by ESR. *Chem. Geol.* **56**, 185-192.

Ikeya M. and Toyoda S. (1991): Thermal effect in metamorphic rock around an intrusion zone with ESR studies. *Appl. Magn. Reson.* **2**, 69-81.

Ikeya M. and Tani A. (1997): Dating of ancient lithic tool factory and geological fault: Electron spin resonance of fractured grains. *Anthropos* **13**, 65-69.

Ildefonse P., Muller J.P., Clozel B. and Calas G. (1990): Study of two alteration systems as natural analogous for radionuclide release and migration. *Engin. Geol.* **29**, 413-439.

Ildefonse P., Muller J.,P., Clozel B. and Calas G. (1991): Records of past contact between altered rocks and radioactive solutions through radiation-induced defects in kaolinite. *Mater. Res. Soc.* **212**, 749-756.

Imai N., Shimokawa K. and Hirota M. (1985): ESR dating of volcanic ash. *Nature* **314**, 81-83.

Imai N. and Shimokawa K. (1988): ESR dating of Quaternary tephra from Mt. Osorezan using Al and Ti centers in quartz. *Quat. Sci. Rev.* **7**, 523-527.

Imai N. and Shimokawa K. (1989): ESR dating of tephra "crystal ash" distributed in Shinshu, Central Japan. *Appl. Rad. Isot.* **40**, 1177-1180.

Imai N., Shimokawa K., Sakaguchi K. and Takada M. (1992): ESR dates and thermal behavior of Al and Ti centers in quartz for tephra and welded tuff in Japan. *Quat. Sci. Rev.* **11**, 257-265.

Imai N., Shimokawa K. and Yamamoto M. (1994): ESR study of radiation centers and thermal behavior in chert. *Quat. Geochr.* **13**, 641-645.

Ioffe V. A. and Yanchevskaya I. S. (1968): Electron paramagnetic resonance and thermoluminescence of irradiated single crystals of aluminosilicates $NaAlSi_3O_8$ and $LiAlSiO_4$. *Soviet Phys. Solid State* **10**, 370-374.

Jones J.P.E., Angel B.R. and Hall P.L. (1974): Electron spin resonance studies of doped synthetic kaolinite II. *Clay Miner.* **10**, 257-270.

Johnson N. M. (1966): Geothermometry from TL of contact metamorphosed limestones. *J. Geol.* **74**, 607-619.

Kasuya M., Furusawa M. and Ikeya M. (1990): Distributions of paramagnetic centers and α-emitters in a zircon single crystal. *Nucl. Tracks.* **17**, 563-568.

Kasuya M., Ikeya, M. and Danhara T. (1992): Single grain ESR analysis of zircon: Comparison with fission-tracks. *Quat. Intern.* **13/14**, 131-134.

Livingston R. and Zeldes H. (1966): Paramagnetic resonance study of liquids during photolysis: hydrogen peroxide and alcohols. *J. Chem. Phys.* **4**, 1245-1259.

Marfunin A.S. and Bershov L.V. (1970): Electron-hole centers in feldspar and their possible crystalchemical and petrological significance (in Russian). *Dokl. Akad. Nauk. SSSR.* **193**, 412-414.

Marfunin A.S. (1979): *Spectroscopy, Luminescence and Radiation Centers in Minerals*

(Springer-Verlag, Heidelberg).

McBride M.B. (1990): Electron spin resonance spectroscopy. In *Instrumental Surface Analysis of Geologic Materials* ed. Perry D.L. (VCH, New York).

Meads R.E. and Malden P.J. (1975): Electron spin resonance in natural kaolinites containing Fe^{3+} and other transition metal ions. *Clay Minerals* **10**, 313-345.

Miyakawa K. and Tanaka K. (1985): An ESR study on the geothermal histories of the Kurobe, Central Japan. *ESR Dating and Dosimetry* 165-173.

Muller J. P. and Calas G. (1989): Tracing kaolinites through their defect centers: kaolinite paragenesis in a laterite (Cameroon). *Economic Geol.* **84**, 694-707.

Naeser C.W., Ergel J.C. and Dodge F.C.W. (1970): Fission track annealing and age determinations of epidote minerals. *J. Geophys. Res.* **75**, 1579-1584.

Nagai S., Ohnishi S. and Nitta I. (1974): ESR study of Cu(II) ion complexes absorbed on interlamellar surface of montmorillonite. *Chem Phys. Lett.* **26**, 517-520.

Nagasawa K. and Noro H. (1987): An electron spin resonance of halloysites. *Clay Sci.* **6**, 261-268.

Nambi K.S.V. and Sankaran A.V. (1985): ESR dating of laterite of basaltic origin. *ESR Dating and Dosimetry* 175-180.

Ogoh K., Toyoda S., Ikeda S., Ikeya M. and Goff F. (1993): Cooling history of the Valles caldera, New Mexico using ESR dating method. *Appl. Radiat. Isot.* **44**, 233-237.

Perry D.L. ed. (1990): *Instrumental Surface Analysis of Geologic Materials* (VCH, New York).

Pinnavaia T.J. (1980): Applications of ESR spectroscopy to inorganic clay systems. In *Advanced Chemical Methods for Soil and Clay Minerals Research* eds. Stucki J.W. and Banwart W.L. (Reidel Pub.).

Poole C.P., Farach H.A. and Bishop T.P. (1978): Electron spin resonance of minerals: part II silicates. *Magn. Reson. Rev.* **4**, 225-289.

Porat N. and Schwarcz H.P. (1991): Use of signal subtraction methods in ESR dating of burned flint. *Nucl. Tracks* **18**, 203-212.

Porat N., Schwarcz H.P., Valladas H., Bar Yosef O. and Vandermeersch B. (1994). Electron spin resonance dating of burned flint from Kebara cave, Israel. Geoarchaeology 9, 175-190.

Radtke U. and Broeckner H. (1991): Investigation on age and genesis of silcretes in Queensland (Australia): preliminary results. *Earth Surf. Processes Landform* **16**, 547-554.

Regulla D.F., Wieser A. and Goeksu H. Y. (1985): Effect of sample preparation on the ESR spectra of calcite, bone and volcanic materials. *Nucl. Tracks* **10**, 825-830.

Rink W.J., Toyoda S., Rees-Jones J., *et al.* (1999): Thermal activation of OSL as a geothermometer for quartz grain heating during fault movements. *Radiat. Meas.* **30**, 97-105.

Rinneberg H. and Weil J. A. (1972): EPR studies of Ti^{3+}-H^+ center in X-irradiated α-quartz. *J. Chem. Phys.* **56**, 2019-2028.

Sasaoka H., Yamanaka C. and Ikeya M. (1994): An ESR study of NH^{3+}, Al center and HO_2 radicals in alkali feldspars. *Appl. Magn. Reson.* **8**, 243-254.

Sasaoka H., Yamanaka C. and Ikeya M. (1996): Is the quartet due to $\bullet CH_3$ and $\bullet C_2H_5$ or $\bullet NH_3^+$ in alkali feldspars? *Appl. Radiat. Isot.* **47**, 1415-1417.

Shimokawa K., Imai N. and Hirota M. (1984): Dating of a volcanic rock by electron spin resonance. *Isotope Geoscience* **2**, 365-373.

Shimokawa K. and Imai N. (1987): Simulataneous determination of alteration and eruption ages of volcanic rocks by electron spin resonance. *Geochim. Cosmochim. Acta* **51**, 115-119.

Shimokawa K., Imai N. and Moriyama A. (1988): ESR dating of volcanic and baked rocks. *Quat. Sci. Rev.* **7**, 529-532.

Skinner AR, Rudolph MN (1996): The use of the E' signal in flint for ESR dating. *Appl. Radiat. Isot.* **47**, 1399-1404.

Slichter C. P. (1960): Principle of Magnetic Resonance (Harper & Row, New York).

Smith J. V. (1974): Lunar mineralogy: a heavenly detective story presidential address,

354   10. Silica and silicates: Geotherm and Volcanism

Part I. *Am. Mineral.* **59**, 231-243.

Solntsev V.P. and Shcherbakova M.Y. (1972): *Zh. Strukt. Chim.* **13**, 924.

Solntsev V.P. and Shcherbakova M.Y. (1973): EPR study of structure defects in irradiated zircons. *Dokl. Akad. Nauk.* **212**, 156-158.

Solntsev V.P., Shcherbakova M.Y. and Dvornikov E.V. (1974): Radicals SiO_2^-, SiO_3^{3-} and SiO_4^{5-} in $ZrSiO_4$ structure according to EPR data. *Zh. Strukt. Chim.* **15**, 217-222.

Stucki J.W. and Banwart W.L. ed. (1980): *Advanced Chemical Methods for Soil and Clay Minerals Research* (Reidel Pub.).

Speit B and Lehmann G. (1976): Hole centers in feldspar sanidine. *Phys. Stat. Sol.* **36**, 471-481.

Speit B. and Lehmann G. (1982): Radiation defects in feldspars. *Phys. Chem. Minerals.* **8**, 77-82.

Taguchi S., Harayama M. and Hayashi M. (1985): ESR signal of zircon and geologic age. *ESR Dating and Dosimetry*, 191-196.

Takashima I. (1985): Thermoluminescence dating of volcanic rocks and alteration of minerals and their applications to geothermal history. *Bull. Geol. Surv. Japan* **36**, 321-336.

Takeuchi A. and Saeki R. (1985): Electron spin resonance (ESR) signals of pelagic sediments in the southern Pacific. *ESR Dating and Dosimetry*, 125-133.

Tamanyu S. and Suto S. (1978): Stratigraphy and geochronology of Tamagawa welded tuff in the western part of Hachimantai, Akita. *Bull. Geol. Surv. Japan* **29**, 159-169.

Tani A., Bartoll J., Ikeya M., Komura K., Kajiwara H., Fujimura S., Kamada T. and Yokoyama Y. (1997): ESR study of thermal history and dating of a stone tool. *Appl. Magn. Reson.* **13**, 561-569.

Taylor L. A. and Cirlin E. H. (1985): A review of ESR studies on lunar samples. *ESR Dating and Dosimetry*, 19-29.

Toyoda S. and Ikeya M. (1989): ESR as a paleothermometer of volcanic materials. *Appl. Radiat. Isot.* **40**, 1171-1175.

Toyoda S. and Ikeya M. (1991a): ESR dating of quartz and plagioclase from volcanic ashes using E', Al and Ti centers. *Nucl. Tracks* **18**, 179-184.

Toyoda S. and Ikeya M. (1991b): Thermal stabilities of paramagnetic defect and impurity centers in quartz: basis for ESR dating of thermal history. *Geochem. J.* **25**, 437-445.

Toyoda S., Ikeya M., Dunnell R. C. and McCutcheon P. T. (1993): The use of electron in spin resonance (ESR) for the determination of prehistoric lithic heat treatment. *Appl. Radiat. Isot.* **44**, 227-231.

Tsay F.D., Chan S.I. and Stanley L.M. (1972): Electron paramagnetic resonance of radiation damage in lunar rock. *Nature* **237**, 121-122.

Vinokurov V.M., Gainullina N.M., Evgrafova L.A., Nizamutdinov N.M. and Suslina A.N. (1971): Zr^{4+}-Y^{3+} isomorphism in zircon. *Sov. Phys. Cryst.* **16**, 262-265.

Yokoyama Y., Falgueres C. and Quaegebeur J. P. (1985): ESR dating of sediment baked by lava flows: comparison of paleo-dose for Al and Ti centers. *ESR Dating and Dosimetry*, 197-204.

Walther R. and Zilles D. (1994): ESR studies of flint with a difference-spectrum method. *Quat. Geochr.* **13**, 635-639.

Warashina T., Higashimura T. and Maeda Y. (1981): Determination of firing temperature of ancient pottery by means of ESR spectroscopy. *Brit. Mus. Occas. Papers* **19**, 117-128.

Wauchope R. D. and Haque R. (1971): ESR in clay minerals. *Nature* **233**, 141-142.

Weeks R. A. (1973): Paramagnetic resonance spectra of Ti^{3+}, Fe^{3+} and Mn^{2+} in lunar plagioclase. *J. Geophys. Res.* **78**, 2393-2401.

Wild M.T., Tabner B.J., Macdonald R. (1999): ESR dating of quartz phenocrysts in some rhyolitic extrusive rocks using Al and Ti impurity centres. *Quat. Sci. Rev.* **18**, 1507-1514.

Woda C., Mangini A., Wagner G.A. (2001): ESR dating of xenolithic quartz in volcanic rocks. *Quat. Sci. Rev.* **20**, 993-998.

Chapter 11

Solid H_2O and CO_2

– Space and Environmental Sciences –

The stability of radiation-induced radicals in solid H_2O, CO_2 and CH_4 indicates the possibility of future ESR dating of comets, meteorite impact craters and volcanoes of those solid materials on the outer planets and their satellites. Comet studies with a light-weight ESR instrument will be a first step for ESR investigations in planetary research.

11.1   Introduction

Our planet Earth and solar system are full of water (H_2O), to which we owe our lives. Unfortunately, radiation–induced radicals in liquids are not stable and decay in a short time. Solid H_2O, snow or ice, is formed on earth but the temperature is still generally too high to stabilize radicals induced by natural radiation in solid H_2O, even at the cold polar areas. However, when we look at the solar system, such an environment with extremely low ambient temperature sufficient to stabilize radicals in solid H_2O is common on the outer planets and their satellites, as well as on parts of comets. Solar wind protons, cosmic radiation and the radiation from the internal radioactivities produce defects.

The visible and infrared properties of frozen materials (H_2O, CO_2 and SO_2) in the solar system have been studied and compared with the reflection spectra of planets and satellites (Fink and Sill 1982). Infrared telescopy clarified that 98% of the surface of *Pluto* is covered with solid N_2 and a mixture of CH_4 and CO as in the case of *Triton* (*Neptune*'s satellite). Table 11.1 shows some properties of the planets and their satellites together with the surface temperatures and materials (Rothery 1992).

"Voyager" has collected tremendous amounts of information on outer planets, which has changed our understanding of the Earth. An ice volcano has been reported on the surface of *Titan* (one of the satellites of Saturn) in addition to ice craters caused by meteorite impacts. The presence of solid CO_2 (dry ice) has been reported at the polar caps of *Mercury* and *Mars*, while solid H_2O has not been detected so far. Solar winds and natural radiation in the low temperature environment might cause radiation effects. Volcanic eruptions and meteorite impacts might raise the temperature and reset the clock time of the accumulated defects in solid H_2O (ice) and CO_2 to zero, which makes it possible to determine the age of events.

The chemical composition of the *comet Halley* has been studied with ultra–violet (UV), visible and infrared (IR) spectroscopy as well as with *in–situ* mass spectroscopy using spacecraft such as Vega I and II and Giotto in the 1980's (Yamamoto 1991). The chemical evolution of a comet may be studied by examining radicals produced by UV and proton irradiation simulating the solar wind. In space and planetary sciences in the future, substances on planets and in interplanetary space such as solid H_2O, CO_2, NH_3, CH_3, N_2 and H_2, might be the objects of ESR dating by taking materials back to Earth or by launching a light–weight, compact ESR spectrometer (see Appendix 3) for remote sensing studies.

ESR spectroscopy, which can detect as few as 10^{10} spins if the signal has a linewidth of 0.1 mT, may detect trace impurities down to the ppb level through impurity–related radicals by irradiation or through converting intrinsic radicals into impurity–related ones. Environmental studies are being done using Antarctic ice and frozen rain irradiated at 77K. While ESR measurements of solid samples at outer planets and their satellites will probably be carried out in the 21st or 22nd century, this chapter describes the possibility of future ESR dating using these materials.

Table 11.1 Some properties of the planets and their satellites mostly after Science Table published by the National Astronomical Observatory, Tokyo (1993).

Planets (n [a]) satellite	Radius (km)	Mass [c] (Earth = 1)	Density (g/cm³)	T (°C)	Remarks [d]
Mercury [b]	2,439	0.055	5.43	430 ~ –160	ig (da ?)
Venus [b]	6,051	0.815	5.24	470	da, gr
Earth [b] (1)	6,378	1.000	5.52	35 ~ –58	H_2O, da, gr
Moon	1,738	0.012	3.55	140 ~ –130	da
Mars [b] (2)	3,397	0.107	3.93	25 ~ –136 (cloud)	con
Phobos	~ 11	1.6×10^{-9}	1.9		con
Deimos	~ 6.5	3.3×10^{-10}	1.5		con
Jupiter (16)	71,398	317.83	1.33	–150	H_2 + He
Io	1,815	0.015	3.53		S, SO_2, volcano
Europe	1,569	0.008	3.03	–133	ice
Ganymede	2,631	0.030	1.93	–119	ice
Callisto	2,400	0.018	1.79	–106	ice (100 km thick)
Saturn (18)	60,000	95.16	0.70	–180 (cloud)	H_2 + He
Mimas	195	7.610×10^{-6}	1.2		ice
Enceladus	250	1.4×10^{-5}	1.2		ice
Tethys	525	1.3×10^{-4}	1.2		ice
Dione	560	1.8×10^{-4}	1.4		ice
Rhea	765	4.2×10^{-4}	1.3		ice
Titan	2,560	2.3×10^{-2}	1.9		? covered with cloud
Iapetus	720	3.2×10^{-4}	1.2		ice + con
Uranus (15)	25,560	14.54	1.27	–210	? H_2 + He (cloud)
Titania	790	1.8×10^{-4}	1.6?		ice + rock
Oberon	762	4.9×10^{-5}	1.5?		ice + rock
Neptune (8)	24,760	17.15	1.64	–220	s–CH_4 + rock ?
Triton	1,350	0.022	2.07	–238	s–N_2, s–CH_4, s–CO
Nereid	170	1×10^{-9}	2.0?	?	
Pluto (1)	1,142	0.0022	2.07	–230	s–N_2, s–CH_4, s–CO
Charon	650	0.0006	1.85?		ice + rock ?

a) n is the number of satellites.
b) Terrestrial planets with basaltic origin.
c) Mass density normalized by the mass of Earth 5.98×10^{24} kg.
d) Abbreviation: s– (solid), ig (igneous rock), da (dacite), gr (granite), con (chondritic rock). "ice" indicates only solid H_2O in this book.

11.2 Solid H$_2$O

11.2.1 *Structures and Models of Defects*

(a) *Structure* :   Solid H$_2$O has eleven morphological structures and one of them is *hexagonal* (I$_h$) with one proton between oxygen atoms as shown in Figure 11.1 (a). A proton forms a covalent bond with the nearest neighbor oxygen atom and a hydrogen bond with another oxygen atom. In the bent molecule H$_2$O, the H–O–H angle is 104.52° and the bond length of O–H is 0.957 Å. The physics and chemistry of ice, frost, snow, etc., have been studied (Fletcher 1970, Maeno and Hondoh 1992). Amorphous ice having a low thermal conductivity has been formed in the laboratory by UV–irradiation of a cubic ice (I$_c$) film and detected using electron diffraction patterns (Kouchi and Kuroda 1990, Kouchi *et al.* 1992).

(b) *ESR spectra and models of defects* :   Low temperature irradiation of solid ice both pure and doped with impurities has been studied with ESR. The following defects are observed in irradiated solid H$_2$O and their ESR parameters are summarized in Table 11.2.

·OH radical :   Figure 11.2 shows ESR spectra of polycrystalline H$_2$O (a) irradiated by 4 MeV X–rays and measured at 4.2K and (b) irradiated at 4.2K and warmed to 98K. The spectrum (c) is a computer simulation using aniso-

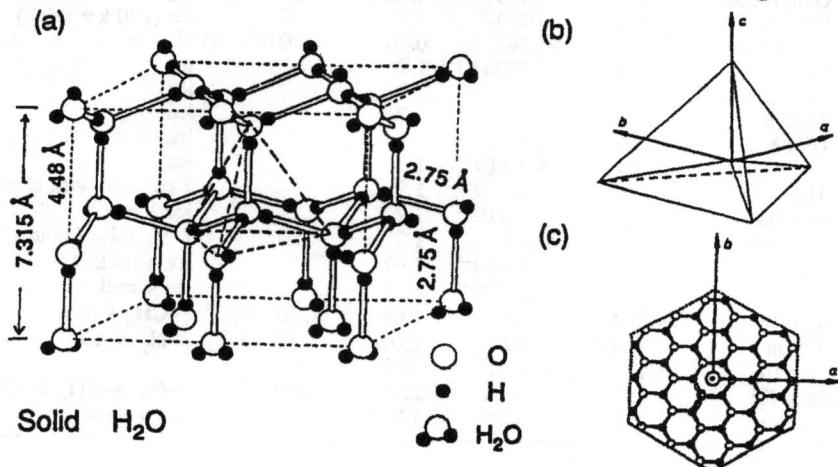

Figure 11.1   (a) Crystal structure of hexagonal ice (solid H$_2$O) lattice with the nearest neighbor oxygen tetrahedron. (b) Coordinate axes of a distorted oxygen tetrahedron. (c) Orientation of the coordinate system. The O–O length is 4.48 and 2.75 Å (after Bednarek and Plonka 1987b).

tropic g factors of three types of ·OH radicals shown in Table 11.2 (Johnson and Moulton 1987).   ENDOR studies using an ice crystal indicated three configurations (I), (II) and (III) with slightly different distances between O and H (Box *et al.* 1970). These give different anisotropic g factors and hf constants. The models for these three ·OH radicals are shown in Figure 11.3 (Bednarek and Plonka 1987b).

The microwave power dependence indicates that the intensity increases up to 100 mW. The ·OH radical is annealed out above 120K. Though poly-crystalline powder was used in the experiment, a slight angular dependence was observed due to crystalline properties (Tsukamoto *et al.* 1993).

·OH radicals in glassy and crystalline ice matrices doped with salts such as NaClO$_4$ and H$_2$SO$_4$ have been studied (Riederer *et al.* 1983).

HO$_2$· radical : Signals of HO$_2$· radicals are observed when ·OH radicals are annealed at 130K. It has been suggested that HO$_2$· radicals are formed by transport of two ·OH radicals to one which remains bound in its original site. Thus, three types of HO$_2$· radicals are formed from these three types of ·OH radicals as shown in Figure 11.3 (Bednarek and Plonka 1987b).

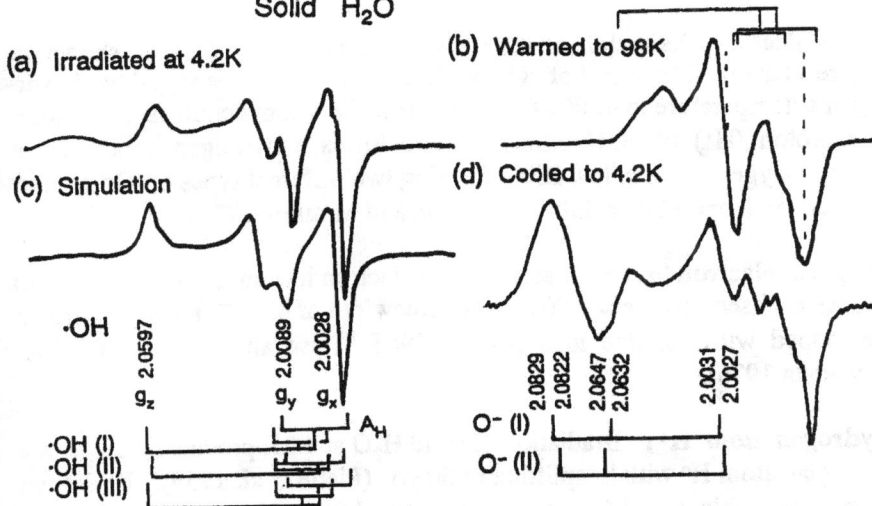

Figure 11.2   ESR spectra of (a) solid H$_2$O irradiated and measured at (a) 4.2K and   (b) the sample subsequently warmed to 98K.   (c) Computer simulation of a powder spectrum using parameters for ·OH I, II and III in a crystal. (d) Cooling again to 4.2K enhances the O$^-$ signal at $g = 2.08$ (after Johnson and Moulton 1987).

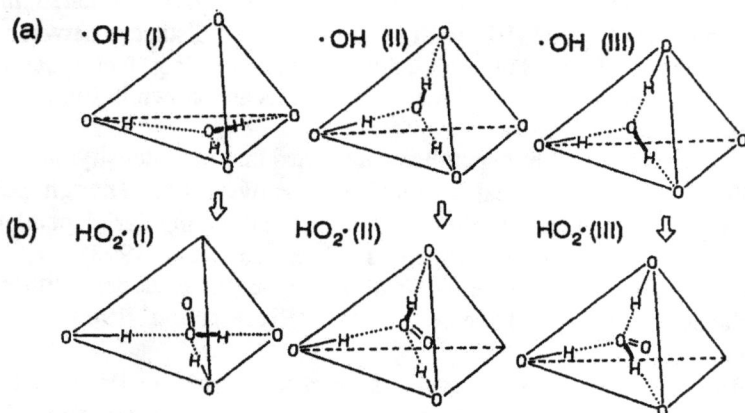

Figure 11.3 (a) Models of three types of ·OH radicals (I), (II) and (III) in the oxygen tetrahedron in Figure 11.1 and (b) conversion into HO_2· after annealing of ·OH above 120K (Bednarek and Plonka 1987b).

O^- radical :   Irradiation at 4.2K gives a signal at $g = 2.08$ as shown in Figure 11.2 (a). The signal of ·OH at 77K changes into this signal by lowering the temperature to 4.2K as shown in (d). This conversion is reversible. The proton (H^+) of ·OH dissociates and forms a hydrogen bond with a nearby oxygen atom below 20K, forming two different types of O^- (I and II) (Bednarek and Plonka 1987a, Johnson and Moulton 1978).

Trapped electron e_t^- :   A self trapped electron like an F center in alkali halides is observed at $g = 2.001$ with a linewidth of 4.3 mT mostly in glassy ice doped with alkaline ions (Kevan 1965, Hase and Kawabata 1976, Kawabata 1976).

Hydrogen atom H^o :   Irradiation of solid H_2O at 4.2K produces signals of a hydrogen atom H^o with hf splitting of 50 mT (Piette et al. 1959). Hydrogen atoms are stable at 4.2K for several hours and form H_2 molecules at 77K (Siegel et al. 1960, 1961).

Table 11.2 ESR parameters of radiation–induced defects in solid H_2O.

Defects	g factor				A tensor in mT			Ref.
	g_{zz}	g_{xx}	g_{yy}	g_{av}	A_{zz}	A_{yy}	A_{xx}	
·OH(I)	2.0597	2.0028	2.0089	2.0240	2.85	4.42	0.32	1
(II)	2.0571	2.0031	2.0089	2.0232	2.80	4.42	0.67	
(III)	2.0581	2.0027	2.0088	2.0234	2.79	4.49	0.54	
	$g_\parallel$=2.0615	$g_\perp$=2.0095		2.0270	$A_\parallel$=0.7	$A_\perp$=4.3		2
	2.0585	2.0050	2.0090	2.0243	0.5	4.4	2.6	3
	2.06	2.005	2.009	2.025	0	2.6	4.4	4
HO_2· (I)	same g factors as D_2O				0.8	1.3	1.6	5
(II)	"				0.5	1.1	1.5	
(III)	"				0.3	1.0	1.4	
DO_2· (I)	2.0376	2.0025	2.0117	2.0173	no hf line			5
(II)	2.0443	2.0022	2.0081	2.0183	"			
(III)	2.0455	2.0023	2.0105	2.0195	"			
O^- (I)	2.0031	2.0647	2.0822	2.0503				6
(II)	2.0027	2.0632	2.0829	2.0499				
			g_{iso}=2.08					7
e^-_t H_2O			g_{iso}=2.001		(ΔH = 4.3)			8
D_2O			g_{iso}=2.001		(ΔH = 3.1)			9

Reference: 1) Box *et al.* 1970; 2) Gunter 1967; 3) Dibdin 1967; 4) Brivati *et al.* 1969; 5) Bednarek and Plonka 1987b; 6) Bednarek and Plonka 1987a; 7) Johnson and Moulton 1978; 8) Kevan 1965; 9) Hase and Kawabata 1976.

11.2.2 Lifetime of Defects at Ambient Temperature

The signal intensity of ·OH is decreased slowly at 100K and rapidly above 120K by isothermal annealing of γ–irradiated ice. The decay follows first–order kinetics, from which the lifetime at respective temperatures is obtained. After the ·OH signal disappears, a weak, broad line remains until the temperature reaches 140K. This signal is derived from HO_2· radicals (Bednarek and Plonka 1987b) and is stable up to 160K, but decays rapidly above 160K. Thermoluminescence is observed around this temperature.

An Arrhenius plot of the lifetime obtained by isothermal annealing of ·OH gives an activation energy of 0.085 eV from the slope as shown in Figure 11.4. The annealing temperature of 120K is too low for radicals to be stabilized at ambient temperature on Earth even in the polar area. However, reported temperatures at the outer planets and their satellites are sufficiently low as to allow the accumulation of defects as indicated in the figure. An

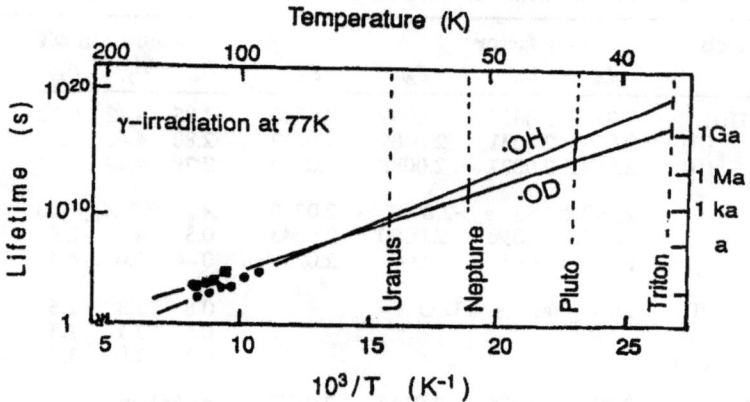

Figure 11.4 Arrhenius plot of the lifetime for radiation–induced ·OH and ·OD radicals in polycrystalline H_2O and D_2O, respectively. The lifetime was extrapolated down to ambient temperature on the outer planets and their satellites to indicate a possibility of ESR dating.

extrapolation gives the lifetime of ·OH radicals at the ambient temperature, to be 10^2 Ga at Triton, a satellite of Neptune. ESR dating is possible based on the assumption that the accumulated radiation effects are erased at *"time zero"* of a dating clock by some events.

11.2.3   Growth Curve of ·OH Radicals

Figure 11.5 shows the growth of ·OH radicals in solid H_2O irradiated by an electron beam (Gunter 1967). A linear increase with the radiation dose and saturation at a high dose (> 1 kGy) are observed. The concentration, n, is expressed as the number of radicals per gram of weight. The theoretical growth curve by artificial irradiation at a dose rate D' is written as

$$n = (N_0/b)[1 - \exp(-abD't'/N_0)] , \qquad (11.1)$$

where N_0 is the initial radical concentration, a is the defect production efficiency and b is the insensitive volume (see Section 3.5.1 of Chapter 3).

Parameters a and b are obtained from the initial growth, $aD't'$, and the saturation concentration, $n_s = N_0/b$, respectively. From the initial growth rate, one can get the G–value (the number of radicals formed per 100 eV). These parameters are calculated using Eq. (3.23) and (3.24) to be

a = 4.6×10^{15} Gy^{-1}g^{-1} = 6.25×10^{15} $G\rho$

$\therefore$ G = $a/6.25 \times 10^{15}\rho$ = 0.8 ($\rho = 0.917$ for solid H$_2$O)

N_o = $6.02 \times 10^{23}/(16+2)$ = 3.34×10^{22}/g

b = N_o/n_s = $3.34 \times 10^{22}/5 \times 10^{18}$ = 6.7×10^3

The reconstructed growth curve using the obtained parameters is shown by a solid curve in the figure. The saturation dose $(D't' = N_o/ab)$ that yields the saturation growth, $n = n_s(1 - e^{-1}) = 0.6321 \times n_s$, gives the saturation lifetime, $\tau_s = n_s/aD$, at an ambient dose rate. If the lifetime τ at ambient temperature is larger than τ_s, the additive dose method may be used. If τ is comparable to or less than τ_s, the real TD must be calculated using the correction factor (see Section 3.5.2 (d) of Chapter 3).

The thermal behavior of ·OH radicals in γ–irradiated crystalline ice has been studied and it was found that the G–value at 77K is nearly half of that at 4.2K due to thermal migration of the hydrogen atom H° in H$_2$O (Siegel *et al.* 1960, 1961).

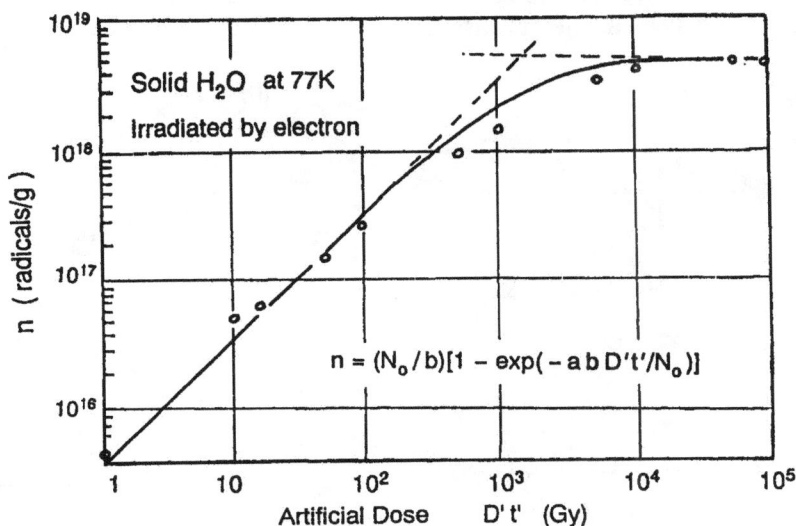

Figure 11.5 Growth curve of ·OH radicals in 1 g solid H$_2$O as a function of electron irradiation dose at 77K. The reconstructed growth curve using the calculated parameters is shown by a solid line (revised from Gunter 1967).

11.2.4 *Environmental Sciences*
(a) SO_3^- *in Solid H_2O : Acid Rain*

Effects of impurity anions on radical formation in solid H_2O have been studied. The signal intensity at $g_{\parallel} = 2.0036$ and $g_{\perp} = 2.0020$ is enhanced considerably in γ–irradiated solid H_2O doped with SO_3^{2-}. O_3^- radicals are stable after ·OH radicals are annealed out. One can detect the concentration down to $10 \sim 100$ ppm of SO_3^{2-} in irradiated solid H_2O.

SO_3^- radicals were detected in *frozen rain* after γ–irradiation of 10 kGy. The rain was collected at a site close to a heavy traffic road in the suburbs of Osaka, Japan. The same signal was observed in bore–hole core *antarctic ice*, presumably due to acid rain caused by volcanic activity.

(b) *Atmospheric Radicals in Ice Matrix : Troposphere*

Radicals present in atmosphere have been collected in a polycrystalline ice matrix for ESR measurements (Mihelcic *et al.* 1990). Air samples were collected on a cooled copper finger together with vapor of H_2O and D_2O. Peroxy radicals such as HO_2 and CH_3O and NO_2 were measured in free troposphere (altitude > 10 km) and atmosphere boundary layer. The procedure may be used in the studies of air polution and planetary atmospheres using a small ESR spectrometer.

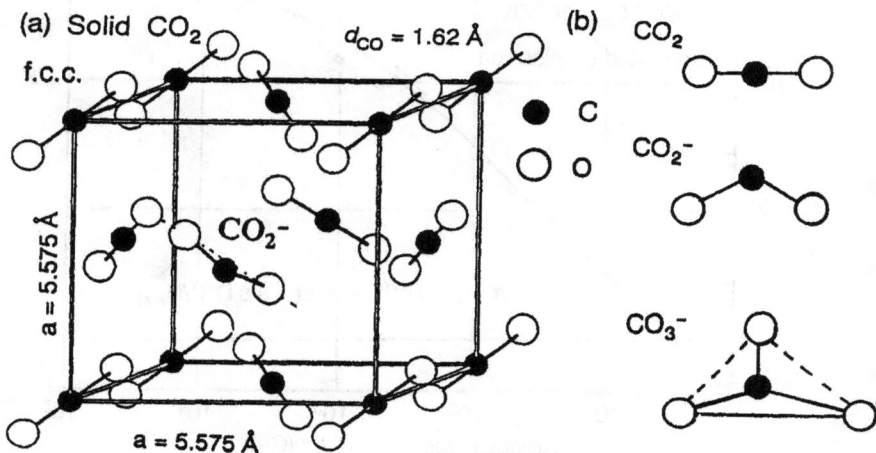

Figure 11.6 (a) The crystal structure of face–centered cubic (fcc) CO_2. The molecular crystal is bound by Van der Waals forces. (b) The linear CO_2, bent CO_2^- and planar CO_3^- molecules.

11.3 Solid CO2

11.3.1 *Structure and Models of Defects*

Molecule of CO2 is a straight rod having the distance between C and O of 0.162nm. The crystal structure of solid CO$_2$ is face-centered cubic (fcc) with a lattice constant of a = 0.5575 nm (Wyckoff 1982) (see Figure 11.6). The defects observed in γ-irradiated solid CO$_2$ are summarized in Table 11.3.

(a) *Intrinsic defects:* Figure 11.7 shows an ESR spectrum of solid CO$_2$ irradiated by g-rays at 77K and measured at 100K. ESR signals are grouped into several sets, which are identified based on radiation-induced defects in carbonates (see Chapter 5). The signals of a hole type-center CO$_3^-$ and an electron center CO$_2^-$ were identified in solid CO$_2$ (Tsukamoto *et al.* 1993; Hirai *et al.*, 1994). A hole type center CO$^+$ was not identified though CO$_3^-$ (an interstitial O$^-$) was observed.

(b) *Impurity-related defects:* Irradiation of solid CO$_2$ doped with water vapor, gaseous SO2 and CH$_4$ produces signals of impurity-related defects. OH radicals with similar g factors but different linewidths, methyl radicals (CH$_3$) and SO$_2^-$ were observed in addition to CO$_2^-$ and CO$_3^-$. Radicals of HO$_2$ generated by electric discharges in CO$_2$ atmosphere with moisture were observed in the dry ice frost condensated on the cold tip from the CO$_2$ atmosphere.

Table 11.3 ESR parameters of radiation-induced defects in solid CO$_2$.

Defects	Material	g factor				Ref
		g_{zz}	g_{xx}	g_{yyy}	g_{av}	
Intrinsic defects						
CO$_3^-$	solid CO$_2$	2.006	2.013	2.016	2.012	1, 8
CO$_2^-$	solid CO$_2$	2.001	2.003	1.998	2.001	1, 8
O$_3^-$	solid CO$_2$	2.014	2.011	1.993	1.993	1, 8
Impurity-related defects						
Li+-CO$_2^-$	solid CO$_2$-Li complex	g_-=1.9972	g_+=2.0024	2.0007		2
_OH	solid CO$_2$ with moisture	2.055	2.0029	1.995	2.0178	
SO$_2^-$	solid CO$_2$ doped with SO$_2$				g_{iso} = 2.006	1
_CH$_3$	solid CO$_2$ doped with CH$_4$				g_{iso} = 2.00251	
			(A = 22.5 mT; quartet)			
HO$_2$	solid CO$_2$ doped with H$_2$O	2.004	2.006	2.036	2.015	8
		(A = 1.3	1.85	1.8 mT)		
c.f.						
CO$_2^-$	HCOONa	2.0014	2.0032	1.9975	2.0007	3
CO$_2^-$	CaCO$_3$	2.0016	2.0032	1.9973	2.0007	4
CO$_2^-$	KHCO$_3$	2.0066	2.0086	2.0184	2.012	5
^{13}CO	MgO	g_l=2.0055	g_+=2.0021		2.0032	6
CO$^+$	CO on H-Y zeolite	g_l=2.0045	g_+=2.0005		2.0018	7

Reference:1) Tsukamoto *et al.*,1993; 2) Borel et al. 1981; 3) Ovenall and Whiffen 1961; 4) Marshall et al. 1965; 5) Chantry et al. 1962; 6) Lunsford and Jayne 1966; 7) Vedrine and Naccache 1973; 8) Hirai *et al.*, 1994. (Table revised in 2001).

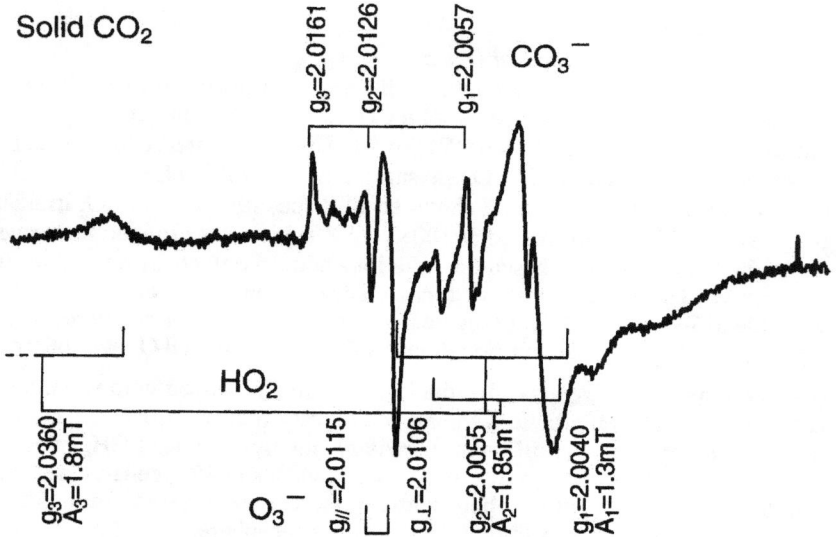

Figure 11.7   ESR spectrum measured at 77 K of polycrystalline solid irradiated with γ-rays at 77K. The microwave power is 10 mW, the magnetic field modulation is 0.1 mT and the time constant is 0.1 s. The signals are identified based on the radiation-induced defects in carbonates.

11.3.2 *Lifetime of defects*

Most atmospheric carbon dioxide (CO_2) on Earth has been changed into $CaCO_3$ biologically during the last 600 millions years. Solid CO_2 (dry ice) is present in nature on several planets. Isothermal annealing experiments indicate that decay of the defects in solid CO_2 follows more or less first- order kinetics from which the lifetimes are obtained. Radicals of HO_2 and CH_3 are more stable in solid CO_2 than in H_2O and D_2O. The lifetimes of several defects are shown in Figure 11.8. Extrapolation of the Arrhenius plot gives the lifetimes of defects at ambient temperature of the outer planets and their satellites. ESR dating is thus possible using the radiation-induced defects.

11.4 Solid Methane (CH_4) and Others in Space
11.4.1 *Solid CH_4*

Methane (CH_4) is a thermodynamically stable molecule. The blue color of Uranus and Neptune is due to the presence of gaseous CH_4 in atmospheres of H_2 and He. Neptune is known to have two satellites, Triton and Nereid. Naturally,

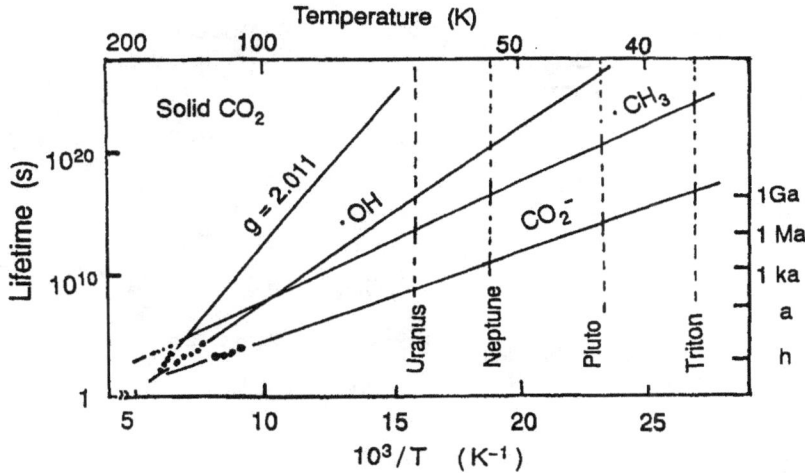

Figure 11.8 Lifetimes of several defects in solid CO_2 (dry ice) as a function of reciprocal temperature in an Arrhenius plot. The ambient surface temperatures are indicated in the figure.

solar system. Voyager II discovered additional 5 satellites of Neptune and craters of solidified CH_4 and N_2 on the surface of Triton in 1989. *Black smoke* (gaseous N_2 and solid CH_4) was observed a few km above the ground volcanic eruptions.

Studies on radical formation in solid CH_4 gave the following defects (Gordy and Morehouse 1966, Timm and Willard 1969):

Hydrogen atoms (H^o) : $g_H = 2.00207;$ $A_H = 50.8$ mT
Methyl radical ($\cdot CH_3$) : $g_{CH_3} = 2.00242;$ $A_{CH_3} = 2.3$ mT; $G = 0.7$
Paired H^o and $\cdot CH_3$: $g_{pair} = (g_H + g_{CH_3})/2;$ $A_{pair} : A_H/2$ & $A_{CH_3}/2$
($\cdot CH_3 ... H^o$)

Figure 11.9 (a) shows an ESR spectrum of solid CH_4 irradiated with X–rays at 4.2K and a simulation spectrum of weak pair signals (Toriyama *et al.* 1979). A doublet due to H^o and a quartet due to $\cdot CH_3$ are clearly ob–served in addition to weak satellite signals due to an $H^o - \cdot CH_3$ pair.

A singlet ($S = 1$) and a triplet ($S = 1$) in the energy levels arise because of the exchange coupling of $S_1 + S_2 = 1$ and $S_1 - S_2 = 0$. Radical pairs in irradiated alkanes have also been studied (Kurita 1964, Toriyama *et al.* 1979,

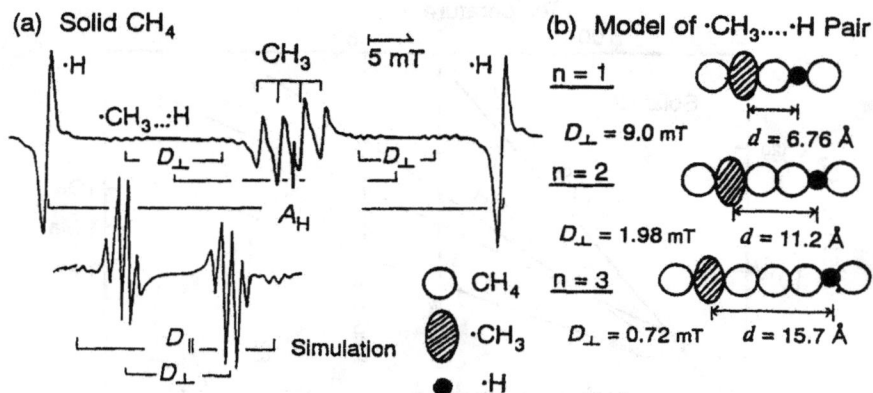

Figure 11.9   (a) ESR spectrum of solid CH_4 irradiated at 4.2K (revised after Toriyama *et al.* 1979). The simulation spectrum of the weak pair signals is shown below.   (b) Models for exchange coupled pairs ($S = 1$) of $H°....·CH_3$. n is the number of CH_4 molecules present between $H°$ and $·CH_3$.

Harrison and Symons 1993). The anisotropic exchange coupling for the triplet ($S = 1$) gives a doublet ($S = 1$: 2S lines) fine structure splitting with a separation of $(D/2)(3\cos^2\theta - 1)$ as shown in Figure 2.18. The angle dependent spectrum is averaged for polycrystalline CH_4 giving prominent signals at $D_{||} = 2D$ ($\theta = 0°$) and $D_\perp = -D$ ($\theta = 90°$) as shown in Figure 2.22 (a) and Section 2.5.2. These lines further split into a hf doublet with a hf constant of $A_H/2$ by $H°$ and a quartet with $A_{·CH_3}/2$ because an unpaired electron is equally shared by both $H°$ and $·CH_3$ in the exchange coupled pair; the hf coupling constant is proportional to the spin density at the nucleus due to the *Fermi contact interaction* (see Section 2.4.4). Models for exchange coupled pairs at several distances are given in Figure 11.9 (b). A mechanism of pairwise trapping is discussed (Toriyama *et al.* 1991).

     Annealing of these coupled pairs and formation of $·C_2H_5$ have been studied in addition to the theoretical estimation of the spectra for pairs separated by CH_4 molecules ($n = 1, 2, 3 \cdots$). Coupled pairs might be formed by heavy irradiation of cosmic rays and UV photons, which may lead to polymerization on the outer planets. Dense excitation by energetic heavy ions on the CH_4 frost produces aromatic materials (Kaiser *et al.* 1994). Hence, petroleum of non–biological origins might be abundant on the outer planets and their satellites where solid CH_4 are present. Neither growth curve nor isothermal annealing has been measured to study the lifetime and the saturation concentration of radiation–induced defects in solid CH_4.

11.4.2   Other Solids

Elements such as H, C, N and O and their compounds which exist as gaseous molecules on Earth are in a solid form on the outer planets and their satellites. They may form "minerals" and "rocks" of various phases and compositions under various temperatures and pressures. A systematic study on phase diagrams has to be carried out to understand the structure of the outer planets. Defects in other solid molecules are as follows.

(a)   *Solid H_2*:   Irradiation of solid H_2 at 4.2K gives only a hf doublet of H^o having a linewidth of 0.04 mT. The saturation level and dose as well as the defect formation efficiency ($G = 0.2$) are reported in early work (Piette *et al.* 1959). The saturation level ($N_s/b = 1.2 \times 10^{19}$/g) gives the *interaction volume*, $b = 2.5 \times 10^4$ H_2 sites. The saturation dose of 72 kGy suggests the saturation lifetime, $\tau_s = 72$ Ma for an annual dose rate of 1 mGy/a. The actual dose rate may be much smaller at Jupiter. No data on isothermal annealing experiment are available to estimate the lifetime.

(b)   *Solid N_2 and NH_3*:   In solid N_2 bombarded by a 1 keV energetic particle, atomic N, N_3 and N_3^- have been detected with ESR and optical spectroscopy (Tian and Michl 1988). Nitrogen molecular dimer cation radicals, N_4^+, are observed at 4K in a matrix of Ne for both ^{14}N and ^{15}N isotopes (Knight *et al.* 1987).

Amino radicals, NH_2 are produced in irradiated solid NH_3 at 77K (Smith and Seddon 1970). It is interesting to note that the gaseous phase was irradiated by 1 MeV He^+ using a "*crossed beam technique*" and deposited. NH_2 radicals are trapped in a solid deposit. This indicates that initially signals might be present. Precaution may be necessary when one thinks of the condensation from natural (radiolysis) species.

Ammonium ions, NH_3^+ are also observed in many solids. Radiolysis of NH_4NO_3 gives signals of NO_3 and NO_3^{2-} (Jarke and Ashford 1975).

(c)   *Hydrocarbons (outer planet petroleum)*:   Hydrocarbons are abundant on outer planets where atoms lighter than Si and Fe coagulate to form the planet. Radio–astronomical work indicates numerous organic molecules and thus solid hydrocarbons may be on the outer planets and their satellite. Both *saturated and unsaturated hydrocarbons* are expected at outer planets. There are accumulated data on radical species in solidified methane, ethane, propane and other n-alkanes. Strangely, no atomic hydrogen, H^o is detected in alkanes except in CH_4 (Timm and Willard 1969).

Solid CH_4 and the organic substances have been exposed to solar winds and cosmic rays. Experimentally, aromatic molecules (Kaiser et al. 1994) and fullerene (C_{60}) and a diamond were formed by high density excitation of energetic particle irradiation. *Outer planet petroleum* of non-biological origins may be abundant at the outer planets and their satellites.

(d) *Solid SO_2 on Io and Europa and frosts of CO_2*
Schwalbe et al., (1975) identified SO_2^+ in γ-rayed solid SO_2, while Moore (1984) indicated SO_3 by proton irradiation. Kanosue (2000; PhD Thesis Osaka Univ.) identified various radical species and their thermal stability in γ-rayed solid SO_2 for ESR, electron spin echo and TL dating. Eleven species (tentative models are marked "?") of SO_3^-, SO_2^-, CO_3^-, CO_2^-, SCO^-(?), O_3^-, S_3^-(?), CS_2^- and others were suggested from ESR spectra in comparison with those in carbonates and sulfates. Norizawa et al., (2000) studied trapping of atmospheric HO_2 in solid CO_2 on icy satellites in outer planet world.

11.4.3 *Magnetic Resonance in Space?*
Microwave photons are commonly observed in space. Radio-astronomers have been studying microwaves and radiowaves from space and solar activities. The hf interaction of hydrogen atoms gives an emission of 21 cm radiowaves. A cosmic background explorer (COBE) satellite launched into the Earth orbit brought information on an almost uniform microwave background using differential microwave radiometers. An isotropic cosmic photon background with a black body temperature of about 3K was expected in a standard "big bang" scenario.

The magnetic field intensity in the universe ranges from 10^{-8} T to 10^8 T and is about 10 mT around the sun. There are plenty of photons in space and it would be no wonder that interstellar atoms, molecular ions, clusters of atoms and solid materials may cause magnetic resonance transitions. Such a natural magnetic resonances in the solar and galactic systems might be used to map the magnetic field intensity profile from the shift of the resonance frequency.

11.5 Summary
Radiation effects on solid H_2O, CO_2, CH_4 aannd SO_2 have been reviewed to establish ESR as a viable dating method for icy bodies like outer planets, their satellites and comets. The lifetime of radicals has been measured at several temperatures and extrapolated to ambient temperature at icy bodies in an Arrhenius plot. ESR dating of volcanoes and craters may be made using radical signals in this century. A light-weight ESR spectrometer should be launched for remote dating, material science and atmospheric measurements in planetary research.

References

Bednarek J. and Plonka A. (1987a): Single-crystal electron spin resonance studies on radiation-produced species in ice Ih: Part 1 The O- radicals. *J. Chem. Soc. Faraday Trans.* I **83**, 3725-3735. *ibid* Part 2 The HO2 radicals. *ibid* I **83**, 3737-3747.

Borel J. P., Faes F. and Pittet A. (1981): Electron paramagnetic resonance of Li-CO2 complexes in a CO2 matrix at 77K. *J. Chem. Phys.* **74**, 2120-2123.

Box H. C., Budzinski E. E., Lilga K. T. and Freund H. G. (1970): ENDOR study of X-irradiated single crystals of ice. *J. Chem. Phys.* **53**, 1059-1065.

Brivati J. A., Symons C. R., Tinling D. A. J. and Williams D. O. (1969): Unstable intermediates: Part LIX Electron spin resonance studies from 4 to 77K of hydrogen-bonded hydroxyl radicals in irradiated ice. *J. Chem. Soc.* (A) 719-720.

Chantry G. W, Horsfield A., Morton J. R. and Whiffen D. H. (1962): The structure, electron resonance and optical spectra of trapped CO3- and NO3. *Mol. Phys.* **5**, 589-599.

Dibdin G. H. (1967): ESR of g-irradiated single crystals of ice at 77K: identification of the hydroxyl radical and its trapping site. *Trans. Faraday Soc.* **63**, 2098-2111.

Fink U. and Sill G.T. (1982): *Infrared spectral properties of frozen volatiles.* Comets. ed. Widkearing L. L. (Univ. Arizona Press, Tueson)

Fletcher N. H. (1970): *The Chemical Physics of Ice* (Cambridge Univ. Press).

Gordy W. and Morehouse R. (1966): Triplet-state electron spin resonance of an H- atom-methyl radical complex in a solid matrix. *Phys. Rev.* **151**, 207-210.

Gunter T. E. (1967): Electron paramagnetic resonance studies of radiolysis of H_2O in the solid state. *J. Chem. Phys.* **46**, 3818-3829.

Harrison N. and Symons M. C. R. (1993): Detection of radical pairs in glassy system. *J. Chem. Soc. Faraday Trans.* 89, 59-61.

Hase H. and Kawabata K. (1976): Trapped electrons in crystalline D_2O ice at 4K. *J. Chem. Phys.* **65**, 64-67.

Hirai M., Ikeya M., Tsukamoto Y. and Yamanaka C. (1994): Radiolysis of solid CO_2 annd future electron spin resonance of outer planets. *Jpn. J. Appl. Phys.* **33**, L1453-L1455.

Ikeya M., Sasaoka H., Toda K., Kanosue K. and Hirai M. (1997): Future ESR and optical dating of outer planet icy materials. *Quatern. Geol.* **36**, 431-436.

F. H. and Ashford N. A. (1975): Electron paramagnetic resonance of NO_3, NO32- and O_2^- in irradiated NH_4NO3. *J. Chem. Phys.* **62**, 2923-2924.

Johnson J. E. and Moulton G. C. (1978): Study of ice irradiated at 4.2K, a thermally stable radical. J. Chem. Phys. **67**, 3108-3111.

Kaiser R. I., Klebs B. and Roessler K. (1994): Cosmic ray simulation in organic frosts. Nucl.Instr. Method B in press.

Kanosue K., Hirai M. and Ikeya M. (1996): Preliminary study for future ESR dating of solid SO_2. *Appl. Radiat. Isot.* **47**, 1433-1436.

Kanosue K., Toda H., Hirai M. Kanamori K. and Ikeya M. (1997): Thermoluminescence (TL) and ESR studies of gamma-irradiated SO_2 frost for future dating in outer planets. *Radiat. Meas.* **27**, 399-403.

Kawabata K. (1976): Electron traps in irradiated crystalline ice. *J. Chem. Phys.* **65**, 2235.

Kevan L. (1965): Cation interaction of trapped electrons in irradiated alkaline ice. J. Am. Chem. Soc. **87**, 1481-1484.

Knight L. B., Johannessen D., Cobranchi D. C. and Earl E. A. (1987): ESR and ab initio theoretical studies of the cation radicals $^{14}N_4+$ and $^{15}N_4+$: The trapping of ion- neutral reaction products in neon matrix at 4K. *J. Chem. Phys.* **87**, 885-897.

Kouchi A. and Kuroda T. (1990): Amorphization of cubic ice by ultraviolet irradiation. *Nature* **344**, 134-135.

Kouchi A., Greenberg J. M., Yamamoto T. and Mukai T. (1992): Extremely low thermal conductivity of amorphous ice: relevance to comet evolution. *Astrophys. J.* **388**, L73.

Kurita Y. (1964): Electron spin resonance study of radical pairs trapped in irradiated single crystals of dimethylglyoxime at liquid nitrogen temperature. *J. Chem. Phys.* **41**, 3926.

Lunsford J. H. and Jayne J. P. (1966): Study of CO radicals on magnesium oxide with electron paramagnetic resonance techniques. *J. Chem. Phys.* **44**, 1492-1496.

Maeno N. and Hondoh T. ed. (1992): *Physics and Chemistry of Ice* (Hokkaido UP).

Marshall S. A., Reinberg A. R., Serway R. A. and Hodges J. A. (1964): ESR absorption spectrum of CO_2^- molecule-ions in single crystal calcite. *Mol. Phys.* **8**, 225-231.

Mihelcic D., Volz-Thomas A., Patz H.W. and Kley D. (1990): Numerical analysis of ESR spectra from atmospheric samples. *J. Atm. Chem.* **11**, 271-297.

Muto H., Matsuura K. and Nunome K. (1992):   Large isotope effect due to quantum tunneling in the conversion reaction of electrons to H and D atoms in irradiated H2O and D2O ice. *J. Phys. Chem.* **96**, 5211-5213

Norizawa K. Kanosue K. and Ikeya M. (2000): Radiation effects in dry ice: models for a peak on the Arrhenius curve. *Appl. Radiat. Isot.* **52**, 1259-1263.

Norizawa K. Hirai M. Kanosue K. and Ikeya M. (2000): Trapping of atmospheric HO_2 in solid CO2 on icy satellitess: simulation using electron spin resonance and thermoluminescence analyses. *Jpn. J. Appl. Phys.* **39**, 6759-6762.

Ovenall D. W. and Whiffen D. H. (1961):   Electron spin resonance and structure of the CO2- radical ion. *Mol. Phys.* **4**, 135-144.

Piette L. H., Rempel R. C., Weaver H. E. and Flournoy J. M. (1959): EPR studies of electron irradiated ice and solid hydrogen. *J. Chem. Phys.* **30**, 1623-1624.

Riederer H., H ttermann J., Boon P. and Symons M. C. R. (1983): Hydroxyl radicals in aqueous glasses: characterization and reactivity studied by ESR spectroscopy. *J. Magn. Reson.* **54**, 54-66.

Rothery D.A. (1992): *Satellites of the Outer Planets* (Oxford Univ. Press, Oxford).

Siegel S., Baum L. H., Skolnik S. and Flournoy J. M. (1960): Observations of the thermal behavior of radicals in γ-irradiated ice. *J. Chem. Phys.* **32**, 1249-1256.

Siegel S., Flournoy J. M. and Baum L. H. (1961): Irradiation yields of radicals in γ-irradiated ice at 4.2 and 77 K. *J. Chem. Phys.* **34**, 1782-1788.

Smith D. R. and Seddon W. A. (1970): Electron spin resonance spectra of ^{15}N-labeled amino radicals. *Canad. J. Chem.* **48**, 1938-1942.

Tian R. and Michl J. (1988): Fast particle bombardment of solid nitrogen.   Faraday Discuss. *Chem. Soc.* no.**86**, 113-124.

Timm D. and Willard J. E. (1969): Absence of hydrogen atom production in radiolysis of solid hydrocarbons. *J. Chem. Phys.* 73, 2403-2408.

Toda H., Sasaokaa H., Kanamori H. Kaanosue K. , Hirai M. and Ikeya M. (1997): The nature of amino radicals in frozen ammonia water for future dating of comets. Proc. 1st Asia-Pacifcic EPR/ESR Symposium, 186-191.

Toriyama K., Iwasaki M. and Nunome K. (1979): ESR studies of irradiated methane and ethane at 4.2K and mechanism of pairwise trapping of radicals in irradiated alkanes. *J. Chem. Phys.* **71**, 1698-1705.

Toriyama K. (1991): ESR studies on cation radicals of saturated hydrocarbons. in *Radical Ionic System* (Kluwer Academic Pub) eds. Lund A. and Shiotani M. 99-124.

Tsukamoto Y., Ikeya M. and Yamanaka C. (1993): Fundamental study on ESR dating of outer planets and their satellites. *Appl. Radiat. Isot.* 44, 221-225.

Vedrine J. C. and Naccache C. (1973): ESR study of CO⁺ radicals on a H-Y type zeolite. *Chem Phys. Lett.* **18**, 190-194.

Wyckoff W.G. (1982): *Crystal Structure* (Wieley & Sons Inc., New York).

Yamamoto T. (1991): Chemical theories on the origin of comets. *Comets in Post- Halley Era* **Vol. 1**, 361-376.

Chapter 12

Chemical ESR Dating

– Organic Materials –

Chemical ESR dating can be extended from potato chips to a dead body. The chemical reaction of radical formation or oxidation can be used to assess the age as in racemization dating. The accuracy of this method depends on the assessment of the temperature.

12.1 Introduction

One advantage of ESR dating is to measure the concentration of trapped electron and/or hole centers at room or low temperature without heating the materials. No combustion of organic materials occurs in ESR measurement: for this reason, fossils that contain organic materials have been chosen as objects of dating by ESR. This advantage is further extended to include organic materials themselves as objects for ESR dating.

Organic radicals are produced by photochemical processes and chemical decomposition or oxidation (Fritsh *et al.* 1974) as well as by radiation (Gordy *et al.* 1955). Although radiation can create radicals, the natural radiation dose is too low in general to compete with the chemical process of radical formation by oxidation in archaeological and historical organic materials. The lifetime of most radical species is relatively short. Hence, organic materials cannot be objects of geological ESR dating, but ESR might be used for historical and forensic substances within the limitation.

Chemical ESR dating following a natural passage of time instead of involving additive irradiation was first used for dating potato chips (Ikeya and Miki 1980) and then rabbit skins using lipid peroxy radicals. The age was assessed by considering the decomposition rate equation (Ikeya and Miki 1986). A method of chemical ESR analysis to determine the age of organic materials by accelerating the speed of time passage at a high temperature is presented as an auxiliary method in archaeometry. Although the limitation of chemical dating depending on *accurate temperature assessment* cannot be avoided, this limitation is usable in cases to assess the age and the past temperature. Pre–Inca furs and mummy were studied to establish an ESR dating method. Dating of blood stains in forensic science was done utilizing the oxidation of hemoglobin (Miki *et al.* 1985, 1987b, Sakurai *et al.* 1989).

In this chapter, principle and several examples of chemical ESR dating are described. Almost no work except ours has been reported in early 1990's since most of ESR dating researchers are geoscientists. There will be many organic materials whose age and origin can be assessed with ESR. Chemists should be involved in archaeological and preservation science

12.2 Principle of Chemical ESR Dating

Chemical reactions such as radical formation and valency change by oxidation can be detected with ESR and used as a method of dating. Photo-chemical reactions or radiolysis produces usually unstable radicals and cannot be used for practical dating.

12.2.1 *Rate Equation*

Radicals are formed in organic materials by chemical reactions in addition to an exposure to radiation or light. The organic materials are thermally converted into the intermediate state of radicals and then form more stable diamagnetic products. The rate equations for radical formation with the rate constant, K_1, and for decay with the rate constant, K_2, are described by

$$dN/dt = -K_1 N \ , \tag{12.1}$$

and

$$dn/dt = K_1 N - K_2 n \ , \tag{12.2}$$

where n and N are the concentration of radicals and host organic materials. Eq. (12.2) is identical to the parent–daughter relation in a radioactive disintegration series. The host material decays exponentially as

$$N = N_0 \exp(-K_1 t) \ , \tag{12.3}$$

and n is expressed as

$$n(t) = [K_1 N_0/(K_2 - K_1)] [\exp(-K_1 t) - \exp(-K_2 t)] \ . \tag{12.4}$$

The calculated formation of radicals is shown in Figure 12.1 (Ikeya and Miki 1985a). In ESR dating, the signal intensity, I, which is proportional to the concentration of radicals, n, is used.

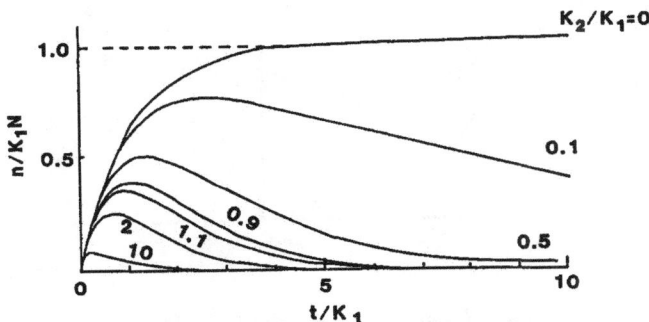

Figure 12.1   Theoretical growth curves of radicals for the rate constants, K_1 and K_2, of radical formation and decay, respectively. A simple saturation curve is obtained for a long lifetime ($K_2 = 0$) of radicals. The initial growth is $K_1 N_0$.

12.2.2 Graphic Analysis of Chemical Formation and Decay

(a) *Initial growth analysis : wait and see method* : The age T_{ESR} can be determined by extrapolating the linear enhancement of the signal intensity back to the zero ordinate using $I(t') = I_0(1 + t'/T_{ESR})$ since Eq. (12.4) leads to $N_0 K_1 t$ for $K_1 t \ll 1$ and $K_2 t \ll 1$. This procedure was demonstrated using potato chips (see Figure 3.3).

(b) *Saturation curve* : $\underline{I = I_s(1 - \exp(-K_1 t))}$: If the growth curve shows saturation ($K_2/K_1 \approx 0$), the difference between the saturation intensity, I_s, and the growth curve, $I(t)$, gives a straight line in a semilogarithmic plot. Extrapolation of the straight line to I_s (i.e., $n = 0$) gives the age T_{ESR} as shown in Figure 12.2 (a). A special case for $K_2 = 0$ (a long lifetime) is oxidation of Fe^{2+} or Cu^+. Impurities such as Fe^{3+} and Cu^{2+} are stable ($K_2 = 0$) and accumulated as the time proceeds. Dating of bloodstains in Section 12.4 belongs to this category.

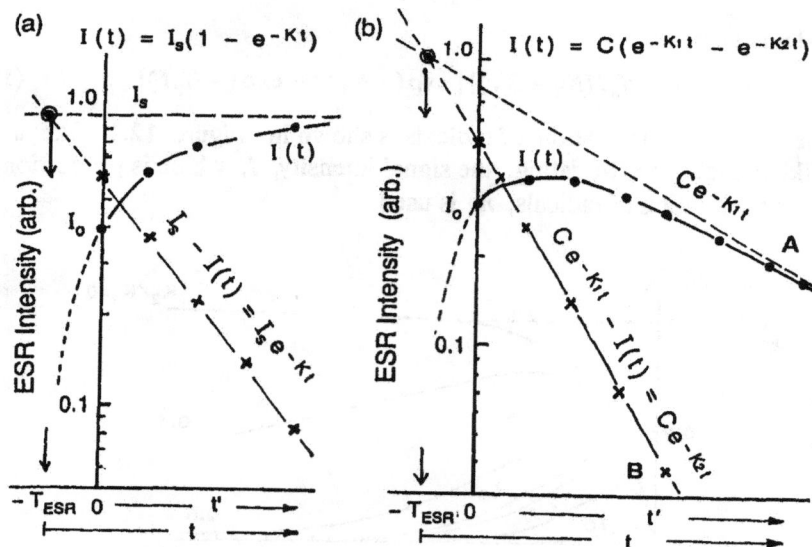

Figure 12.2 Analysis of formation and decay curves.   (a) A saturation curve. The difference from I_s is plotted in a semilogarithmic graph to obtain a straight line. The age T_{ESR} is obtained by extrapolating to I_s. (b) Decay analysis. The difference between line A with a slope of K_1 and the growth curve is plotted to obtain a straight line B with a slope of K_2. The age T_{ESR} is obtained from the coordinate where the two lines meet.

(c)   *Decay analysis* : *age where two lines meet* :   Saturation and decay are sometimes observed at the late stage of radical formation if K_2 is not negligible $(K_2/K_1 > 0)$, as shown in Figure 12.2 (b).   A method of analysis based on Eq. (12.4) is as follows:

(1) A semilogarithmic plot of the intensity as a function of time gives a growth curve $I(t)$. Extrapolation of the growth curve at the late stage gives a straight line A of $C \exp(-K_1 t)$, where $C = K_1 N_0 / | K_2 - K_1 |$.

(2) Subtraction of the growth curve $I(t)$ from the straight line A gives another straight line B of $C \exp(-K_2 t)$ in the same semilogarithmic plot. The age is obtained from the coordinate where the two straight lines intersect.

12.2.3   *Time Machine to Accelerate a Reaction*

(a)   *Laboratory   heating* :   Since a chemical reaction depends on the temperature,   its time scale is shortened by raising the temperature.   Elevated temperature is hence a kind of "*time machine*" to accelerate the time passage. One has to solve Eq. (12.4) under boundary conditions of $n(T_{ESR}) = n_0$ at $t' = 0$ with new rate constants K_1' and K_2' at an elevated temperature.   The solution $n(t')$ normalized by n_0 is given as

$$n(t')/n_0 = \exp(-K_2' t') + (K_1'/K_1)(K_2 - K_1)/(K_2' - K_1') \exp(K_1 T_{ESR})$$
$$\times [\exp(-K_1' t) - \exp(-K_2' t)] / [\exp(-K_1' T_{ESR}) - \exp(-K_2' T_{ESR})] \quad (12.5)$$

K_1' and K_2' can be obtained experimentally from the slope in a semilogarithmic plot as described in the decay analysis. An Arrhenius plot gives the rate constants K_1 and K_2 at the ambient temperature as shown later in Figure 12.5. The age is obtained from computer fitting to the theoretical curve.

(b)   *Oxidation in* O_2 *atmosphere* :   One way to accelerate the passage of time without raising the temperature is to increase the *oxygen concentration* in the atmosphere. The rate constant for formation K_1 is simply increased, while that for decay K_2 may or may not depend on the oxygen concentration. High–pressure experiments may be useful to see the diffusion of oxygen.

12.2.4   *Light–Induced Reaction* : *Photochemical ESR Dating*

Yellowing of wools, silks and cottons occurs by a prolonged exposure to light. ESR dating of organic materials based on *light–induced defects* is analogous to the ESR dating using signals produced by natural radiation.

Figure 12.3 shows the growth of the signal intensity of a silk textile by

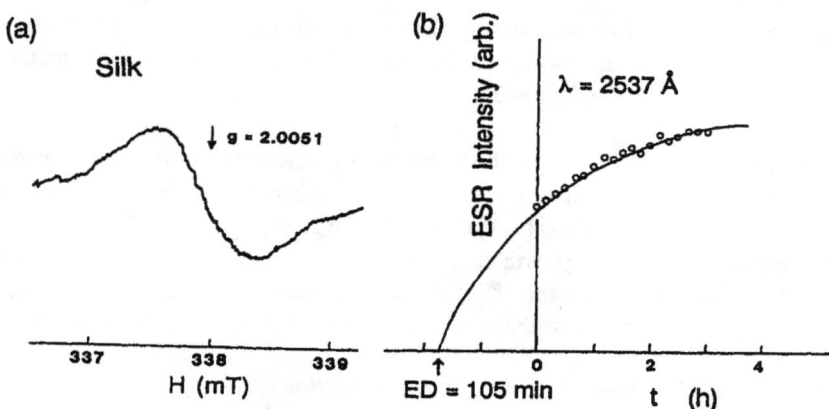

Figure 12.3 (a) ESR spectrum and (b) formation of radicals by UV–irradiation of a silk textile in air. Yellowing of silk remains though radicals decay after the termination of UV–irradiation.

UV–irradiation. A *"time machine"* of the light exposure can be used to obtain the *"equivalent light dose (ED)"* as was done for natural radiation. The assessment of the *annual or daily light dose (D)* must be done separately. The dependence of the signal intensity on the wavelength of both natural and additive artificial light must be evaluated to obtain the *ED* just like the radiation quality effect in the dose rate estimation. The photon energy dependence of radical formation for wools has been studied on textiles (Smith 1973, 1976, Smith *et al.* 1979, Leaver 1968, Leaver and Ramsey 1969).

The effective efficiency, $k(\lambda)$, is calculated as

$$k(\lambda) = I(\lambda) I_a(\lambda_a) ,$$ (12.6)

where $I(\lambda)$ and $I_a(\lambda_a)$ are the intensity of the natural and artificial light, respectively. The annual or daily light energy, D in Gy/a, must be the average throughout the age of the materials. The age, T_{ESR} [a], is then obtained as

$$T_{ESR} = kED/D .$$ (12.7)

UV–light and radiation cause *fiber breakdown, chain rupture, cross–linking, gelatin formation* and *color change* in addition to radical formation in fibrous proteins. *Glycyl radicals* showing a doublet were observed in wool (Cosgrove *et al.* 1971). Quantum efficiencies are of the order of 10^{-6} for radical production at several wavelengths for proteins such as keratin, insulin, egg white, fibroin, casein and gelatin (Dunlop and Nichols 1965). Radicals produced by UV–light occur densely at the surface and are stable only for a few hours. Presumably second–order annealing at the high con-centration surface gives a short effective lifetime as described in Chapter 3.

12.3   Organic Radicals
12.3.1   *Manufactured Date of Potato Chips*

The manufactured date determination of potato chips based on lipid peroxidation was demonstrated as *"wait and see method"* in Chapter 3. Figure 12.4 (a) shows ESR spectra of potato chips kept at 60°C for different periods. The spectrum is composite, consisting of two or more signals (Ikeya and Miki 1980, 1986). The radical formation at several temperatures is shown in (b). The rate constants can be obtained from the growth.

We have encountered a drastic increase in the signal intensity around $g = 2.0055$ which is due to fungal colony formation on potato chips, but the intensity drastically decayed when the temperature dropped.

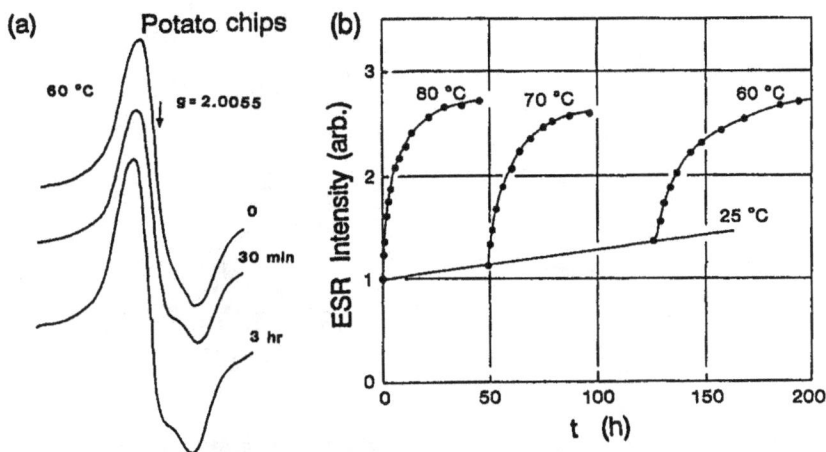

Figure 12.4 (a) ESR spectra of potato chips kept at 60°C. (b) Formation of lipid peroxy radicals in potato chips at several temperatures as a function of time. The formation rate is enhanced at high temperatures.

12.3.2 *Dead Body and Fur*: *Forensic Science*

Lipid peroxidation products are stable in human skin (Meffert *et al.* 1976). Radical formation in a rabbit skin dried outside has been studied in order to evaluate chemical ESR dating for a dead body in forensic science (Ikeya and Miki 1985a). The ESR signal intensity increased as the time passes presumably due to lipid peroxy radical formation and protein deterioration as shown in Figure 12.5 (a). The content of lipid peroxy radicals, n, may be expressed by Eq. (12.2) as a simple case. The radical formation rate constant K_1 was obtained from the initial increase and the decay constant K_2 was estimated from the saturation, i.e., $dn/dt = 0$. When the rabbit skin was kept at constant temperatures of 35°C and 50°C, a linear enhancement was observed up to 20 and 10 days, respectively.

In another model experiment using a rabbit body, oxidation proceeded both from the surface and the back of the skin when a part of the skin was excised from the body. Light also produces radicals at the surface (Miki *et al.* 1985). Thus, precaution is necessary for actual dating of a dead body.

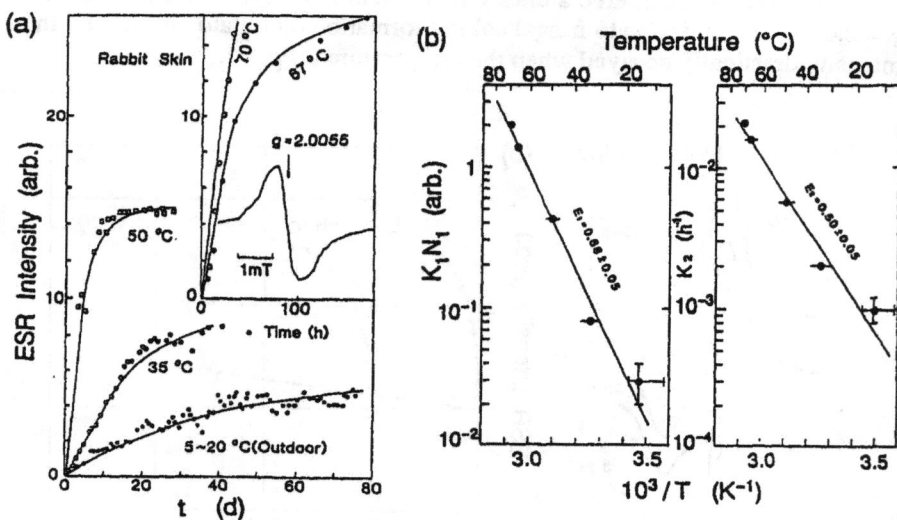

Figure 12.5 (a) The growth of the radical signal intensity in rabbit skins as a function of time at several storing temperatures. The ESR spectrum is inserted in the figure. (b) The rate constants K_1 and K_2 are obtained from the initial growth and the saturation behavior. Curves are the theoretical ones assuming a constant $K_1 N$ or $N \gg n$.

12.3.3 Leather and Mummy : Archaeometry

(a) *Leather* : Leather manufactured from skins of various animals has been used for thousands of years. The composition of skins (protein, fat and carbohydrate) is modified by artificial curing, tanning, pickling and bacterial actions. Treatment of leather with grease, petroleum, paraffin or bee wax gives additional ESR signals in some cases.

Figure 12.6 shows ESR spectra of some historical leathers. It is noticed that the broad signal associated with Fe^{3+} is appreciably large for old furs. The lifetime of radicals is usually not so long as to give a dating range of hundreds of years. The growth of the signal intensity shows a tendency to saturate. However, the signal of Fe^{3+} at $g = 4.3$ and broad signals $g = 2.25$ are intense in old furs but not in 2 year–old leather. These signal are stable and formed by oxidation and may be used for dating (Ikeya and Miki 1986).

(b) *Pre–Inca mummy* : Although the stability of lipid peroxy radicals at $g = 2.0055$ in rabbit skins is about a few months at room temperature, radicals in old leather, skin and mummy would be bound more to proteins after decay of the radicals close to the surface. The late Mr. Amano at Museum Amano, Peru supplied pieces of pre–Inca mummy, fur, cotton,

Leather

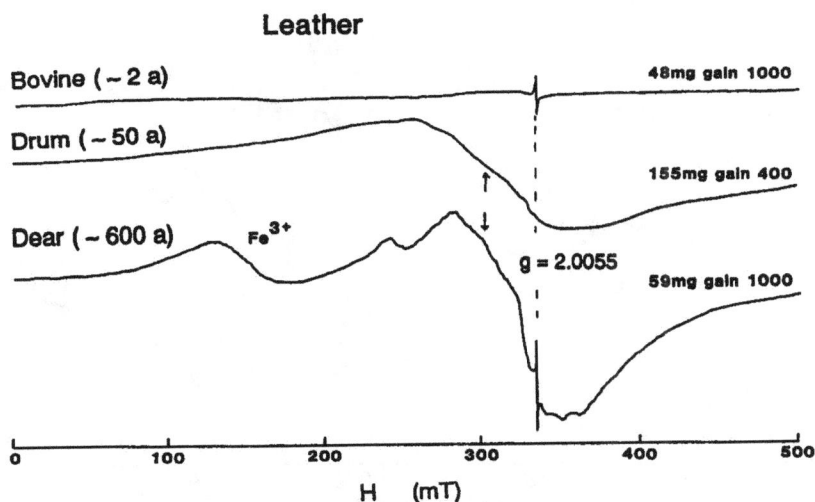

Figure 12.6   ESR spectra of modern and archaeological leathers. The signal intensities of Fe^{3+} ($g = 4.3$) in rhombic field and $g = 2.3$ presumably associated with low–spin hemoprotein–peroxide complexes increase as the age increase due to gradual oxidation reaction.

feather, bean and popcorn as well as pieces of bone needle and shell ornaments for a trial experiment with ESR. A mummy in Egypt is a dead body
preserved with salt (NaCl), natron (Na_2CO_3) and asphalt. On the other hand,
mummies in north and south America were considered to be natural due to
the arid climate, but later they were found to be artificially desiccated.

Figure 12.7 shows an ESR spectrum of the pre–Inca mummy. The
broad signal at $g = 2.4$ is more intense than that at $g = 2.0055$. This broad
signal is due to heme proteins associated with iron impurities or to decomposition of proteins. The signal at $g = 2.4$ may be associated with a low ferric
spin state of the heme protein–peroxide complex (see Section 12.4; Sym-
mons and Peterson 1978, Tajima et al. 1986). Radical signals around $g = 2.005$ were detected in proteins or amino acids exposed to lipid peroxides
(Karel et al. 1975). Oxidation of hemoprotein using Fe^{3+} signal may be used
for dating when the growth of lipid peroxy radical ($g = 2.005$) saturated.

Feather clothes and hairs at pre–Inca days as well as modern bird
feathers and some insects showed signals due to paramagnetic species.
Zero–age (background) signals make ESR dating difficult. In Asia, people
reserve umbilical cords of babies, hoping their happiness. However, signals
due to chemicals (talc powder for example) mask radical signals.

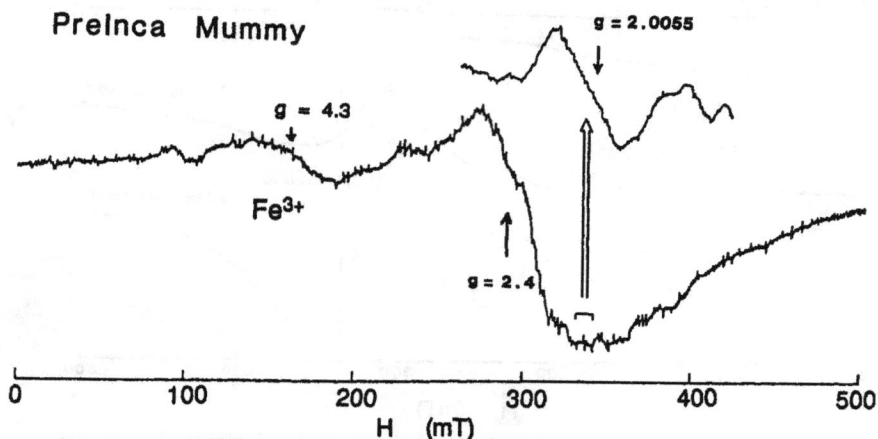

Figure 12.7 ESR spectrum of a pre–Inca mummy. The Fe^{3+} signal at $g = 4.27$ and an unidentified signal at $g = 2.4$ are dominant. The lipid peroxy radical at $g = 2.0055$ is still observed in a 2,000 year–old mummy.

12.4 Forensic Science: Oxidation of Bloodstain

The date of blood after bleeding is vitally important in forensic science. Color of the bloodstain has been inspected and evaluated using optical spectroscopy by measuring oxy–hemoglobin (HbO_2) and methemoglobin (met–Hb) (Kind and Watson 1973, Lins and Blazek 1982). ESR dating of bloodstains has been developed using signals of Fe^3 and radicals for forensic science (Miki *et al.* 1985, 1987a, b) Bloodstain dating has also been investigated by measuring the sample at 77K rather than at room temperature for ages up to two years (Sakurai *et al.* 1989). Although there are a number of biochemical ESR studies on hemoglobin, its decomposition in air has not been fully investigated (Banerjee and Stetzkowski 1970).

Figure 12.8 (a) shows an ESR spectrum of the bloodstain and (b) the growth of the intensity of some prominent signals as a function of time at 70°C after bleeding. The observed species are summarized in Table 12.1. The signal intensity at $g = 6.0$ $(I_{6.0})$ saturates at an early stage and then that of radicals at $g = 2.0057$ (I_R) saturates around 20 hours at 70°C,

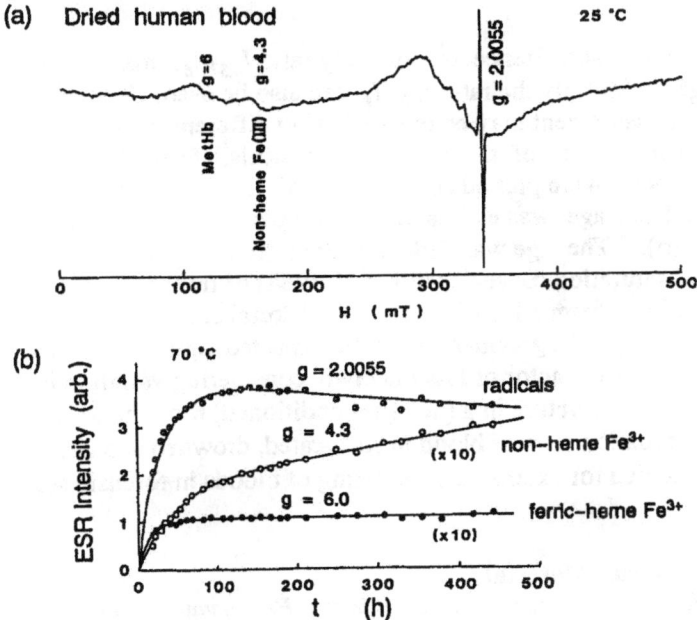

Figure 12.8   (a) ESR spectrum of dried human blood measured 7 days after sampling. (b) The growth of the signal intensities at $g = 6.0$ and 4.3 due to Fe^{3+} and at $g = 2.0055$ as a function of time at 70°C (Miki *et al.* 1987b).

Table 12.1 ESR parameters of paramagnetic ferric species and radicals observed in human bloodstains.

Species	g factor			Comments	Ref.
	g_1	g_2	g_3		
High–spin heme Fe^{3+}	$g = 6.0$			large tetragonality: DS_z^2	1,2
High–spin nonheme Fe^{3+}	$g = 4.3$			large rhombicity $E(S_x^2 - S_y^2)$	1,2
Low–spin heme Fe^{3+}					
$N - Fe^{3+} - N$	2.93	2.27	1.54 [a]		2
	2.92	2.26	1.53 [b]		
$S^- - Fe^{3+} - N$	2.43	2.27	1.90 [a]		2
	2.45	2.26	1.90 [b]		
Lipid peroxy type radical	$g = 2.0055$			early increase and late decay	1

Blood samples on (a) the first day and (b) 4 months after extraction.
Reference: 1) Miki *et al.* 1985, 1987b;   2) Sakurai *et al.* 1989.

while $I_{4.3}$ increases. Hence, the intensity ratio $I_{4.3}/I_{6.0}$ may be an indicator of the age. Similarly the ratio $I_{4.3}/I_R$ can also be used. The ambiguity of temperature assessment may be reduced using different intensity ratios.

Intensity ratios of the signals of radicals, ferric heme and ferric nonheme species were plotted as a function of time in Figure 12.9 (a), from which the ESR age was estimated and compared with the actual ages as shown in (b). The age was deduced either from the theoretical intensity ratios using saturation curves (Miki *et al.* 1988) or from empirical regression curves obtained from a logarithmic plot (Sakurai *et al.* 1991). Both results show a fairly good agreement with the expected age. ESR age may be determined within a factor of less than two considering variation in intensity ratios and the temperature in a (not air–conditioned) room up to 2 – 3 years.

The oxidation state of blood in suffocated, drowned and charred bodies should be studied for extensive ESR dating of bloods in forensic science.

12.5 Historical Materials
12.5.1 *Historical Papers and Cellulose : Preservation Science*

Ordinary pieces of wood show an ESR spectrum of *lignin radicals* (Rex 1960). White cellulose is manufactured for paper production by boiling

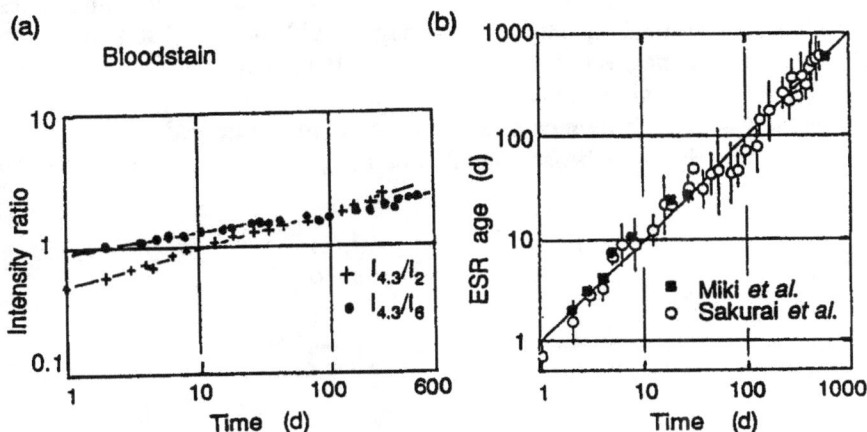

Figure 12.9 (a) The relation between the intensity ratios $I_{4.3}/I_6$ for signals at $g = 4.3$ and $g = 6.0$ and $I_{4.3}/I_R$ for $g = 4.3$ and $g = 2$ (radicals) as a function of the time. (b) the ESR age and the actual age of bloodstains at room temperature (replotted from Sakurai *et al.* 1991 and Miki *et al.* 1988).

wood or pulp in a NaOH solution and removing lignin and the ESR signal of lignin radicals disappears at this stage. Addition of carbonates and glue gives somewhat complicated ESR signals depending on paramagnetic impurities involved. Papers manufactured on an industrial scale with addition of aluminum sulfate and pine tree resin, degrade due to hydration of cellulose in less than 100 years. Plant resin gives an intense ESR signal (Robins *et al.* 1985) as amber in which particle tracks are registered (Uzgris and Fleischer 1971).

Figure 12.10 shows the structure of cellulose consisting of polyglucose. Addition of water molecules or H^+ and OH^- results in decomposition of cellulose. Unpaired electrons in lignin radicals formed by the photochemical process are transferred to cellulose leading to decomposition. The peroxidation of glue organics is also a reason for radical formation, though UV–light exposure in air considerably enhances radical formation. Radical formation at elevated temperatures is reduced considerably by heating in vacuum and in an inert atmosphere (Ikeya and Miki 1985c). This clearly indicates that oxidation is the main cause of radical formation. Thus, the preservation of papers should be in an inert atmosphere without contact with oxygen in a cool and dark site just like foods in refrigerator or packed in a light shutting envelope with oxygen active chemicals.

Some papers show ESR signals due to paramagnetic impurities. Figure 12.11 is ESR spectra of book pages published in 1804 in Berlin, Germany for the magnetic field perpendicular to the paper surface. The low field ($g = 9.0$) signal which is somewhat similar to that in magnetite (Fe_3O_4) is less intense in the brown colored edge than in the white inside. The brown colored edge shows a broad signal around $g = 2.3$, which is somewhat

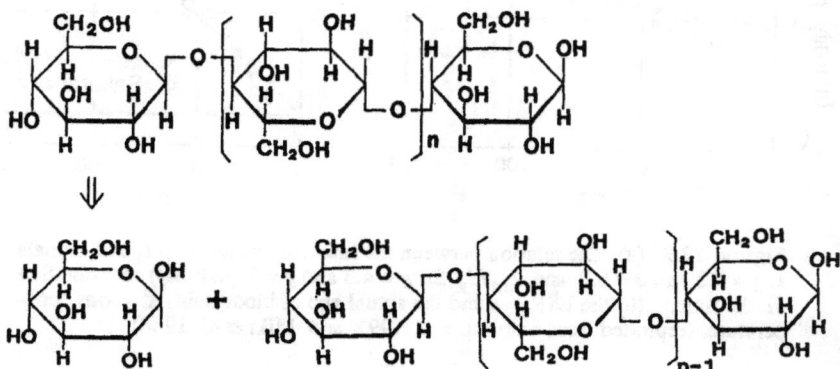

Figure 12.10   Structure of cellulose and hydration decomposition. Woods and papers are more complex since lignin and other materials are involved.

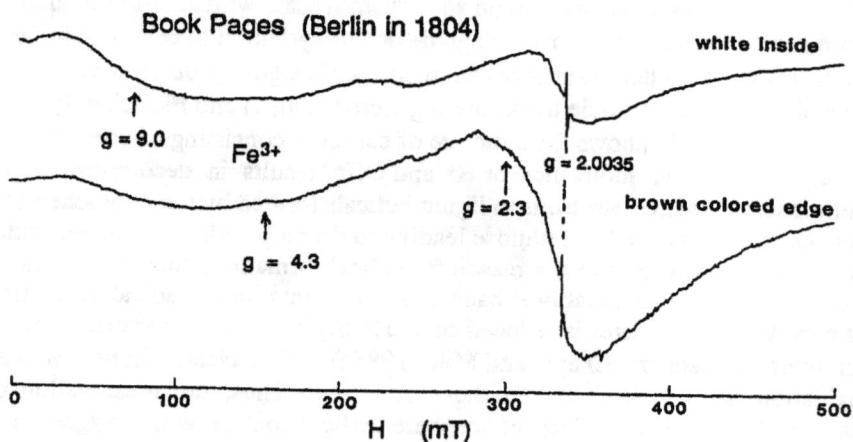

Figure 12.11   ESR spectra of the white inside and brown colored edge parts of the book published in 1804.

similar to that in more oxidized form, hematite (Fe_2O_3) and may be associated with oxidation of Fe^{2+} to Fe^{3+}, i.e., presumably oxidation of magnetite to hematite. The spectra of these minerals are shown in Figure 10.17 (see Chapter 10). The spectrum changes as the direction of the magnetic field is changed from perpendicular to parallel to the paper surface suggesting that the orientation of microcrystal minerals along the pressing direction gives angle dependent ESR signals. Dollar or yen notes also give ESR signals due to radicals in the paper and to printing pigments. Art pigments and their oil oxidation are objects of studies usina a cavity with an aperturein Appendix 5.

Figure 12.12 (a) shows the radical formation curves of three different aged papers by heating at 45°C. The intensity is normalized by the initial one. A slight decrease in the intensity at the initial stage was observed for the young papers, but the increase is a general trend. The rate of increase is small in sample III (old Japanese papers of the Edo era (200 a). ESR ages were obtained from the analysis of radical formation at different temperatures and were compared with known ages in (b) and overall agreement was obtained though the ambiguity by a factor of two is present.

Cotton is also cellulose and shows an enhancement of the signal intensity at the initial stage in an aging experiment. It saturates soon presumably because the charge transfer from lignin and other radicals at the late stage is absent for pure cellulose materials. The obtained ED by additive γ–irradiation is 100 Gy, which corresponds to 100 ka for $D = 1$ mGy/a.

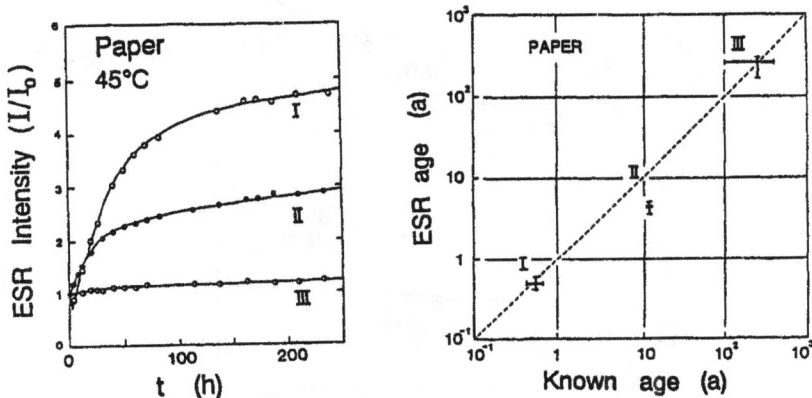

Figure 12.12 (a) Aging experiment for papers with three different ages. Sample I, II and III are a recent newspaper, magazine pages of about ten years ago and old Japanese papers of about 200 years ago, respectively. (b) Ages estimated with ESR are compared with the known ages.

12.5.2 *Silk, Wool and Japanese Lacquer*

Silk is an important historical textile material as is seen from the historic *"Silk Road"*. Silk and wool become yellowed with ages in the same way
as does cotton. ESR studies of these materials were performed earlier by
Gordy *et al* (1958). *Glycyl radicals* are formed by a hydrogen extraction
reaction in glycine residues of fibroin during a photochemical reaction and
further react with oxygen to become hydroperoxide (Patten and Gordy 1960).
Figure 12.13 (a) shows ESR spectra of historical silk textiles prepared in
about A.D. 1700 and A.D. 1400. The overall spectra show a Fe^{3+} signal at
$g = 4.3$. A broad Fe^{3+} signal around $g = 2.0$ overlaps the radical signal at
$g = 2.0035$. Thermal formation of radicals is shown in (b).

Japanese lacquer is a resin from a particular tree and its preservation
in museums is of special interest. Black lacquer is a mixture with iron
particles. An ESR spectrum of black lacquer shows broad lines around
$g = 2.0$ and $g = 4.3$ due to Fe^{3+}. Natural raw lacquer shows a radical signal
alone around $g = 2.0$ (Ikeya and Miki 1986).

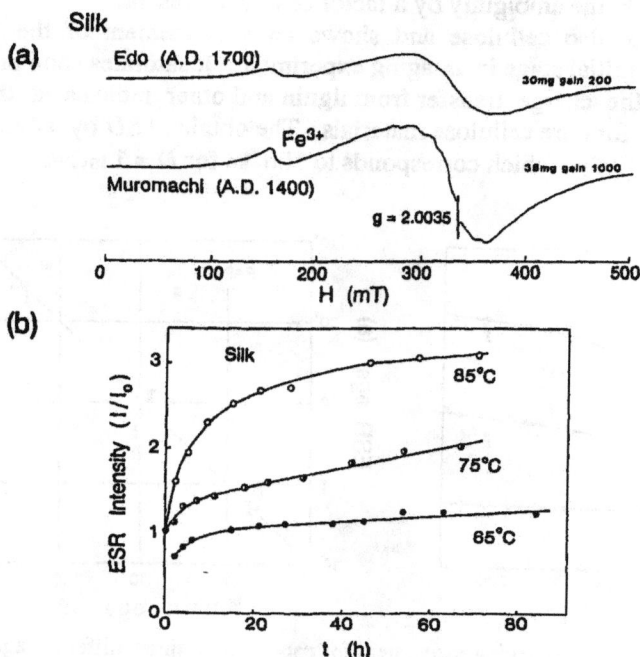

Figure 12.13 (a) ESR spectra of silk clothes and (b) radical formation of at
elevated temperatures as a function of time.

12.6   Agricultural Products

12.6.1   *Coffee Beans : Quality Assessment*

Some materials such as tea leaves and coffee beans show an increase in the signal intensity at $g = 2.0057$ after harvest. Figure 12.14 (a) shows an ESR spectrum of coffee beans taken from a Brazilian farm. The Mn^{2+} and radical signals were observed (Ikeya *et al.* 1989). Although radical species might be different from those produced by natural radiation, they were also enhanced by γ–irradiation. The *ED* was obtained by an additive dose as was done in ESR dating. The relation between the *ED* and the quality of coffee beans is shown in (b). It is clear that the quality is better for coffee beans with a small *ED*, i.e., the flavor or taste inversely correlates to the *ED*. If the signal intensity is plotted instead of the *ED*, fluctuations are much larger indicating a fluctuation in the original lipid abundance or composition in beans. Peroxy radicals would be indicative of aroma decomposition.

12.6.2   *Thermal History of Cereal Grains : New Archaeology*

In new archaeology, a main concern is people's life in history rather than the materials they left. Thermal history of archaeological cereal grains has been studied with ESR to know the ancient cooking recipe (Hillman *et al.*

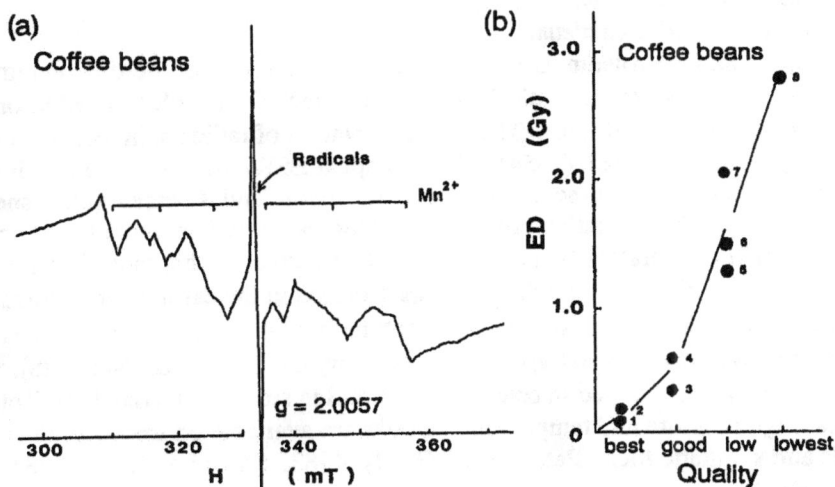

Figure 12.14   (a) ESR spectrum of coffee beans half a year after harvest from the tree. (b) The equivalent dose (*ED*) obtained by the additive dose method is related to the grade of the coffee beans.

1983, 1985, Sales *et al.* 1987). Scattering of *g* factors, the intensity and linewidth of pyrolytically produced radicals have been plotted in two- or three-dimension. The *g* factor shift by heating indicates that different types of radicals are produced by oxidation, and that the signal is composite (Robins *et al.* 1984). Although temperature dependence of the *g*-shift was reported, no attempt to date has been made using the *g*-shift.

Chemical ESR dating using the g-shift may be done as follow:

(1) Plot the *g*-shift as a function of time at various high temperatures.
(2) Estimate the *g*-shift at ambient temperature by plotting the rate in an Arrhenius plot. Draw the expected *g*-shift versus time curve.
(3) One can obtain the age from the *g*-shift using the above curve.

Peroxy radicals are detected in a variety of materials. For instance, several crops such as wheat and rice show ESR signals of radicals whose concentration increases day by day. The radical signal saturates or rather decreases for old samples. Archaeological old crops show signals associated with Fe^{3+} due to oxidation. Detailed studies of other organic materials would lead to a dating method or provenance study in history and forensic science.

12.7 Coal and Petroleum (Engine Oil)
12.7.1 *Coal and Petroleum*

Coal and petroleum are fossil fuels on which most of our energy depends. Various organic radicals have been detected with ESR in addition to paramagnetic metal ions. The first observation of radicals in coals was made by Uebersfeld *et al.* (1954). An attempt at ESR dating was made using radicals produced by geothermal heat in coal, *lignite* and *kerogen* (Duchsne *et al.* 1961). Carbonization process for biological materials (wood, cotton, and sugar) in a laboratory by raised temperature indicated that radical formation in coals is identical with the pyrolisis occurred in the earth crust at lower temperature. Estimation of the geological time or temperature for coal formation was made by extrapolating laboratory data (Morozowski 1988).

ESR has been used in coal chemistry and in studies on coal liquidization at high pressure and temperature. Books are available on free radicals in coals and synthetic fuels (Petrakis and Gandy 1983, Shick and Kevan 1984).

Petroleum-related materials such as *heavy oils, asphalt, pitch, coal tar, tar sands, kerogen* and *oil shale* have been studied with ESR. The signal intensity and the linewidth of kerogen radicals ($g = 2.0030$, $\Delta H =$

0.4 ~ 0.6 mT) were plotted as a function of depth (geothermal heat) to assess the maturity of petroleum (Pusey 1973), but the results were not persuasive because of the scattering of data due to heterogeneous structure of kerogens. Kerogen treated with pyridine gave a good correlation (Bark *et al.* 1990).

Paramagnetic ions, especially vanadyl compounds (VO^{2+}) abundant in heavy oil, give ESR signals due to the anisotropic g factor and hf splitting. Characterization of crude oil has been made using this signal. Petroleum asphaltene contains three fundamental types of vanadyl complexes with aromacity as well as organic radicals ($g = 2.0027 ~ 2.0037$) of 10^{18} spins/g (Yen *et al.* 1969).

A single peak dignal of radicals without hf splitting makes an analysis of radical species in petroleum difficult. ESR lineshapes and relaxation times (spin–echo) using both CW– and pulsed ESR can select out species of particular relaxation time (Singer *et al.* 1987) and might be used for dating.

12.7.2 *Engine Oil* : *Mileage Determination*

The degradation rate of engine oil during the car use has been studied with infrared spectroscopy (Zieba 1985) and with ESR (Ikeya 1985). Figure 12.15 (a) shows an ESR spectrum of engine oil out of a car. ESR signals of lipid peroxy radicals ($g = 2.0055$) overlaps the signal of carbon radicals ($g = 2.0035$). The intensities increase as a function of mileage at a steady driving condition (driving distance and speed) up to 1000 km as shown

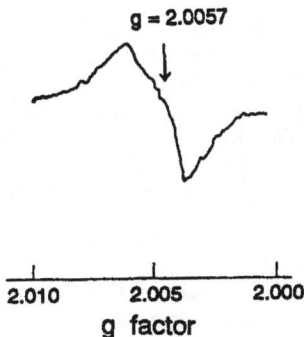
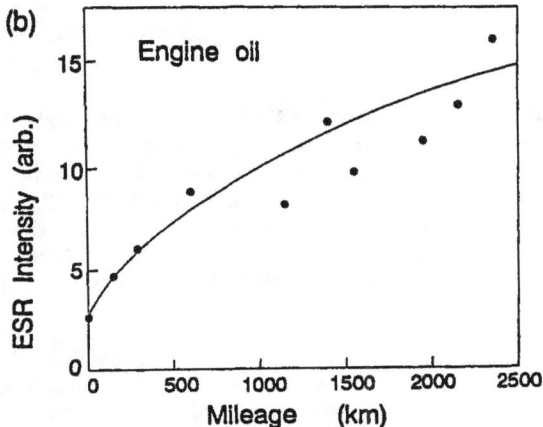

Figure 12.15   (a)   ESR spectrum of engine oil and (b) the increase of the peroxy radical as a function of mileage.

in (b).   However, the radical formation efficiency is no longer constant
under different driving conditions (driving on a highway, etc.). This is
because the engine temperature is not constant; chemical rate constants K_1
and K_2 depend on the driving conditions. The signal intensity of engine oil
may be   used to assess the quality of engine oil and engine itself. Signals
associated with additive antioxidants are also observed in some commercial
engine oil.

12.8   Summary

Principles of chemical ESR dating based on radical formation and
valency change by oxidation are described together with the results of some
applications in our laboratory. Procedures should be developed and refined.
Geoscientists are little interested in chemical ESR dating where temperature
variation results in a large error, a factor of two or three as known in
racemization dating. Measurements of a day–by–day increase in the signal
intensity used for dating of potatochips can be applied to dating of bloodstains in forensic science. Archaeological objects and historical materials can
be dated within the limitation of temperature variation. Relevant references
cited in this chapter are not complete since the related field is too wide for
the author to cover all. The scope of chemical ESR dating may be expanded
into forensic sciences, history and archaeology.

References

Banerjee R. and Stetzkwski F. (1970):   Heme transfer form hemoglobin and ferrihe-
     moglobin to new ligands and its implication in the mechanism of oxidation of
     ferrihemoglobin by air. *Biochim. Biophys. Acta.* **221**, 636–639.
Bark M. Y., Akiyama M. and Sanada Y. (1990):   ESR assessment of kerogen matura-
     tion and its relation with petroleum genesiss. *Org. Geochem.* **15**, 595–599.
Cosgrove M. A., Collins M. A., Grant R. A. and Allock B. J. (1971):   Some effects of
     radiation on glycyl peptides in solid states. *New Zealand J. Science* **14**, 599–
     607.
Duchsne J., Depireus J. and Van der Kaa J. M. (1961):   Origin of free radicals in
     carbonaceous rocks. *Geochem. et Cosmochim. Acta,* **23**, 209–218.
Dunlop J. I. and Nichols C. H. (1965):   Electron spin resonance studies of ultraviolet
     irradiated keratin and related protein. *Photochem. Photobiol.* **4**, 881–890.
Fritsch G., Lopez C. and Rodriguez L. (1974):   Generation and recombination of free
     radicals in organic materials studied by electron spin resonance. *J. Magn. Reson.*

16, 48–55.

Gordy W., Ard W. B. and Sheilds H. (1955): Microwave spectroscopy of biological substances: I paramagnetic resonance in X–irradiated amino acids and proteins. *Proc. Nat. Acad. Sci.* **41**, 983–996.

Hillman G. C., Robins G. V., Oduwole D., Sales K. D. and McNeil D. A. C. (1983): Determination of thermal history of archaeological cereal grains by electron spin resonance spectroscopy. *Science* **222**, 1235.

Hillman G. C., Robins G. V., Oduwole D., Sales K. D. and McNeil D. A. C. (1985): The use of electron spin resonance spectroscopy to determine the thermal history of cereal grains. *J. Archaeol. Sci.* **12**, 49

Ikeya M. and Miki T. (1980): A new dating method with a digital ESR. *Naturwissenschaften* **67**, 191–192.

Ikeya M. (1985): Car mileage determination with ESR signal of engine oil: a case of organic ESR dating. *ESR Dating and Dosimetry* 453–457.

Ikeya M. and Miki T. (1985a): Principle of chemical ESR dating for organic materials using radical formation. *J. Speleol. Soc. Jpn.* **10**, 32–37.

Ikeya M. and Miki T. (1985b): ESR dating of organic materials: from potato chips to a dead body. *Nucl. Tracks* **10**, 909–912.

Ikeya M. and Miki T. (1985c): ESR dating and preservation of papers. *Naturwissenschaften* **72**, 32–33.

Ikeya M. and Miki T. (1986): Organic ESR dating in archaeology. *J. Archaeol. Chem.* **4**, 1–10.

Ikeya M., Baffa F. O. and Mascarenhas S. (1989): Quality assessment of coffee beans with ESR. *Radiat. Isotop.* **40**, 1219–1222.

Karel M., Schaich K. and Roy R. B. (1975): Interaction of peroxidizing methyl linoleate with some proteins and amino acids. *J. Agr. Food Chem.* **23**, 159–163.

Leaver I. H. (1968): Studies in wool yellowing Part XXII : ESR investigations of the action of UV radiations. *Textile Res. J .* **38**, 729–734.

Leaver I. H. and Ramsay G. C. (1969): Studies in wool yellowing Part XXVI : Photoyellowing of fluorescent–brightened wools–wavelength dependence and ESR studies. *Textile Res. J.* **39**, 722–729.

Meffert H., Diezel W. and Sonnichse N. (1976): Stable lipid peroxidation products in human skin. *Experientia* **32**, 1937.

Miki T., Yahagi T., Ikeya M., Sugawara N. and Furuno J. (1985): ESR dating of organic substances: corpse for forensic medicine. *ESR Dating and Dosimetry* 440–450.

Miki T., Kai A. and Ikeya M. (1987a): ESR dating utilizing change in valency or ligand of transition metals. *Jpn. J. Appl. Phys.* **26**, 972–973.

Miki T., Kai A. and Ikeya M. (1987b): Electron spin resonance of bloodstain and its application to the estimation of the time after bleeding. *Forensic Sci. Intern.* **35**, 149–158.

Miki T., Kai A. and Ikeya M. (1988): ESR dating of organic substance utilizing paramagnetic degradation products. *Nucl. Tracks* **14**, 253–258.

Morozowski S. (1988): ESR studies of carbonization and coalification process, Part I. Carnonaceous compound; Part II. Biological materials. *Carbon* **26**, 521–529; *Carbon* **26**, 531–541.

Patten F. and Gordy W. (1960): Temperature effects on free radical formation and electron migration in irradiated proteins. *Proc. Nat. Acad. Sci.* **46**, 1137–1144.

Petrakis L. and Gandy D. W. (1983): *Free Radicals in Coals and Synthetic Fuels* (Elsevier, New York).

Pusey W.C. (1974): How to evaluate potential gas and oil source. *World Oil* **176**, 71–73.

Rex R. W. (1960): Electron paramagnetic resonance studies of stable free radicals in lignins and humic acids. *Nature* **188**, 1185–1186.

Robins G. V., Sales K. D. and McNeil D. A. (1984): Ancient spins. *Chemistry in Britain* **84**(II), 896–899.

Robins G. V., Sales K. D. and Oduwole A. D. (1985): Electron spin resonance of plant resins: an assessment of ESR dating possibilities. *ESR Dating and Dosimetry* 435–438.

Sakurai H., Tuchiya K., Fujita Y. and Okada K. (1989): Dating of human blood by electron spin resonance spectroscopy. *Naturwissenschaften* **76**, 24–25.

Sakurai H., Tuchiya K., Fujita Y. and Okada K. (1991): ESR dating of human bloods (II). *Advances in ESR Applications* **7**, 38–39 (in Japanese).

Sales K. D., Oduwole A. D., Convert J. and Robins G. V. (1987): Identification of jet and related materials with ESR spectroscopy. *Archaeometry* **29**, 103–109.

Shick S. and Kevan L. (1984): *Introduction, Advanced Topics and Applications to Fossil Energy* In *Magnetic Resonance* eds. Petrakis L. and Fraissard J. P. (Reidel Pub. Co., Dordrecht, the Neitherlands).

Singer L.S., Lewis I.C. Riffel D.M. and Doetschman D.C. (1987): EPR studies of separated fractions of mesophase pitches. *J. Phys. Chem.* **91**, 2408–2415.

Smith G. J. (1973): Triplet state in wool. *Photochem. Photobiol.* **18**, 243–244.

Smith G. J. (1976): Effect of light at different wavelength on electron spin resonances in wool. *Textile Res. J.* **46**, 510–512.

Smith G. J., Claridge R. F. C. and Smith C. J. (1979): The action spectra of free radicals produced by irradiation of keratin containing bound iron (III) ions. *Photochem. Photobiol.* **29**, 777–779.

Symmons M. C. R. and Peterson R. L. (1978): Electron capture by oxyhemoglobin: an e.s.r. study. *Proc. Roy. Soc.* **B201**, 285–300.

Tajima K., Ishizu K., Sakurai H. and Ohya–Nishiguchi H. (1986): A possible model of hemoprotein–peroxide complexes formed in an iron–tetraphenyl porphyrin system. *Biochem. Biophys. Res. Commun.* **135**, 972–978.

Tajima K., Jinno J., Ishizu K., Sakurai H. and Ohya–Nishiguchi H. (1989): Direct evidence of heme–tert–butyl peroxide adduct formation demonstrated by simulataneous ESR and optical measurements. *Inorg. Chem.* **28**, 709–715.

Uebersfeld J., Etienne A. and Combrisson J. (1954): Paramagnetic resonance: a new property of coal like materials. *Nature* **174**, 6114.

Uzgris E. E. and Fleischer R. L. (1971): Charged particle registration in amber. *Nature* **234**, 28–29.

Yen T. F., Boucher L. J., Dickie J. P., Tynan E. C. and Vaughan G. B. (1969): Vanadium complexes and porphyrins in asphaltens. *J. Inst. Petrol.* **55**, 87–99.

Zieba J. (1985): Examination of the rate of engine oil degradation during car use by infrared spectroscopy. *Forensic Sci. Intern.* **29**, 269–274.

Chapter 13

ESR Dosimetry

– Dosimeter, Accident Dose and Foodstuffs –

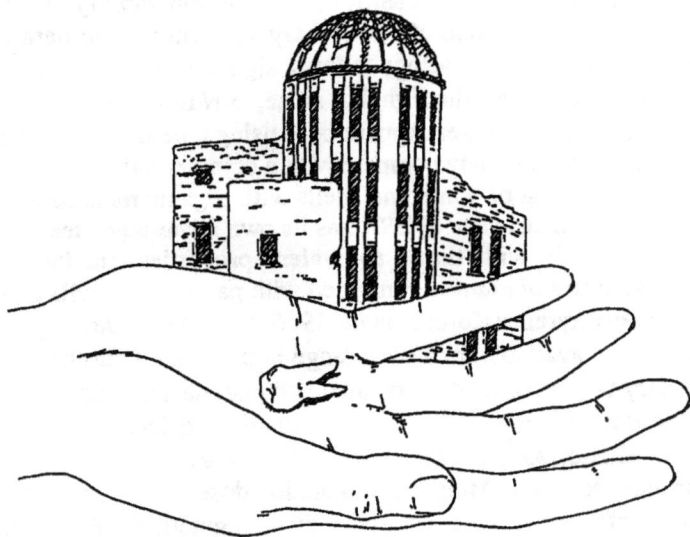

ESR can be used as a radiation dosimeter in nuclear facilities using alanine, sugar and other tissue equivalent organic materials. A new material with a high G-value and a sharp linewidth is strongly hoped for. Accident dosimetry of direct exposure to human beings can be made by measuring human tooth enamel and shell buttons. Some food irradiation doses can also be checked with ESR.

13.1   Introduction

Radiation effects such as fission tracks (FT), thermoluminescence (TL), phosphorescence and optical absorption have been used as methods for radiation dosimetry. The *TL dosimeter* (TLD) has been developed considerably for personal monitoring and is used for assessment of the environmental radiation dose (Mahesh and Viji 1985, McLaughlin *et al.* 1982, 1989). The concentration of radiation–induced radicals can be used in radiation dosimetry as we have seen in ESR dating of various materials.

The information on the cumulative radiation dose is unaltered in ESR dosimetry by repeated measurements, but it is destroyed in TL dosimetry. Unfortunately, the sensitivity of ESR at the moment is several orders of magnitude less than that of TL; only the intermediate to high dose ranges can be measured with ESR. However, the recent development of the ESR spectrometer has increased its sensitivity and reproducibility dramatically and has allowed the detection of a few mGy exposure. Furthermore, the development of digital technology allows signal averaging by repeated measurements and enhances the signal to noise, *S/N* ratio. A compact and a portable ESR spectrometer (see Appendix 3) using a permanent magnet will be a dosimeter reader and may compete with a commercial TL reader.

Just as ESR dating of bones and shells with organic remains has a great advantage, the dosimetric use of ESR has its own advantage; teeth (bones) and organic materials with tissue equivalent properties can be dosimeter elements. Dosimeters of *alanine* combined with paraffin (Regulla and Deffner 1982) or polystyrene (Kojima *et al.* 1986) have been developed; both are commercially available for a dose range higher than radiotherapy level. ESR dosimetry will be used for personnel and materials in nuclear facilities as well as for atomic bomb (A–bomb) and radiation accident survivors.

Food irradiation to kill bacteria and to keep the shelf–life long is being studied for practical uses. Monitoring radiation dose to foods must be made to check the safe upper dose endorsed by a regulation. Post–irradiation dosimetry of foodstuffs is a promising field of ESR applications (McLaughlin *et al.* 1982, Rossi *et al.* 1992).

In this chapter, the methods and limitations of ESR dosimeters such as alanine and sugar are described together with ESR dosimetry of A–bomb survivors in Hiroshima and Nagasaki and of residents close to the Chernobyl reactor accident using *tooth enamel, shell buttons* and *sugar*. Food irradiation dosimetry is also reviewed briefly.

13.2 Principle of ESR Dosimetry

13.2.1 ESR Radical Dosimeters

(a) *Dose–response function* : Radiation dosimeters are classified into following two different types according to a response to radiation. One utilizes formation and the other utilizes decay of radicals, as shown schematically in Figure 13.1.

(1) *Formation by radiation* : An ESR intensity, I, produced by radiation is proportional to the radiation dose rate, D, and the irradiation time, t. The dose–response curve is simply described by

$$I(t) = cDt , \qquad (13.1)$$

where c is a constant. "*Radical dosimeters*" belong to this category.

(2) *Decay by radiation*: Radiation–induced decay of some quantities such as stored charges, impurity centers, etc., follows a simple rate equation, $dn/dt = -cn$, and so to

$$I(t) = I_o e^{-cDt} . \qquad (13.2)$$

An *ionization pocket chamber, electret dosimeter, leucodye* in a polymer (Uribe *et al.* 1981, Pasenkiewicz and Knapoczyk 1981), etc., belong to this category. No practical dosimeter to detect decay of paramagnetic centers has been developed though a large G–value (see Item (d) of this section) is common in a valency change dosimeter as in a "*Fricke dosimeter*" which detects a valency change from Fe^{2+} to Fe^{3+} by optical absorption measurement ($G = 16$ for γ–rays) (Pejuan and Kuhn 1981). ESR detection of Fe^{3+}, however, is not sensitive.

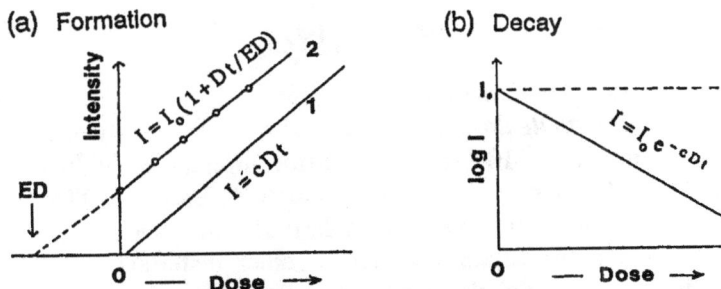

Figure 13.1 ESR dosimetry utilizing (a) increase (curve 1) and (b) decay of the signal intensity after irradiation. The equivalent dose (ED) is obtained by additive irradiation in ESR dosimetry (curve 2).

(b) *Radical dosimeter* : The principle of conventional ESR dosimetry belongs to category (1). Electron and/or hole centers are directly detected by ESR. The ESR intensity of some defect centers is proportional to the absorbed radiation dose. A material with a specific impurity or impurities is used as a dosimeter element. The sensitivity heavily depends on the production yield of unpaired spins and the linewidth of the ESR signal (see Chapter 2). The radiation dose is determined from the internal calibration curve (curve 1 in Figure 13.1 (a)). Accident and A–bomb radiation dosimetry belongs to this category, where an additive dose of Q Gy is used to calibrate the intensity in an additive dose method (curve 2) $(Q = D't')$.

(c) *Lifetime of a defect for dosimetry* : One requirement for a dosimeter is that the lifetime of defect centers must be longer than a few years. If annealing temperature is above 200°C for inorganic solids, the lifetime of defects is longer than a few years. Results of TLD and ESR dating can be used for the development of ESR dosimeters.

(d) *Radiation sensitivity* : <u>G–value</u> : The efficiency of a dosimeter is expressed by a "G–value", i.e., the number of radicals or valency change of ions produced by the absorbed radiation per 100 eV. The G–value of ethanol or propylalcohol is around 8.3 at liquid nitrogen temperature. Aromatic molecules such as benzene and naphthalene have generally small G–values of 0.1 ~ 0.2. The absorbed radiation energy of 1 J/kg or 6.3×10^{18} eV/kg is defined as the absorbed dose of 1 Gy (gray) in the SI unit (see Chapter 4).

The number of radicals produced by the absorbed dose of 1 Gy is about 6.3×10^{16}/kg for a material with $G = 1.0$ (Ikeya 1985). Thus, the number of radicals, n, for 1 g dosimetric materials is written as

$$n = 6.3 \times 10^{13} G \quad [Gy^{-1} g^{-1}] , \tag{13.3}$$

and is shown as a function of dose for different G–values in Figure 13.2. The *minimum detectable dose* is evaluated with the sensitivity of a commercial ESR spectrometer, ~ 10^{10} spins/0.1 mT for different linewidths as shown in Table 13.1. In practice, inorganic compounds used for TLD ($CaSO_4$, $CaCO_3$, etc.) give a minimum detection limit of about 1 Gy ($G < 1.0$) and is only used for an intermediate dose range as other materials. A high–dose range dosimeter has been developed using the signal intensity of the E'_1 center in synthetic quartz (Wieser and Regulla 1989).

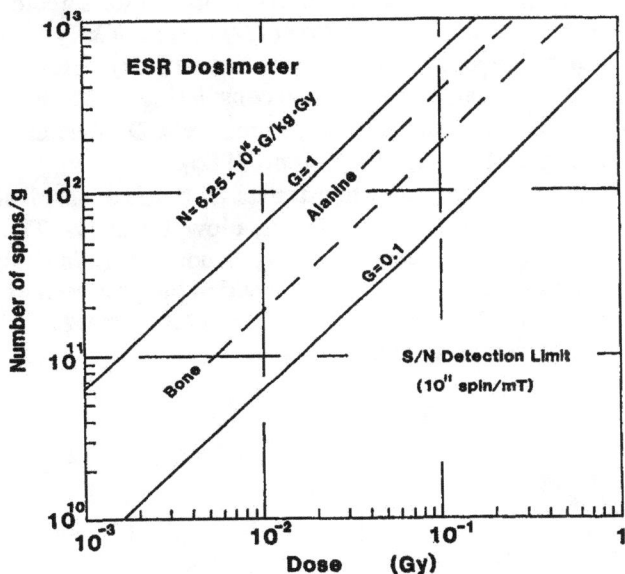

Figure 13.2 Theoretical estimation of the ESR dosimeter response for materials with different G-values. The responses for alanine and bone are shown by dashed lines.

Table 13.1 Detection limit (mGy) per 100 mg of materials with different G-values using an ESR spectrometer with the sensitivity of 10^{10} spins/0.1 mT.

G-value	spins/Gy·g	ΔH (mT)	Detection limit *) (mGy/100 mg)	Comment
20	1.3×10^{15}	0.05	0.04	ideal
10	6.3×10^{14}	0.1	0.16	probable goal
1	6.3×10^{13}	~1.0	~16	apatite
1	6.3×10^{13}	~10	~160	alanine
0.1	6.3×10^{12}	1~10	160~1600	most materials

*: The detection limit is reduced by a factor of four by adopting a sample of 300 mg, a cylindrical cavity of TE_{011} mode and the present sensitivity of 7×10^9 spins/0.1 mT.

13.2.2 *Dosimetry with Natural Rock*

(a) *Theoretical Background* : *AD and TD* : Natural rocks have been
exposed to both internal and external natural radiation for a geologically long
time. The *total dose of natural radiation (TD)* of tens of kGy is higher than
artificial dose of an A–bomb or accident radiation (AD). However, the *AD*
given to natural rocks can be determined considering defect formation at a
high dose rate, D', of A–bomb and a low dose rate, D, in nature as schematically shown in Figure 13.3 (a) (Ikeya *et al.* 1986).

The point in *rock dosimetry* to estimate A–bomb or accident radiation
is to use a relatively unstable signal having a low intensity. The saturation
level of the relatively unstable signal, $aD\tau$, is not so high in nature as in an
accident or A–bomb radiation. If a sample had in the past been exposed to a
radiation dose *AD* at a high dose rate by some reason, the *ED* obtained for
defects with first–order decay is described by

$$ED = AD + D\tau ,\tag{13.4}$$

and thus,

$$AD = ED - D\tau .\tag{13.5}$$

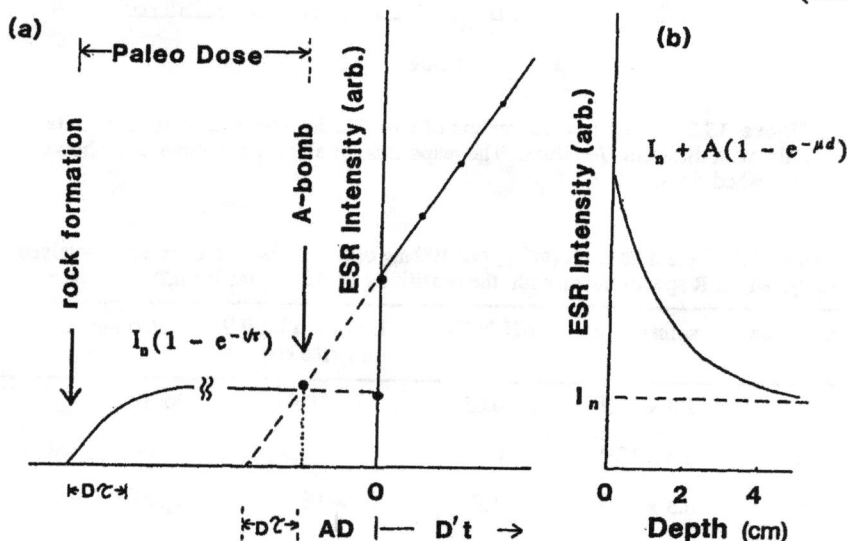

Figure 13.3 Principle of A–bomb or accident dose (*AD*) estimation using the
ESR signal of natural rock minerals. (a) Additive dose method to obtain the
equivalent dose (*ED* = *AD* + *D*τ). (b) Self–attenuation studies using the signal
intensity versus depth profile. The natural intensity, I_n, may be subtracted to
obtain the *AD* directly in (a).

The presence and absence of a relatively unstable signal is a good indication of an accident dose AD. The lifetime of relevant defects must be appropriate so that one can estimate the AD beyond the background TD level.

(b) *Depth profile* : Another way to separate the AD from the TD is to detect a *self–absorption effect* using thick samples, as shown in Figure 13.3 (b), since a natural dose is fairly uniform. Dependence of the signal intensity on the distance from the surface (depth) may be analyzed by estimating self-absorption and by considering the energy spectrum of accident radiation (see Section 13.5.4). ESR imaging of the defect concentration (Chapter 14) will be useful for measurement of the depth profile.

13.3 Gamma- and X-ray Dosimeters

13.3.1 *Alanine, Sugar and Other Organics*

(a) *Alanine* : Early use of ESR as a radiation dosimeter was done by Bradshaw *et al.* (1962) using the signal intensity of alanine radicals (CH_3– $\cdot CHCOOH$; see Figure 13.4 (a)) and their radiolysis was studied extensively (Shields and Gordy 1958, Minegishi *et al.* 1967, Sinclair and Hanna 1967, Simmons and Bewley 1976, Akaboshi *et al.* 1979, Simmons 1989). An alanine dosimeter was then improved considerably by forming a uniform pellet of alanine microcrystals embedded in *paraffin* (Regulla and Deffner 1981, 1982, 1983) or in *polystyrene* (Aminogray; see Figure 13.4 (b)) (Kojima *et al.* 1986), and by using a compact reader (Kojima and Tanaka 1989).

An ESR spectrum at 293K of a commercial *alanine–polystyrene dosimeter* after γ–irradiation and the signal positions at 293K and 77K are shown in Figure 13.4 (c). The signal intensity of alanine radicals increases linearly with the dose from about 3 Gy to 10^4 Gy. The wide range of linearity and the tissue–equivalent energy response are advantages over some TLD elements that show a supralinearity at high radiation dose ranges (Van Laere *et al.* 1989). High–level radiation dosimetry of alanine has been compared with TL and the optical density in solid state dosimetry (Göksu *et al.* 1989). The G–value of alanine is about 0.65 (Katsumura *et al.* 1980).

The background signal intensity, which is equivalent to about 3 Gy in an alanine pellet and 0.1 Gy in synthetic recrystalized powder, limits the minimum detectable dose. This is mainly due to radiolysis by light exposure or by mechanical grinding. The sensitivity of L–alanine is higher than that of DL–alanine and the background intensity is small. The dependence of the

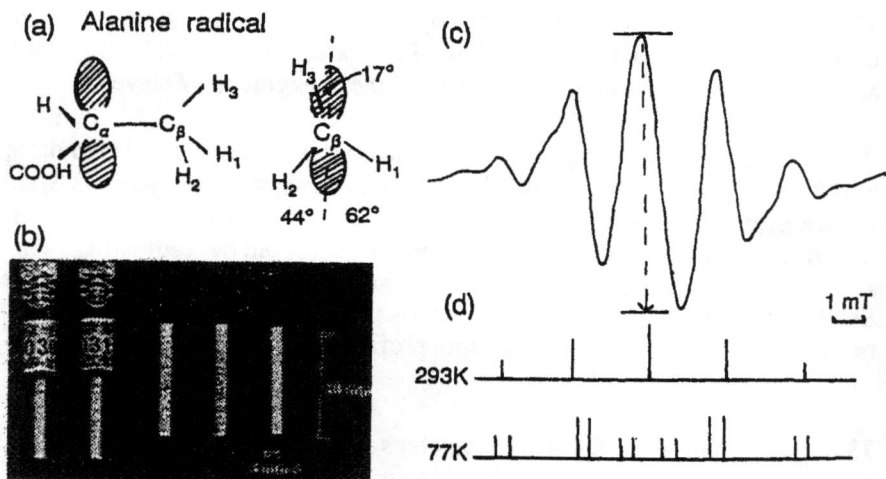

Figure 13.4 (a) Structure of alanine radicals (Miyagawa and Itoh 1962) and (b) A commercial dosimeter element with polystyrene binder (Aminogray) (Kojima *et al.* 1986). (c) ESR spectrum at 293K of an alanine dosimeter after γ–irradiation and (d) the signal positions of alanine radicals at 293K and 77K. A triplet due to H_1 and H_2 splits into a doublet by H_3, which further splits by H_α at 77K. Free rotation at room temperature gives a quintet spectrum.

response for the *alanine–paraffin dosimeter* on the irradiation temperature has been shown in the range of +50 ~ –60°C (Wieser *et al.* 1989).

Alanine has been studied as a dosimetric system in radiation processing plants (Bartolotta *et al.* 1989, Nam and Regulla 1989, Olsen *et al.* 1989, Sollier *et al.* 1989). ESR alanine dosimetry of high–energy electrons has been studied in radiotherapy (Chu *et al.* 1989) as well as in an accelerator radiation environment (Coninckx *et al.* 1989) and in dose control in industrial plants (Dorda 1989). *Alanine film dosimeters* developed for electron dosimetry (Janovsky *et al.* 1988, Miller *et al.* 1989, Hjortenberg *et al.* 1989, Kojima *et al.* 1992) will also be useful to study the spatial distribution of radiation dose as an image (see Chapter 14).

(b)   *Sugar and Other Organics* : Organic materials other than alanine can also be used as a tissue–equivalent dosimetric material. Sugar has been used for *lyoluminescence dosimetry* in which emission of light is observed by dissolving the substance in solution (Azorin *et al.* 1989, Oommen *et al.* 1989). Sugar is found to be sensitive to radiation so as to detect 50 mGy and down to 1 mGy by repeated ESR measurements of 20 ~ 30 times and subsequent signal averaging (Nakajima 1988, Nakajima and Ohtsuki 1990). An ESR spectrum of irradiated sugar is shown in Figure 13.5 (a) and the linear enhancement of the intensity by irradiation is shown in (b). Sugar has been used as an emergency accident ESR dosimeter in the Chernobyl reactor accident (see Section 13.6.2).

Organic materials have in many cases no energy dependence (tissue equivalence) during radical formation by ionizing radiation, however, the thermal stability of radicals in organic substances is a major problem. Transfer of electrons from unstable radicals to some impurities which give ESR signals is preferable for development of dosimeter materials. Hence, synthetic materials doped with impurities are suitable for a dosimeter element. The fading property of an organic polymer (acrylic) material, "*lucite*", indicated that the signal intensity decays to 88% of the initial value after 2 days (Nakajima and Watanabe 1974, Nakajima 1982). The minimum detectable dose was about 1 Gy with ESR at that time, but radicals decay in a few hours. The situation is the same for some natural and synthetic fabrics (Barthe *et al.* 1989, Kamenopoulou *et al.* 1986).

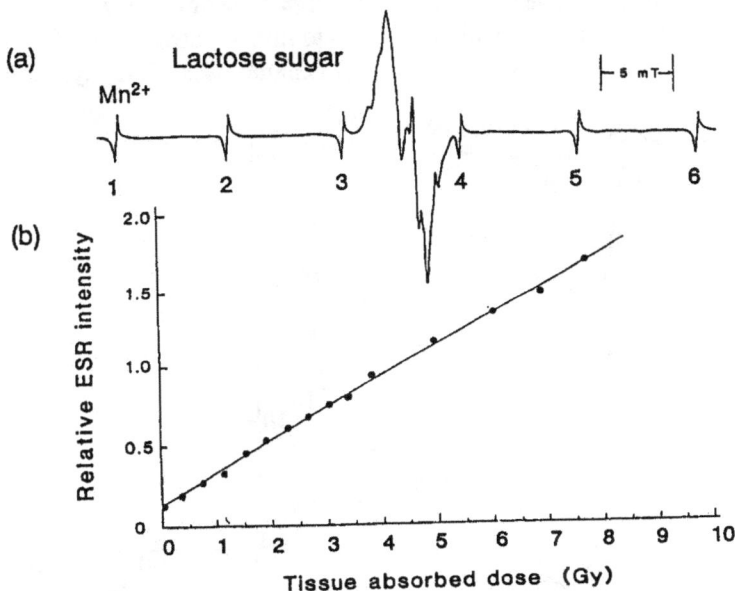

Figure 13.5 (a) ESR spectrum of lactose sugar with six hyperfine lines of Mn^{2+} as standard signals. (b) The relationship between the signal intensity and the radiation dose (revised from Nakajima 1988).

13.3.2 *Apatite and Biological Materials*

(a) *Synthetic apatite* : Tooth enamel and bone have stable CO_2^- *radicals* in their mineral part, hydroxyapatite. *Hydroxyapatite* doped with CO_3^{2-} was synthesized for ESR dosimetry (Ostrowski *et al.* 1971). The sensitivity increases as the CO_3^{2-} content increases. A dosimetric use of synthetic hydroxyapatite is promising, but the material is not tissue equivalent, as the constituent elements include Ca^{2+} which has inner K–shell atomic orbitals leading to a high mass absorption coefficient below a photon energy of 30 keV. CO_2^- radicals in fluorapatite give a sharper line than hydroxyapatite due to a different mode of hindered rotation (see Chapter 8). The optimum width of 100 kHz field modulation is checked as is shown in Figure 13.6 for the ESR signal of synthetic hydroxyapatite powder. The signal intensity is increased monotonously as the modulation field is increased up to 2 mT, though the separation of signals is not clear due to overmodulation.

(b) *Tooth enamel and bone* : Tooth enamel is the best *biological dosimeter* (Ikeya *et al.* 1984, Pass and Aldrich 1985, Aldrich *et al.* 1986a, b). ESR of tooth enamel has been applied to dosimetry of cancer patients under radiation therapy (Hoshi *et al.* 1985), to the cumulative dose of dentists (Tatsumi and Okajima 1985, Shimano *et al.* 1989) and to human dosimetry of radiation accidents. Since signals of CO_2^- in tooth enamel is sufficiently stable, the

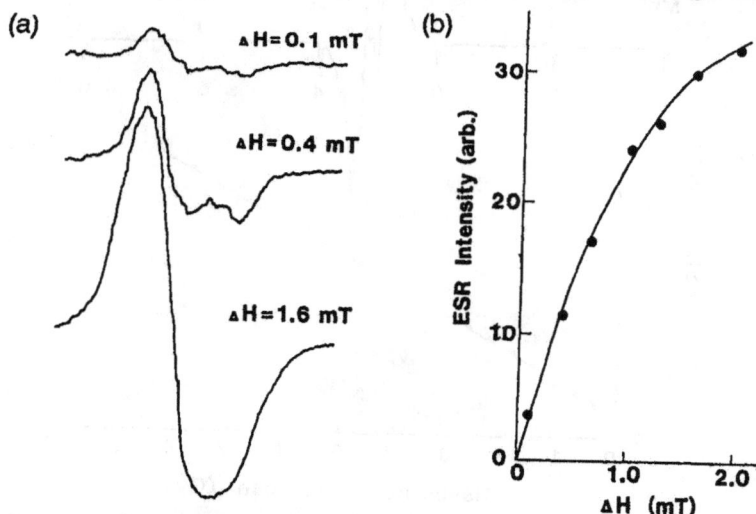

Figure 13.6 (a) ESR spectra of irradiated synthetic hydroxyapatite with different 100 kHz modulation field intensities. (b) The ESR intensity versus modulation field for an optimum dosimeter condition using hydroxyapatite.

lifetime cumulative radiation dose of personnel working in nuclear facilities will be determined using teeth.

The ESR signal of biological materials such as bone tissue and rat tail has been used to determine the radiation dose (Swartz 1965, Brady *et al.* 1968, Stachowicz *et al.* 1974). The sensitivity (minimum measurable dose) was about 1 Gy for canine, human and bovine bones (Koberle *et al.* 1973), now reduced to 50 mGy though the subtraction of a broad organic signal is needed. Sensitivity enhancement was made by detecting rapid passage out–of–phase (second harmonic) signal at 77K (Galtsev *et al.* 1993).

The absorbed doses to legs of two workers in a radiation accident at a San Salvador ^{60}Co–irradiation facility (Desrosiers 1991) and to a piece of tooth and finger bones in British radiation facilities (Lloyd *et al.* in preparation) were determined with ESR using the CO_2^- signal. Radioactive holmium–166 (^{166}Ho) in a form of phosphonate was injected into a beagle to study the uptake and the radiation dose to bone (Desrosiers *et al.* 1991).

(c) *Shell Button* : Shells are marine carbonates which are sensitive to radiation (see Chapter 6). *Shells*, especially *shell buttons* are ideal dosimeters. A–bomb radiation dosimetry using a piece of shell button is described in Section 13.5.3. Plastic buttons did not show a stable signal after irradiation.

13.3.3 *Radiation Quality Effect*

(a) *Energy Dependence* : The tissue–equivalent property of organic substances is a major focus of research on alanine for an ESR dosimeter (Regulla and Deffner 1982, Hoshi *et al.* 1985, Tatsumi and Okajima 1985, Aldrich and Pass 1988b). Figure 13.7 shows an apparent energy dependence of the signal intensity for alanine and hydroxyapatite relative to water by radiation exposure at a dose of 1 Gy. The response of an alanine dosimeter shows little energy dependence as expected. On the other hand, the signal intensity of hydroxyapatite shows a high sensitivity to X–ray photons below 100 keV. The response to X–rays at 30 keV is about eight times higher than to γ–rays from ^{60}Co (1.3 MeV) at the same absorbed dose expressed as a dose to water. This effect cannot be neglected, especially when the *ED* for tooth enamel is obtained using γ–rays in the estimation of A–bomb radiation (see Section 13.5.2).

Dosimetry using inorganic materials involves such energy dependence as is well known for TLD of $CaSO_4$ doped with rare earth elements. The inner shell ionization of heavy elements is the cause of the energy dependence. Similar energy dependence was observed for a shell button ($CaCO_3$).

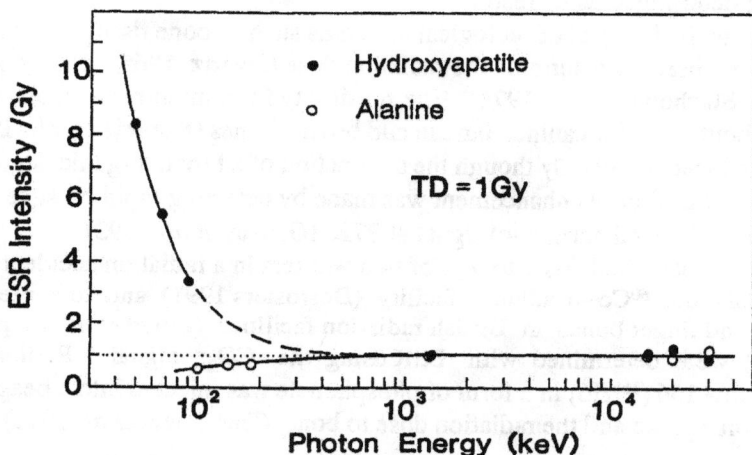

Figure 13.7   Energy dependence of the apparent ESR signal intensity for alanine and hydroxyapatite. If the defect formation efficiency considering the mass absorption coefficient of hydroxyapatite is taken into account, the efficiency is less than 1.0 at the low energy (based on the figure from Hoshi *et al.* 1985, Tatsumi and Okajima 1987).

(b) *Separation of X- and γ-ray exposure* : A mass attenuation coefficient at photon energy of about 60 keV is so high compared with γ-ray photons from ^{60}Co. X-rays give the ESR signal intensity at the front several times higher than at the back of a tooth. The back surface is not irradiated due to self-attenuation for X-ray exposure. This phenomenon has been used for separating nuclear accident or A-bomb radiation doses from dental X-ray doses (Aldrich and Pass 1986, 1988a, Shimano *et al.* 1989).

Figure 13.8 shows ESR spectra of the front and back (tongue) side of tooth enamel irradiated by (a) X-rays and (b) γ-rays. It is apparent that X-rays produces ESR signals only at the front surface of enamel, while γ-rays creates signals much more uniformly (only about 10% less at the back). The effect of secondary electron build-up is also seen with and without acrylic plates attached to the surface in (c) and (d).

The presence of metallic filling for dental caries gives a higher radiation dose in tissues close to the metal. Spatial images of the ESR signal intensity for human teeth irradiated by X or γ-rays and with metal fillings are studied with a microwave scanning ESR microscope (see Figure 14.19).

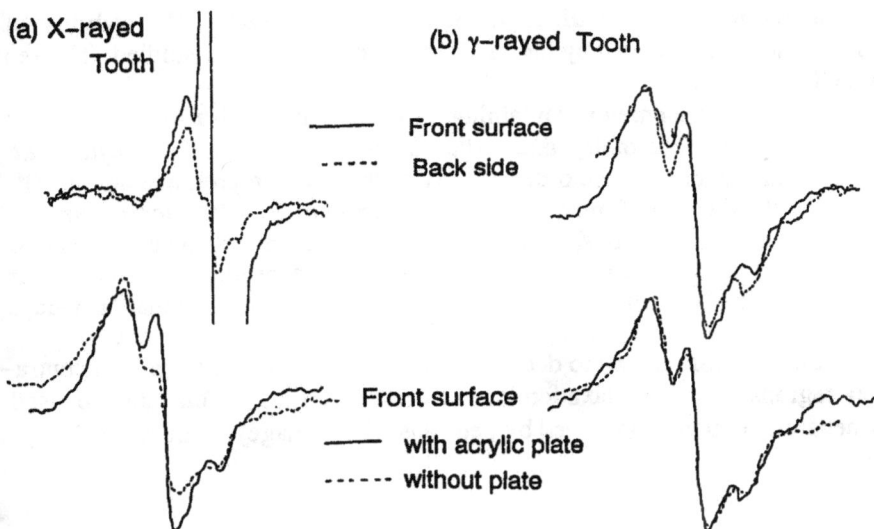

Figure 13.8 ESR spectra of tooth enamel at the front surface and the back (tongue) side of a tooth by (a) X–irradiation and (b) γ–irradiation. The effect of a surface acrylic plate is shown in (c) and (d) to indicate the secondary electron build–up effect (Shimano *et al.* 1989).

13.4 Neutron and High Linear Energy Transfer (LET) Dosimetry

Fast neutrons knock off protons (H^+) or light nuclei as a result of energy transfer; nuclei with a mass close to that of the projectile neutron have a high kinetic energy transfer from neutrons and can be displaced from the lattice site. The ionization caused by the displaced nucleus produces radicals in a material. The relative response to fast neutrons as compared with γ– or X–rays has been studied using alanine and other organic compounds having sufficient protons. The linear response up to 10^4 Gy is confirmed, although the relative sensitivity to neutrons is low, the ratio of the G–value for neutrons (G_n) relative to γ–rays (G_γ) being $G_n/G_\gamma = 0.65$ for 14 MeV neutrons and 0.4 for 1 MeV neutrons (Katsumura *et al.* 1986).

The high *linear energy transfer* (LET) effect by the *recoil protons* allows to detect the radical localization, which has been confirmed by comparing the microwave power saturation behavior in γ– and neutron–irradiated organic materials (Waligorski *et al.* 1981, 1989, Miyagawa and Itoh 1962, Katsumura and Tabata 1985, Ogawa *et al.* 1985), especially an alanine probe (Ikezoe *et al*. 1984, Schraube, *et al*. 1989). The stability of alanine radicals induced by high LET radiation have been studied (Hansen and Olsen 1989).

Dosimeter elements containing ^6Li and ^{10}B could also be used for thermal neutron dosimetry due to the (n, α) nuclear reaction (Ikeya *et al*. 1972). Since energy of the order of MeV in the reaction produces about 10^4 radicals, the number of reactions would be around $10^5 \sim 10^6$ for detectable spins of $10^9 \sim 10^{10}$. Thus, thermal neutron measurement may also be done utilizing a dosimeter containing Li or B. The LET dependence is involved in this case. The G-value for neutrons G_n could be smaller than that of γ-rays G_γ if the energy production by the (n, α) reaction is taken into account.

Pulsed ESR allows to detect alanine radicals at locally high concentration regions from the relaxation time (Höfer *et al*. 1989). This may be used to separate radicals produced by neutrons (H^+-damage) from those by γ-rays.

13.5.   A-Bomb Radiation Dosimetry
13.5.1   *A-bomb Dose Reevaluation*
A-bomb radiation dose as a function of distance from the epicenter in Hiroshima and Nagasaki, Japan has been studied using a bare reactor on a tower (*Ichiban Assembly*) and in a series of A-bomb tests in USA (Roesch 1987). The *tentative radiation dose (T65D)* determined at Oak Ridge National Laboratory in 1965 (Auxier *et al*. 1966), was doubtful according to theoretical calculation of neutron spectrum and the dose estimate. Neutron doses in Hiroshima were reestimated as approximately 1/10 of *T65D* (Löwe and Mendelson 1985). This indicates that the *maximum permissible dose* has to be lowered by one order of magnitude for neutron doses. Radiation effects on human beings therefore were previously underestimated (Okajima 1985).

The radiation dose of A-bomb survivors was determined using the shielding effect of houses and the *T65D*. The conventional standards such as the maximum permissible dose are partly based on radiation effects on human beings at Hiroshima and Nagasaki. The radioactive phosphor (^{32}P) induced in the sulfur of ceramic insulators was measured soon after the A-bombing. The residual radioactivities of ^{60}Co in iron materials and recently ^{152}Eu in rocks have been measured by γ-ray spectroscopy (Nakanishi *et al*. 1983, Hasai *et al*. 1987, Hoshi *et al*. 1989).

The γ-ray radiation dose was determined from TL of *roof-tiles* (Higashimura *et al*. 1963). A considerable improvement of TL dosimetry of A-bomb radiation has been made using *quartz grains* separated from ceramics in the area within 2 km from the hypocenter (Ichikawa and Nagatomo 1987). This procedure contributes greatly to the determination of TL dose in A–bomb radiation dosimetry and of TL ages in archaeology. ESR may provide further information on A–bomb doses, though there is concern whether the additive dose method with a dose rate of 10 Gy/h can be used for A–bomb radiation dosimetry of 10 Gy in 10^{-7} s ($= 3.6 \times 10^{10}$ Gy/h).

Two volumes of reports on the reevaluation of the A–bomb dose system (*DS86*) were issued in 1987 by the Radiation Effects Research Foundation (RERF) (Roesch 1987). A new *dose system, DS86,* was determined as a function of distance from the epicenter. The radiation dose was estimated, considering the *shielding factors* depending on the building type (wooden, concrete, brick, etc.). The correlation with *chromosomal aberrations* was also studied and it was found that chromosomal aberration is sufficiently sensitive to detect an exposure of 0.1 Gy. ESR detection of radiation effects on *blood* is not sensitive (Kawano *et al.* 1985a, b).

13.5.2 *Tooth Enamel of A–bomb Survivors*

Mascarenhas *et al.* (1973) have extended the study of bone dosimetry to the measurement of bones from victims in Hiroshima. However, the sensitivity of ESR spectrometer was not sufficient at that time. The signal at $g = 2.0045$ associated with organic radicals seems to have been measured resulting in an overestimate of A–bomb radiation dose (Duran *et al.* 1985).

Extensive studies on A–bomb dosimetry has been performed using teeth, in collaboration with the groups at Hiroshima and Nagasaki (Ikeya *et al.* 1984, 1986, Tatsumi and Okajima 1985, 1987), based on the experience in dating of anthropological bones and teeth (Ikeya and Miki 1980). Figure 13.9 (a) shows an ESR spectrum of tooth enamel extracted by dental treatment from A–bomb survivors. The dentine and cementum were removed since they give a broad organic signal at $g = 2.0045$. The signal intensity due to CO_2^- is proportional to the weight of the sample up to $300 \sim 400$ mg as shown in (c) (Iwasaki *et al.* 1990).

The enamel was irradiated by γ–rays to obtain an *ED*. The signal intensity at $g = 1.9973$ due to CO_2^- was taken as indicative of the radiation dose and an *ED* of 3 Gy was obtained as shown in Figure 13.10. Thus, human exposure can be determined directly from the ESR signal of tooth enamel. A–bomb radiation dose estimated from the ESR signal intensity of tooth enamel was compared with the *DS86* (Tatsumi and Okajima 1987). The assessment of shielding factors is not simple.

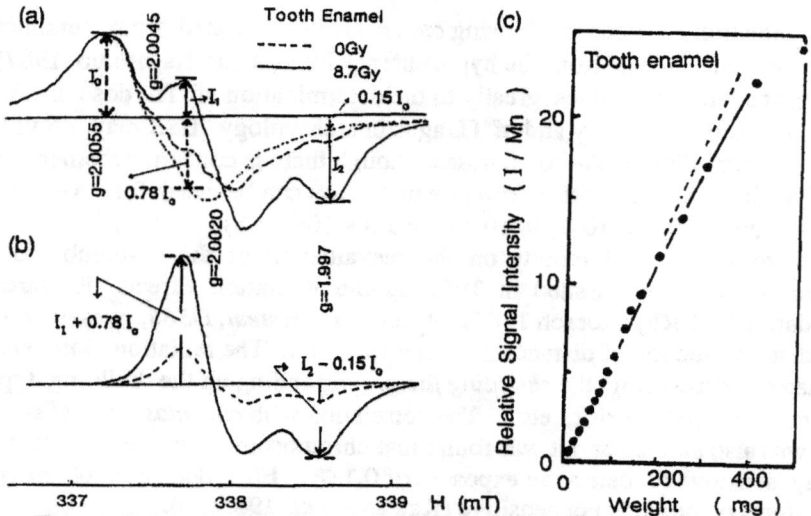

Figure 13.9   (a) ESR spectrum of tooth enamel extracted by dental treatment from an A–bomb survivor. The contribution of the broad signal at $g = 2.0045$ due to organic radicals must be subtracted as in (b) (Shimano *et al.* 1989, Ishii and Ikeya 1990). (c) The relation between the signal intensity and the weight of the sample (Iwasaki *et al.* 1990).

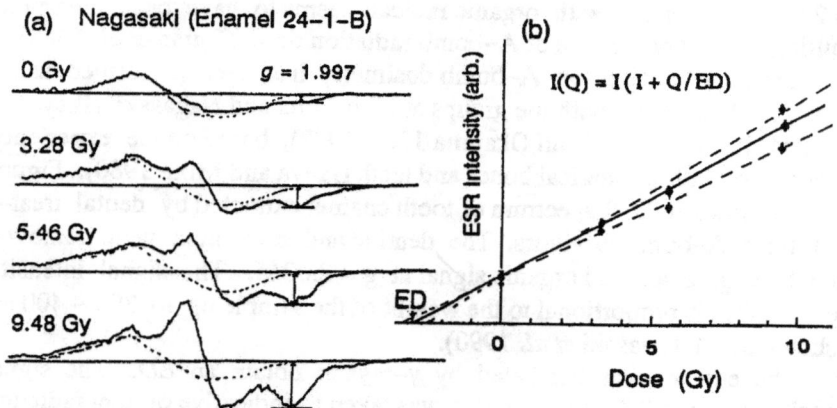

Figure 13.10   (a) ESR spectra of a human tooth by A–bomb radiation at Nagasaki and the enhancement of the signal intensity by γ–irradiation. (b) The A–bomb radiation dose is estimated by a linear extrapolation of the intensity enhancement. A broad signal around $g = 2.0045$ must be subtracted with a computer to determine the low radiation dose of tens of mGy, as in ESR dating of Holocene bones.

13.5.3 *Shell Buttons*

Figure 13.11 shows the city map of Nagasaki indicating the epicenter of the A–bomb explosion and the medical school hospital of Nagasaki University. The black dots in the insertion are those who were killed and the white ones are survivors. A medical doctor exposed to A–bomb radiation in

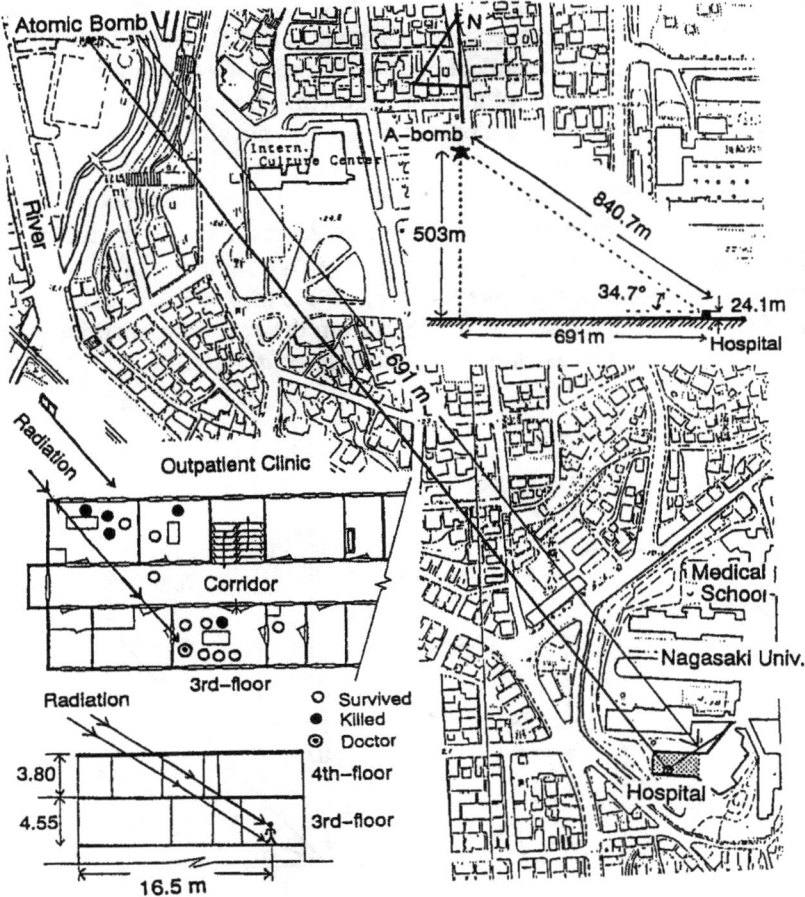

Figure 13.11 A city map of Nagasaki indicating the site of the A–bomb explosion and the hospital of Nagasaki University where a doctor was exposed to A–bomb radiation. The black and white dots in a hospital building indicate those who were killed and who survived, respectively. ESR of a tooth piece and a shell button from the doctor gave an A–bomb dose of 2.1 Gy and 1.9 Gy, respectively.

the hospital is shown as a double circle. He was in the building at the ground distance of 506 m through the ceiling and 4th floor. No other wall shielded the radiation. The doctor, who suffered the symptoms of radiation disease, provided samples of a tooth and a shell button from his clothes.

Figure 13.12 shows ESR spectra of a half piece of *shell button* as received and irradiated by γ–rays. The same spectra as for fossil shells are observed. The signals at $g_C = 2.0007$ and $g = 1.997$ are due to isotropic and orthorhombic CO_2^-, respectively (signals C and D; see Chapter 6). The ESR signal is angular dependent because of the preferential orientation of *c*–axis along the direction of the shell growth. Hence, measurements after additive doses must be made by setting the sample always the same relative to the magnetic field direction.

The *ED* of 1.9 ± 0.2 Gy obtained from the enhancement of the signal intensities was concordant with the *ED* of 2.1 ± 0.2 Gy for the tooth enamel. The absorbed dose of 2 Gy indicates an effective shielding of about 33 cm in concrete thickness for a γ–ray energy of 1.3 MeV (^{60}Co), using *T65D* of 32.46 Gy at a ground distance of 506 m from the epicenter. Fabrics were not good ESR dosimeters as claimed by Alekhin *et al*. (1982).

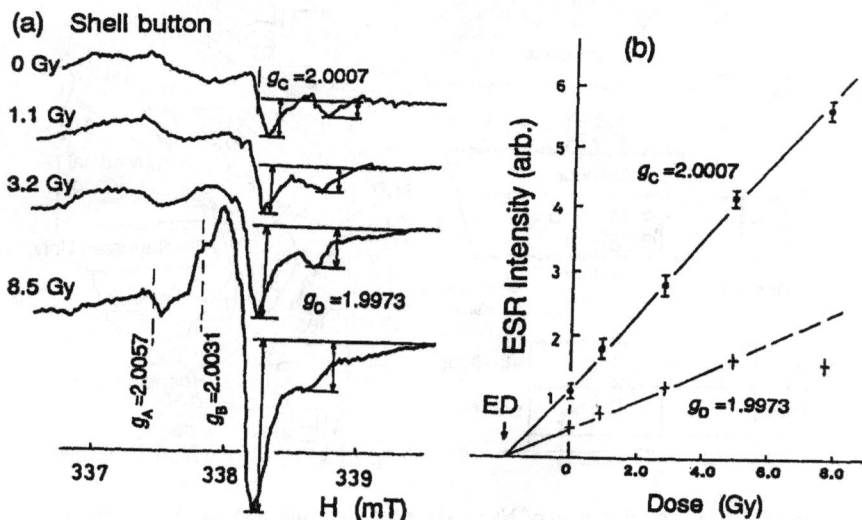

Figure 13.12 (a) ESR spectra of a half piece of shell button supplied by the medical doctor. (b) The enhancement of the signal intensity at $g = 2.0007$ and $g = 1.997$ as a function of additive dose. The A–bomb dose equivalent to γ–ray dose from ^{60}Co is 1.9 ± 0.2 Gy (Ikeya *et al*. 1984).

13.5.4 *Dosimetry with Quartz in Granite Bridge Pillar*

(a) *ESR spectra* : *Roof–tiles* and *brick ceramics* once fired have been used for A–bomb dosimetry using TL at Hiroshima and Nagasaki (Higashimura *et al.* 1963, Ichikawa and Nagatomo 1987). Fine *quartz grains* separated from these materials shows ESR signals of Mn^{2+} and Fe^{3+}, presumably due to diffusion from the surface. Unfortunately, the amount of the material was too small to etch the surface region where impurity might diffuse. In addition, the minimum dose detected using quartz grains is 10 Gy with a present day ESR spectrometer. Hence, *rock dosimetry* of granite bridge pillar was attempted based on the principle described in Section 13.2.2.

Figure 13.13 (a) and (b) show the ESR spectra of quartz derived from granite drilled from the granite pillar of the Motoyasu bridge in Hiroshima. These spectra are typical of quartz grains in nature. The signals at $g = 2.0009$ (E'_1 center) and at $g = 1.997$ (Ge center; see Chapter 9) are more clearly observable in the surface sample than in the inside (Ikeya *et al.* 1986). The signals associated with oxygen hole centers (OHC) or the peroxy center

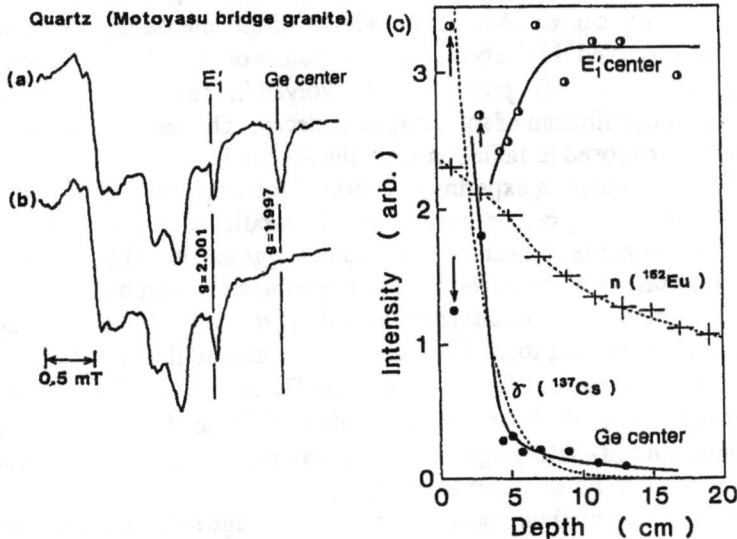

Figure 13.13 (a) and (b) ESR spectra of quartz grains in a granite pillar of the Motoyasu bridge with a ground distance of 120 m from the epicenter of A–bomb at Hiroshima. The signal of Ge centers at $g = 1.997$ is more intense at the surface (a) than inside (b). (c) The depth profile of ESR signal intensity of the E'_1 and Ge centers. The activities of ^{152}Eu determined by γ–ray spectroscopy and the calculated relative γ–ray intensity, considering the self–absorption of γ–rays (662 keV, ^{137}Cs), are also shown by dashed curves.

produced by α-particles or α-recoil atoms and also by A–bomb fast neutrons should be more intense at the surface than the inside; however, these
were not observed. These centers created by natural radiation must have
been abundant.

(b) *Depth profile* : The depth profile of the signal intensity for the Ge and
E'_1 center is shown in Figure 13.13 (c). The calculated *self–shielding effect*
for γ–rays from ^{137}Cs at the incident angle of *A–bomb radiation* is shown by
a dashed curve. The activity of ^{152}Eu in the same rock is also shown to
indicate the neutron dose for comparison (Hasai *et al.* 1987, Hoshi *et al.*
1989). Except for the surface sample, the intensity of the *Ge center* decreases as a function of the depth. The intensity of the E'_1 *center* is reduced
at the early stage of artificial irradiation but that of the Ge center is enhanced.
       The absorption effect of A–bomb radiation in the bridge pillar granite
rock at Hiroshima is clear. The intensity of the Ge center at the surface is
relatively smaller than that in slightly deeper layers. On the other hand, the
intensity of E'_1 centers is large at the surface, suggesting a charge transfer
from the Ge to E'_1 center. A similar tendency of the intense E'_1 signal at the
surface was observed in the building wall granite of the Fukoku Insurance
Company. Charge transfer processes are involved in the defect formation of
old rocks in an equilibrium of formation and decay. The results indicate that
the granite was exposed to radiation from the A–bomb.
       A thermal annealing experiment at 100°C for a few days indicated that
the intensity of the E'_1 center is enhanced, in parallel with the decrease of
the relatively unstable Ge center. The result suggests that the intense E'_1
signal at the surface of the granite is due to partial annealing by the surface
heating. Heat from the A–bomb and sunlight as well as the light bleaching of
the Ge center in the past forty years may have enhanced the intensity of the
E'_1 center at the surface at the expense of the Ge center. Sunlight bleaches
the signal intensity of the Ge center in irradiated SiO_2 as described in Chapter 9. Thus, the self–shielding effect in natural rocks can yield information
about accident or A–bomb doses by using a relatively unstable signal.
       The proposed method, utilizing the results of the self–shielding effect
from the depth profile, can give an accident or A–bomb radiation dose. For
quartz, the Ge center with a less saturation concentration can give an accident dose. However, electron transfer to the E'_1 center occurs by heating or
light bleaching as time proceeds (Ikeya and Ishii 1989).
       Garrison *et al.* (1981) have used additive irradiation by fast neutrons to
estimate the accumulated dose in ESR dating of flints. The fast neutron dose

can be obtained relatively in the depth profile of the ESR signal intensity, if the natural saturation dose is either comparable to or much smaller than the A–bomb dose. The effect was not observed though an additive neutron dose from a ^{252}Cf source was used to estimate the minimum detectable neutron dose in this granite.

13.5.5 High Dose Rate Effect : A–bomb or Accident Radiation

It has often been considered whether a high dose rate of artificial irradiation (D') can actually simulate the natural radiation dose rate.

(1) The natural radiation dose rate is usually of the order of mGy/a.
(2) The artificial dose rate of $10 \sim 100$ Gy/h $(= 8.76 \times 10^4 \sim 10^5$ Gy/a) is about 10^8 times of the natural dose rate.
(3) In case of the A–bomb radiation dose rate, the absorbed dose of 1 kGy close to the epicenter corresponds to the prompt dose rate of 3.6×10^{13} Gy/h $(= 3.15 \times 10^{17}$ Gy/a).

In case of A–bomb radiation dosimetry, the A–bomb dose rate is extremely high since the exposure time is of the order of 0.1 μs. The *dose rate effect* on defect formation has already been given in Figure 3.13. The intensity, I_0, relative to the saturation intensity, I_s, or the *ED* relative to the saturation dose is a valuable parameter to estimate the dose rate effect. The A–bomb dose at the Motoyasu bridge 110 m away from the hypocenter was 200 Gy or so, still less than the saturation dose. Hence, the high dose rate effects on human tooth may be neglected and the correction factor for the real dose TD_{real}/ED is close to 1.0 for I_0/I_s, being less than 0.1 (see Figure 3.14).

13.6 Accident Dosimetry

13.6.1 Chernobyl Resident Teeth : Hot Spot or Artifact ?

The disastrous accident of the nuclear reactor at Chernobyl is another tragedy of the 20th century civilization following the dropping of the A–bombs at Hiroshima and Nagasaki. Our limited information on the radiation effect on human beings has been based on the data at Hiroshima and Nagasaki. This information is not adequate to predict such effects at Chernobyl, as the situation is completely different. Although the major problem of radiation exposure to residents at Chernobyl arises from *internal exposure* due to radioactive elements ingested mainly from food, the *cumulative external radiation exposure* is also important in assessing the *health hazards*.

The ESR spectrum before additive irradiation shows a broad line for all except one of the eight samples since an organic radical signal ($g = 2.0045$) is dominant and CO_2^- signals due to the radiation effect ($g = 2.0020$ and $g = 1.997$) are masked. Separation of the radiation–induced signal from the organic radical signal is necessary to obtain an accurate intensity as described in Section 13.5.2. The *ED*'s by an additive dose method are shown in Figure 13.14 (a) in which those tabulated by Brillibit *et al.* (1990) are shown as a histogram.

The results of *accidental radiation dose* assessment using human tooth enamel are clear for the samples from the state of Gomel due to the Cherno–byl accident and Timsk in Siberia due to accidents in a nuclear processing plant. It is surprising that a few samples showed a very high radiation dose, though the low–dose exposure of other samples is consistent with our results. The problem is that they have just checked the signal intensity of enamel and obtained the dose from the calibration curve. If some tooth enamel contains an extraordinary high content of CO_3^{2-} ions, the signal intensity would be

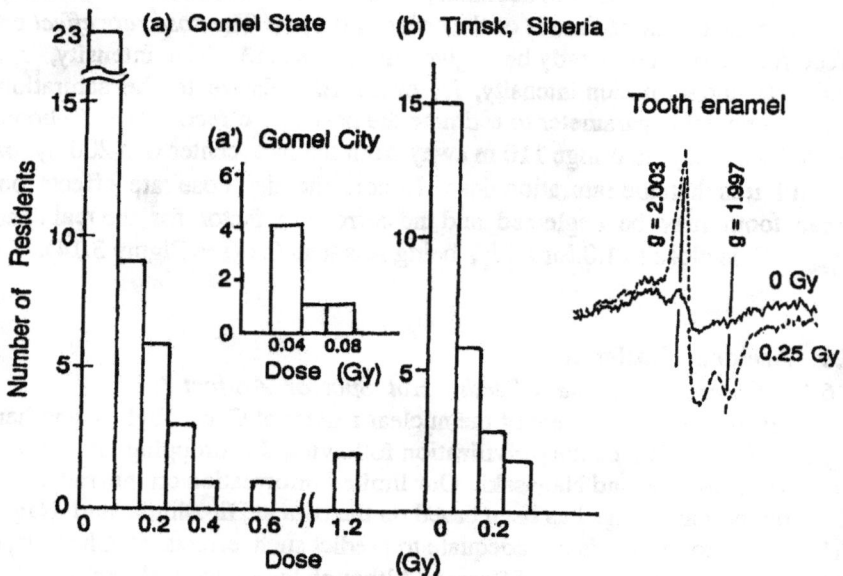

Figure 13.14   ESR dosimetry of human tooth enamel. (a) The accident dose received by the residents in Gomel city (insertion: Ishii and Ikeya 1990) and in the state of Gomel close to the Chernobyl reactor. (b) Timsk, Siberia (from the table by Brillibit *et al.* 1990).

too high. An *additive dose method* should be used for those particular samples to check the efficiency of signal formation using a calibrated source of radiation. Our results indicates that the radiation dose was <u>about 40 mGy</u>.

If 3 ~ 4% of residents were really exposed to high radiation at 1.2 Gy, it might be due to radiation doses from *hot spots*. The contamination is so highly nonuniform as to raise doubt whether the routine radiation detector is working correctly in my radiation survey measurement at Chernobyl area. The problem, whether the high dose is due to a high content of CO_3^{2-} or to the hot spots, was noted during a private talk with Dr. G. A. Klevezal at the Gomel Hematology Workshop in 1991.

Precautions are still necessary for studying the effect of *metallic dental fillings*, due to intense damage at the surface in contact with the metal. This is demonstrated as an image in Chapter 14. Whether the high–dose exposure is due to a high radiation dose to residents from hot spots or to the high sensitivity produced by a high content of CO_3^{2-} impurity or by attached dental metal is not certain at present. Systematic studies on residents, especially on school children using *milk teeth*, are needed to investigate radiation dose equivalents to residents. Teeth at the contaminated area should be collected to know the radiation dose. *A piece of tooth is a life–long cumulative radiation dosimeter* and hence <u>extracted teeth should be preserved</u> for evaluation of personal radiation exposure.

13.6.2 *Dosimetry of Sugar at Chernobyl Home*

Sugar is an ESR dosimeter material sensitive to radiation. Granulated sugar found in most homes and restaurants can act as an *accident dosimeter* of external radiation dose at particular sites inside houses or buildings. Sugar samples collected from a hotel, schools and polyclinic at Pripyat city, 3 km north–west of the Chernobyl nuclear reactor were measured with ESR (Nakajima *et al.* 1991). External radiation to evacuees at the Chernobyl accident was estimated from the sugar samples to be 55 ~ 91 mGy in good agreement with our result described above. Hence, these results and tooth dosimetry of residents indicate that the radiation dose to ordinary residents was not so high, except for a few percent who might have been exposed to much higher radiation from presumably hot spots.

Sugar–coated drug tablets have also been used in accident dosimetry (Regulla and Deffner 1989, Kai and Ikeya 1988). The drug tablet "trapizil" for heart disease showed an intense signal at the boundary between the sugar coat and the inner drug. The origin of this signal is not clear but might be associated with some mineral intentionally added.

13.7 Dosimetry for Food Irradiation

Radiation sterilization technologies are practiced in various fields with standards and recommended practices. Microbial inactivation by radiation is common in sterilizing medical devices (McLaughlin *et al.* 1989). This is the reason why the alanine dosimeter for an intermediate–dose range has been developed to monitor the absorbed dose. Food spoilage and contamination of foods with pathogens can be prevented by radiation processing (Slager *et al.* 1964). Traditional methods of processing and preserving foods have been drying, fermentation, pickling, pasteurization and canning in addition to using preservatives, inert gas and oxygen absorptive chemicals. In radiation processing, γ–rays from a source of ^{60}Co or ^{137}Cs are used in addition to high–energy X–rays (for penetration) and electron beams from an accelerator. The *safe upper absorbed dose* endorsed by the World Health Organization (WHO) is 10 kGy and US Food & Drug Administration (FDA) has set 1 kGy. *Post–irradiation dosimetry* for various foods must be developed to check the proper dose.

ESR dosimetry is simple, without requiring sample pretreatment, and is essentially a non–destructive method for the detection of unpaired electrons created by ionizing radiation. Detection of irradiated foods has thus been investigated with ESR relatively recently (Dodd *et al.* 1988a, b). ESR spectra of some irradiated foodstuffs are shown in Figure 13.15.

Bones and calcified tissues such as egg shells, mollusk and some other organic materials, which have been used for ESR dating (Ikeya and Miki 1980, Ikeya and Ohmura 1981), are used for dosimetry of some foodstuffs (Rossi *et al.* 1992). So far, CO_2^- radicals in bones (Chapter 8) and in shells (Chapter 6) have been used for dosimetry of poultry, fish and frog bones (Lea *et al.* 1988) and shellfish.

Meats and vegetables are processed by radiation (Desrosiers and Simic 1988, Desrosiers and McLaughlin 1989), in addition to spices. Fruits such as mango and strawberries (Raffi *et al.* 1988) or vegetables like onions (Desrosiers and McLaughlin 1989, 1990), although investigated, have not yet been the subject of ESR dating or dosimetry. Organic radicals are also produced thermally by oxidation (see Chapter 12) and are detected in non-irradiated foods. Hence, some precaution is required so as not to include the signal intensity from chemical reactions. Organic radicals produced by radiation are generally unstable and cannot be used as a radiation dosimeter.

Monitoring foods processed by radiation is important to know whether they are within the safe upper dose endorsed by regulation. However, not much work has been performed. Although monitoring an absorbed radiation dose with ESR for foods is within the framework of this textbook, details are included in "IAEA Newsletters", etc. Radicals are unstable due to reactions. The diamagnetic products were changed into paramagnetic ones by UV-irradiation (**D-P Conversion**) *as* shown in Figure 13.16 in the *Symp. 2001-ESRDD-Osaka.*

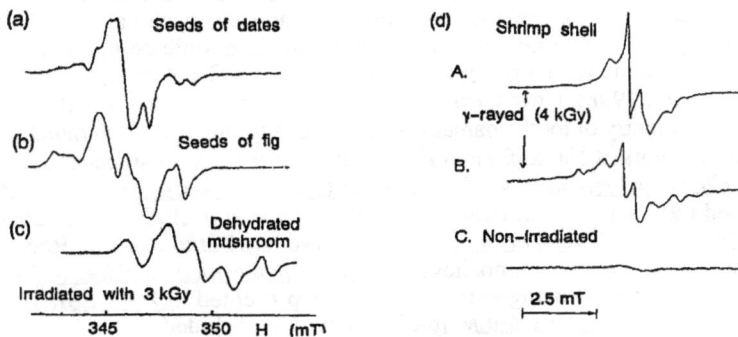

Figure 13.15 ESR spectra of some irradiated foodstuffs. (a) Seeds of dates, (b) seeds of fig and (c) dehydrated mushrooms irradiated with 3 kGy (Stachowicz et al. 1993). (d) Pink shrimp shell; γ-irradiated (4 kGy) (A) before and (B) after cleaned, dried and ground, and (C) non-irradiated control (Morehouse and Desrosiers, 1993).

Figure 13.16 ESR spectra of gynger (a) as received, (b) after gamma irradiation and (c) and (d) subsequent UV-irradiation as a new method of checking irradiated foods. Formation of UV-iduced signal intensity is strongly supressed in gamma-rayed sample. Heavi irradiation can be monitired simply by UV-irradiation.

13.8   Summary

Advantages of cumulative dose measurements with a wide range of linearity in the dosage without destroying the information by reading is in contrast to TLD showing supralinearity. A linear response in an intermediate dose range around $1 \sim 10^4$ Gy can be obtained from common materials. The sensitivity at a low radiation dose may be improved by signal averaging with a computer and by development of some new dosimeter materials like ionic compounds of alkali ion-organic acids with a sharp linewidth and a high G-value (Ikeya et al., 1998; Hassan *et al.*, 1998). New sensitive materials is required for dosimeter elements.

Dosimetry of tooth enamel was made in estimating A-bomb radiation doses of the survivors at Nagasaki and Hiroshima, accident doses to people and workers at _Chernobyl reactor_ (WHO, 1996) and _Techa river waste disposal plant_ which released radioactivities at Ural area (Romanyukha *et al.*,1996; 1999; 2000a,b,c) and _Semipalatinsk nuclear bomb test site_ (Pivovarov *et al.*, 2000). Recently, radiation dose of victims in unbelievable _JCO-Toukai critical accident_ in 2000 were evaluated with ESR using extracted tooth as presented in _2001-ESRDD-Osaka_ (http://quartz.ess.sci.osaka-u.ac.jp). These will be included in laer revision.

Bones, sugar, shell buttons and gultamic soda for cooking can be used for ESR dosimetric materials in daily lives. A technique to measure the radiation dose accurately is to select enamel carefully by removing dentine with organic materials from a tooth. An attempt to build an apparatus to measure the ESR signal in tooth without extraction is in Appendix 6. A-bomb radiation effects on rock minerals can be detected with an ESR signal depth profile of the Ge center in quartz grains.

The dosage of food irradiation to kill pathogens and to increase the shelf-life of the product should be monitored. Post-irradiation dosimetry should be developed using ESR of foodstuffs. A new method to use UV-irradiation to convert diamagnetic species to paramagnetic ones (D-P Conversion) was proposed in _2001-ESRDD-Osaka_ We will see the capability of this method in various fields. Compact ESR spectrometers using a permanent magnet of Nd-B-Fe alloy (Neomax) for a dosimetry reader are now commercially available (see Appendix 3).

References

Aldrich J. E. and Pass B. (1986): Dental enamel as in-vivo radiation dosimeter: separation of X-ray dose from the dose due to natural sources. *Radiat. Prot. Dos.* 17, 175-179.
Aldrich J. E., Mailer C. and Pass B. (1986a): The quantification of the ESR signals in dental enamel. *Int. J. Radiol.* 61, 433-437.
Aldrich J. E., Scallion P. L. and Pass B. (1986b): An analysis of paramagnetic centers in irradiated dentin using spin resonance. *Calcif. Tissue Int.* 46, 166-168.
Aldrich J. E. and Pass B. (1988a): Determination of the radiation exposure from nuclear accidents and atomic bomb tests using dental enamel. *Health Phys.* 54, 469- 471.
Alekhin L. A., Babenko S. P., Kraitor S. N., Kushnereva K. K., Barabanova A. V., Ginzburg S.R., Drutman R.D. and Petushkov V. N. (1982): Experience from applica tion of radioluminescence and electron spin resonance to dosimetry of accident irradiation. *Atomnaya Energiya* 53, 91-95.
Auxier J.A., Cheka J.S., Haywood F. F., Jones T.D. and Thorngate J.H. (1966): Free-field

radiation dose from Hiroshima and Nagasaki bombings. *Health Phys.*12, 425-429.

Azorin J., Gutierrez A., Munoz E. and Gleason R. (1989): Correlation of ESR with lyoluminescence dosimetry using some sugars. *Appl. Radiat. Isot.* 40, 871- 874.

Azorin J., Rivera T. and Solis J. (2000): A new ESR dosimeter based on polytetrafluoroethylene. *Appl. Radiat. Isot.* 52 No.5, 1243-1246.

Barthe J., Kamenopoulou V.,Cattoire B. and Portal G. (1989): Dose evaluation from textile fibers: a post-determination of initial ESR signal.*Appl. Radiat. Isot.* 40,1029-1034.

Bartolotta A., Onori S. and Pantaloni M. (1989): Intercomparison of the dosimetric system used in industrial irradiation plants in Italy. *Appl. Radiat. Isot.* 40, 871-874.

Birk A. Baraboy V., Atamanenko O., Shevehenko Yu and Brik V. (2000): Matabolism in tooth enamel and reliability of retrospective dosimetry. *Appl. Radiat. Isot.* 52, 1305-1310.

Bogushevich S.E. and Ugolev I.I. (2000):   Inorganic EPR dosimeter for medical radiology. *Appl. Radiat. Isot.* 52 ,1217-1220.

Bradshaw W.W., Cadena D.G., Crawford G.W. Jr. and Spetzler H.W.W. (1962): The use of alanine as a solid dosimeter. *Radiat. Res.*171, 11-21.

Brady J.M., Aarestad N.O. and Swartz H.M. (1968): In vivo dosimetry by electron spin resonance spectroscopy. *Health Phys.* 15, 43-47.

Brillibit M.D., Klevezal G.A., Mordvintoev P.I., Sukhovskaya L.I., Rzhenkov V.A., Voevodskaya N.V. and Vanin A.V. (1990): On the use of dental enamel as in- vivo dosimeter to g-rays. *Hematologia* 35, No.12, 11-15.

Bugay A., Kolesnik S., Metha K., Nagy V. and Desrosiers M. (2000):   Temperature stablization of alanine dosimeters used for food processing and sterilization. *Appl. Radiat. Isot.* 52, 1371-1374.

Bugay A.A., Onischuk V.A., Petrenko T. L. and Teslenko V. V. (2000): The mechanisms of radical formation in L- α -alanine. *Appl. Radiat. Isot.* 52,1189-1194.

Chen F., Covas D.T. and Baffa O. (2001): Dosimetry of blood irradiation using an alanine/ESR dosimeter. *Applied Radiation and Isotopes* 55, 13-16.

Chu S., Wieser A., Feist H. and Regulla D.F. (1989): ESR/alanine dosimetry of high-energy electrons in radiotherapy. *Appl. Radiat. Isot.* 40, 993-996.

Chumak V., Likhtarev I., Sholom S., Meckbach R. and Krjuvhkov V. (1998): Chernobyl experience in field of retrospective dosimetry: Reconstruction of doses to the population and liquidators involved in the accident. *Radiat. Protec. Dosiim.* 77, 91-98.

Coninckx F., Sch+nbacher H., Bartolotta A., Onori S. and Rosati A (1989): Alanine dosimetry as the reference dosimetric system in accelerator radiation environments. *Appl. Radiat. Isot.* 40, 977-983.

De Angelis C., Mattacchini A., Onori S., Aragno D., De P.U. and P.V. (2000): Electron arc therapy treatment planning verfication with alanine/EPR dosimetry. *Appl. Radiat. Isot.* 52, 1203-1208.

De Angelis C., Fattibene P., Onori S., Petetti E., Bartolotta A. and Sansone S. A. (2000): Transferability of ASTM/NIST alanine-polyethylene recipe at ISS. *Appl. Radiat. Isot.* 52, 1197-1202.

Desrosiers M. F. and Simic M. G. (1988): Post-irradiation dosimetry of meat by electron spin resonance spectroscopy of bones. *J. Agric. Food Chem.* 36, 601-603.

Desrosiers M.F. and McLaughlin W.L. (1989): Examination of g-irradiated fruits and vegetables by electron spin resonance spectroscopy. *Radiat. Phys. Chem.* 34, 895-898.

Desrosiers M. F. and McLaughlin W. L. (1990): Onion skin as a radiation monitor. Radiat. *Phys. Chem.* 35, 321-323.

Desrosiers M. F. (1991): In vivo assessment of radiation exposure. *Health Phys.* 61, 859.

Desrosiers M.F., Coursey B.M., Avila M. J. and Parks N. J. (1991): Radiation doses. *Nature* 349, 287-289.

Dodd N. J. F., Swallow A. J. and Ley F. J. (1988a): Use of ESR to identify irradiated food. Radiat. Phys. Chem. 26, 451-453.

Dodd N. J. F., Lea J. S. and Swallow A. J. (1988b): ESR detection of irradiated food. *Nature* 334, 387.

Dorda D. M. (1989):   Dose control in the semi-industrial irradiation plant at Ezeiza Atomic Center. *Appl. Radiat. Isot.* 40, 1009-1012.

Duran J. E. R., Panzeri H. and Mascarenhas S. (1985):   ESR dosimetry of irradiation human teeth. *ESR Dating and Dosimetry*, 391-396.

Flores J., Cabrera B.E., Calderon T., Munoz P.E., Adem E., Hernandez A.Z., Boldu J.L., Ovalle M.P. and Murrieta S.H. (2000): ESR and optical absorption studies of gamma- and electron irradiated sugar ctystals. *Appl. Radiat. Isot.* **52**, 1229-1234.

Garrison E. G., Rowlett R. M., Cowan D. L. and Holroyd L. V. (1981): ESR dating of ancient flints. *Nature* **290**, 44-45.

Galtsev V.E., Grinberg O.Ya., Lebedev Ya.S. and Galtseva E.V. (1993): EPR dosimetry sensitivity enhancement by detection of rapid passage signal of the tooth enamel at low temperature. *Appl. Magn. Reson.* **4**, 331-333.

G+ksu H.Y., Wieser A., Waibel A., Vogenauer A. and Regulla D.F. (1989): Comparing measurements of free radicals, optical density and thermoluminescence in solids for high-level dosimetry. *Appl. Radiat. Isot.* **40**, 905-909.

Hansen J.W. and Olsen K.J. (1989): Predicting decay in free-radical concentration in L-_-alanine following high-LET radiation exposures. *Appl. Radiat. Isot.* **40**, 935-939.

Hasai H., Iwatani K., Shizuma K., Kosako T. and Morishima H. (1987): Europium- 452 depth profile of a stone bridge pillar exposed to the Hiroshima atomic bomb: 152Eu activities for analysis of the neutron spectrum. *Health Phys.* **53**, 227-239.

Haskell E.H., Hayes R. B., Romanyukha A. A. and Kenner G. H.(2000): Preliminary report on the development of a virtually nondestructive additive dose technique for EPR dosimetry. *Appl. Radiat. Iso.* **52**, 1065-1070.

Hassan G.M.and Ikeya M. (1997): ESR dosimetry of Li-lactate. *J. Nucl. Sci. Techn.* **47**, 1185-1187.

Hassan G.M.,Uesugi A.and Ikeya M. (1999): Radical formation in Li- and Mg-citrates for electron spin resonance dosimetry. *Jpn. J. Appl. Phys.* **38** 6703-6710.

Hassan G.M. and Ileya M. (2000): Metal ion-organic compound for high sensitive ESR dosimetry. *Appl. Radiat. Isot.* **52**, 1247-1254.

Higashimura T., Ichikawa Y. and Sidei T. (1963): Dosimetry of atomic bomb radiation in Hiroshima by thermoluminescence of roof-tiles. *Science* **139**, 1284-1285.

Hjortenberg P., Hansen J.W. and Wille M. (1989): Measurement of dose rate from106Ru/106Rh ophthalmic applications by means of alanine-polymer foils. *Appl. Radiat. Isot.* **40**, 997-1001.

H+fer P., Holczer K. and Schmalbein D. (1989): Characterization of g-ray irradiated powder alanine samples by pulsed EPR. *Appl. Radiat. Isot.* **40**, 1233-1236.

Hoshi M., Sawada S., Ikeya M. and Miki T. (1985): ESR dosimetry for A-bomb survivors. *ESR Dating and Dosimetry*, 407-414.

Hoshi M., Takada J., Kim R. aand Nitta Y. ed. (1996): *Effects of Loe Level Radiation for Residents Near Semipalatinsk Nuclear Test Site.* Proc. 2nd Hiroshima Intern. Symposium.

Hochi A., Hurusawa M. and Ikeya M. (1993): Apploication of microwave sacanning microscope: Human tooth with metal. *Appl. Radiat. Iso.* **44**, 401-405.

Ichikawa Y. and Nagatomo T. (1987): Thermoluminescent measurements. In U.S.- *Japan Workshop for Reassessment of Atomic Bomb Radiation Dosimetry* (Radiation Effects Foundation).

Ignatiev E.A., Lyubashevskii N. M., Shishkina E.A. and Romanyukha A.A (1999): EPR dose reconstruction for bone-seeking 90Sr. *Appl. Radiat. and Iso.* **51**, 151-159

Ikeya M., Ishibashi M. and Itoh N. (1972): Thermoluminescence response of the mixture of CaSO$_4$ and Li$_2$SO$_4$ to thermal neutrons and X-ray fields. *Health Phys.* **21**, 407-412.

Ikeya M. and Miki T. (1980): ESR dating of animal and human bones. Science 207, 977-979.

Ikeya M. and Ohmura K. (1981): Dating of fossil shells with electron spin resonance. *J. Geology* **89**, 247-250.

Ikeya M., Miyajima J. and Okajima S. (1984): ESR dosimetry for atomic bomb survivors using shell buttons and tooth enamel. *Jpn. J. Appl. Phys.* **23**, L699-L701.

Ikeya M. (1985): *Techniques of Radiation Dosimetry*, ed. by Mahesh K. and Viji D.R. (Wiley Eastern, New Delhi), Chapter 15 Electron spin resonance dosimetry.

Ikeya M., Miki T., Kai A. and Hoshi H. (1986): ESR dosimetry of A-bomb radiation using tooth enamel and granite rocks. *Radiat. Protect. Dos.* **17**, 181-184.

Ikeya M. and Ishii H. (1989): Atomic bomb and accident dosimetry with ESR: natural rocks and human tooth in-vivo spectrometer. *Appl. Radiat. Isot.* **40**,1021- 1027.

Ikeya M., Sumitomo H., Yamanaka C., Lloyd DC and Edwards AA (1996): ESR dosimetry of a deaseased radiation worker. *APPl. Radiat. Iso.* **47**, 1341-1344.

Ikeya M., Hassan G.M., Sasaoka H, Kinoshita Y., Takaki S. and Yamanaka C.(2000): Strategy for finding ESR dosimeter materials. *Appl. Radiat. Isot.* **52**, 1209-1215.

Ikezoe Y., Sato S., Onuki K., Morishita N., Nakamura T., Katsumura Y. and Tabata Y. (1984): Alanine dosimetry of 14 MeV neutrons in FNS target room. *J. Sci. Techn.* **21**, 722-724.

Ishii H., Ikeya M. and Okano S. (1990): ESR dosimetry of teeth of residents close to Chernobyl reactor accident. *J. Nucl. Sci. Tech.* **27**, 1153-1155.

Iwasaki M., Miyazawa C. and Shimano T. (1990): The relation between the weight of human tooth enamel and the CO_3^{3-} signal intensity on the ESR dosimetry. *Radioisotopes* **17**, 39-44.

Iwasaki M., Miyazawa C. and Uesawa T. (1995): Difference in radiation sensitivity of human tooth enamel in an individual and among indivisuals in dental ESR dosimetry. *Radioisotopes* **44** 785-788.

Janovsky I., Hansen J. W. and Cernoch P. (1988): A polymer-alanine film for measurement of radiation dose distribution. *Appl. Radiat. Isot.* **39**, 651-657.

Kai A. and Ikeya M. (1989): ESR accident dosimetry using medicine tablets coated with sugar. *Radiat. Protect. Dos.* **34**, 307-310.

Kamenopoulou V., Barthe J., Hickman C. and Portal G. (1986): Accidental g-irradiation dosimetry using clothing. *Radiat. Protect. Dos.* **17**, 185-188.

Katsumura Y., Hamamoto M., Yanagi H. and Tabata Y. (1980): Spatial distribution of radicals produced by fast neutron irradiation. *Rad. Phys. Chem.* **6**, 255-262.

Katsumura Y. and Tabata Y. (1985): Radical yield in alanine dosimeters irradiated by fast neutron. *ESR Dating and Dosimetry*, 415-421.

Katsumura Y., Tabata Y., Seguchi T., Morishita N. and Kojima T. (1986):Fast neutron irradiation effects - III. Sensitivity of alanine systems for fast neutrons having an energy of ca 1 MeV. *Radiat. Phys. Chem.* **28**, 337-341.

Kawano K., Nakata R. and Sumita, M. (1985a): Effect of X-ray irradiation on transition metal ions in human blood proteins. *ESR Dating and Dosimetry* 341-434.

Kawano K., Nakata R. and Sumita M. (1985b): Detection of radiation damage on vertebrate erythrocytes irradiated by X-ray. *ESR Dating and Dosimetry*, 423- 429.

Kinoshita A., Francisco Braga J.H.N., Graeff CFO and Baff O. (2000): ESR dosimetry of ^{89}Sr and [153]Sm in bone. *Applied Radiation and Isotopes* **54**: 269-274.

Koberle G., Terrile C., Panepuchi C. and Mascarenhas S. (1973): On the paramagnetism of bones irradiated in-vivo. *Acad. Brazil. Cienc.* **45**, 157-160.

Kojima T., Tanaka R., Morita Y. and Seguchi T. (1986): Alanine dosimeter using polymers as binders. *Appl. Radiat. Isot.* **37**, 517-520.

Kojima T. and Tanaka R. (1989): Polymer-alanine dosimeter and compact reader. *Appl. Radiat. Isot.* **40** , 851-857.

Kojima T., Chen L., Haruyama Y., Tachibana H. and Tanaka R. (1992): Fading characteristics of an alanine-polystyrene dosimeter. *Appl. Radiat. Isot.*   **43**, 863-865.

Koshta A.A., Wieser A., Ignatiev E.A., Bayankin S., Romanyukha A.A. and Degteva M.O. (2000) New computer procedure for routine EPR - dosimetry on tooth enamel description and verification. *Appl. Radiat. and Iso.* **52**, 1287-1290

Lea J. S., Dodd N. J. F. and Swallow A. J. (1988): A method of testing for irradiation of poultry. J. Food Sci. Tech. 23,

Lloyd D.C, Edward A.A, Fitzsimons EJ Evans C.D, Raailton R., Jeffrey P., Williams TG, White A.S., Ikeya M. and Sumitmo H. (1994): Death of a classified worker probably caused by overexposure to gammaa-irradiation. *Occup. Environ. Medic.* **51**, 713-718.

Loewe W.E. and Mendelson E. (1985): Revised dose estimate at Hiroshima and Nagasaki. *Health Phys.* **1**, 663-666.

Mahesh K. and Viji D.R. (1985): *Techniques of Radiation Dosimetry* (Wiley Wastern, New Delhi).

Mascarenhas S., Hasegawa A. and Takeshita K. (1973): ESR dosimetry of bones from Hiroshima A-bomb site. *Bull. Am. Phys. Soc.* **18**, 579.

McLaughlin W. L., Jarnett R. D. and Olejnik T. A. (1982): *Preservation of Food by Ionizing Radiation*, ed. Josephson E. S. and Peterson M. S. (CRC Press, Boca Raton FL) Vol. 1, Ch.8 Dosimetry, 189-245.

McLaughlin W. L., Boyd W. A., Chadwick K. H., McDonald J. C. and Miller A. (1989): *Dosimetry for Radiation Processing* (Tayler & Francis, London)

Mehta K. and Girzikowsky R. (2000): IAEA high-dose intercomparison in ^{60}Co field. *Appl. Radiat. Isot.* **52** No.5, 1179-1184.

Miller A., Kovacs A., Wieser A. and Regulla D. F. (1989): Measurements with alanine and film dosimeters for industrial 10 MeV electron reference dosimetry. *Appl. Radiat. Isot.* 40, 967-970.

Minegishi A., Shinozaki Y. and Meshitsuka G. (1967): Radiolysis of solid L-alpha alanine. *Bull. Chem. Soc. Japan* **40**, 1271-1272.

Miyagawa I. and Itoh J. (1962): Electron spin resonance of irradiated single crystal of alanine: hindered rotation of the methyl group in a free radical. *J. Chem. Phys.* **36**, 2157-2163.

Morehouse K. M. and Desrosiers M. F. (1993): Electron spin resonance investigations of g-irradiated shrimp shell. *Appl. Rad. Isotop.* **44**, 429-432.

Nagy V. (2000):Accuracy considertion in ESR dosimetry.*Appl. Radiat. Isot.* 52,1039-1050.

Nakajima T. and Watanabe S. (1974): New method for estimating g-ray exposure sustained in radiation accidents: possibility of using organic substance as monitor. *J. Nucl. Sci. Tech.* **11**, 575-582.

Nakajima T. (1982): The use of organic substance as emergency dosimeter. *Appl. Radiat. Isot.* **33**, 1077-1084.

Nakajima T. (1988): Sugar as an emergency populace dosimeter for radiation accident. *Health Phys.* **55**, 951-955.

Nakajima N. (1989): Possibility of retrospective dosimetry for persons accidentally exposed to ionizing radiation using electron spin resonance of sugar and mother-of-pearl. *Br. J. Radiol.* **62** 143-148.

Nakajima T. and Ohtsuki T. (1990): Dosimetry for radiation emergencies: radiation- induced free radicals in sugar of various countries and effect of pulverizing on the ESR signal. Appl. Radiat. Isot. **41**, 359-365.

Nakajima T., Ohtsuki T. and Likhtariov I. (1991): Tentative dose estimation in house at Pripyat-city on Chernobyl accident with sugar. *J. Nucl. Sci. Techn.* **28**, 71-73.

Nakanishi T., Imura T., Komura K. and Sakanoue M. (1983): Eu-152 in samples exposed to the nuclear explosions at Hiroshima and Nagasaki. *Nature*   **302**, 132- 134.

Nam J. W. and Regulla D. F. (1989): The significance of the international dose assurance service for radiation processing. *Appl. Radiat. Isot.* **40**, 953-955.

Nishiwaki Y. and Shimano T. (1990): Uncertainties in dose estimation under emergency conditions and ESR dosimetry with human teeth. *Radiat.Prot. Dos.* **34**, 295-297.

Ogawa M., Hirashima S., Ishiguro K. and Oshima K. (1985): ESR study of yields and distribution of radicals in organic compounds irradiated with fast neutrons. *Rad. Phys. Chem.***16**, 15-19.

Oka T., Ikeya M., Sugawara N. aand Nakanishi A. (1996): A high sensitivity portable spectrometer for ESR dosimetry. *Appl. Radiat. Isot.* **47**, 1589-1594.

Okajima S. (1985): Reassessment of A-bomb radiation dosimetry. *ESR Dating and Dosimetry*, 381-389.

Olsen K. J., Hasen J. W. and Waligorski M.PR. (1989): ESR dosimetry in calibration intercomparisons with high-energy photons and electrons. *Appl. Radiat. Isot.* **40**, 985.

Olsson S.K., Lund E. and Lund A. (2000): Development of ammonium tartrate as an ESR dosimeter materials for clinical purposes. *Appl. Radiat. Isot.* **52**, 1235-1342.

Oommen I. K., Nambi K. S. V., Sengupta S., Gundu Rao T. K. and Ravikumar M. (1989): Lactose and "Tris" lyoluminescence dosimetry systems and ESR correlation studies. *Appl.Radiat.Isot.***40**, 879-883.

Ostrowski K., Goclawska A. D., Stachowicz W., Michalik J.,Tarsoly E. and Komender A. (1971): Application of the electron spin resonance technique for qualitative evaluation of the resorption rate of irradiated bone grafts. *Calc. Tiss. Res.* **7**, 58- 66.

Fainstein C., Winkler E. and Saravi M. (2000): ESR/Alanine γ-dosimetry in the 10-30 Gy range. *Appl. Radiat. Isot.* **52** No.5, 1195-1196.

Pasenkiewicz M. and Knapoczyk J. (1981): Radiation Protection (*Proc. 4 Symp. Neutron Dosimetry* ed. G. Burger and H. G. Ebert, EUR 7448, 489.

Pass B. and Aldrich J. E. (1985): Dental enamel as in-vivo radiation dosimeter. *Med. Phys.* **12**, 305-307.

Pejuan A. and Kuhn H. (1981): The G-value of the ferrous sulphate dosimeter for 3 and 14 MeV neutrons. *Phys. Med. Bio.* **26**, 163-169.

Pivovarov S., Rukhin A. and Seredavina T. (2000): ESR of environmental objects from Semipalatinsk nuclear test site. *Appl. Radiat. Isot.* **52**, 1255-1258.

Raffi J., Angel J. P. L., Buscarlet L. A. and Martin C. C. (1988): Electron spin resonance identification of irradiated strawberry. *J. Chem. Soc. Faraday Trans.* **84**, 3359-3362.

Regulla D. F. and Deffner U. (1981): Standardization in high-level photon dosimetry based on ESR transfer metrology. Biomedical Dosimetry: Physical Aspects, Instrumentation, Calibration. *IAEA-STI/PUB/567*, 391-401.

Regulla D. F. and Deffner U. (1982): Progress in alanine/ESR transfer dosimetry in high dose dosimetry. Trends in Radiation Dosimetry ed. Mclaughlin, W. L. (Pergamon Press, Oxford); *Appl. Radiat. Isot.* **33**, 1101-1114.

Regulla D. F. and Deffner U. (1983): A system of transfer dosimetry in radiation processing. *Radiat. Phys. Chem.* **22**, 305-309.

Regulla D. F. and Deffner U. (1989): Dose estimation by ESR spectroscopy at a fatal radiation accident. *Appl. Radiat. Isot.* **40**, 1039-1043.

Regulla D. F. (2001): From dating to biophysics - 20 years of progress in applied ESR spectroscopy. *Appl. Radiat. Isot.*, **52**, 1023-1030.

Roesch W. C. ed. (1987): US-Japan Joint Assessment of Atomic Bomb Radiation Dosimetry in Hiroshima and Nagasaki: Final Report vol. 1 and vol. 2.

Romanyukha A.A. and Regulla D. (1996): Aspect of retrospective ESR dosimetry. *Appl. Radiat. Isot.* **17**, 1293-1298.

Romanyukha AA, Ignatiev EA, Degteva MO, Kozheurov VP, Wieser A. and Jacob P. (1996): Radiation dose from Ural region. *Nature* **381**, 199-200.

Romanyukha A.A., Hayes R.B., Haskell E.H. and Kenner G.H.(1999) Geographic variations in the EPR spectrum of tooth enamel. *Radiat. Prot. Dom.* **84**,No.1, 445-449

Romanyukha A.A, Ignatiev EA, Vasilenko EK, Degteva MO, Wieser A., Jacob P., K-Markus IB, Kleschenko ED, Nakamura N. and Miyazawa C. (2000): EPR dose reconstruction of Russian nuclear workers. *Health Phys.* **78**, 15-20.

Romanyukha A.A, Desrosiers MF and Regulla D. (2000): Curreent issues on EPR dose reconstruction in tooth enamel. *Appl. Radiat. Isot.* **52**, 1265-1273.

Romanyukha A.A., Ignatiev E.A., Ivanov D.V. and Vasilyev A.G.(2000): The distance effect on the individual exposures evaluated from the soviet nuclear bomb test at totskoye test site in 1954. *Radiat. Prot. Dom.* **86**,No.1, 53-58

Romanyukha A.A., Ignatiev E.A., Vaslenko.K., Vasilenko.K.,Drozhko G.,Wieser A.,Jacob P.,Keirim-Markus I.B., Kleschenko E.D., Nakamura N. and Miyazawa C.(2000): EPR dose reconstruction for Russian nuclear workers. *Helth Physics* **78**,15-20

Rossi A. M., Poupeau G., Chaix O., Raffi J., Angel J. P. and Jeunet A. (1992): Paramagnetic species induced in bioapatites by foodstuff ionization. Electron spin resonance (ESR) *Applications in organic and bioorganic materials* Ed. Catoire, B. (Springer-Verlag, Berlin) 151-166.

Schraube H., Wietzeneggar E., Wieser A. and Regulla D. F. (1989): Fast neutron response of alanine probes. Appl. Radiat. Isot. **40**, 941-943.

Sharpe P. and Stephton J. (2000): An automated system for the measurement of alanine/EPR. *Appl. Radiat. Isot.* **52** No.5, 1185-1188.

Shields H. and Gordy W. (1958): Electron spin resonance studies of radiation damage to amino acids and protein. Proc. Natl. Acad. Sci. U. S. **41**, 983-966.

Shimano T., Iwasaki M., Miyazawa C., Miki T., Kai A. and Ikeya M. (1989): Human tooth dosimetry for g-rays and dental x-rays using ESR. *Appl. Radiat. Isot.* **40**, 1035-1038.

Silveira e O.and Baffa (1995): Lyoluminescence and ESR Measurements on alanine and sucrose dosimeters. *Appl. Radiat. Isot.* **46**, 827-830.

Simmons J.A. and Bewley D.K. ((1976): The relative effectiveness of fast neutrons in creating stable free radicals. *Radiat. Res.* **65**, 197-201.

Simmons J.A. (1989): Thermal effects in irradiated amino acids *Appl. Radiat. Isot.* **40**, 901-904.

Sinclair J. W. and Hanna M. W. (1967): Electron paramagnetic resonance of L-alanine irradiated at low temperature. *J. Phys. Chem.* **71**, 84-88.

Slager U. T., Zucker M. J. and Reilly E. B. (1964): The persistence of electron spin resonance in bone grafts sterilized by ionizing radiation. *Radiat. Res.* **22**, 556- 563.

Sollier T. J. L., Mosse D. C., Chartier M. M. T. and Joli J. E. (1989): The LMRI ESR/alanine dosimetry: description and performance. *Appl. Radiat. Isot.* **40**, 961-966.

Stachowicz W., Michalik J., Goclawska A. D. and Ostrowski K. (1974): Evaluation of absorbed dose of g-ray and X-ray radiation using bone tissue as a dosimeter. *Nucleonika* **19**, 845-850.

Stachowicz W., Burlinska G., Michalik J., Goclawska A. D. and Ostrowski K. (1993): Applications of EPR spectroscopy to radiation treated materials in medicine, dosimetry and agriculture. *Appl. Rad. Isotop.* **44**, 423-427.

Swartz H. M. (1965): Long-lived electron spin resonance in rats irradiated at room temperature. *Radia. Res.* **24**, 579-586.

Tatsumi M. J. and Okajima, S. (1985): ESR dosimetry using human tooth enamel. *ESR Dating and Dosimetry* 397-405.

Tatsumi M. J. and Okajima S. (1987): ESR dosimetry for atomic bomb survivors and radiologic technologists. Nucl. Instr. Meth. A257, 417-422.

Uribe R M., McLaughlin W.L., Miller A., Dunn T.S. and Williams E.E. (1981): Possible use of electron spin resonance of polymer film containing leucodyes for dosimetry. *Radiat. Phys. Chem.* **18**, 1011-1016.

Van Laere K., Buysse J. and Berkvens P. (1989): Alanine in high-dose dosimetry: spectrophotometric and electrochemical readout procedures compared with ESR. *Appl. Radiat. Isot.* **40**, 885-895.

Vanhaelewyn G., Sadlo J., Callens F., Mondelaers W., De Frenne D. and Matthys P. (2000): A decomposition study of EPR spectrum of irradiated sucrose. *Appl. Radiat. Isot.* **52**, 1221-1228.

Waligorski M.P.R., Katz R., Byrski S.T., Gierula P.M. and Knapczyk J. (1981): Mixed field dosimetry in a cyclotron-induced fast neutron radiotherapy beam with special emphasis on a newly developed alanine system. *Proc. 4th Symp. on Neutron Dosimetry* ed. Burger G. and Ebert H.G. (EUR 7448) 489-498.

Waligorski M.P.R., Denialy G., Kim S.L. and Katz R. (1989): The response of the alanine detector after charged-particle and neutron irradiation.*Appl. Radiat. Isot.* **40**, 923-933.

WHO (1981): *Wholesomeness of irradiated food.* WHO Technical Reports Ser. **659**.

WHO (1996): *Health consequences of the Chernobyl accident.* V.2.3.3 ESR dosimetry, 232-242 (WHO, Geneva).

Wieser A. and Regulla D. F. (1989): ESR dosimetry in the "Gigarad" range. *Appl. Radiat. Isot.* **40**, 911-913.

Wieser A., Siegele R. and Regulla D. F. (1989): Influence of the irradiation temperature on the free-radical response of alanine. *Appl. Radiat. Isot.* **40**, 957-960.

Wieser A., Goeksu HY, Regulla D.F. and Vogenauer (1994): Limits of retrospective accident dosimetry. *Radiat. IMeas* **23**, 509-516.

Wieser A., Romanyukha A.A, Degteva MO, Kozherov V.P., and Petzoldt D. (1996): Tooth enamel as a natural beta dosimeter for bone seeking radionuclides. *Radiat. Prot. Dos.* **65**, 413-416.

Wu K. Guo L., Cong JB. Sun CP, Hu JM, Zhou ZS, Wang S., Chang X., and Shi YM (1998): Researches and applications of ESR dosimetry for radiation accident dose assessment. *Radiat. Prot. Dosim.* **77**, 65-67.

Yordanov ND, Novakova E. and Lubenova S. (2001): Consecutive estimation of nitrate and nitrite ions in vegetables and fruits by electron spin resonance spectroscopy. *Analy. Chemica Acta* **21**, 1-8.

Chapter 14

ESR Microscopy

– Scanning ESR Imaging of Spin Density –

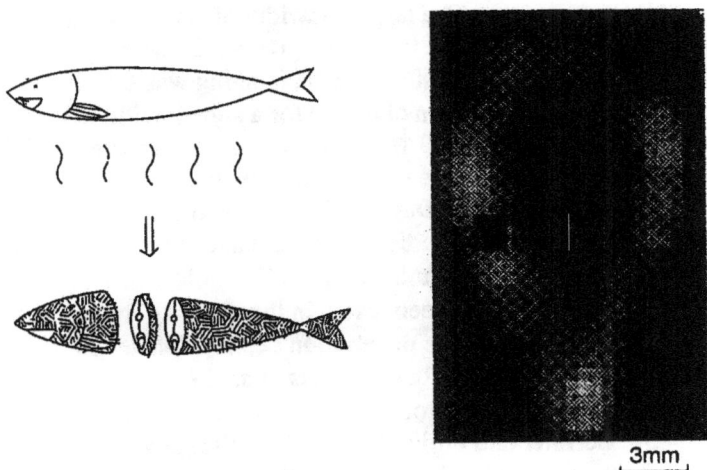

3mm

Microscopic spatial imaging (microscopy) of the concentration of unpaired electron spins or paramagnetic ions has been achieved (a) by a magnetic field gradient method used in the conventional magnetic resonance imaging (MRI) as well as by scanning (b) the localized magnetic or modulation field and (c) the localized microwave field. Heavily broiled fish or meat is hazardous to health as free radicals might cause cancer. The distribution of radicals in broiled dry sardine is imaged.

14.1   Introduction

Many analytical methods visualize the spatial distribution of some physical parameters as an image since the picture provides us insights for understanding the physical and chemical nature of the material. *Magnetic resonance imaging* (MRI) as used in hospitals is a *computer tomography* (CT) using *nuclear magnetic resonance* (NMR). One of the advocated advantages of MRI is the safety in comparison with X–ray CT–imaging in which the medical radiation dose is high. CT–NMR using pulsed NMR employs a magnetic field gradient so that the spatial resolution is realized (Lauterbur 1973). The resolution is improved as to image a single cell in NMR microscopy (Aguayo *et al.* 1986). Principles and procedures of NMR microscopy are described in some books (Callaghan 1991, Blümich and Kuhn 1992).

ESR or EPR imaging has been developed following the field gradient method in NMR imaging. The large linewidth of an ESR signal and the difficulty in the pulse microwave technique prevent the development of ESR imaging. The theoretical possibility of ESR imaging was foreshadowed with a maximum resolution of 10 μm obtained for a signal with a linewidth of 0.1 mT and a field gradient of 10 T/m (Karthe and Wehrsdorfer 1979). The defect concentration in diamonds was actually imaged with a magnetic field gradient of 0.5 T/m (Hoch and Day 1979). Trials to develop ESR imaging based on the magnetic field gradient method have been done (Eaton and Eaton 1986a, b, 1987, 1990, Ohno 1986, 1987). A low frequency micro-wave (L–band: 1.5 GHz) has been used in the field of biology since the microwave loss is small and the penetration depth is plausible at this fre-quency. ESR imaging of a rat brain (Ogata *et al.* 1986) and a plant stem with a stable nitrogen radical involvement has been demonstrated (Fujii and Berliner 1985, Berliner and Fujii 1987, 1991, Nishikawa *et al.* 1985a, b). Most of ESR imaging studies have been at this frequency using a loop gap resonator (Froncisz and Hyde 1982).

As a new method, the field gradient was modulated at a certain fre-quency (Herrling *et al.* 1982, Ewert and Herrling 1986a, b, 1989). ENDOR imaging was also studied for spin labeled compounds (Janzen and Kotake 1986). Spectral spatial imaging method has been developed and FT–EPR is being used for imaging (Eaton *et al.* 1987, 1989, Sueki *et al.* 1993). Books on EPR imaging and *in–vivo* EPR were published by pioneers in this field (Ohno 1990, Berliner and Reuben 1991, Eaton *et al.* 1991).

The requirements to determine the spatial age distribution for the growth of stalactites or to assess the accumulated radiation dose at the surface and inside of fossils or minerals force us to carry out ESR dating or ESR radiation dosimetry using imaging (Ikeya *et al.* 1990b). I hesitate to add this chapter considering EPR imaging books published by specialists, but a simple method by scanning the spatially localized magnetic or microwave field (Ikeya 1989, 1991, 1992, Ikeya and Miki 1992) may be useful for material sciences and radiation dose imaging described in this book.

14.2   CT–ESR Imaging
14.2.1   *Field Gradient Method*
The conventional technique of ESR imaging employs a field gradient in addition to the uniform static magnetic field. Figure 14.1 shows a schematic diagram of the apparatus. The anti–Helmholtz coil pairs produce a magnetic field gradient along the uniform field direction (z direction). The ∞–shaped coils create the field gradient along x or y direction.

The magnetic field along z direction at the site (x, y, z) is expressed as

$$H(x, y, z) = H_s + (\partial H/\partial x)x + (\partial H/\partial y)y + (\partial H/\partial z)z \, , \qquad (14.1)$$

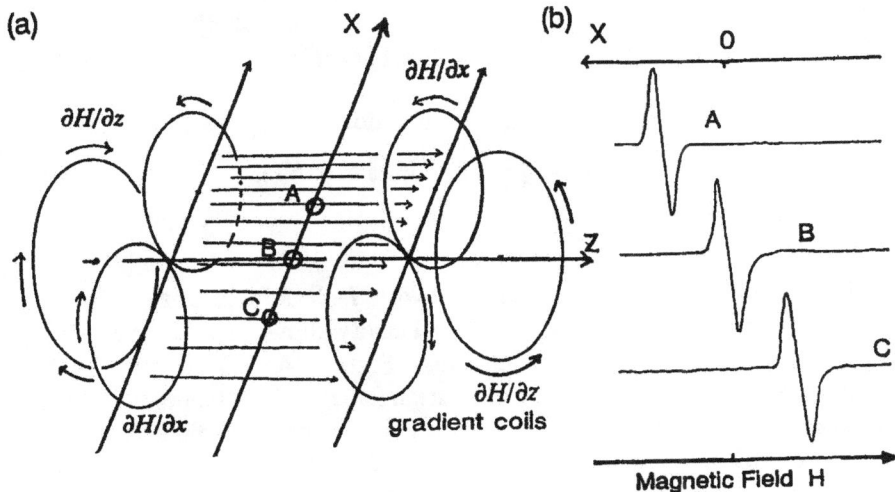

Figure 14.1 (a) Magnetic field gradient ($\partial H/\partial x$) created by the ∞–shaped coil pair. (b) The shift of the ESR signal by the application of the magnetic field gradient. The shift of the resonance field is directly related to the position.

where $\partial H/\partial x$, $\partial H/\partial y$ and $\partial H/\partial z$ are the field gradients produced by the field gradient coils and H_s is the uniform field produced by the electromagnet.

Resonance occurs at a magnetic field, H_0, at a microwave frequency of ν ($h\nu = g\beta H_0$) without a field gradient. Hence, the signal position obtained by sweeping the uniform magnetic field H_s is indicative of the spin position. The distribution of the spin concentration, $f(x)$, can be expressed as $f(H)$, i.e., as a function of the magnetic field H using $H = H_0 + (\partial H/\partial x)x$ (only x–component is considered here). The ESR spectrum under a constant field gradient, $g(H)$, is expressed by convolution as

$$g(H) = \int r(H-H')f(H')\,dH' , \qquad (14.2)$$

where $r(H)$ is the shape function under the uniform magnetic field. The Fourier transform of the convolution integral in Eq. (14.2) is

$$G(\omega) = R(\omega)F(\omega) + N(\omega) , \qquad (14.3)$$

where $G(\omega)$, $R(\omega)$ and $F(\omega)$ are the Fourier transforms of $g(H)$, $r(H)$ and $f(H)$, respectively, and $N(\omega)$ is an additive noise term. The distribution function $f(H)$ of interest can be obtained by deconvolution as

$$f(H) = \int [G(\omega)W(\omega)/R(\omega)]\exp(i\omega H)\,d\omega . \qquad (14.4)$$

Actual integration must include a filter function, $W(\omega)$, as

$$W(\omega) = \|R(\omega)\|^2/(\|R(\omega)\|^2 + K) , \qquad (14.5)$$

since $R(\omega)^{-1}$ disperses for $R(\omega) = 0$.

14.2.2   Field Gradient Coils in a Cavity : 1–D Microscope

A simple field gradient coil pair in a cavity allows us to observe a one-dimensional (1–D) distribution of spins. Figure 14.2 (a) shows a typical example of the 1–D ESR microscope arrangement with a commercial cylindrical microwave cavity TE_{011}. A test sample of organic diphenyl–picryl-hydrazyl (DPPH) particles separated about 50 μm by a mylar foil and sandwiched with other polystyrene is used to check the field gradient. The ESR spectral change on applying current to the field gradient coils and the clear separation of 50 μm for the two dots are shown in (b). A field gradient of

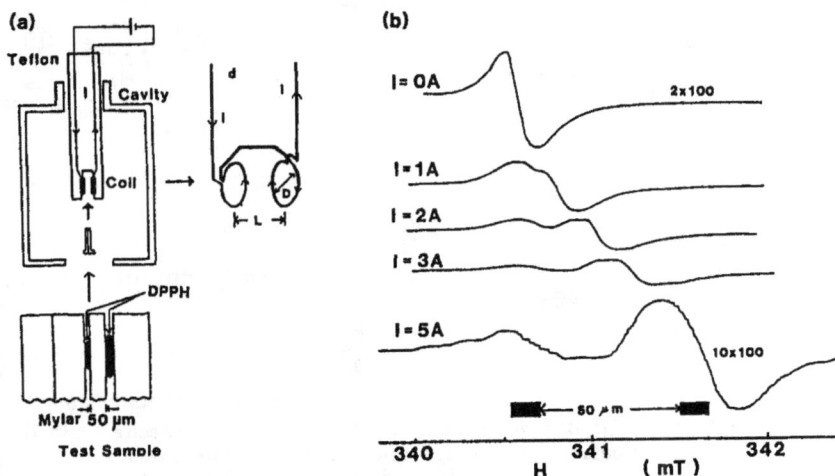

Figure 14.2 (a) An ESR cylindrical cavity of TE$_{011}$ mode and a sample holder with a test sample of DPPH separated by 50 μm mylar sheet. (b) ESR spectra with the current to the anti–Helmholtz coil.

20 T/m is produced judging from the signal separation. A resolution of a few μm is practical in this case without deconvolution (Ikeya and Miki 1987).

The main difficulty of the field gradient method for microscopy is the necessity of an intense field gradient for an ESR signal having a wide line–width. The effort to create an intense field gradient has been made by cooling the field gradient coils using insulator oil and by employing pulse operation to avoid heating (Ohno 1986). A small helix cavity was fabricated at the tip of the semirigid coaxial cable to make a small gradient coil pair for an intense field gradient of 30 T/m (Ohno and Murakami 1988). The problem was solved by making a sample holder with small field gradient coils that can be inserted into a microwave resonator (cavity). The advantage of the method is that a low cost current and voltage source are sufficient to produce a field gradient of 20 T/m at the pinhole site though the microwave loss is caused by the insertion of the coils. The sample is rotated to get a 2–D or 3–D image using the sample holder shown in Figure 14.2 (a).

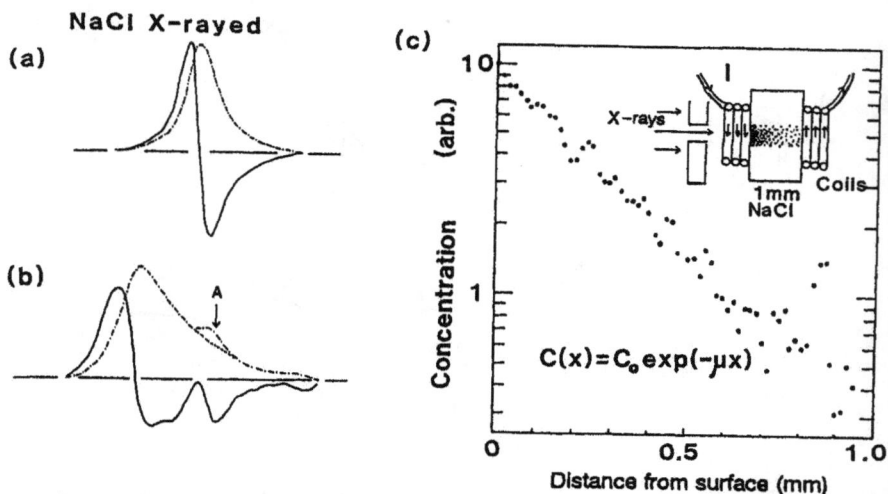

Figure 14.3   ESR derivative (solid curve) and integrated spectra (dashed curve) of NaCl irradiated with X–rays from one side (a) in the absence [$r(H)$] and (b) in the presence of the field gradient [$g(H)$].   The small absorption peak indicated by A was subtracted.   (c) The 1–D distribution of the spin concentration.

An example of 1–D ESR imaging for the study of microdosimetry is shown in Figure 14.3. A single crystal of NaCl was irradiated by soft X–rays at the position under the field gradient.   The signal intensity of defects without and with the field gradient is shown in (a) and (b), respectively, and the obtained 1–D distribution in (c).   An exponential decay of the spin concentration from the surface was observed following Fourier transforma-tion, which indicates the attenuation of X–rays (Ikeya and Miki 1987).

An intense field gradient of 60 T/m was established by using a super-conducting field gradient coil pair of Nb (Ti) inside the cavity at a current of 5 A (Ikeya and Furusawa 1989).   Increasing current up to the critical current of 15 A will make a gradient of 180 T/m.

14.2.3   *Field Gradient Wires in a Cavity : 2–D Microscope*

Standard Maxwell and Anderson coils produce a linear gradient only at the small central region of the coils (Anderson 1961).   Large–size coils and large driving power are necessary to produce a large gradient in a large space using these coils.   Insertion of small gradient coils into a microwave cavity to produce a large gradient causes a considerable microwave loss and limits the sample size due to the low linearity of the gradient. Several techniques to produce a linear gradient in a large space have been developed in the field of NMR imaging using quadruple coils and straight wires (Webster and Marsden 1974, Zupancic and Pirs 1976, Bangert and Mansfield 1982, Poltz and Kuffel 1985, Momo et al. 1988) or saddle–type coils (Siebold 1990). Bird cage coils and magnetic field screening coils (Mansfield and Chapman 1986) were developed for gradient and *radio frequency* (r. f.) coils as described in a textbook on NMR microscopy (Callaghan 1991). Uniformity of the field gradient and r. f. field in a small space is the main interest.

Straight wires parallel to the cylinder axis cause much less microwave loss in a TE_{011} –mode cylindrical cavity. The straight four–wire gradient system is appropriate for insertion into a cavity. The configuration was generalized to a cylindrically placed infinite number of parallel straight wires whose current values and directions can independently be controlled to obtain a highly uniform magnetic field or magnetic field gradient (Furusawa and Ikeya 1991). A magnetic field at (y, z) produced by a single infinite–length wire at (a, b) carrying current, I (Zupancic and Pirs 1976) is

$$H_z(y, z) = \frac{I}{2\pi} \cdot \frac{a - y}{(y - a)^2 + (z - b)^2} \cdot \qquad (14.6)$$

Eq. (14.6) represents the real part (Re) of the complex function,

$$H_z(y, z) = \frac{I}{2\pi} \, \mathrm{Re} \, \frac{1}{(a + bi) - (y + zi)} \cdot \qquad (14.7)$$

Eq. (14.7) can be expanded to a power series of $y + zi$ at the inside of the circle of radius, r, where $|y + zi| < |a + bi|$. Rewriting $y + zi = \xi$ and $a + bi = re^{i\theta}$, the magnetic field is

$$H_z(y, z) = \frac{I}{2\pi r} \, \mathrm{Re} \sum_{n=1}^{\infty} \left(\frac{\xi}{\rho}\right)^{n-1} e^{-in\theta} \,, \qquad (14.8)$$

where (r, θ) is the polar coordinate of the wire position. The result is generalized to an infinite number of current lines on a cylinder parallel to the x–axis. The current, I, is then generalized to $I(\theta)$, which denotes the current density (per unit angle) of a small arc region located at the angle, θ. In this case, the magnetic field at (y, z) produced by all current lines is expressed as

$$H_z(y, z) = \frac{1}{2\pi r} \int_{-\pi}^{\pi} I(\theta) \operatorname{Re} \sum_{n=1}^{\infty} \left(\frac{\xi}{r}\right)^{n-1} e^{-i n\theta} d\theta \quad . \tag{14.9}$$

Expanding $I(\theta)$ in a Fourier series with the Fourier coefficients, a_n and b_n,

$$I(\theta) = a_0 + \sum_{m=1}^{\infty} (a_m \sin m\theta + b_m \cos m\theta) , \tag{14.10}$$

and expressing $e^{-in\theta}$ as $\cos n\theta - i \sin n\theta$, Eq. (14.9) is rewritten as

$$H_z(y, z) = \frac{1}{2\pi r} \operatorname{Re} \int_{-\pi}^{\pi} [a_0 + \sum_{m=1}^{\infty} (a_m \sin m\theta + b_m \cos m\theta)]$$
$$\times \sum_{n=1}^{\infty} \left(\frac{\xi}{r}\right)^{n-1} (\cos n\theta - i \sin n\theta) d\theta \quad . \tag{14.11}$$

Executing the integral in Eq. (14.11), all terms vanish except the terms of $\cos^2 n\theta$ and $\sin^2 n\theta$ because the function series $\{ \sin n\theta, \cos n\theta \mid n = 1, 2, 3, \cdots \}$ is orthogonal. Thus, Eq. (14.11) can be reduced to

$$H_z(y, z) = \frac{1}{2r} \operatorname{Re} \sum_{n=1}^{\infty} \left(\frac{\xi}{r}\right)^{n-1} (b_n - i a_n) \quad . \tag{14.12}$$

The first term ($n = 1$) in Eq. (14.12) represents the uniform magnetic field and the second one ($n = 2$) the linear gradient. All terms except the first-order term of ξ must be eliminated to generate a linear gradient. This is accomplished when all the Fourier coefficients a_n and b_n are zero except a_2 and b_2, giving the current density,

$$I(\theta) = a_2 \sin 2\theta + b_2 \cos 2\theta \quad . \tag{14.13}$$

In this case, the magnetic field inside the cylinder is exactly expressed as

$$H_z(y, z) = \frac{1}{2r^2} (a_2 z + b_2 y) \quad . \tag{14.14}$$

Eq. (14.13) and (14.14) can be rewritten in practical forms as

$$I(\theta) = I_0 \cos(2\theta - \phi) , \tag{14.15}$$

and

$$H_z(y, z) = \frac{1}{2r^2} (y \cos\phi + z \sin\phi) \ , \tag{14.16}$$

respectively, where $I_0 = \sqrt{a_2^2 + b_2^2}$, $\cos\phi = b_2/I_0$ and $\sin\phi = a_2/I_0$.

Eq. (14.16) indicates that a linear magnetic field gradient is generated in the entire region inside the cylinder. The direction and the magnitude of the gradient are ϕ and $I_0/(2r^2)$, respectively. The other component, $\partial H_z/\partial x$, can similarly be generated using another set of wires; the $\pi/2$ rotation of the current configuration producing $\partial H_z/\partial y$ around the z–axis forms the configuration for $\partial H_z/\partial x$.

A practical gradient apparatus based on the above theory was designed as shown in Figure 14.4. Sixteen copper wires are attached to the inner quartz tube using epoxy resin and the cooling gas can be flowed between the inner and outer quartz tubes. The parallel wires were connected to each other and to the current sources, forming four independent current loops.

The magnetic field produced by the N wires of infinite length was calculated numerically. Results are presented in Figures 14.5 using a con-

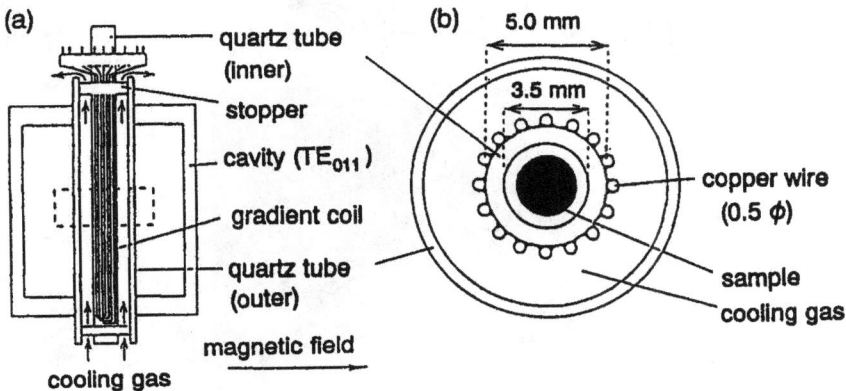

(a) quartz tube (inner) — stopper — cavity (TE$_{011}$) — gradient coil — quartz tube (outer) — magnetic field — cooling gas

(b) 5.0 mm — 3.5 mm — copper wire (0.5 ϕ) — sample — cooling gas

Figure 14.4 (a) Illustration of experimental assembly of the gradient coil inserted into a TE$_{011}$ mode cavity. (b) The cross sectional view of the central part of the gradient coil in the y–z plane.

tour map of field strength, $B_z(y, z)$; (a) – (c) and (d) – (f) represent the z–gradient ($\phi = \pi/2$) and the y–gradient ($\phi = 0$), respectively. The number of wires, N, is 8 in (a) and (d), 16 in (b) and (e), and 32 in (c) and (f). A photograph of a home–made sample holder with wires is shown in (g).

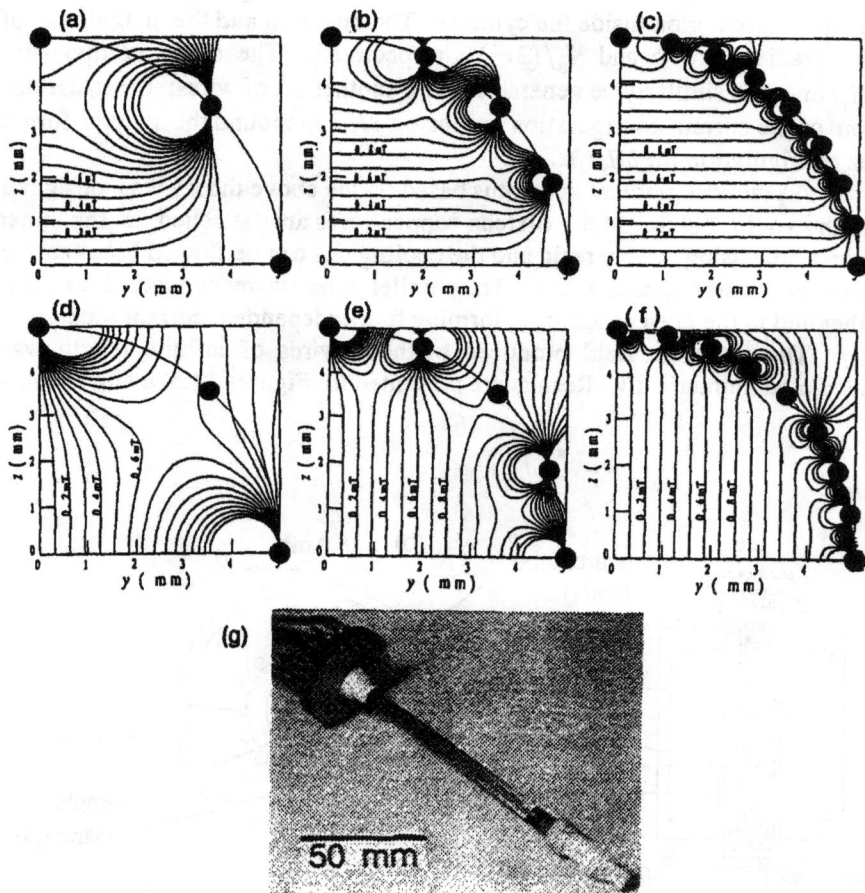

Figure 14.5   The contour map of field strength produced by straight current lines: (a) $N=8$, $I_{max}=10$ A, $\phi=90°$, (b) $N=16$, $I_{max}=5$ A, $\phi=90°$, (c) $N=32$, $I_{max}=2.5$ A, $\phi=90°$, (d) $N=8$, $I_{max}=10$ A, $\phi=0$, (e) $N=16$, $I_{max}=5$ A, $\phi=0$, (f) $N=32$, $I_{max}=2.5$ A, $\phi=0$. The contour lines are drawn in 0.1 mT steps, illustrating a gradient of $G=0.32$ T/m. Closed circles denote the positions of wires. (g) A photograph of a home–made holder with field gradient wires (Furusawa and Ikeya 1993).

An imaging experiment was carried out for a test sample consisting of DPPH powder filling 16 holes located at 4 × 4 lattice points of a teflon plate as shown in Figure 14.6 (a). The gradient coil was operated with a maximum current of 5 A per single wire without additional cooling to produce a gradient of 1.05 T/m. Varying the direction of the field gradient in the y–z plane by 5–degree steps, 36 ESR spectra were obtained as the projection data. The projection spectra were integrated after subtracting the background spectrum caused by the gradient coil inserted into the cavity. Then, the ESR image was reconstructed onto a 128 × 128 pixels grids using filtered back projection (FBP) (Sotgiu *et al.* 1987) as shown in (b).

Another experiment was performed using a teflon tube irradiated by γ–rays (30 kGy) as shown in Figure 14.7 (Furusawa 1992, Furusawa and Ikeya 1993). Projection spectra were measured for 18 directions by 10–degree steps. ESR spectra of the irradiated teflon were much broader than those of DPPH. Thus, signal deconvolution before FBP was performed for all projections using a spectrum without the field gradient to improve the resolution. An ESR–CT image is clearly seen in Figure 14.7 (d).

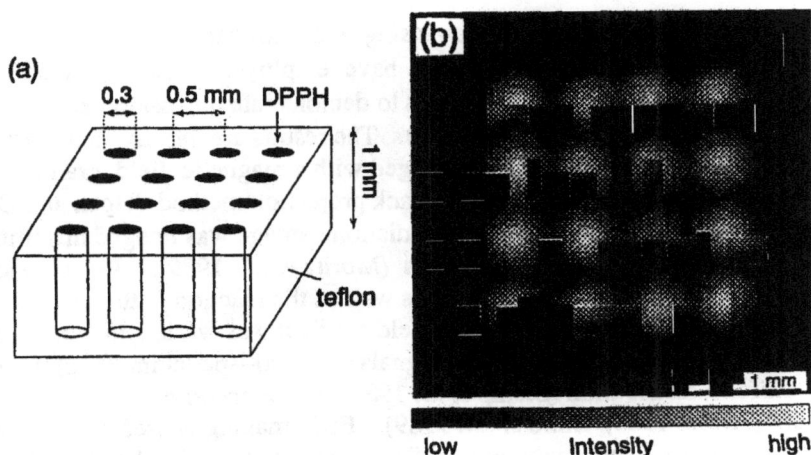

Figure 14.6 (a) Illustration of a DPPH test sample ; DPPH powder filling 16 holes of 0.3 mm in diameter located at 4 × 4 lattice points separated by 0.5 mm. (b) An ESR–CT image of the test sample consisting of 128 × 128 pixels. This image was reconstructed from 36 projection data of 5–degree steps using the filtered back projection (Furusawa 1992).

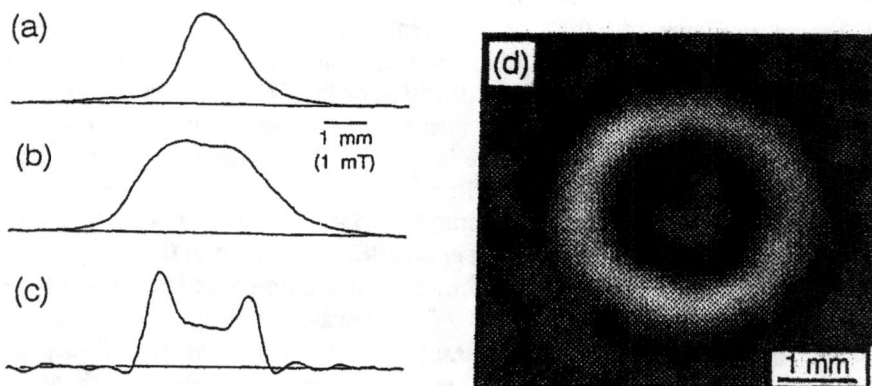

Figure 14.7   ESR spectra of an irradiated teflon tube (a) without and (b) with the field gradient of 1 T/m. (c) Deconvoluted spectrum of (b) using the spectrum (a) as a point spread function. (d) An ESR–CT image of a γ–irradiated teflon tube reconstructed using FBP from 18 sets of projection data of 10–degree steps apart.

14.2.4   *ESR Imaging Application using Field Gradient*

Most of ESR imaging studies have employed phantoms made of chemicals having intense ESR signals to demonstrate the usefulness of the methods (both hardware and software). The results are briefly summarized here. Two small diamonds were imaged with a magnetic field gradient of 0.5 T/m using the deconvolution and back projection method (Hoch and Day 1979). Non–uniform distribution of radiation damage was imaged in sulfuric acid ice (Ohno 1981) in organic solids (Morita *et al.* 1989). Photobleaching and thermal annealing of radicals as well as the reaction with oxygen can be imaged. The magnitude of the field gradient was varied to obtain the image of some overlapping radical signals (spectral–spatial imaging) in the non–uniform distribution (Stemp *et al.* 1987, Maltempo *et al.* 1988, Eaton *et al.* 1987, Ohno 1985, Ohno *et al.* 1989). ESR imaging is useful for characterization of catalysts, studying on diffusion (Shin *et al.* 1991) and profiling of radiation effects on a microscopic scale. Imaging of electromagnetic fields in a cavity was made using paramagnetic gas (Marshall *et al.* 1988).

The optimum frequency for biomedical studies is L–band (1 ~ 2 GHz) using a loop–gap resonator instead of a cavity (Froncisz and Hyde 1982). The dielectric loss of microwave for water–involved materials is reduced at low frequency, while the ESR sensitivity is increased at high frequency. A spectrometer for imaging a small animal at 26 MHz was designed for biomedical studies (Galstewa *et al.* 1983). The low sensitivity of this spectrometer is a major obstacle for the practical use in ESR imaging. Hence, most of the applications to biomedical systems are limited to the detection of the distribution of standard compounds for spin labeling (Ogata *et al.* 1986). Soaking in a nitroxyl radical solution was reported for zeolite powder (Ulbricht *et al.* 1984) and more interestingly celery stalks (Fujii and Berliner 1985). The permeable paths or sites can be imaged clearly. Similar methods may be used to observe the porosity in ceramic materials by infiltration of paramagnetic chemicals (Kordas and Kang 1991).

Remarkable studies for biological applications are the detection of nitroxyl radicals in a living tumor (Nishikawa *et al.* 1985b) and the 3–D imaging of the head of a living rat receiving an administration of nitroxyl radicals (Ishida *et al.* 1992). The radical concentration is also reduced by the presence of oxygen, hence, the oxygen concentration was obtained from the spatial variation of the radical concentration in fatty tissues, seeds and cell aggregates. The method may be used in ESR microscopy in material science though only the non–active sites can be imaged using radical–bearing standard chemicals. 2–D ENDOR imaging was made to study the oxygen concentration (Kotake *et al.* 1988) and radical diffusion in porous media (Sukhoroslov *et al.* 1991). We should continue to follow the development in biomedical fields since most ESR imaging studies are oriented to biomedicine. High field ESR imaging using a frequency of 140 GHz and a high field gradient of 78 T/m produced by two ferromagnetic discs was developed to obtain the resolution of 1 μm using standard stable radicals (Smirnov *et al.* 1992).

14.3 Scanning the Localized Field

14.3.1 *Mechanical Scanning of Modulation Field*

The idea of creating the magnetic field at a pinhole region with microfabricated coils in a microwave cavity has lead to the invention of "**scanning ESR microscopy**". Either the modulation or static resonance field can be localized and scanned to obtain a surface image of the sample just like a scanning tunneling microscope (Binning *et al.* 1982).

Magnetic field modulation of 100 kHz is applied to the ESR measurement to enhance the sensitivity by means of a phase–sensitive lock–in–amplifier system. If the modulation is localized at a pinhole region, one can detect the magnetic resonance at this site. The scanning of the modulation field at the surface of the sample gives a magnetic resonance image (Miki and Ikeya 1987). Figure 14.8 (a) shows a schematic illustration of the scanning ESR microscope. The microcoil for 100 kHz field modulation is mechanically scanned to obtain a surface image using the stepping motor and X-Y stage. Scanning is actually made by moving the sample rather than the modulation coil so that the frequency change of the resonant cavity ' by moving the coil is avoided.

In the differential modulation method shown in (b), the surface modulation wire produces a field along the z direction, but each side of the wire is placed under the field in the opposite direction; the phase is 180–degree different. Hence, what one can obtain by scanning the wire along the x direction is the difference or a differential form $df(x, y)/dx$. The spin distribution must be obtained by integrating along the x direction (Miki 1989).

Figure 14.8 (c) shows an example of microdosimetry using alanine

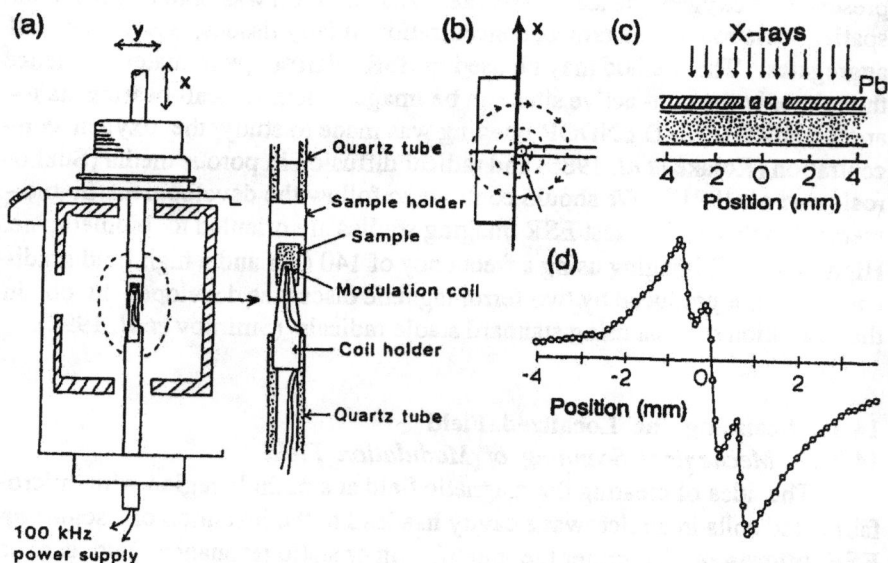

Figure 14.8 (a) An illustration of the scanning field ESR microscope using a small modulation coil and a computer controlled X-Y stage. The sample is scanned in a cavity. (b) A differential modulation method using a single wire of $2a$ in diameter and the sample. The direction of the modulation magnetic field is opposite. (c) Microdosimetry of an alanine pellet X-rayed through a pinhole and (d) the differential distribution of spins obtained (Miki and Ikeya 1987, 1988).

powder cast in paraffin, which is widely used as an ESR dosimeter element. The alanine pellet was exposed to X–rays through pinholes of a lead plate and ESR imaging was obtained in (d). An X–ray image may be obtained using an imaging plate though the radiation sensitivity is too low to be practical in comparison with X–ray film using silver bromide.

14.3.2   *Electronic Scanning of Static Magnetic Field*

An additional magnetic field of a few tenths of 1 mT at a pinhole region shifts the resonance magnetic field. A more refined method of electronically scanning the local static field is achieved using microfabricated wire arrays (Ikeya and Ishii 1990). Figure 14.9 shows a schematic illustration of the wire arrays. The magnetic field at the wire spacing is enhanced or reduced by ΔH due to the two–way current flow at the $i–$ and $(i+1)$–th wires. A typical 1–D imaging sample holder as well as the magnetic field slightly above the wire array are shown in (a) and wire arrays for 2–D

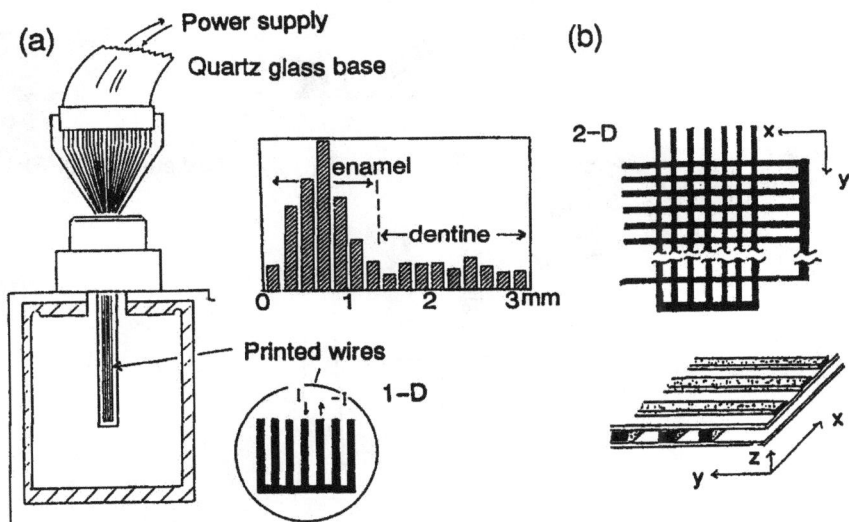

Figure 14.9   (a) The 1–D microfabricated wire array for creation of the local magnetic field in a cavity. The quartz plate sample holder with the printed thin wire arrays. The tip part is enlarged. The obtained 1–D image of the enamel and dentin boundary of a shark tooth is shown.   (b) The 2–D wire arrays for 2–D images. A local static or modulation field can be scanned by switching on and off using a computer (Ikeya and Ishii 1990, Miyamaru and Ikeya 1993)

imaging in (b). Similar currents along the x direction produce a field of H_z between the $j-$ and $(j+1)$-th wires. The intersection square is under the magnetic fields of $(\Delta H_1 + \Delta H_2)$ in 2-D arrays. If the uniform static magnetic field is kept constant so that $H + \Delta H_1 + \Delta H_2$ is at resonance and H is at "off-resonance", the ESR signal intensity at the square site is obtained. By switching "on" and "off" the current in the wire, 2-D ESR imaging can be obtained. A test sample holder for 1-D microwire arrays is printed on a quartz sample holder with the modern resist-masking method. The wire width is about 50 μm and one-sided ends of those wires are short circuited.

Figure 14.10 (a) shows ESR spectra with the current in two adjacent wires, $i-$ and $(i+1)$-th, using 1-D wire arrays. The test sample consists of two dots of DPPH on wire arrays and the ESR signal intensity estimated from the height at the shifted field position shown by an arrow in (a) gives the 1-D image shown in (b). 1-D arrays can thus give the spin distribution of two dots and the spin concentration can be obtained with a resolution equal to the wire spacing, d. A higher resolution of d/n may be obtained

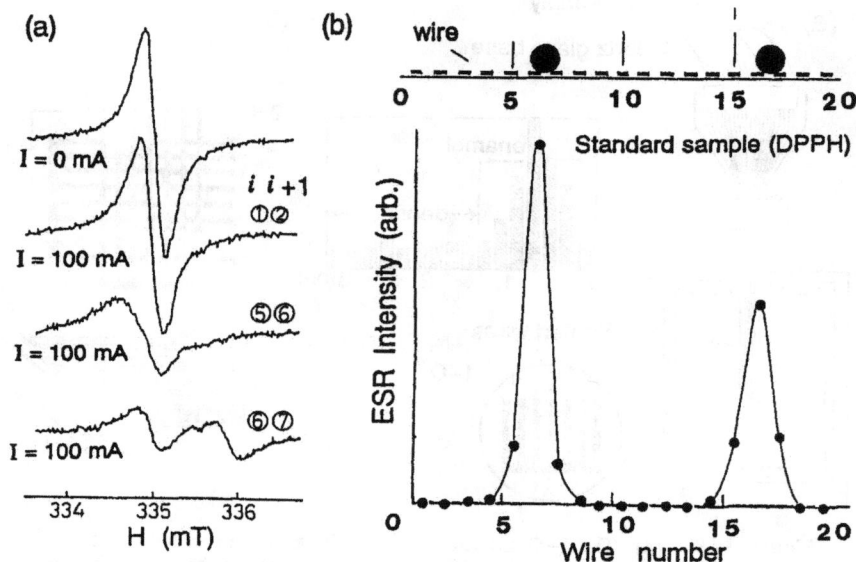

Figure 14.10   (a) ESR spectra of the DPPH test sample. The signal intensity is shifted by the application of the voltage to an adjacent wire (from $i-$ to $(i+1)$-th). (b) The intensity profile of the shifted ESR signal as a function of the wire number, i, for a two point test pattern of DPPH.

if measurements are made $2n$ times by moving sample by steps of d/n. The intensity function may be deconvoluted considering the magnetic field distribution or point spread function as described in 14.4.2.

The advantage of electronic scanning of the static magnetic field is the ability to measure the concentration of specific radicals by tuning the magnetic field. The construction of an ESR image only at the surface, i.e., 2–D imaging at the surface region rather than in the bulk material, is possible since the magnetic field intensity is strongest at the surface but decreases inside the sample.

A modulation field can also be applied so that the local intensity is detected. The static and modulation fields may be applied between the i-th and $(i+1)$-th, and j-th and $(j+1)$-th wires to obtain the surface image for 2–D microwire arrays. Modulation is applied at two different frequencies for 2–D wire arrays so that the signal intensity at the intersection can be detected with two lock–in amplifiers. Experimentally, one frequency is at 100 kHz and the other is at 120 Hz so that the 100 kHz narrow–band lock–in amplifier of the spectrometer is used with an additional lock–in amplifier. The ESR spectrum at the intersection with a double lock–in–amplifier detection is naturally a second derivative curve.

Typical sample holders are shown in Figure 14.11 (a) for 1–D and (b)

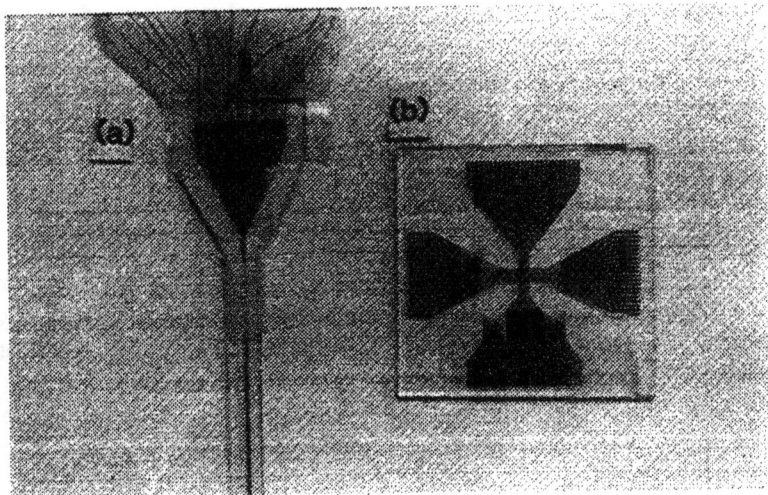

Figure 14.11   (a) 1–D microwire arrays using a computer flat cable and a printed circuit on quartz plate.   (b) Two–layered printed microwire arrays.

for 2–D microwire arrays using a computer flat cable. The simplest wire array for a 1–D scanning ESR microscope is a commercial flat cable of a personal computer attached to a standard glass sample holder. Two 24–channel relay boards for a personal computer switch on and off the microwire arrays.

14.4   Localized Microwave Field

14.4.1   *Simple Microwave Scanning ESR*

There are two physical quantities in magnetic resonance spectroscopy: the static magnetic field and electromagnetic waves. Scanning microscopy is possible whether it is NMR, ESR or other methods by localizing spatially and scanning one of the parameters. The present system employs a microwave magnetic field for resonance at a pinhole localized region. The simplest one is the system with a TE_{102} rectangular cavity with an aperture shown in Figure 14.12 (a). The microwave cavity with a small hole from which the microwave magnetic field along the x direction is leaking was used for ESR measurements. The sample is mechanically scanned using a mechanical X–Y stage with stepping motors controlled by a computer.

The S/N ratio was improved by using a cylindrical TE_{111} mode cavity rather than a simple TE_{102} rectangular one and by inserting a small loop–gap below the aperture. The apparent intensity was enhanced by a factor of 5 ~ 7. The distribution of the microwave magnetic field or actually the power around the pinhole was measured with a DPPH dot sample. The image obtained by scanning it is shown in Figure 14.13 (a). This image may be considered as a *point spread function* (PSF) due to the microwave diffraction through the pinhole.

14.4.2   *Deconvolution Method*

The resolution of a microwave scanning ESR microscope is determined not by the size of the hole but by the step length of the position scanning. If the microwave power intensity inside the hole region, $r(z, x)$, is calibrated with a small DPPH particle, one can get the system function or PSF, $r(z, x)$. The convolution integral with the distribution function of the spin concentration, $f(z, x)$, within the pinhole region gives an experimental image of the microwave absorption, $g(z, x)$, as

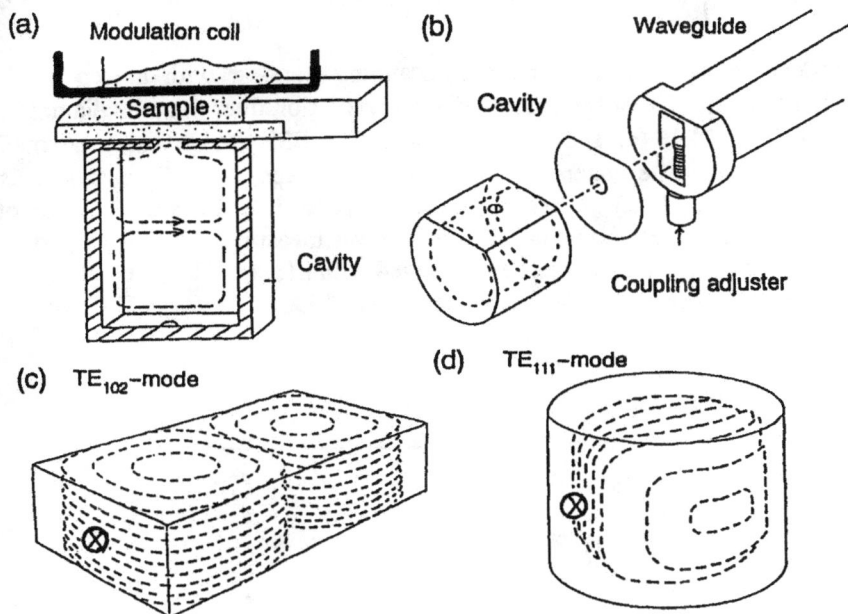

Figure 14.12 Microwave cavity with an aperture for the microwave scanning microscope. The sample over the aperture is mechanically scanned using an X-Y stage and stepping motors. (a) The TE_{102} rectangular cavity (Furusawa and Ikeya 1988) and (b) TE_{111} cylindrical cavity with an aperture (Hochi et al. 1993). (c) and (d) Microwave magnetic field for (a) and (b), respectively (Alger 1968).

$$g(z, x) = \int \int r(z - z', x - x') f(z', x') \, dx' \, dz' \ . \tag{14.17}$$

For a dot sample of DPPH, $f(z, x) = \delta(z, x)$, the system function is obtained as $g(z, x) = r(z, x)$. The 2-D Fourier transform gives

$$G(\omega_1, \omega_2) = R(\omega_1, \omega_2) F(\omega_1, \omega_2) \ . \tag{14.18}$$

Using this relation, $F(\omega_1, \omega_2) = G(\omega_1, \omega_2) / R(\omega_1, \omega_2)$ can be transformed with appropriate filtering to

$$f(z, x) = \int \int [G(\omega_1, \omega_2)/R(\omega_1, \omega_2)] \exp(i(\omega_1 z + \omega_2 x)) \, d\omega_1 \, d\omega_2 \quad . \quad (14.19)$$

The deconvolution of the image by scanning z and x with a step much smaller than the size of the pinhole gives a resolution finer than the diameter.

Figure 14.13 (a) shows an example of an ESR image using a small DPPH particle. The system function, $r(z, x)$, is experimentally obtained as $g(z, x)$ by using $f(z, x) = \delta(z, x)$ for the DPPH particle. The ESR image of three DPPH particles shown as $f(z, x)$ in (b) was measured in (c) as an apparent image, $g(z, x)$, which was deconvoluted with $r(z, x)$. The three particles can clearly be seen in (d) (Furusawa and Ikeya 1990a, Ikeya *et al.* 1990a).

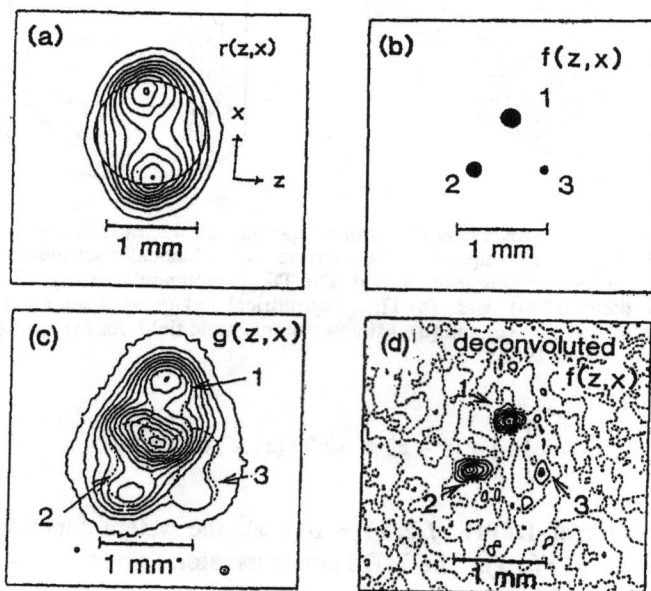

Figure 14.13 (a) A point spread function (PSF) of the aperture, $r(z, x)$, was obtained by scanning a DPPH dot sample on the aperture (circled), (b) an image function, $f(z, x)$, of a test sample of three DPPH dots, (c) an apparent image function, $g(z, x)$, of the three dots and (d) a resolved contour map image, $f(z, x)$, obtained by deconvoluting $g(z, x)$ with $r(z, x)$.

14.5 Distribution of Paramagnetic Ions

14.5.1 *Fossils of Crinoids and Ammonites*

Some carbonates contain a considerable amount of Mn^{2+} substitutionally located at Ca^{2+} sites in the $CaCO_3$ lattice. The distribution of Mn^{2+} can easily be imaged with an ESR microscope. Images of carbonate fossils such as a crinoid and an ammonite have been obtained (Furusawa and Ikeya 1988). The Jurassic Crinoid is a flower like echinoderm characterized by five–fold radial symmetry of the columnar stem section. The spectrum shows six prominent lines associated with a hf structure of Mn^{2+} and a signal due to radicals created by natural radiation.

A crinoid slice was scanned by 0.2 mm step movement over an aperture of 1 mm in diameter. The distribution of Mn^{2+} in the crinoid sample is shown in Figure 14.14 (a). The outer part of the crinoid has a higher Mn^{2+} concentration than the inner part with five–fold symmetry. In the central pentagonal part, no Mn^{2+} signal was detected, but those due to radiation-induced radicals were observed and imaged separately. The image of crinoid fossils was analyzed to obtain a better resolution. An appropriate filter function in Eq. (14.19) gives a clear resolution of the edge.

Ammonite fossils contain much Mn^{2+} similar to the crinoid. A cross sectioned ammonite of about 25 mm in diameter was scanned by 0.25 mm-step movement. The ESR image of Mn^{2+} is concordant with the pattern of chambers in the ammonite sample. The chambers with intense Mn^{2+} are filled with tiny crystals of carbonate, while the other part (pyrite) gives no signal as shown in (b).

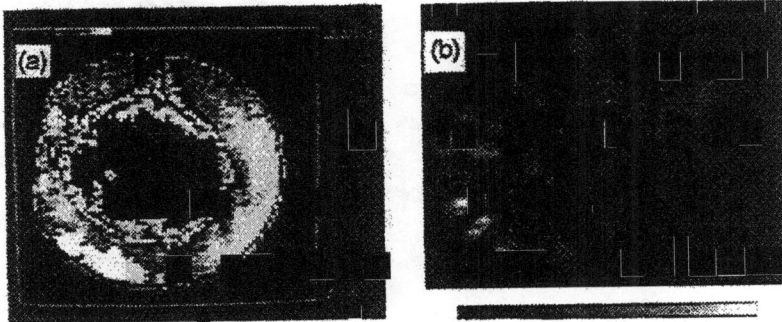

Figure 14.14 ESR microscopic imaging of Mn^{2+} ions in (a) a carbonate fossil of Jurassic crinoid and (b) an ammonite on a CRT display (Furusawa and Ikeya 1988, 1990a).

14.5.2 Reactive Centers on Catalysts : Valency Changes

Paramagnetic metal ions have been used at reactive sites in heterogeneous catalysis. The valency and the state before and after catalytic reactions have been investigated by several analytical techniques. Tablets and pellets of automotive catalysts, sometimes used in chemical industries, employ multilayer structures or concentration gradients of reactive ions. The spatial distribution of these ions and their valency change by catalytic reactions at the active site give us information on their efficiency, appropriate size, proper preparation, poisoning effects and lifetime (O'Reilly 1960).

The tablet formed from the precipitated powder by drying a mixture of $H_3PMo_{12}O_{40} \cdot H_2O$ and silica gel was heated at 460°C for 4 hours in a reducing gas stream of N_2 and H_2 flowed from the bottom of the tablet as shown in Figure 14.15 (a). The ESR spectrum in (b) shows Mo^{5+} signals with Mn^{2+} standards (Xu et al. 1990). The intensity of Mo^{5+} signals is high at the outer edge of upper surface and was gradually decreased toward the center as shown in (c). The reducing gas did not reach the center of the

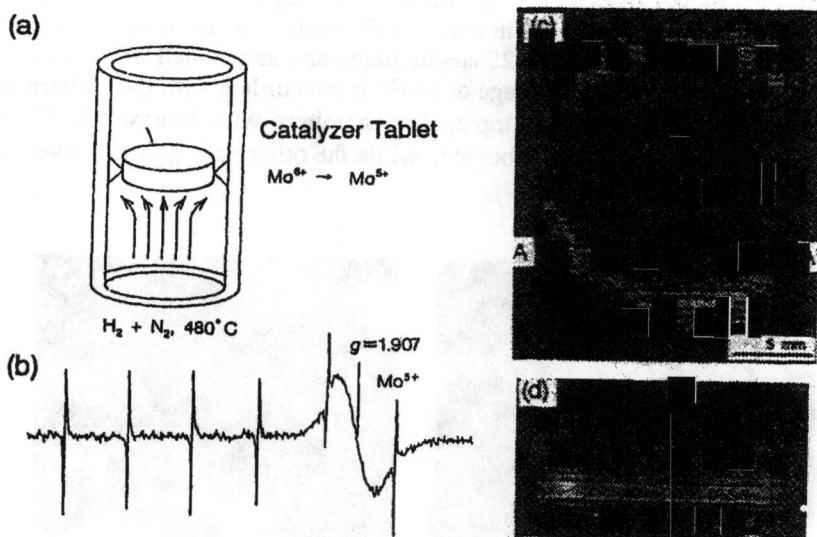

Figure 14.15 (a) A tablet of silica gel catalyst was heated at 460°C from the bottom in the reducing atmosphere. (b) ESR spectrum shows signals of Mo^{5+} and Mn^{2+} ions. Distribution of Mo^{5+} ions in (c) the upper surface and (d) the cross section of the tablet (Xu et al. 1990).

upper surface and presumably the reduction proceeded both from the bottom surface and the edge as shown in the cross sectional image (d). The intensity at the lower surface is high and uniform. Both V^{4+} and Mo^{5+} signals as well as an unidentified signal at $g = 2.0036$ were observed in other tablets. Monitoring the distribution of active species during the catalytic reaction will help to clarify the mechanism and to design a catalytic reactor.

14.6   Crystals : Impurity Distribution

14.6.1   *Synthetic Diamond : Growth Sector*

Optical devices such as the blue light emitting diodes (LED) or lasers are being developed using synthetic diamond crystals with color centers induced by radiation. Two diamond particles were imaged using ESR (Zommerfelds and Hoch 1986). The ESR microscope can determine the spatial distribution of paramagnetic defects in diamond crystals. Nitrogen (N) and nickel (Ni) are common impurities in synthetic diamonds (Samilovich *et al.* 1971, Hagiwara *et al.* 1988). Optical and luminescence studies indicate that the concentration of impurities depends on crystal growth sectors (Woods and Lang 1975).

A diamond grown from a seed crystal in Fe–Co–Ni solvent under a high temperature and high pressure using the temperature gradient method was provided by Sumitomo Electric Industries (Kanda *et al.* 1981). It has a cubo–octahedral shape and is cut along the horizontal line as shown in Figure 14.16 (a) and (b). All surfaces are as–grown except the bottom (111) surface which is cut and polished. The growth sectors of the bottom surface are shown in (c). The contents of N and Ni in the diamond are about 100 and 10 ppm, respectively. The Ni signal was measured using a special sample holder with a small dewar filled with liquid nitrogen placed on the microwave cavity (Furusawa and Ikeya 1990b). An isotropic line of $g = 2.032$ due to Ni (Isoya *et al.* 1990) was observed below 150K in addition to the hf signals of single substitutional N (P1) centers.

The Ni signal intensity measured at 77K with the magnetic field parallel to the [110] direction is mapped in (d). Ni was detected only in the portion of the N–rich (111) sectors. Almost no Ni was observed in as–grown (111) surface. ESR signal intensity due to Ni impurities increases only for a certain period of the crystal growth. The preferential inclusion of Ni only at a portion of (111) sectors would be due to the variation in the growth conditions such as temperature, the growth velocity and the local concentration in the solvent. This explains the absence of Ni in the as–grown surface.

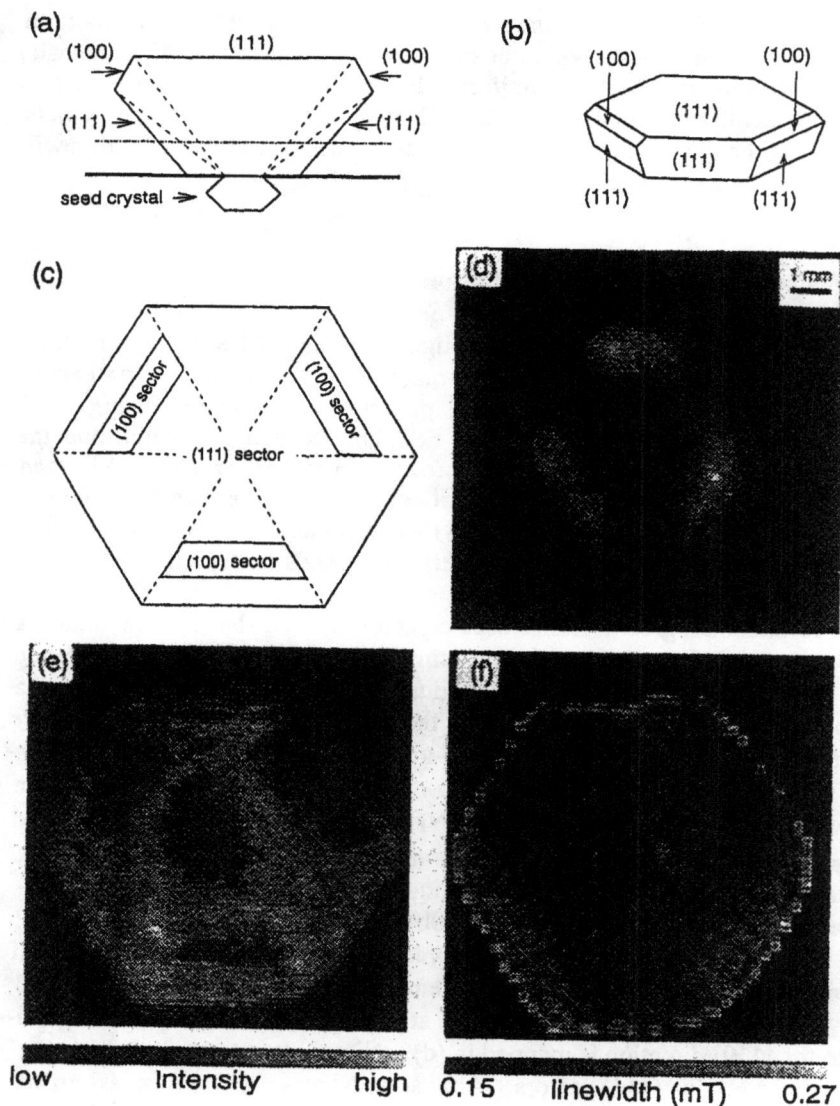

Figure 14.16 (a) A synthetic diamond grown from a seed crystal in Fe–Co–Ni solvent. (b) A diamond piece cut at the dashed line in (a). (c) The bottom surface of the piece where growth sectors are indicated. (d) The distribution of Ni impurities at the bottom surface. (e) The distribution and (f) the line width mapping of substitutional N (P1) centers (Furusawa 1992).

The intensity map of N in the bottom (111) surface at room temperature is shown in (e). The four N-poor regions at the center and three satellite parts showed the concentration of about half of the other part. The three N-poor satellite regions correspond to cubic (100) growth sectors. All other N-rich parts were octahedral (111) sectors. The concentration was low at the central part, i.e., the region close to the (111) surface of a seed crystal. In the top as-grown (111) surface, the distribution of N was almost uniform.

The linewidth of the N signal at room temperature varied spatially and its linewidth map had almost the same pattern as the intensity map of Ni. Although the concentration of N was almost uniform in the whole region of the (111) sectors, the linewidth was narrow in the region where Ni was present. These observations indicate that the line broadening is mainly due to dipole-dipole and exchange interactions among N centers. The irregular narrowing of the linewidth for the N center at the Ni-rich part may be due to cross relaxation between N and Ni centers (Shul'man and Podzyarei 1972, 1975). High-resolution ESR microscopy on the isotope abundance of C and N in natural diamonds would clarify their origin in the inner mantle of the earth (Boyd et al. 1988).

14.6.2   Distribution of Gd^{3+} and Radioactive Elements

Zircon ($ZrSiO_4$) is a mineral which has relatively high concentrations of uranium and rare earth ions. ESR signals of Gd^{3+} and the hole center at $g = 2.011$ are observed in a zircon crystal. The distribution of Gd^{3+} and the hole center can separately be imaged as shown in Figure 14.17 (a) and (b). These images are almost identical and correspond to the zonal structure of the crystal (Furusawa et al. 1991). The fission track map of the same crystal showed similar images, which indicates that the distribution of the hole center is proportional to that of uranium. Alpha-ray tracks on a CR-39 plastic film attached to the zircon crystal showed nearly the same image as that of the fission track (Kasuya et al. 1991).

The intensity of the hole center was nearly proportional to the square of that of Gd^{3+}. This is because the intensity used is the central line due to isolated Gd^{3+} ions. The association-dissociation reaction of Gd^{3+} and some monovalent cations, $Gd^{3+} + M^+ \leftrightarrow Gd^{3+}-M^+$, gives the rate equation,

$$[Gd^{3+}-M^+] = K[Gd^{3+}][M^+] , \qquad (14.20)$$

where $[Gd^{3+}]$, $[M^+]$ and $[Gd^{3+}-M^+]$ are the concentration of the isolated Gd^{3+}, isolated monovalent cations introduced as a charge compensator and

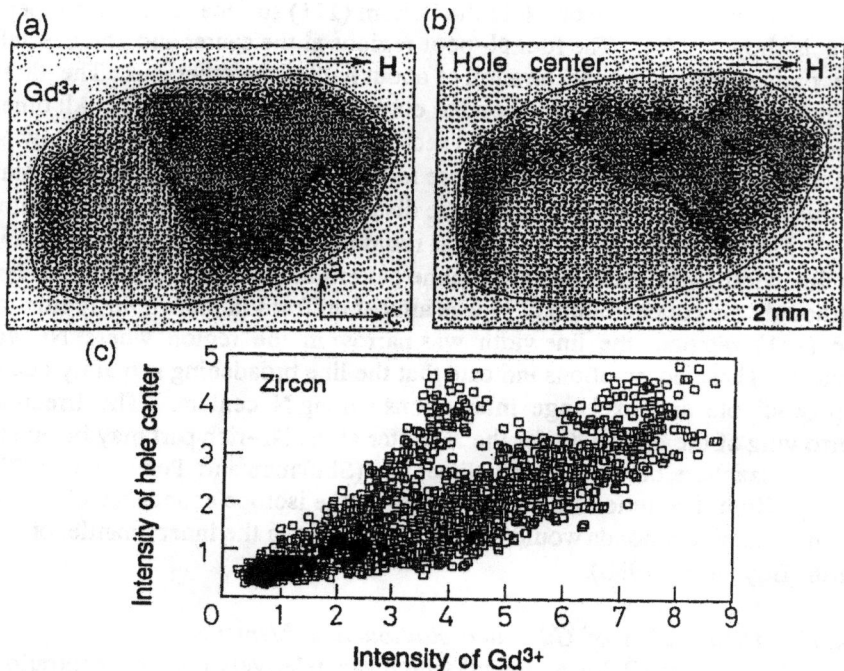

Figure 14.17   The distribution of (a) Gd^{3+} and (b) the hole center in a single crystal of zircon ($ZrSiO_4$) with zone structures. The direction of the magnetic field is parallel to the c–axis of zircon.   (c) A quadratic relation between the hole center and Gd^{3+} intensities (Furusawa *et al.* 1991).

the complex between Gd^{3+} and M^+, respectively.   If the content of uranium is proportional to the total Gd^{3+} concentration, the intensity of the hole center is proportional to the square of that of the isolated Gd^{3+} ions using the relations of $[Gd^{3+}] = [M^+]$ and $[Gd^{3+}-M^+] = N_{Gd}$.

14.6.3   *Crystal Axis Orientation*

The g factor depends on the direction of the magnetic field relative to crystalline axes.   Local tilting of the crystalline axis due to lattice distortion or twinning results in a locally different spectrum.   Two samples were cut from an aragonite crystal from Aragon, Spain, as shown in Figure 14.18 (a). The direction of the magnetic field is perpendicular to the c–axis in the sample A and the image is shown in (b).   The images of the sample B for the direction of the magnetic field parallel and perpendicular to the c–axis are shown in (c) and (d), respectively, which indicate the high concentration of CO_2^- close to the surface presumably due to the external β–ray damage.

The ESR spectral shape changed from area to area indicated by 1, 2 and 3 in (d). The spectrum was identical in each area and for the magnetic field parallel to the c–axis. Thus, we conclude that the a– and b–axes are slightly disoriented though the c–axes coincide. A similar result was obtained for an apparent single crystal of apatite [$Ca_{10}(PO_4)_6F_2$], where a portion was crystallized in a different direction, as in local twinning (Ishii et $al.$ 1991). The signal linewidth of an area was somewhat large, but the intensity was almost uniform. ESR microscopy may be useful in evaluating crystal perfection using anisotropic ESR signals.

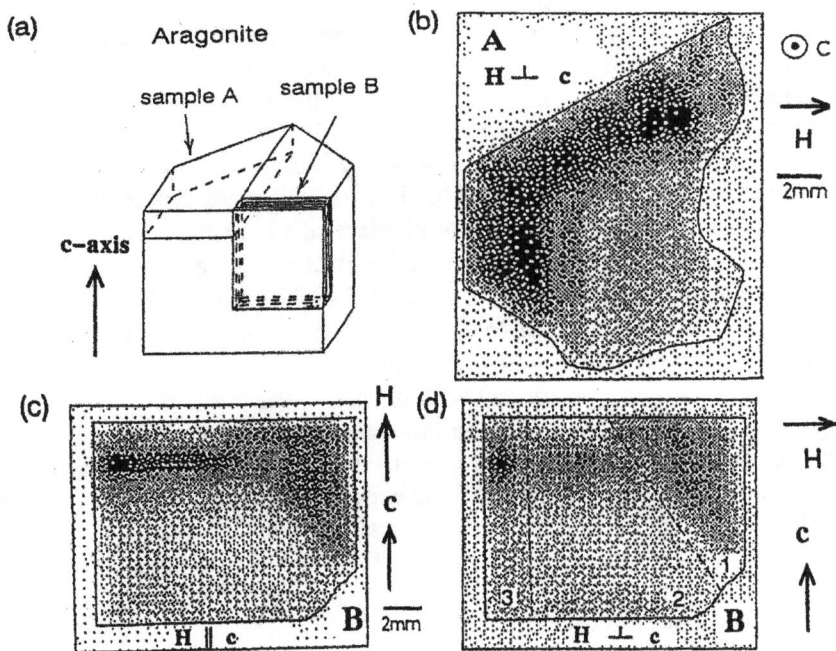

Figure 14.18   (a) The positions of samples A and B taken out of a single crystal of aragonite ($CaCO_3$) from Aragon, Spain. (b) ESR image of aragonite crystal of sample A. (c) and (d) ESR images of sample B for the direction of the magnetic field parallel and perpendicular, respectively, to the c–axis. Different spectra were observed in regions 1, 2 and 3 due to the slightly different crystallographic orientations (Furusawa et $al.$ 1991).

14.7 Dosimetric Image

14.7.1 *Microdosimetry of Minerals*

Signal intensities of electron and hole centers in a single crystal of barite ($BaSO_4$) from Sterling, Colorado gave two different images. When the polished surface of the (100) plane was observed, the signal at $g = 2.019$ was only at the surface. The surface of the crystal had a high concentration of electron or hole centers either due to external $\alpha-$ and $\beta-$radiation to the surface or due to zoning of radioactive or electron trapping impurities related to this defect. This was also observed in the aragonite crystal shown in Figure 14.18 (b) (Furusawa *et al.* 1991). Electron and hole centers related to some impurities in irradiated semiconductors and ceramics would be useful for assessing the distribution of diamagnetic impurities.

14.7.2 *Radiation Dose Image in Human Teeth*

The spatial distribution of CO_2^- in X– and γ–irradiated teeth has been studied. The CO_2^- signal is observed in the enamel part and less so in the dentine of the γ–ray exposed tooth, while only the X–ray exposed surface of the enamel shows the intense signal (Ikeya and Furusawa 1988, Furusawa 1992). Thus, radiation quality can be deduced from the spatial dosimetry. Unfortunately, the sensitivity is still too low to distinguish the radiation dose of accidents and dental X–rays in a tooth piece in an image.

The signal intensity in tooth enamel in contact with a metal is high due to the secondary electron from the metal (Hochi *et al.* 1993). This was demonstrated using extracted tooth half covered with aluminum foils and irradiated by γ–rays. The irradiation conditions are drawn in Figure 14.19 (a) and (b). The intensity in contact with the metal was high as shown in the image (c), where the spin distribution along the horizontal line is shown below each image. Apparently the intensity at the edge is extraordinarily high. We might overestimate radiation exposure by taking enamel close to dental amalgam embedded during the dental treatment. Precaution is necessary in human dose estimation using extracted tooth enamel (see Chapter 13).

14.7.3 *ESR Imaging Plate*

An ESR imaging plate composed of a radiation sensitive material with a high G–value and a sharp signal will be useful since the quality effect is different from that of X–ray photographic films. Recently, Ishii *et al* (1993) have developed an ESR imaging plate using CO_2^- radicals in a sintered fluorapatite disc doped with CO_3^{2-}. The radiation–induced ESR signal due

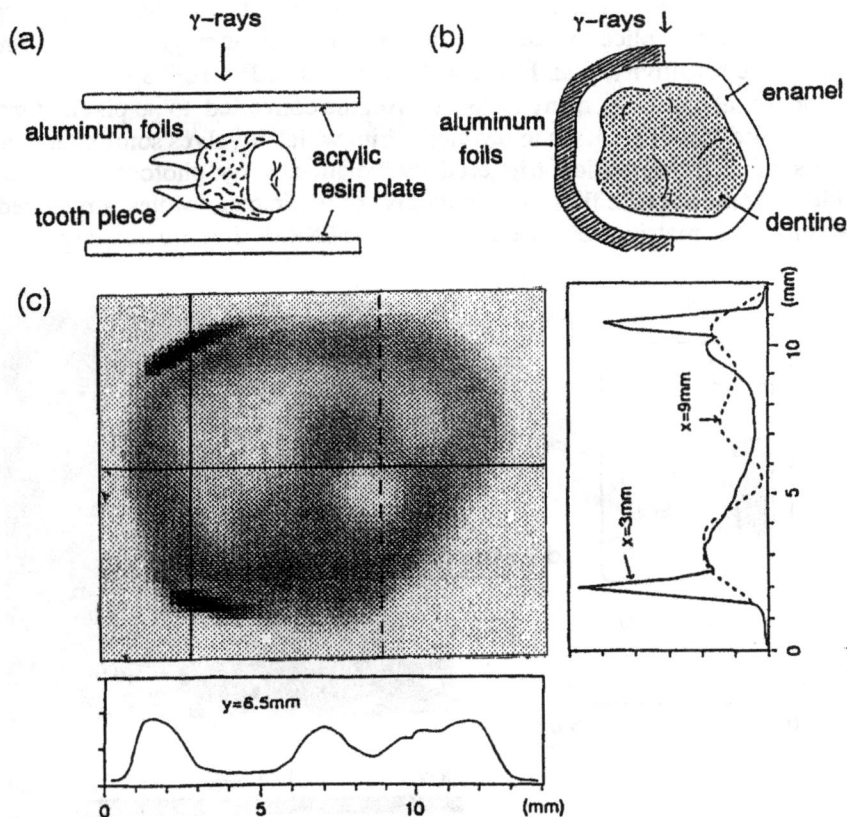

Figure 14.19 Microdosimetric imaging of γ–ray induced CO_2^- in a human tooth half covered by aluminum foil. (a) and (b) Drawings of irradiation conditions. (c) ESR image of the tooth cross section and spin distribution along the central horizontal line and vertical line at the bottom and the right.

to CO_2^- in apatite and its application to an ESR dosimeter have already been described in Chapters 8 and 13. Figure 14.20 shows an example of a disc exposed to X–rays in a commercial X–ray film badge case. The open window portion, central cross letter and shielded portion have been clearly imaged. Irradiation by γ–rays gives an uniform image of the disc (Ishii and Ikeya 1993). Alanine and other films with a high sensitivity to radiation should be developed as imaging plates for ESR dosimetry.

One cannot compare radiation sensitivity of an ordinary ESR dosimeter with that of a photographic film since the process of the photon imaging involves the multiplication to 10^7 times in a form of aggregation of silver atoms. The sensitivity of an ESR spectrometer is $10^9 \sim 10^{10}$ spins/0.1 mT. One needs at least this many spins in a volume converted to a pixel. One way to avoid this dilemma is to develop a film which involves some chemical process of spin formation triggered by radiation or by photons as silver halides. Some specific film using polymers or organic molecules embedded in an inorganic matrix should be developed for ESR dosimetric imaging.

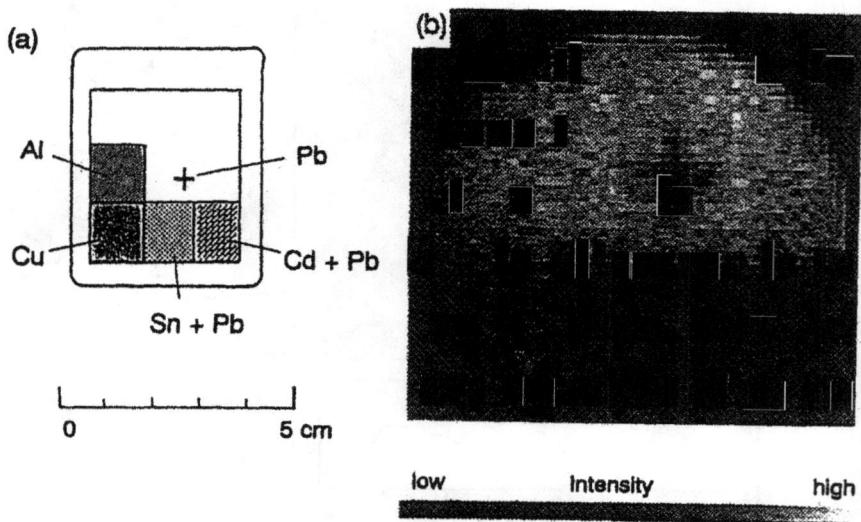

Figure 14.20   An image of radiation dose with an imaging plate of sintered fluoroapatite disc. (a) The disc in a film badge case with filters of Al, Pb, Cu, Sn + PB and Cd + Pb, (b) an image of the plate irradiated by X–rays (Ishii *et al.* 1993).

14.8. Semiconductor, Bioscience and New Apparatus

14.8.1 *Dangling Bonds on Poly–Si CVD Film : Annealing Image*

ESR microscope at the X–band frequency has only limited applications to biological materials but has a large applicability in material sciences. One such example is the distribution measurement of dangling bonds in a thin

film formed by chemical vapor deposition (CVD). Figure 14.21 shows an
image of the distribution of dangling bonds on a CVD poly–Si film on Si
wafer. Two positions are locally heated to anneal out the dangling bonds.

The activation energy and pre–exponential factor of defect centers can
be determined using an annealing image. Each pixel, i, around the heated
region shows the different signal intensity, I_i, from the one before annealing, I_0. The decay time, τ_i, of the dangling bonds can be calculated as
$\tau_i = t_a \log_e(I_0/I_i)$, where t_a is the anealing time. τ_i is related to the local
temperature T_i as $\tau_i = (1/\nu_0)\exp(E/k_B T_i)$. Hence, each intensity ratio is
used as a data point for T_i. An Arrhenius plot gives the activation energy E
and pre–exponential factor ν_0.

An annealing image of defects with an image of the temperature allows
to analyze the thermal activated process in a short time. This can be used for
the lifetime determination of defects in ESR dating.

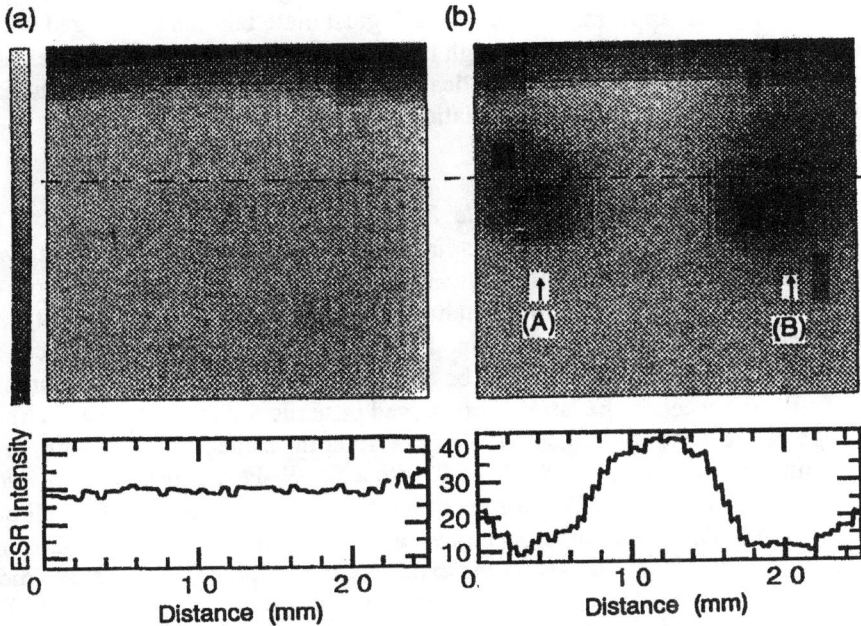

Figure 14.21   (a) An ESR imaging of the distribution of dangling bonds
using signal at $g= 2.0052$ on a part of poly–silicon CVD film on a Si wafer.
(b) An image of the same wafer by locally heating the sites A and B using a
pen–torch. The annealing image together with the temperature profile gives
the activation energy and the preexponential factor in a short time.

14.8.2   Biological Application

(a) *Broiled Sardine* :   Fish or meat broiled too much is said to cause cancer. The heavily broiled part may contain lipid peroxy and other organic radicals. The radical distribution in broiled sardine has been obtained as shown in the front page of Chapter 14. Apparently, the broiled part has radical species. Broiled coffee beans and chestnuts were also imaged: all showed a high concentration of radicals at the heavily broiled parts.

(b) *Melanoma* :   Biological materials contain water which causes microwave loss beyond the detection of an ESR signal at the X–band frequency. However, frozen samples cooled to 77K show no microwave loss. The same apparatus with a similar dewar used for imaging the Ni signal in a synthetic diamond at 77K was applied to the study of melanoma, a skin cancer (Katsuta *et al.* 1990). Additional rapid scan coils allow the acquisition of ESR signals in a few seconds for the magnetic field region where the particular signal of interest appears. A slice of biological materials can be imaged with the present–day apparatus although the signal intensity of most samples is not sufficient for a practical application. Decoration with paramagnetic standard chemicals or heavy irradiation is needed to obtain an image at the moment.

14.8.3   Can One Detect a Single Spin ? (STM–ESR)

The detection limit, 10^{9-10} spins/0.1 mT, is a barrier as long as we measure microwave absorption (Ewert 1991). If one could detect the precession of a single spin with tunneling current (Manassen *et al.* 1989), a scanning microscope system combined with a scanning tunneling microscope (STM) (Binning *et al.* 1982) will be an ultimate goal in the development of an ESR microscope: the latter is often used in semiconductor studies. Although trials have been made to detect ESR at the stage of STM with the tunneling current using a portable ESR with a Nd–B–Fe magnet (Sagawa *et al.* 1984, Ikeya *et al.* 1991), the result has not been successful so far (Hirai *et al.* 1993). The same tunneling current at the microwave frequency (Cohen and Wolf 1991) was observed in the experiment for STM–ESR. An atomic force microscope (AFM) and ESR or electron spin–echo using pulsed ESR may be combined. If one could detect the distribution of single spins in an image, ESR microscope dating based on a day–by–day increase (potato chips method) may become possible on an atomic scale. This may be an ultimate goal in the future.

14.9 Summary

Applications of ESR microscopy to some problems in material sciences are described following the description of principles of MRI and several types of ESR microscopes. The conventional field gradient method used in ESR imaging may be useful if a large field gradient is produced by coils in a cavity. Applications of CT-ESR are still few due to the complexity of the software (FFT, back projection, etc.) and the necessity to use a computer to obtain a clear image in a short time.

Scanning ESR microscopes,though they may look like "macroscope ?" at the moment, have already gone beyond the wavelength and the diameter of the localized area. Scanning three physical parameters for ESR, i.e., (1) the local static magnetic field, (2) the local modulation field and (3) the localized microwave field spatially on the surface of the sample is a new development that could be used in magnetic resonance imaging in general. Commercial ESR spectrometers and personal computers can give an ESR image on a microscopic scale. Simple 2-D imaging methods offer some advantages over the CT-ESR imaging using the field gradient. The potential use of both CT-ESR and scanning ESR microscopes is enormous for physics, chemistry, biology, mineralogy as well as for ceramic and semiconductor technology.

References

Aguayo J.B., Blackband S.J., Schoeniger J., Mattingly M.A. and Hintermann M. (1986): Nuclear magnetic resonance imaging of a single cell. *Nature* **322**, 190-191.

Alger R.S. (1968): *Electron Paramagnetic Resonance: Technique and Application* (John Wiley & Sons Inc., New York).

Anderson W.A. (1961): Electrical current shims for correcting magnetic fields. Rev. Sci. Instr. **32**, 241-250.

Bangert V. and Mansfield P. (1982): Magnetic field gradient coils for NMR imaging. *J. Phys.* **15**, 235-239.

Berliner L.J. and Fujii H.(1987): Magnetic resonance imaging of biological specimens by electron paramagnetic resonance of nitride spin labels. *Science* **227**, 317-318.

Berliner L. J. and Fujii H.(1991): *In Vivo Spectroscopy Biological Magnetic Resonance* Vol. **11**, Chapt. 7 Some Applications of ESR to in Vivo Animal studies and EPR Imaging.

Berliner L.J. and Reuben J. (1991): In Vivo Spectroscopy Biological Magnetic Resonance Vol. **11** (Plenum Press, New York).

Binning G., Rohrer H., Gerber C. and Wiebel E. (1982): Surface studies by scanning tunneling microscopy. Phys. Rev. Letters **49**, 57-61.

Blumich B. and Kuhn W. ed. (1992): *NMR Microscopy* (VCH Publishers, Weinheim).

Boyd S.R., Pillinger C.T., Milledge H.J., Mendelssohn M.J. and Seal M. (1988): Fluctuation of nitrogen isotopes in a synthetic diamond of mixed crystal habit. *Nature* **331**, 604-607.

Callaghan P.T. (1991): *Principles of Nuclear Magnetic Resonance Microscopy* (Clarendon

Press, Oxford).

Cohen L.F. and Wolf E.L. (1991): Microwave-coupled cryogenic STM. *Meas. Sci. Technol.* **2**, 83-85.

Colacicchi S., Onori S., Petetti E. and Sotgiu A. (1993): Application of low frequency EPR imaging to alanine dosimetry. *Appl. Radiat. Isot.* **44**, 391-395.

Eaton G.R. and Eaton S.S. (1986a): Electron spin-echo-detected EPR imaging. *J. Magn. Reson.* **67**, 73-77.

Eaton G.R. and Eaton S.S. (1986b): EPR imaging using flip-angle gradients. a new approach to two-dimensional imaging. J. Magn. Reson. 67, 561-564.

Eaton G.R. and Eaton S.S. (1987): EPR imaging using T1 selectivity. *J. Magn. Res.* **71**, 271-275.

Eaton S.S., Maltempo M.M., Stemp E.D.A. and Eaton G.R. (1987): Three dimensional EPR imaging with one spectral and two spatial dimensions. *Chem. Phys. Lett.* **142**, 567-569.

Eaton G.R., Eaton S.S. and Maltempo M.M.(1989): Three approaches to spectral spatial ESR imaging. *Appl. Radiat. Isot.* **40**, 1227-1231.

Eaton G.R. and Eaton S.S. (1990): *Modern Pulsed and Continuous-Wave Electron Spin Resonance* ed. Kevan L. and Bowman M. K. (John Wiley, New York) 405-435.

Eaton G.R., Eaton S.S. and Ohno K. ed. (1991): *EPR Imaging and In-vivo EPR* (CRC Press, Boca Raton).

Ewert U. and Herrling T. (1986a): Numerical analysis in EPR zeugmatography with modulated field gradient. *J. Magn. Reson.* **61**, 11-21.

Ewert U. and Herrling T. (1986b): Spectrally resolved EPR tomography with stationary gradient. *Chem. Phys. Lett.* **129**, 516-520.

Ewert U. and Herrling T. (1989): EPR imaging: progress in methodological development. *Electron Magnetic Resonance in Disordered Systems* , ed. Yordanov N.D. (World Scientific, Singapore) 205-219.

Ewert U. (1991): Sensitivity in EPR imaging. In *EPR Imaging and In-vivo EPR*, 127-134.

Froncisz W. and Hyde J.S. (1982): The loop-gap resonator: a new microwave lumped circuit ESR sample structure. *J. Magn. Reson.* **47**, 515-521.

Fujii H. and Berliner L.J. (1985): One- and two-dimensional ESR imaging studies on phantoms and plant specimens. *Magn. Reson. Med.* **2**, 275-282.

Furusawa M. and Ikeya M. (1988): ESR microscopic imaging of fossil crinoid utilizing localized microwave field. *Analyt. Sci.* **4**, 649-651

Furusawa M. and Ikeya M. (1990a): Electron spin resonance imaging utilizing microwave magnetic field. *Jpn. J. Appl. Phys.* **29**, 270-276.

Furusawa M. and Ikeya M. (1990b): Distribution of nitrogen and nickel in a synthetic diamond crystal observed with scanning ESR imaging. *J. Phys. Soc. Jpn.* **59**, 2340-2343.

Furusawa M. and Ikeya M. (1991): A method of producing high quality linear field gradient for magnetic resonance imaging using straight current lines. *Jpn. J. Appl. Phys.* **30**, L1682-L1685.

Furusawa M., Kasuya M., Ikeda S. and Ikeya M. (1991): ESR imaging of minerals and its application to dating. *Nucl. Tracks* **18**, 185-188.

Furusawa M. (1992): Development of ESR Imaging and Its Application Ph. D. Thesis (Physics Department, Osaka University).

Furusawa M. and Ikeya M. (1993): ESR imaging of irradiated teflon tube using a highly linear field gradient by straight wires inserted in a cavity. *Appl. Radiat. Isot.* **44**, 381-

384.

Furuta H. Yamanaka C. and Ikeya M. (1996): ESR imaging of alanine dosimeter annealed under thernak gardient. *Appl. Radiat. Isot.* **47**, 1611-1614.

Furuta H. Yamanaka C., Okigawa M., Nakata H. and Ikeya M. (1996): Correlation between photoluminescence and Pb-like center of porpous silicon: ESR image analysis for porpous silicon. *Modern Applications of EPR/ESR from Geophysics to Material Science* (Proc. 1st Asia-Pacific EPR/ESR Symposium), 167-174.

Galstewa E.V., Yakimchenko O.Y. and Lebedev Y.S. (1983): Diffusion of free radicals as studied by tomography. *Chem. Phys. Lett.* **99**, 301-304.

Hagiwara M., Uemura T., Chiba Y. and Date, M. (1988): Electron spin resonance of N1 center in diamond. *J. Phy. Soc. Jpn.* **57**, 741-743.

Herrling T., Klimes N., Karthe W., Ewert U. and Ebert B. (1982): EPR zeugmatography with modulated field gradient. *J. Magn. Reson.* **49**, 203-211.

Hirai M., Yamanaka C. and Ikeya M. (1993): Some trials of a new method of ESR detection using tunneling current. *Appl. Radiat. Isot.* **44**, 385-389.

Hoch M. J. R. and Day A. R. (1979): Imaging paramagnetic centers in diamond. *Solid State Commun.* **30**, 211-213.

Hochi A., Furusawa M. and Ikeya M. (1993): Applications of microwave scanning ESR microscope: Human tooth with metal. *Appl. Radiat. Isot.* **44**, 401-405.

Ikeda S., Furusawa M. and Ikeya M. (1994): Spatial variation of CO_2^- and SO_3^- radicals in massive coral from Ishigaki Island, Japan and its implication. *Proc. 29th Intern. Geol. Congr.* Part **B**, 225-228.

Ikeya M. and Miki T. (1987): ESR microscopic imaging with microfabricated field gradient coils. *Jpn. J. Appl. Phys.* **26**, L929-L931.

Ikeya M. and Furusawa M. (1988): Microdosimetric imaging of tooth irradiated by X- and g-rays with ESR surface scanning microscope. *Oral Radiol.* **5**, 5-12.

Ikeya M. (1989): Use of electron spin resonance spectrometry in microscopy, dating and dosimetry. *Analytical Sciences* **5**, 5-12.

Ikeya M. and Furusawa M. (1989): A portable spectrometer for ESR microscopy, dosimetry, and dating. *Appl. Radiat. Isot.* **40**, 845-850.

Ikeya M. and Ishii H. (1990): Scanning ESR microscopy using microwire arrays on quartz sample holder. *J. Magn. Reson.* **88**, 130-134.

Ikeya M., Furusawa M. and Kasuya M. (1990a): Near field scanning electron spin resonance microscopy. *Scanning Microscopy* **4**, 45-48.

Ikeya M., Furusawa M., Ishii, H. and Miki T. (1990b): ESR microscopy. *Appl. Magn. Reson.* **1**, 70-84.

Ikeya M. (1991): Electron spin resonance (ESR) microscopy in material science. *Ann. Rev. Mater. Sci.* **21**, 45-63.

Ikeya M., Meguro K., Miyamaru H. and Ishii H. (1991): Educational experiments on ESR imaging with a portable spectrometer. *Appl. Magn. Reson.* **2**, 663-673.

Ikeya M. (1992): Scanning and CT-ESR microscopes. In *NMR Microscopy* ed. Blumich B., (VCH Publisher, Weinheim).

Ikeya M. and Miki T. (1992): *ESR Microscopy- New Development of Applied Electron Spin Resonance-* (Springer-Verlag Tokyo, Tokyo) in Japanese.

Ikeya M., Yamamoto M. and Ishii H. (1994): Nondestructive measurements of large objects with electron paramagnetic resonance: Pottery, sculpture and jewel ornaments. *Rev. Sci. Instr.* **65**, 3670-3672.

Ikeya M. (1994): ESR and ESR Microscopy in geosciences and radiation dosimetry. Appl. Mahn. Reson. 7, 237-255.

Ikeya M. and Yamamoto M. (1994): Electron spin resonance image of annealing under thermal gradient: Chemical vapor deposition poly-silicon film. *Jpn. J. Appl. Phys.* **33**, 1087-1089

Ikeya M. (1994): ESR Microscopy: scanning ESR imaging of spin density. *EPR News Lett.* **6**, No.4, 12-14.

Ikeya M., Oka T., Omura T., Okawa M. and Takeno S. (1997): Evaluation of environment using electron spin resonqance (ESR): Microscopic images of gypsum ($CaSO_4$ $2H_2O$) microcrystals in borehole cores at Konya Basin, Turkey. *Japan Review* **8**, 193-208.

Ishida S., Matsumoto S., Yokoyama H., Mori N., Kumashiro H., Tsuchihashi N., Ogata T., Yamada M., Ono M., Kitajima T., Kamada H. and Yoshida E. (1992): An ESR-CT imaging of the head of a living rat receiving an administration of a nitroxide radicals. *Magn. Reson. Imaging* **10**, 109-114.

Ishii H., Ikeya M., Furusawa M. and Kasuya M. (1991): ESR, TL and FT measurements of a natural apatite. *Nucl. Tracks* **18**, 189-192.

Ishii H., Hochi A. and Ikeya M. (1993): An ESR imaging plate using sintered fluorapatite doped with CO_3^{2-}. *Jpn. J. Appl. Phys.*   in press.

Isoya J., Kanda H., Norris J. R., Tang J. and Bowman, M.K. (1990): Fourier transform and continuous wave EPR studies of nickel in synthetic diamond: site and spin multiplicity. *Phys. Rev.* **41**, 3905-3914.

Janzen E. G. and Kotake Y. (1986): ENDOR imaging based on differences in viscosity or oxygen concentration in aqueous solution of nitroxides. *J. Magn. Reson.* **69**, 567-571.

Kanda H., Ohsawa T., Fukunaga O. and Sunagawa I. (1981): Effect of solvent metals upon the morphology of synthetic diamond. *J. Cryst. Growth* **94**, 115-124.

Karthe W. and Wehrsdolfer E. (1979): The measurement of inhomogeneous distributions of paramagnetic centers by means of EPR. *J. Magn. Reson.* **33**, 107-111.

Kasuya M., Furusawa M. and Ikeya, M. (1991): Distribution of paramagnetic centers and α-emitters in a zircon single crystal. *Nucl. Tracks* **16**, 563-568.

Katsuta H., Kobayashi T., Saito H., Matsunaga T. and Ikeya M. (1990): Electron spin resonance imaging of mouse B16 melanoma. *Chem. Pharm. Bull.* **38**, 2838- 2840.

Kordas G. and Kang Y. H. (1991): Three dimensional electron paramagnetic resonance imaging technique for mapping porosity in ceramics. *J. Am. Ceramic. Soc.* **74**, 709-713.

Kotake Y., Oehler U.M. and Janzen E.G. (1988): Two dimensional ENDOR imaging based on differences in oxygen concentration. *J. Chem. Soc. Faraday Trans.* **184**, 3275-3278.

Lauterbur P. C. (1973): Image formation by induced local interactions: examples employing nuclear magnetic resonance. *Nature* **242**, 190-191.

L'vov S.G., Cherkasov F.G., Vitol A.Ya. and Silaev V.A. (1996): ESR and ESR-imaging of heavily irradiated alkali halide crystals. *Appl. Radiat. Isot.* **47**, 1615-1619.

Maltempo M. M., Eaton S. S. and Eaton G. R. (1988): Reconstruction of spectral spatial two dimensional EPR images from incomplete sets of projections without prior knowledge of the component spectra. *J. Magn. Reson.* **77**, 75-83.

Manassen Y., Hamers R.J., Demuth J.E. and Castellano A.J. (1989): Direct observation of the precession of individual paramagnetic spins on oxidized silicon surfaces. *Phys. Rev. Letters* **62**, 2531-2534.

Mansfield P. and Chapman B. (1986): Active magnetic field screening of gradient coils in NMR imaging. *J. Magn. Reson.* **66**, 573-576.

Marshall S.A., Suits B.H., Umlor M.T. and Zhang Y.N. (1988): Imaging electromagnetic fields using magnetic resonance absorption spectrum of a paramagnetic gas. *J. Magn.*

Reson. **76**, 494-503.

Miki T. and Ikeya M. (1987): Electron spin resonance microscopy by localized field modulation. *Jpn. J. Appl. Phys.* **26**, 1495-1498.

Miki T. and Ikeya M. (1988): A method of EPR imaging for differential detection of paramagnetic species distribution utilizing field modulation. *J. Magn. Res.* **80**, 502-508.

Miki T. (1989): ESR Spatial dosimetry using localized magnetic field modulation. *Appl. Radiat. Isot.* **40**, 1243-1246.

Miki T., Murata T., Kumai, H. and Yamashiro A. (1996): A high resolution EPR-CT microscope using cavity-resonators equipped with small field gradient coils. *Appl. Radiat. Isot.* **47**, 1599-1603.

Miyamaru H. and Ikeya M. (1993): One dimensional ESR microscope using microwire array. *Appl. Radiat. Isot.* **44**, 397-405.

Momo F., Adriani O., Gualtieri G. and Sotgiu A. (1988): Generalized Anderson coils for magnetic resonance imaging. *J. Phys. E* **21**, 565-568.

Morita Y., Ohno K., Ohashi K. and Sohma J. (1989): ESR imaging investigation on depth profiles of radicals in organic solid dosimetry. *Appl. Radiat. Isot.* **40**, 1237-1242.

Nishikawa H., Fujii H. and Berliner J.L. (1985a): Helix and surface coils for low field in vivo ESR and EPR imaging application. *J. Magn. Reson.* **62**, 72-86.

Nishikawa H., Fujii H. and Berliner J.L. (1985b): L-band ESR and ESR imaging. *ESR Dating and Dosimetry*, 307-314.

Ogata T., Ono M., Fujisawa T., Ooshida E. and Kamade H. (1986): An example of in vivo analysis of L-band ESR technique using a loop gap resonator. *Chem. Lett.* 1681-1684.

Ohno K. (1981): A method of EPR imaging: Application of spatial distributions of hydrogen atoms trapped in sulfuric acid ices. *Jpn. J. Appl. Phys.* **20**, L179-182.

Ohno K. (1985): Two dimensional ESR imaging for paramagnetic species with anisotropic parameters. *J. Magn. Reson.* **64**, 109-114.

Ohno K. (1986): ESR Imaging and Its Applications. *Appl. Spectro. Rev.* 22, 1-56.

Ohno K. (1987): ESR Imaging. *Magn. Reson. Rev.* **11**, 275-310.

Ohno K. and Murakami T. (1988): Microscopic ESR imaging using micro coil system. *J. Magn. Reson.* **79**, 343-347.

Ohno K., Morita Y. and Sawamura S. (1989): Spectral-spatial studies on radicals in polymers irradiated with electron beam using ESR imaging. *Electron Spin Resonance in Disordered System* ed. Yordanov, N. D. (World Scientific, Singapore) 221-234.

Ohno K. ed. (1990): *ESR Imaging* (IPC Publisher, Tokyo) in Japanese.

Oka T., Ikeya M.,Sugawara N. and Nakanishi A. (1996): A high-sensitivity portable spectrometer for ESR dosimetry. *Appl. Radiat. Isot.* **47**, 1589-1594.

Oka T., Grun R., Tani A., Yamanaka C., Ikeya M., and Huang PH (1997): ESR microscopy of fossil teeth. *Radiat. meas.* **27**, 331-337.

Oikawa K., Ogata T., Togashi, H. Yokoyama H.,Ohya-Nishiguchi H.and Kamada H. (1996): A 3D- and 4D-ESR imaging system for small animals. *Appl. Radiat. Isot.* **47**, 1605-1609.

Omura T. and Ikeya M. (1995): Evaluation of the ambient environment of minerals (gypsum (CaSO$_4$ 2H$_2$O)) growth by ESR microscope. *Geochem. J.* **29**, 317-324.

O'Reilly E. (1960): Magnetic resonance detection in catalytic research. *Advances in Catalysis* **12**, 31-60.

Poltz J. and Kuffel E. (1985): A method of magnetic field synthesis for medical NMR imaging. *J. Appl. Phys.* **57**, 3869-3871.

Sagawa M., Fujiwara S., Togawa N., Yamamoto H. and Matsuwa Y. (1984): New material for P.M. on a base of Nd and Fe. *J. Appl. Phys.* **55**, 2083-2087.

Samilovich M. I., Bezrukov G. N. and Butuzov V. P. (1971): Electron paramagnetic resonance of nickel in synthetic diamond. *Sov. Phys. JETP Lett.* **14**, 379-381.

Shin Y.K., Ewert U., Budil D.E. and Freed J.H. (1991): Microscopic versus macroscopic diffusion in model membranes by electron spin resonance spectral-spatial imaging. *Biophys. J.* **59**, 955-957.

Shul'man L.A. and Podzyarei G.A. (1972): Exchange broadening of the hyperfine component of the spectrum of nitrogen in diamond. *Sov. Phys. Solid State* **16**, 1377-1378.

Shul'man L. A. and Podzyarei G. A. (1975): Cross relaxation in ESR of synthetic diamond. *Sov. Phys. Solid State* **16**, 1377-1378.

Siebold H. (1990): Gradient field coils for MR imaging with high spectral purity. *IEEE Trans. Magn.* **26**, 897-900.

Smirnov A.I., Poleuctov O.G. and Lebedev Y. (1992): High field EPR imaging. *J. Magn. Reson.* **97**, 1-12.

Sotgiu A., Gazzillo D. and Momo F. (1987): ESR imaging: spatial deconvolution in the presence of an assymetric hyperfine structure. *J. Phys. C.* **20**, 297-304.

Stemp E.D.A., Eaton G.R., Eaton S.S. and Maltempo M.M. (1987): Reconstruction of spectral-spatial two dimensional electron paramagnetic resonance imaging and transport of radicals in non-uniform media. *J. Chem. Phys.* **91**, 6467-6469.

Sueki M., Eaton S.S. and Eaton G.R. (1993): Spectral-spatial EPR imaging of irradiated silicon dioxide. *Appl. Radiat. Isot.* **44**, 377-380.

Minoru Sueki, Sandra S. Eaton and Gareth R. (1996): EPR imaging of irradiated silicon dioxide: increased concentrations of E' defects near the surface. *Appl. Radiat. Isot.* **47**, 1595-1598.

Sukhoroslov A. A., Samoilova R. I. and Milov A. D. (1991): Application of electron spin echo ESR-tomography to study radical diffusion in porous media. *Appl. Magn. Reson.* **2**, 577-586.

Ulbricht K., Herrling T., Ewert U. and Ebert B. (1984): Studies on spatial distribution of di-t-butyl nitroxide in zeolite powders by means of EPR zeugmatography. *Colloids and Surfaces* **11**, 19-29.

Webster D.S. and Marsden K.H. (1974): Improved apparatus for NMR measurement of self-diffusion coefficients using pulsed field gradients. *Rev. Sci. Instr.* **45**, 1232-1234.

Woods G.S. and Lang A.R. (1975): Cathodeluminescence, optical absorption and X- ray topographic studies of synthetic diamonds. *J. Cryst. Growth* **28**, 215-226.

Xu Y., Furusawa M., Ikeya M., Kera Y. and Kuwata K. (1990): ESR microscopic imaging of the spatial distribution of paramagnetic reactive centers on catalysts. *Chem. Letters* **12**, 293-236.

Yamamoto M. and Ikeya M. (1994): Electron spin resonance- computer tomography with magnetic field gradient wires inside the cavity: Images of alanine dosimeter and some materials. *Jpn. J. Appl. Phys.* **33**, 4887-4890.

Yamamoto M. and Ikeya M. (1994): Electron spin echo imaging using a field gradient wire inserted into a cavity. *Jpn. J. Appl. Phys.* **34**, 115-118.

Yamanaka C., Hosokawa K., Ikeya M., Nishijima J. and Okada T. (1995): Scanning positoron microscopy. *Jpn. J. Appl. Phys.* **34**, 6528-6529.

Zommerfelds W. and Hoch M.J.R. (1986): Imaging of paramagnetic defect centers in solids. *J. Magn. Reson.* **67**, 177-178.

Zupancic I. and Pirs J. (1976): Coils producing a magnetic field gradient for diffusion measurements with NMR. *J. Phys. E* **9**, 79-80.

Appendix

Appendix 1: Theory of Defect Formation and Decay

A1.1 *First− and Second Mixed−Order Annealing*

First− and second−order annealing kinetics are described in Chapter 3 (Sections 3.4.1 and 3.4.2). The defect concentration n in 1st− and 2nd−order *mixed−order annealing kinetics* is described as

$$dn/dt = -n/\tau - \lambda n^2 \ , \tag{A1.1}$$

where τ is the lifetime and λ is the 2nd−order decay constant. The exact solution for $n = n_0$ at $t = 0$ is

$$n = \frac{n_0 e^{-t/\tau}}{1 + \lambda \tau n_0 (1 - e^{-t/\tau})} \ . \tag{A1.2}$$

This equation is reduced to the exponential decay, $n = n_0 e^{-t/\tau}$ in 1st−order annealing ($\lambda = 0$) and to $n = n_0/(1 + \lambda n_0 t)$ in 2nd−order annealing ($1/\tau = 0$). Isothermal annealing curve must be fitted to a general mixed−order annealing equation to estimate τ and λ. A linear extrapolation in an Arrhenius plot of these parameters to the ambient temperature is made to estimate λ and τ.

Figure A1.1 shows isothermal annealing curves of 1st−order, 2nd−order and mixed annealing as a function of time t normalized by τ for three different initial concentrations, $n_0 = 10^{18}$, 10^{17} and 10^{16} cm^{-3}. The 2nd−order decay parameter λ is taken as $10^{-17}/\tau$ so as to make initial decay for the concentration of 10^{17} cm^{-3} the same as 1st−order decay. When the initial concentration is high, 2nd−order annealing predominates. When the concentration is low or decreases at a late stage, 1st−order exponential decay is dominant. Hence, in the *analysis of an isothermal annealing curve,*

(1) the late stage *exponential decay is subtracted* using a semilogarithmic plot and

(2) the *difference* is plotted in 1/n *versus* t for the 2nd−order decay analysis.

A1.2 *General Solution for Natural Growth : Formation and Decay*

A general equation for growth of the defect concentration accompanying mixed−order annealing is written using the effective lifetime τ_e in Eq. (3.31) as

$$dn/dt = aD - n/\tau_e - \lambda n^2 \ . \tag{A1.3}$$

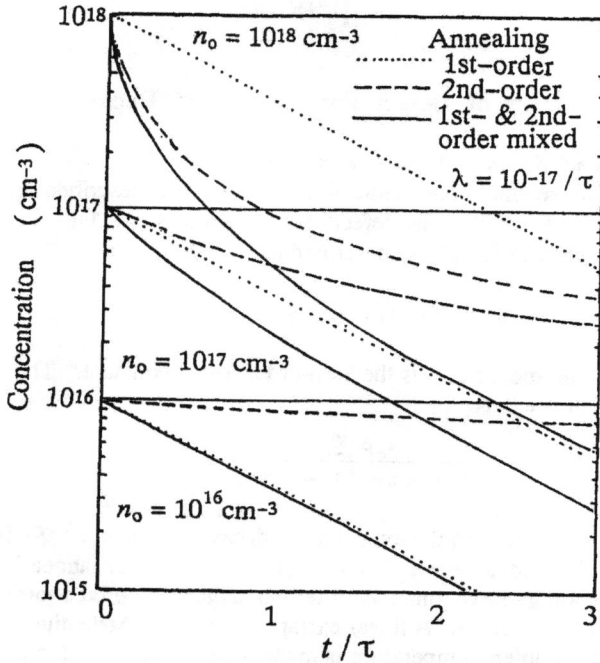

Figure A1.1 Theoretical isothermal annealing for (a) 1st–order, (b) 2nd–order and (c) 1st and 2nd–order mixed annealing for three different initial concentrations, 10^{18}, 10^{17} and 10^{16} cm^{-1}. The 2nd–order rate constant λ is taken as $10^{-17}/\tau$ so that initial decay is nearly the same for both 1st–order and 2nd–order annealing at the concentration of 10^{17} cm^{-3}.

The final solution for the boundary condition of $n = 0$ at $t = 0$ is

$$n = \frac{2aD\tau_e \tanh\left[(1 + 4aD\lambda\tau_e^2)^{1/2}(t/2\tau_e)\right]}{(1 + 4aD\lambda\tau_e^2)^{1/2} + \tanh\left[(1 + 4aD\lambda\tau_e^2)^{1/2}(t/2\tau_e)\right]} \ . \tag{A1.4}$$

The solution is equal to a saturation curve of Eq. (3.32) for $\lambda = 0$ and to Eq. (3.40) for $\tau_e = \infty$. The initial growth is $n = aDt$ for $t/\tau_e \ll 1$. Thus, the growth curve is no longer a simple saturation curve but follows to Eq. (A1.4). The real age T_{real} for $n = n_0$ can be determined directly as

$$T_{real} = \frac{\tau_e}{(1 + 4aD\lambda\tau_e^2)^{1/2}} \cdot \ln \frac{2aD\tau_e - (1 + 4aD\lambda\tau_e^2)^{1/2}n_0 + n_0}{2aD\tau_e - (1 + 4aD\lambda\tau_e^2)^{1/2}n_0 - n_0} \ . \tag{A1.5}$$

This equation is rewritten using the saturation intensity $I_s = aD'\tau_s' = aD\tau_s$ by the additive dose and the initial intensity I_0 before irradiation as

$$T_{real} = \frac{\tau_e}{(1 + 4aD\lambda\tau_e^2)^{1/2}} \cdot \ln \frac{(2\tau_e/\tau_s) + [1 - (1 + 4aD\lambda\tau_e^2)^{1/2}](I_0/I_s)}{(2\tau_e/\tau_s) - [1 + (1 + 4aD\lambda\tau_e^2)^{1/2}](I_0/I_s)} . \quad (A1.6)$$

Thus, if the intensity ratio I_0/I_s and the annealing parameters λ and τ as well as τ_s are available, the real age can be calculated. A simple graphic solution can be obtained as shown in Figure 3.12, where the calculated curve is shifted to fit I_0/I_s, from which the growth curve by an additive dose starts ($t' = 0$).

A correction factor for the ED, $TD_{real}/ED = T_{real}/T_{add}$, is obtained using $T_{add} = \tau_s \ln(1 - I_0/I_s)$ and is shown in Figure A1.2 as a function of I_0/I_s for $\tau = \infty$ ($\tau_e = \tau_s$) and different λ. A growth curve for the ^{238}U–series disequilibrium can be calculated using Eq. (A1.3) and the dose rate $D(t)$ in Eq. (4.28).

A1.3   A New Method of Additive Dose : Growth Curve

Eq. (A1.5) is an exact age solution considering 1st & 2nd mixed–order annealing of defects at the dose rate D. Thermal annealing experiments give the parameters τ and λ at the ambient temperature in the past. Parameters a and b are experimentally obtained using the additive dose method. An additive irradiation

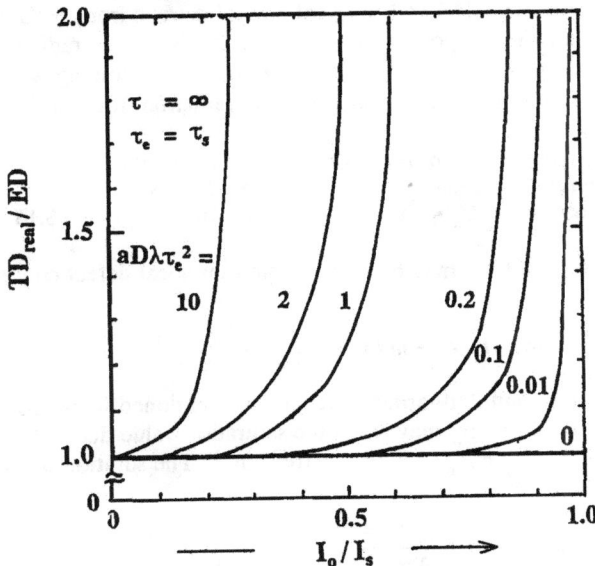

Figure A1.2   Correction factor to additive dose ED as a function of the initial intensity (I_0/I_s) for the 2nd–order natural annealing parameters.

is carried out at a high dose rate, $\underline{D' = 1 \sim 10 \text{ Gy/h} = 8.76 \times 10^3 \sim 8.76 \times 10^4 \text{ Gy/a}}$ which is about $\underline{10^8 \text{ times higher than the natural dose rate.}}$

The growth curve by an additive dose with the dose rate D' can be obtained by solving Eq. (3.39) with D' and τ_e' instead of D and τ and a new boundary condition of $n = n_o$ at $t = T$ or $t' = 0$. The growth curve in a laboratory is

$$n = \frac{(2aD'\tau_e' - n_o)\tanh[(1 + 4aD'e\lambda\tau_e'^2)^{1/2}(t'/2\tau_e')] + n_o(1 + 4aD'\lambda\tau_e'^2)^{1/2}}{(1 + 4aD'\lambda\tau_e'^2)^{1/2} + (1 + 2\lambda\tau_e'n_o)\tanh[(1 + 4aD'\lambda\tau_e'^2)^{1/2}(t'/2\tau_e')]} \quad . \tag{A1.7}$$

The time t' required for additive irradiation is usually of the order of hours, which is much shorter than the lifetime τ. Hence, the *saturation behavior* in the growth curve after irradiation is mostly due to the *interaction among defects* rather than annealing during irradiation except for unstable defects.

A1.4   Defect Formation by Natural α–Rays
A1.4.1   Defect Concentration in α–Tracks

Internal α–rays from radioactive elements produce defects along the α–ray track. The recoil atom by α–disintegration also produces an α–recoil track. The local defect concentration in fission tracks or tracks by α–rays or α–recoil atoms is extremely high and may follow natural annealing at ambient temperature.

The local dose rate can be estimated using the energy deposition of a single α–particle with an energy of 4 MeV (typical natural α–rays from ^{238}U–series disintegration in Chapter 4). Supposing that the track size is 2 nm in radius and 10 μm in length and the density of the material is 2.6 g/cm^3, the following values are obtained assuming a time interval of a single α–ray radiation of the order of μs:

The cylinder volume of an α–ray track : $V_\alpha = 4\pi \times 10^{-17}$ cm^3
The local energy deposition   : 1.96×10^6 Gy
The α–ray dose rate : $D_\alpha = 2 \times 10^{12}$ Gy/s $= 6.3 \times 10^{19}$ Gy/a (1 a $= 3.15 \times 10^7$ s)

Natural annealing of defects may be written using the local defect concentration in a track, m, as

$$dm/dt = -m/\tau - \lambda m^2 \quad , \tag{A1.8}$$

considering both 1st– and 2nd–order annealing as mentioned in Section A1.1. The initial local concentration m_0 may reach the saturation value due to the interaction among defects, i. e., $m_0' = N_0/b = 10^{19} \sim 10^{20}$ cm^{-3}. The solution using the initial condition of $m = m_0$ at $t = 0$ is written as

$$m = \frac{m_0 e^{-t/\tau}}{1 + \lambda\tau m_0(1 - e^{-t/\tau})} \quad . \tag{A1.9}$$

The defect concentration in each track is different and depends on the age of the track. What is measured, however, is the total defect concentration.

A1.4.2   *Growth of α-Tracks and Natural Annealing*

If the α-ray dose rate D_α is time dependent, the total concentration n_α is obtained by the convolution integral as

$$n_\alpha(t) = \int_0^t a_\alpha V_\alpha D_\alpha(t') m(t - t') \, dt' \, , \tag{A1.10}$$

where $n_\alpha(t)$ is the concentration of defects produced by α–rays, $m(t-t')$ is the local concentration of α–ray–induced defects between time t' and t, and a_α is the number of α–rays per 1 Gy. If the D_α is constant, the integral becomes

$$n_\alpha(t) = (a_\alpha V_\alpha D_\alpha/\lambda) \log_e [1 + \lambda \tau m_0 (1 - e^{-t/\tau})] \, . \tag{A1.11}$$

For 1st–order annealing, this equation is reduced to

$$n_\alpha(t) = a_\alpha V_\alpha D_\alpha \tau (1 - e^{-t/\tau}) \, , \tag{A1.12}$$

and for 2nd–order annealing,

$$n_\alpha(t) = (a_\alpha D_\alpha V_\alpha/\lambda) \log_e (1 + \lambda m_0 t) \, , \tag{A1.13}$$

where V_α is the volume of an α–track.

If the interaction among α– or α–recoil tracks is considered, an approximate calculation can be made by multiplying the factor $[1 - (b m_0/N_0) n]$ to the convolution integral of Eq. (A1.10) as

$$n_\alpha(t) = \int_0^t a_\alpha V_\alpha D_\alpha(t') (1 - b n_\alpha/N_0) m(t-t') \, dt' \, . \tag{A1.14}$$

More precisely, assuming a constant dose rate in radioactive equilibrium, the overlap of α tracks can be taken into account using the accumulated number of α–rays, $N_\alpha(t') = a D_\alpha t'$ at time t' as

$$n_\alpha(t) = \int_0^t a_\alpha D_\alpha V_\alpha (1 - V_\alpha)^{a D_\alpha (t - t')} m(t - t') \, dt' \, . \tag{A1.15}$$

Numerical calculation should be made for radioactive disequilibrium. Typical growth curves by α–rays are given in Figure A1.3 for several annealing parameters.

A1.4.3   *Theoretical k–value*

The defect concentration in an α–track may be equal to a saturation level, $m_0 = N_0/b$, at which there is no space of defect creation immediately after the track formation. The α–ray defect production efficiency relative to γ–rays is called the "α–ray efficiency (k–value) (see Section 3.5.1 in Chapter 3). The k–value is described using the G value (the number of stable defects per 100 eV) as

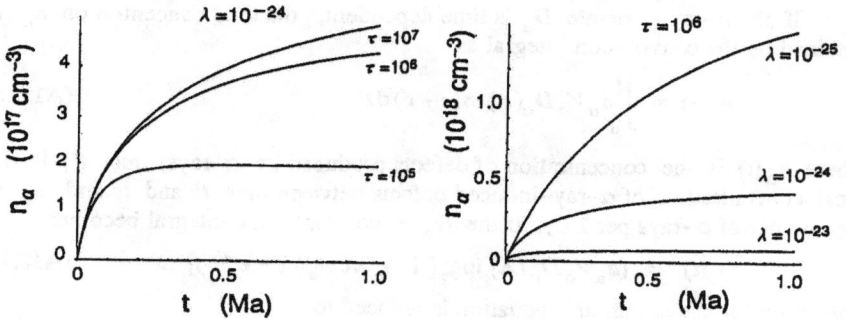

Figure A1.3 Growth of the defect concentration by α-ray irradiation based on 2nd-order annealing of defects within a track. The theoretical functional form is $n_\alpha = (aD_\alpha/\lambda)\log(1 + \lambda m_0 t)$ in Eq. (A1.13).

$$k = (m_0 V_\alpha/E_\alpha)/(G/100) .$$ (A1.16)

For an α-particle with the energy $E_\alpha = 4$ MeV, this equation becomes

$$k = (N_0 V_\alpha)/(b G \times 40,000) .$$ (A1.16')

Simply, if the following parameters are assumed or obtained,
the volume of an α-track   : $V_\alpha = 1.26 \times 10^{-16}$ cm³ ,
the interaction volume   : $b = 10^3$ and
the number of the available lattice site : $N_0 = 2.6 \times 10^{22}$ cm⁻³ ,

then, the following values are calculated for the G-value : $G = 0.5$:
the α-ray efficiency   : $k = 0.164$,
the local defect concentration   : $m_0 = N_0/b = 2.6 \times 10^{19}$ cm⁻³ and
the number of defects in a track : $m_0 V_\alpha = 3.27 \times 10^3$.

More precise V_α and b must be determined experimentally. Empirically obtained k-values may be used for practical ESR dating and dosimetry.

Appendix 2: Molecular Orbital Schemes
– AB$_2$–, AB$_3$– and AB$_4$–Type Molecules –

Molecular orbital (MO) schemes for illustration of ESR g–shift and hf constant, A, of simple molecules from the spin–orbit interaction are shown together with the closed shell MO energy levels calculated based on the unrestricted Hartree–Fock approximation (UHFMO) using MNDO–PM3 (modified neglect of diatomic overlap: parametric method 3) with the MOPAC program (Ver. 6.0: PM3) (Stewart 1989).

A2.1 AB$_2$–Type Molecules

The linear AB$_2{}^{16}$ molecular ions with 16 electrons, bent AB$_2{}^{17}$ radicals with 17 electrons and AB$_2{}^{19}$ radicals with 19 electrons are considered:

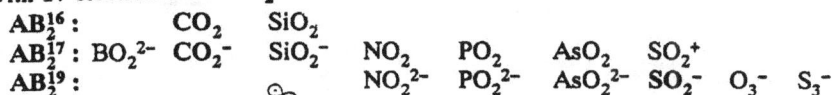

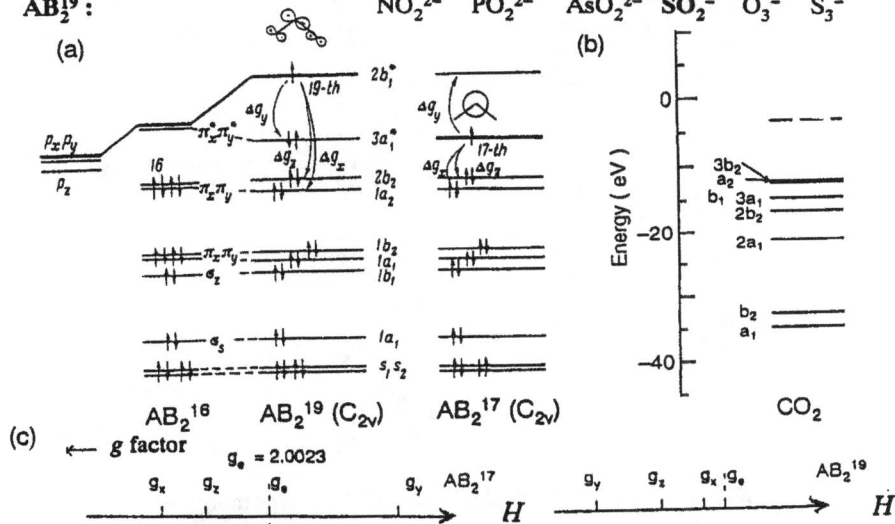

AB$_2^{16}$:   CO$_2$ SiO$_2$
AB$_2^{17}$: BO$_2^{2-}$ CO$_2^-$ SiO$_2^-$ NO$_2$ PO$_2$ AsO$_2$ SO$_2^+$
AB$_2^{19}$:   NO$_2^{2-}$ PO$_2^{2-}$ AsO$_2^{2-}$ SO$_2^-$ O$_3^-$ S$_3^-$

Figure A2.1   (a) Molecular orbital schemes of AB$_2$–type molecules with 16, 17 and 19 electrons. Arrows indicate the spin–orbit interaction which leads to g–shift (Marfunin 1979). (b) Energy levels of CO$_2$ using MOPAC program. (c) g–shift diagram for AB$_2{}^{17}$ and AB$_2{}^{19}$

AB$_2{}^{16}$ (CO$_2$) : Linear molecules.
AB$_2{}^{17}$ (CO$_2^-$) : The unpaired (17th) electron enters the antibonding $3a_1^{\,\bullet}$ from sp orbitals of the A atom, resulting in bent radicals with C$_{2v}$ symmetry. A large isotropic hf A_{iso} and considerable **anisotropy**. Negative g–shift; $-\Delta g_y > +\Delta g_x > \pm \Delta g_z$, ($\Delta g = g - g_e$: $g_e = 2.0023$; see Figure A2.3 (c)).
AB$_2{}^{19}$ (SO$_2^-$) : Bent radicals (C$_{2v}$ symmetry). The unpaired (19th) electron occupies the antibonding $2b_1^{\,\bullet}$ orbital formed from p orbitals of the A atom. A large anisotropic hf A (small A_{iso}); $+\Delta g_y > +\Delta g_z > +\Delta g_x \geq 0$.

A2.2 AB_3 – Type Molecules

The triangle planar AB_3^{24} molecular ions (D_{3h} symmetry) with 24 electrons, planar AB_3^{23} radicals (D_{3h}) with 23 electrons and pyramidal AB_3^{25} radicals (C_{3v} symmetry) with 25 electrons are considered (for the highest occupied molecular orbitals (HOMO) of CO_3^-, CO_3^{2-} and CO_3^{3-}, see Figure 5.4) :

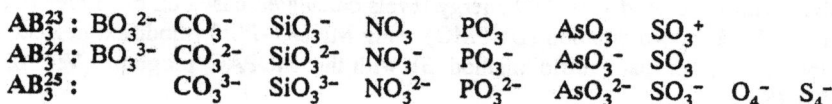

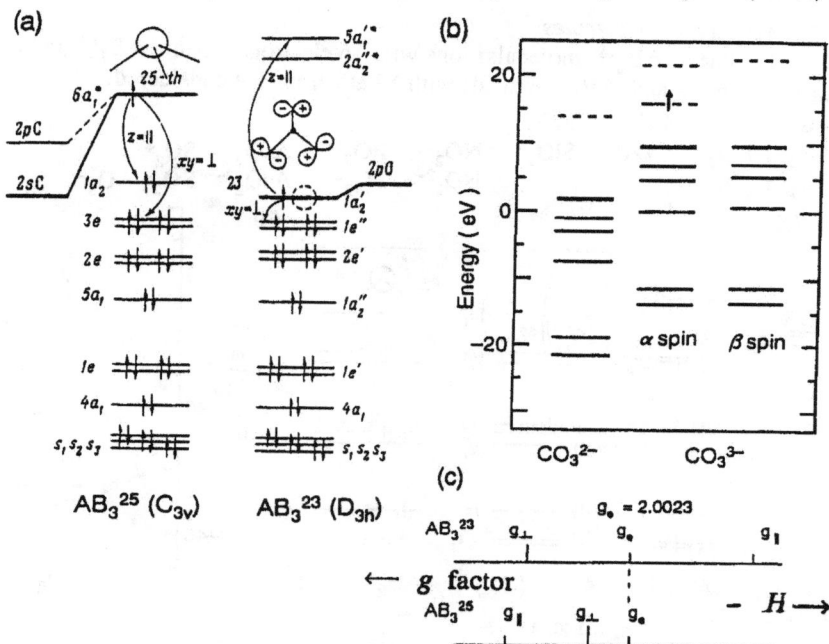

AB_3^{23} : BO_3^{2-} CO_3^- SiO_3^- NO_3 PO_3 AsO_3 SO_3^+

AB_3^{24} : BO_3^{3-} CO_3^{2-} SiO_3^{2-} NO_3^- PO_3^- AsO_3^- SO_3

AB_3^{25} : CO_3^{3-} SiO_3^{3-} NO_3^{2-} PO_3^{2-} AsO_3^{2-} SO_3^- O_4^- S_4^-

Figure A2.2 (a) Molecular orbital schemes of AB_3 type molecules with 23 and 25 electrons (Marfunin 1979). (b) Energy levels of CO_3^{2-} and CO_3^{3-} using MOPAC program. CO_3^{3-} with α and β spins is shown (UHFMO). (c) g–shift diagram of AB_3^{23} and AB_3^{25} molecules.

AB_3^{23} (CO_3^-) : Planar radicals with D_{3h} symmetry. The unpaired (23th) electron is in the nonbonding $1a_2'$ orbital formed from the atomic $2p$ orbital of the three oxygens leading to a **small** A and $\Delta g_\perp > 0$, $\Delta g_\parallel \leq 0$.

AB_3^{24} (CO_3^{2-}) : Planar molecular ions with D_{3h} symmetry.

AB_3^{25} (CO_3^{3-}) : Pyramidal radicals with C_{3v} symmetry. The unpaired (25th) electron occurs in the $6a_1^*$ orbital formed mainly from sp^n (mixture of sp^2 and sp^3) orbitals of the central A atom, leading to a **large** A_{iso} and **g factor with small anisotropy**: $\Delta g_\parallel \geq \Delta g_\perp \geq 0$.

A2.3 AB_4–Type Molecules

The following three types of molecules are considered for AB_4 molecules:

AB_4^{31} : BO_4^{4-} SiO_4^{3-} NO_4^{2-} PO_4^{2-} AsO_4^{2-} SO_4^{-}
AB_4^{32} : SiO_4^{4-} NO_4^{3-} PO_4^{3-} AsO_4^{3-} SO_4^{2-}
AB_4^{33} : SiO_4^{5-} NO_4^{4-} PO_4^{4-} AsO_4^{4-} SO_4^{3-}

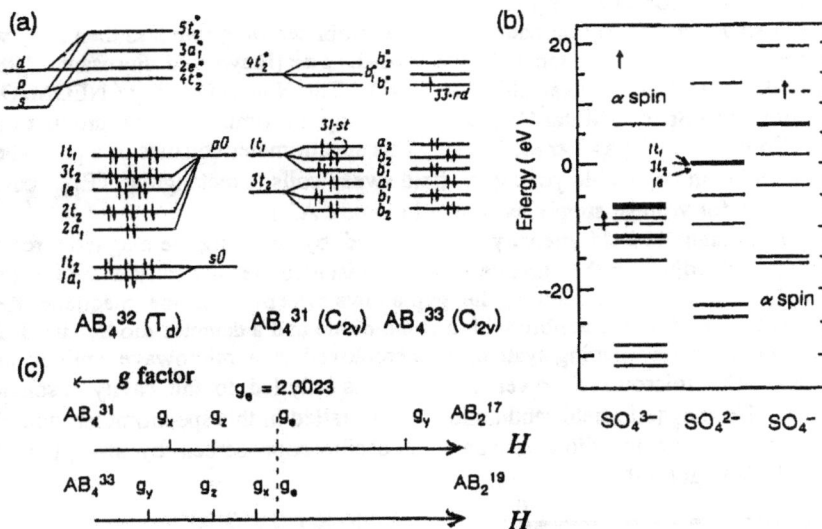

Figure A2.3 (a) Molecular orbital schemes of AB_4–type molecules with 32, 31 and 33 electrons. The T_d symmetry of AB_4 is distorted into C_{2v}. The orbital mixing scheme is the same as for AB_2 (Marfunin 1979). (b) Energy levels of SO_4^{3-}, SO_4^{2-} and SO_4^{-} using MOPAC program (c) g–shift diagram for AB_4^{31} and AB_4^{33} (AB_2^{17} and AB_2^{19}).

AB_4^{31} (SO_4^{-}) : Distorted tetrahedral radicals with C_{2v} symmetry. The unpaired (31st) electron (self–trapped hole center) occurs in the a_2 orbital formed mainly from non–bonding oxygen orbitals; A small hf A value.
$\Delta g = g - g_e$ determined by the spin–orbit interaction and so by splitting of the t_1 orbital; $-\Delta g_y > +\Delta g_x > \pm\Delta g_z$.
AB_4^{32} (SO_4^{2-}) : Tetrahedral oxyanions with T_d symmetry.
AB_4^{33} (SO_4^{3-}) : Distorted tetrahedral radicals with C_{2v} symmetry. The unpaired (33rd) electron (trapped electron) enters the antibonding b_1^* orbital formed by splitting of the lower $4t_2^*$ orbital, which splits into $3b_2$ by distortion of the tetrahedron into C_{2v} symmetry.
Large hf A value: the level consists of the sp^3 orbital of the central A atom, due to the large probability of electron clouds, i.e., $|\Phi(0)|^2$ at the nucleus. Anisotropic g factor shifts due to mixing of b_1^* orbitals by spin–orbit interaction with a_1, b_2 and a_2; $+\Delta g_y > +\Delta g_z > +\Delta g_x \geq 0$.

Appendix 3   Portable ESR Spectrometer for Teaching

A3.1   *A Spectrometer with a Permanent Magnet*

ESR spectrometers used in physics and chemistry researches employ a large and heavy electromagnet, which produces an external magnetic field around 340 mT for an X–band spectrometer. Some light–weight (portable) and low cost ESR machines are strongly hoped for routine ESR dosimetry, dating and microscopy as well as for a teaching program.

A small magnet for an educational spectrometer weighing less than 2 kg was fabricated with a new magnet alloy of Nd–B–Fe (Ikeya and Furusawa 1989, Yamanaka *et al.* 1991, Nakanishi *et al.* 1993). The Nd–B–Fe alloy (NEOMAX–30; Sumitomo Special Metal Co., Ltd.) has a high maximum energy product than a Sm–Co magnet (Sagawa *et al.* 1984). This spectrometer consists of a permanent magnet with an adjustable yoke gap, field sweep coils, a rectangular TE_{102} cavity and circuits for voltage supply as shown in Figure A3.1.

The magnetic field intensity can be varied by changing the magnetic resistance through adjusting the variable yoke gap distance mechanically with a screw. An additional Helmholz coil in the gap allows sweeping of the magnetic field electrically. A diode assembly with a Gunn diode and a detector diode, used for an automatic door opening system, was employed as a microwave emitter and detector. The microwave power of 15 mW is fully led to the cavity resonator directly. No magnetic field modulation was installed in this spectrometer and the output of the detector diode is seen on an oscilloscope screen by sweeping the magnetic field at 60 Hz.

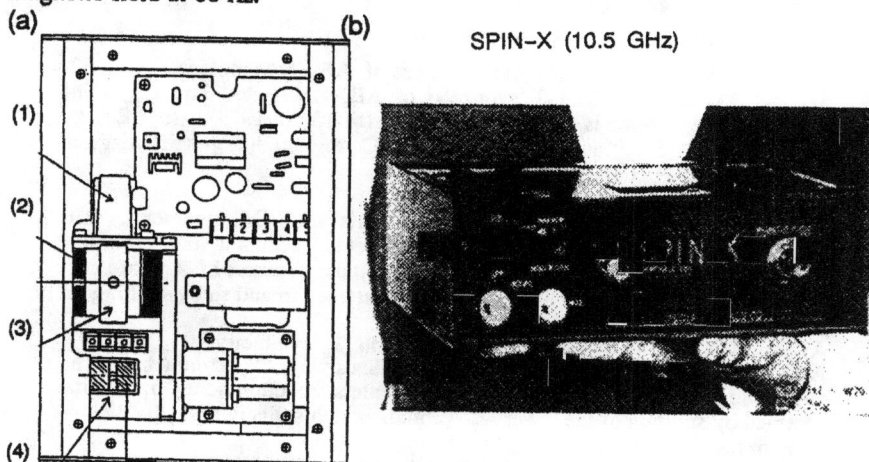

Figure A3.1   (a) A drawing of a light–weight (2 kg) portable ESR spectrometer use using (1) X–band (10.5 GHz) Doppler–radar Gunn diode, (2) a Nd–B–Fe (Neomax) permanent magnet, (3) a cavity and adjustable yoke gap. (b) A photograph of the spectrometer SPIN–X (from the catalog).

ESR spectra for DPPH, TEMPOL and coals are shown in Figure A3.2. The sensitivity of this spectrometer without a phase sensitive lock–in–amplifier system is about 10^{15} spins/0.1 mT; about five orders of magnitude less than that of commercial spectrometers. Nevertheless, there are several teaching programs for students using this simple spectrometer.

An oscilloscope display of hyperfine (hf) splitting with intensity ratios of $1:4:6:4:1$ for semiquinone radicals with four protons is shown in Figure A3.3. If hydroxyapatite powder is grown from a solution containing unstable semi-quinone radicals, stable radicals showing a broad signal are formed in the micro-crystal of hydroxyapatite. Such powder crystals are colored. Chemistry experiments can be done with this spectrometer.

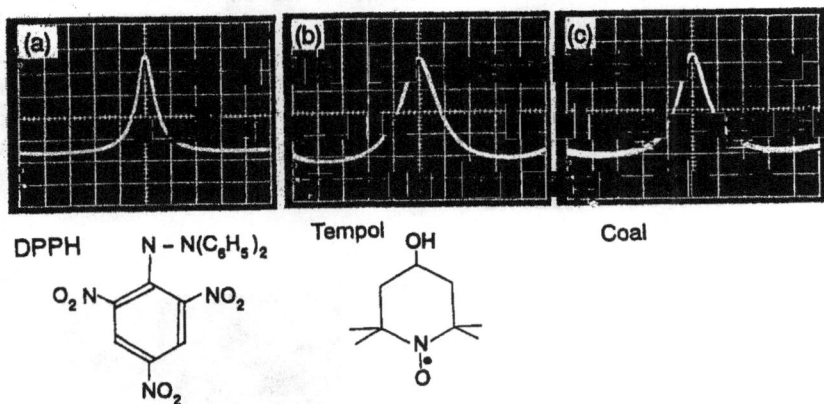

Figure A3.2   ESR spectra of some standard samples with a simple ESR spectrometer for a student experiment.

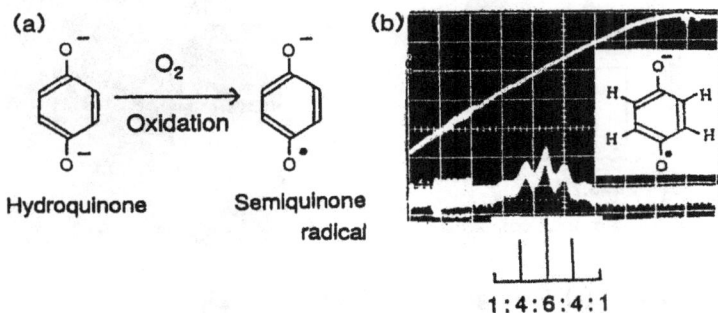

$$1:4:6:4:1$$

Figure A3.3   (a) Formation of semiquinone by oxidation of hydroquinone, (b) An oscilloscope output of the quintet ESR spectrum of semiquinone. The typical hf splitting with intensity ratios of $1:4:6:4:1$ is observed.

A3.2 Magnetic Resonance Imaging (MRI) Experiment
A3.2.1 Field Gradient Method for CT–ESR

The principle of MRI can be demonstrated with a present–day spectrometer by inserting small field gradient coils into a microwave cavity (see Chapter 14). Figure A3.4 (a) shows an anti–Helmholtz coil pair wound with an ordinary enameled wire with a diameter of 0.1 mm. The radius is about 3 mm with a separation of 3 mm. The magnetic field intensity produced by these coils is calculated based on the theory of electromagnetism. The field gradient is uniform when the separation of the coils, d, is nearly equal to the diameter of the anti–Helmholz coils. A sample of two DPPH dots separated by a plastic plate 0.57 mm in thickness was inserted into the coil pairs as shown in (b) (Ikeya et al. 1991).

Figure A3.4 (c) shows the ESR spectra for the current to the coils(0 A, 0.1 A and 0.3 A). The absorption spectrum with a linewidth of 0.2 mT is broadened and split. One can calculate the induced field gradient in mT/cm as an exercise in electromagnetism and compare with the splitting of the signal peaks in mT for a spatial separation of 0.75 mm for a DPPH sample separated by a plastic. The exponential decrease of the spin concentration from the surface in X–irradiated solids can be measured in an advanced course, if the students can process the data with the Fast Fourier Transform (FFT).

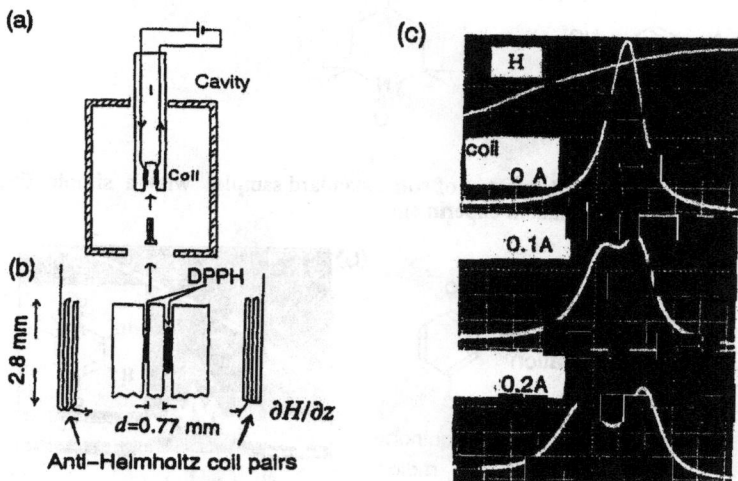

Figure A3.4   (a) Field gradient anti–Helmholz coil pair inserted into a microwave cavity.   (b) A test sample of DPPH particles separated by a plastic spacer with a thickness of 0.75 mm. (c) No field gradient is observed for 0 A but splitting occurs under a magnetic field gradient created by the increasing currents of 0.1 A and 0.2 A.

A3.2.2 *1–D Scanning ESR Microscopy with Wire Array*

In scanning ESR microscopy, the surface distribution of paramagnetic species was imaged by scanning the localized magnetic field (static or modulation field) or the localized microwave field. A localized static field was scanned by placing microwire arrays in a cavity and by shifting the wire position of the two-way current flow from the i–th to the $(i+1)$–th wire (Ikeya *et al.* 1991).

An example using this portable spectrometer is shown in Figure A3.5 for 5 wires. DPPH particles were placed between wire b and c and also between wires d and e. Calculation of ΔH by the wire current was done by taking the distance from the wire plane as a parameter. The DPPH signal without a current to the wires in Figure A3.5 (a) splits and shifts to the negative side when the current flows from wire a to wire b because of the presence of DPPH between wires b and c. This is because a slightly low external H is sufficient for resonance as there is a local static magnetic field between wires b and c. The signal is slightly shifted to the opposite side when the current is from wire c to wire d as a weak magnetic field in an opposite direction is produced at the DPPH position. For the current from wire d to wire e, a negative shift of the field is again detected. ESR micros-copy experiments using DPPH, tempol, pitch, coal and heavily irradiated ceram-ics can be done using this spectrometer.

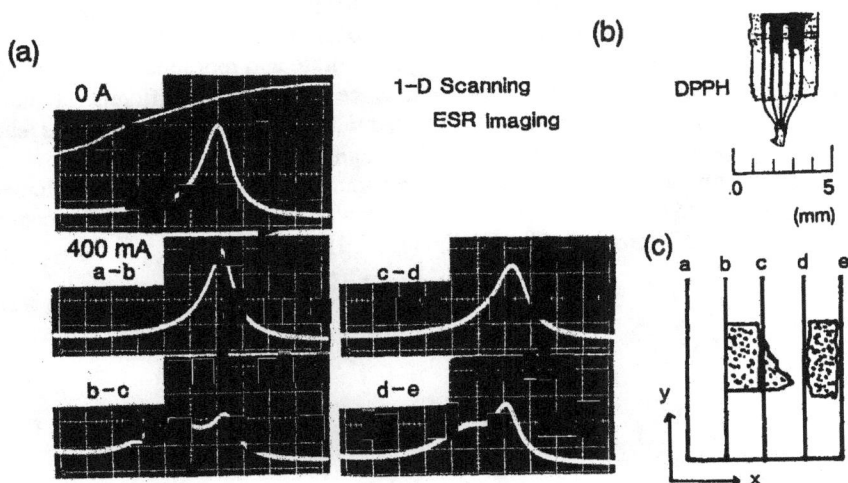

Figure A3.5   (a) ESR signal of DPPH with no two–way current (0 A) to the inserted 1–D wire arrays, two–way current (400 mA) to the wire a to b, b to c and c to d as shown in the microwire array of a photograph (b) and schemat-ic drawing with positions of DPPH particles in (c).

Appendix 4 :
ESR Reader for Dosimetry and Automatic Sample Changer

A4.1 *Compact ESR Reader*

Manufacturers of ESR spectrometers have begun to realize the importance of low–cost spectrometers for specific applications such as dosimetry, food irradiation monitoring, coal characterization and assessments of engine oil deterioration. The sensitivity of such ESR machines may not be as high as those for research. Attempts to use a low frequency ESR (Franconi *et al.* 1993), a permanent magnet of Nd–B–Fe (Ikeya and Furusawa 1989, Yamanaka *et al.* 1991, Nakanishi *et al.* 1993) have been made. Several manufacturers are selling computer–equipped compact ESR spectrometers dedicated for dosimetry systems and fully automated for the measurement of the signal intensity (Kojima *et al.* 1993, Mailer and Schmalbein 1993). A convenient method is to measure a standard sample of Mn^{2+}, and to take the intensity ratio as an indicative of the radiation dose. The names of manufacturers are JEOL, Microdevice and Sumitomo (Japan), Bruker (Germany), Micronow (USA) and Minsk (Belorussia).

A4.2 *Automatic Sample Changer*

Changing samples and measuring ESR repeatedly are tedious labor unless one is interested in the ESR spectrum. Measurements of the dosimeter elements or dating samples should be done with an automatic sample changer. A trial apparatus was made in our laboratory for full automatic spectrum measurements as shown in Figure A4.1. The sample setting is accurate and reproducible. Researchers prefer to measure manually, however, since materials are different in many cases. Only when measurements are repeated with the same kind of samples such as dosimeter elements, an automatic sample changer has a great advantage.

An automatic sample changer to set a commercial alanine dosimeter element into a cavity and back to the turn table sample was made by a pneumatic control similar to an NMR sample holder (Kojima *et al.* 1989).

Automatic sample changer

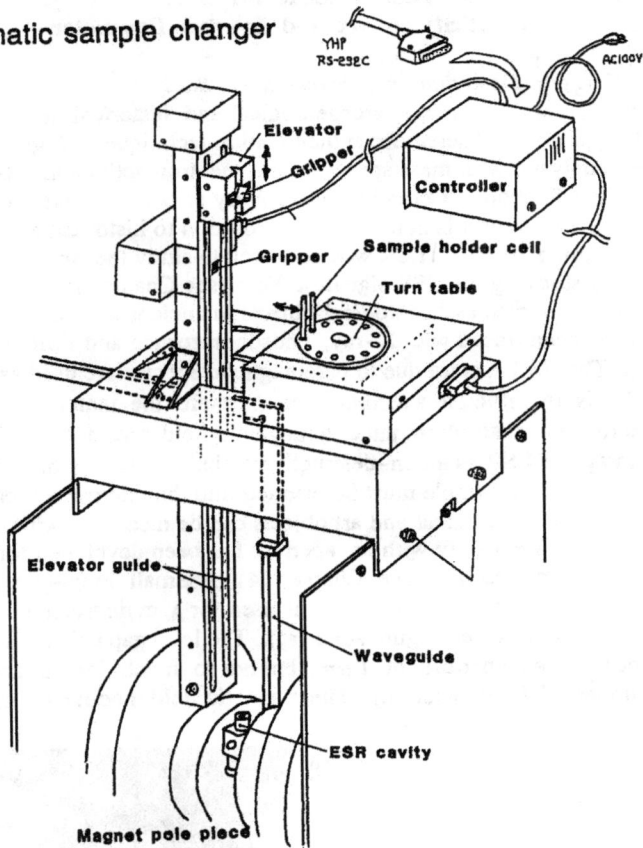

Figure A4.1 A test machine of an automatic sample changer. The sample setting is more reproducible than a manual setting.

Appendix 5:   Nondestructive ESR for a Large Object
– Authenticity of Art and In-vivo Dosimeter –

A5.1   Forgery Detection and Provenance Study

The forgery of art and archaeological and historical treasures have been detected by means of scientific authentication techniques. Pigments in painting have been analyzed with mass spectroscopy, neutron activation, X–rays, infrared and optical reflection.   Ceramics like pottery or porcelain are objects of TL, as rough age determination is in many case a solution to historical and archaeological objects (Fleming 1975).   ESR was also used to study the ancient technology of ceramics by studying the ESR signal of Fe^{3+} (see Chapter 10).   Provenance studies were done to trace archaeological objects in ancient trades. The origin of jade in Japan has been traced with X–ray, phosphorescence and ESR (Higashimura et al. 1985). The ESR spectra due to paramagnetic species are used as a finger–print.

ESR is in principle a technique to measure the material nondestructively since microwave absorption under a magnetic field does not affect the material. The advantage of ESR as a non–destructive technique has not been fully utilized as a small piece of the sample must be inserted into the microwave cavity. A large object such as archaeological and art objects can be measured without breaking.

A microwave cavity with an aperture has been developed for a microwave scanning ESR microscope (see Chapter 14). A small loop–gap resonator in a cavity system (Anderson et al. 1985) is used for a cylindrical cavity of TE_{111}– mode with a hole and (see Figure A5.1 (a)). The loop-gap enhances the sensitivity three times for an aperture of 1 mm, but not so much for an aperture of 3 mm (Sumitomo et al. in preparation). The magnetic field modulation at 100 kHz was

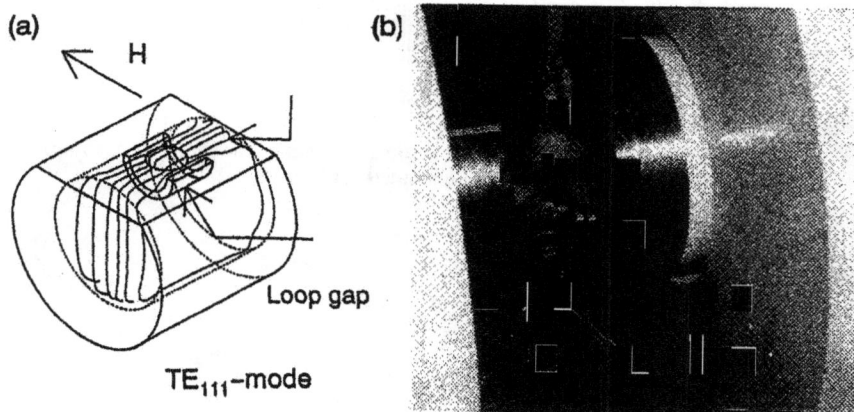

Figure A5.1   (a) ESR cavity with a metal ring at the aperture to enhance the signal sensitivity.   (b) A large object can be measured with ESR using a cavity with an aperture. A pottery of Mexican bird whistle during measurement.

three times for an aperture of 1 mm, but not so much for an aperture of 3 mm. The magnetic field modulation at 100 kHz was applied from either inside or outside the cavity.   A flat art object is placed on the cavity or inserted in the space be-tween the cavity and the pole piece. The penetration depth of the microwave is 0.1 ~ 0.2 mm for the present cavity. The object size must be less than the pole-piece gap, 70 ~ 100 mm in one direction.

A5.2   Applications
A5.2.1   Ceramic Wares : Pottery and Porcelain

ESR spectra of pottery or porcelains are mainly associated with Fe^{3+} in clay minerals (see Chapter 10). The shape or pattern of ESR spectra depends on clay materials, glaze as well as on oxidation state as described in the studies of ancient technology for pottery manufacturing (Warashina et al. 1981). The paramagnetic impurities and their state in minerals give a clue to the origin of clays and glaze. Figure A5.2 (a) shows an ESR spectrum of a pottery bird whistle from Mexico (see Figure 5.1(b)). A signal around $g = 2.0$ associated with Fe^{3+} is characteristic fea-tures of pottery. Archaeological Japanese Jomon potteries show the same signal.

A traditional Japanese pottery, "Bizen pottery vase" was measured nonde-

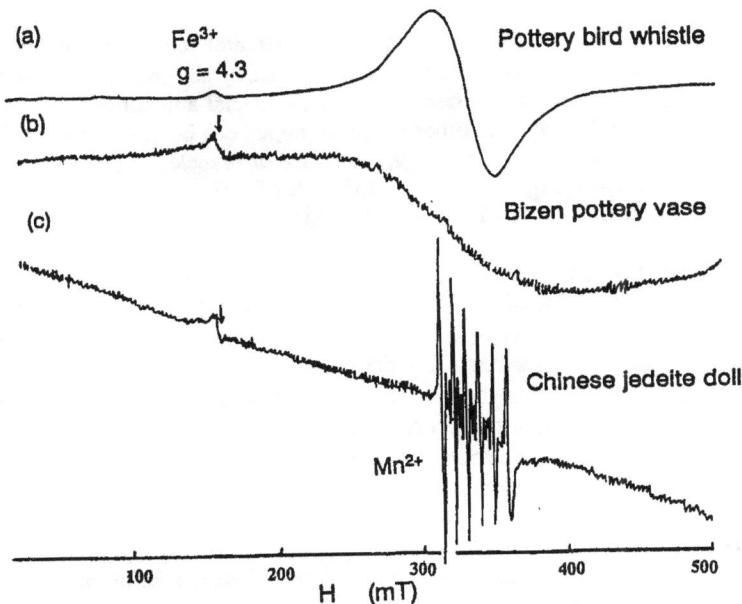

Figure A5.2 ESR spectra of (a) pottery bird whistle (Mexico), (b) Japa-nese traditional pottery "Bizen". (c) Chinese jade sculpture.

structively as shown in (b). The signal at $g = 4.3$ associated with Fe^{3+} and a broad line are observed. A spectrum of pottery differs from one site to the other due to subtle change in color caused by different degrees of oxidation. Although no glaze is used in manufacturing Bizen pottery, dominant signals are from the surface rather than that of the inside.

A5.2.2   Sculptures and gemology

(a)   *Sculptures* : *Jade* is a gemstone mineral of the clinopyroxene group with a chemical form of $Na(Al, Fe)Si_2O_6$. It occurs in unevenly distributed color from dark green to greenish white. It has long been used for jewelry, carved articles and ornament of jade by cutting and polishing. Jade stones and archaeological objects in Japan have been studied with ESR to determine the sources of objects and to establish the ancient transportation (Higashimura *et al.* 1985). Jade beads with holes and crescent jades. The source of jade is mainly Burma and China. Figure A5.2 (c) shows an ESR spectrum of the modern Chinese jade sculpture. The signals of Fe^{3+} and a hf sextet due to Mn^{2+} are observed.

    Marbles are calcite ($CaCO_3$) having ESR signals characteristic of Mn^{2+} with or without the signal of Fe^{3+} at $g = 4.3$. Allocation of marbles is difficult unless a systematic study on the quarry and products is carried out.

(b)   *Gemology* :   Precious gems and ornaments or archaeological materials are closely related with the history and appeal us the poetry and romance. Gemology which adopt physics techniques can distinguish artificial and natural gem stones. Jewels like diamond, ruby and other precious stones can be measured using the present system whether they are in rings, necklace or bracelet.

    Pearls contain a large amount of Mn^{2+}. Artificial pearls covering the shell sphere has a thickness of $0.5 \sim 1$ mm which is thick enough for ESR measurement.

A5.2.3   Color Pigments in Art Objects

    Color pigments contain Fe^{3+} and other paramagnetic ions. An ESR spectrum of pigments in paintings without removing pigments can be measured with a cavity with an aperture. A series of modern color pigments by different manufacturers were measured to identify the pigment manufacturers. The cavity system was slightly modified so that the flat surface with an aperture in the cavity was in the $x–y$ plane parallel to the surface of the pole piece of the electromagnet. Authenticity tests will be done by measuring ESR spectra of oil pigments.

A5.3   In–vivo Oral Dosimeter

    If a human dosimeter for measuring radiation dose at a front molar without extraction were available, many people worrying about radiation exposure in the Chernobyl accident and late effects of A–bomb radiation could know the accurate dose and be at least free from some psychological stress. Those who received high

In–vivo tooth dosimetry using a cavity with an aperture has been studied. A low–cost permanent magnet of Nd–B–Fe alloy was used to apply a magnetic field at the tooth as shown schematically in Figure A5.3 (a). A lower front molar is attached to a hole of a microwave cavity. The 100 kHz modulation field is applied from inside the cavity (Ikeya and Ishii 1989, Ishii and Ikeya 1990).

If a good magnet with a pole–piece gap of about $250 - 300$ mm were avail–able as in magnetic resonance imaging (MRI), monitoring down to 10 Gy would be possible with such a TE_{102} cavity. Improvement of the ESR signal sensitivity of nearly one order of magnitude has been made using a TE_{111} cylindrical cavity. The minimum detectable dose for a whole extracted tooth attached to the cavity is $1 - 0.5$ Gy at best (Yamanaka *et al.* 1993). Improvement by one or two order of the magnitude in sensitivity is desirable for a practical use. The uniformity of the magnetic field and the sensitivity of the ESR apparatus (mostly the Q–value of the cavity) are obstacles for practical use of this dosimeter. The effect of the magnetic field and its gradient on human brains must be considered for a practical use of this dosimeter. A large magnet used for MRI is needed for in–vivo tooth dosimetry. A magnet for tooth dosimetry is being designed in Minsk, Belorussia, where con–cern over the Chernobyl accident is very high.

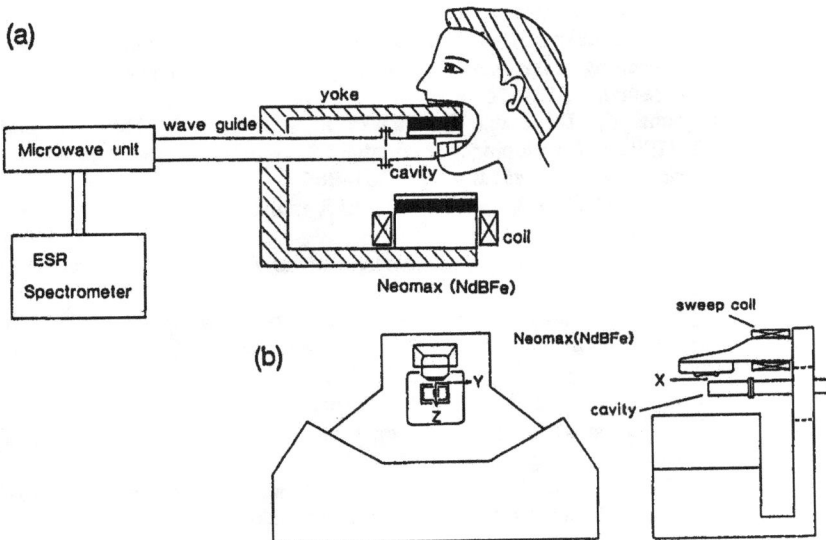

Figure A5.3   (a) A test ESR in–vivo tooth dosimeter to measure the dose of human tooth enamel using a cavity with an aperture and an oral magnet. (b) An oral magnet designed using an intense permanent magnet Nd–B–Fe.

References

Anderson W. A., Venters R. A., Bowman M. K., True A. E. and Hoffman B. M. (1985): ESR and ENDOR applications of loop–gap resonators with distributed circuit coupling. *J. Magn. Reson.* **65**, 165–168.

Fleming S. J. (1975): *Authenticity in Art* (The Institute of Physics, London and Bristol).

Franconi C., Holowacz J., Bonori M., Ettinger K. V. and Laitano R. F. (1993): Investigation of a special S–band spectrometer for ESR dosimetry. *Appl. Radiat. Iso.* **44**, 351–355.

Higashimura T., Warashina T. and Nakano Y. (1985): Allocation of ancient jadeite objects in Japan. *ESR Dating and Dosimetry* (Ionics, Tokyo) 459–468.

Ikeya M. and Furusawa M. (1989): A portable spectrometer for ESR microscopy, dosimetry and dating. *Appl. Radiat. Isot.* **40**, 845–850.

Ikeya M. and Ishii H. (1989): Atomic bomb and accident dosimetry with ESR: natural rocks and human tooth in–vivo spectrometer. *Appl. Radiat. Isot.* **40**, 1021–1027.

Ikeya M., Megro H., Ishii H. and Miyamaru H. (1991): Educational experiments on ESR imaging with a portable ESR spectrometer. *Appl. Magn. Reson.* **3**, 667–673.

Ishii H. and Ikeya M. (1990): An electron spin resonance system for in–vivo human tooth dosimetry. *Jpn. J. Appl. Phys.* **29**, 871–875.

Kojima T., Uematsu T., Tanaka R. and Aramaki S. (1989): An apparatus for continuous setting and resetting of dosimetric elements to an ESR spectrometer. *Japanese Patent* No. 1–138484, 533–536.

Kojima T., Haruyama Y., Tachibana H., Tanaka R., Okamoto J., Hara H. and Yamamoto Y. (1993): Development of portable ESR spectrometer as a reader for alanine dosimeter. *Appl. Radiat. Isot.* **44**, 361–365.

Mailer D. and Schmalbein D. (1993): A dedicated EPR analyzer for dosimetry. *Appl. Radiat. Isot.* **44**, 345–350.

Marfunin A. S. (1979): *Spectroscopy, Luminescence and Radiation Centers in Minerals* (Springer–Verlag, Berlin).

Nakanishi A., Sugawara N. and Furuse A. (1993): Portable ESR spectrometer with Neomax (Nd–Fe–B) permanent magnet circuit. *Appl. Radiat. Isot.* **44**, 357–360.

Sagawa M., Fujimura S., Togawa N., Yamamoto H. and Matsuwa Y. (1984): New material for P.M. on a base of Nd and Fe. *J. Appl. Phys.* **55**, 2083–2087.

Stewart J.J.P. (1989): Optimization of parameters semiempirical methods. I. Methods ; II. Applications. *J. Comp. Chem.* **10**, 209–220; 221–264.

Warashina T., Higashimura T. and Maeda Y. (1981): Determination of firing temperature of ancient pottery by means of ESR spectroscopy. *Brit. Mus. Occas. Papers* **19**, 117–128.

Yamanaka C., Ikeya M., Meguro K. and Nakanishi A. (1991): A portable ESR spectrometer using Nd–B–Fe permanent magnets. *Nucl. Tracks* **18**, 279–282.

Yamanaka C., Ikeya M. and Hara H. (1993): ESR cavities for in–vivo dosimetry of tooth enamel. *Appl. Radiat. Isot.* **44**, 77–80.

Index